Proceedings of the
Specialty Conference on

COMPUTER AND PHYSICAL MODELING IN HYDRAULIC ENGINEERING

Sponsored by the Hydraulics Division of
the American Society of Civil Engineers
McCormick Inn • Chicago, Illinois
August 6-8, 1980

Co-Sponsored by the
Illinois Section, ASCE
Illinois Institute of Technology

George Ashton, Editor

AMERICAN SOCIETY OF CIVIL ENGINEERS FOUNDED 1852

Published by the
American Society of Civil Engineers
345 East 47th Street
New York, New York 10017

Library of Congress Catalog Card Number 80-67878
ISBN 0-87262-252-5
Manufactured in the United States of America

DEDICATION

On this, his 90th birthday year, the Proceedings of this Conference are dedicated to Dr. Garbis Keulegan in grateful recognition of his many outstanding contributions to the fields of hydraulics and hydrodynamics, recognizing these have been in both theory and application, and that his lucidity and clearness of concept have often served a much needed bridge between the two.

CONTENTS

---

*Manuscript not available at time of printing

SESSION 5: GROUNDWATER MODELING

SESSION 6: MODELING THERMAL DISCHARGES II

SESSION 7: NUMERICAL ANALYSIS OF OPEN CHANNEL FLOWS I

SESSION 8: MODELING SEDIMENT TRANSPORT II

SESSION 9: WATER DISTRIBUTION SYSTEMS

---

*Manuscript not available at time of printing

---

*Manuscript not available at time of printing

SESSION 15: POLLUTION MODELING

SESSION 16: PHYSICAL MODELS

USE OF A SCALE MODEL TO IMPROVE POND CIRCULATION

by

William A. Rozeboom[1]

ABSTRACT

The paper describes a physical model used to improve circulation patterns for a portion of a proposed cooling pond system. The objective of the study was to reasonably optimize the design of a spur dyke required to prevent short circuiting of flow across the study area.

A series of qualitative tests were performed to investigate the performance of several dyke design alternatives with the condition of water flow only. The selected final design was examined while under the influence of wind conditions.

Overall, the study exemplifies the role which a qualitative scale model can play in the economical solution to a practical engineering problem.

INTRODUCTON

This paper presents the case study of a qualitative hydraulic model which involved an investigation of cooling pond circulation patterns. Specific results of the study are not intended for general application; however, they may be of some value to those interested in reservoir circulation and wind induced currents.

The objective of the paper is to demonstrate the capability of a low-budget model to provide an economical solution to a practical engineering problem.

STUDY BACKGROUND

The Keephills Thermal Plant is to be constructed by Calgary Power Ltd. approximately 70 km west of Edmonton, Alberta, Canada; the project will include a 12 km^2 cooling pond system which will be created in natural topography.

1 Staff Engineer, Northwest Hydraulic Consultants Ltd., Edmonton, Alberta, Canada

Northwest Hydraulic Consultants Ltd. was engaged by the project design engineers, Montreal Engineering Company Ltd., to construct and test a scale model of the 2 km^2 end portion of the cooling pond. The modelled area in relation to a portion of the pond arrangement is shown in Figure 1; the overall pond scheme will occupy roughly twice the area shown in the figure.

The objective of the model study was to reasonably optimize the length and orientation of a spur dyke designed to prevent short circuiting of flow across the study area. Also, the effect of wind shear on circulation patterns was to be investigated.

The study was performed over an 8 week period, with a total budget of approximately $25,000.

MODEL DESCRIPTION

A floor model was constructed at scales of 1:250 horizontal and 1:100 vertical. In order to reproduce flow patterns where determined by the contours of the pool, topography was reproduced for the upstream portion of the pond, starting with a natural channel constriction 150 m wide and 5 m deep upstream of the main study area. The deeper downstream portion was represented at a constant depth of 13 metres. The invert of the pond outlet (plant intake) canal was 5 metres below the full supply reservoir elevation.

The design flow modelled was 56.6 m^3/s. A closed flow system was used, with circulation provided by a submersible pump situated beyond the pond outlet. Flow was returned to a baffled headbox to produce a uniform entrance flow condition. Flow rates were controlled by a calibrated valve on the return flow line.

An elevated wire grid, with spacing at 200 metres prototype, was used both to provide co-ordinates for photography and to support a 10 mil clear PVC canopy which sealed the model during wind shear tests.

Wind shear was produced in the model by two 2.4 m^3/s fans, which drew air through a well rounded entrance across the model surface and into plenum chambers depressurized by the fans. Desired wind shear was produced by the adjustment of vents on the plenums.

All dimensions and quantities quoted subsequently herein are expressed in prototype units.

MODEL SIMILITUDE/OPERATING PARAMETERS

No attempt was made to simulate the effects of a bouyant plume, which is to say that the model was operated with water of constant density. Practical considerations (time and cost of the model study) enhanced the justification for neglecting thermal effects: the density difference between pond inflow and receiving water in the study area was predicted to be very small.

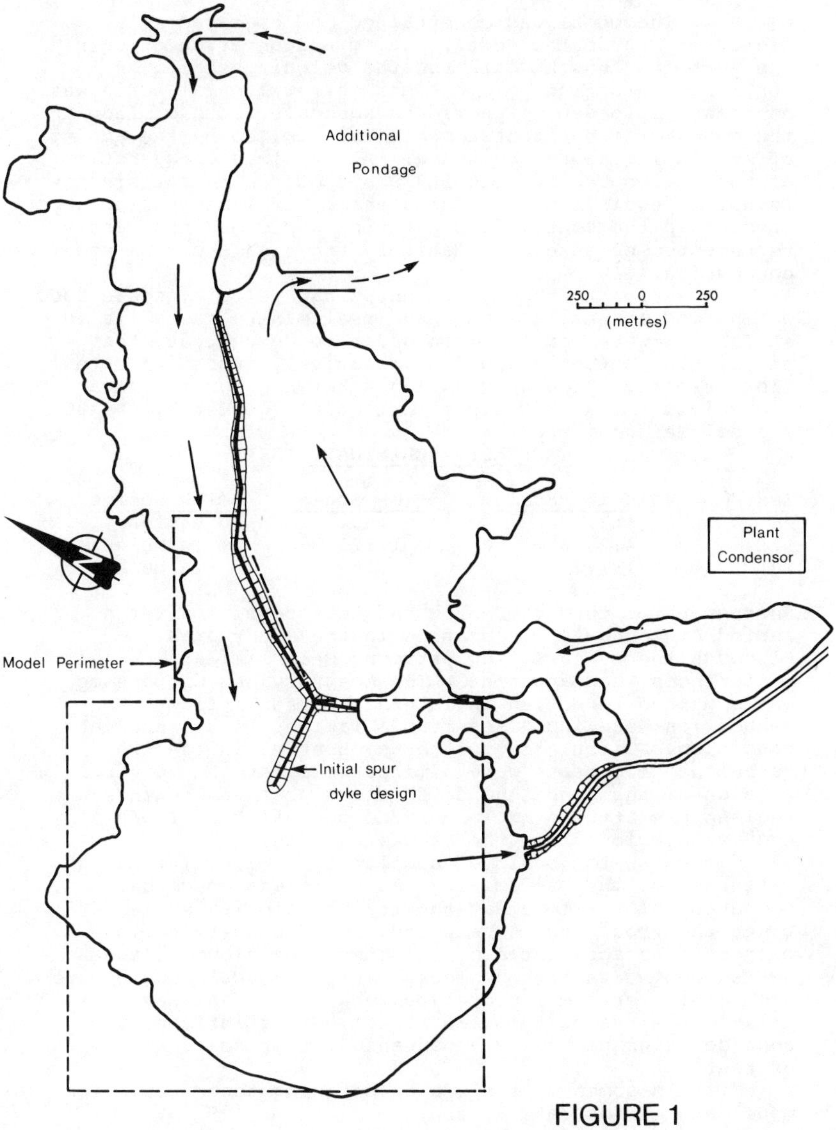

FIGURE 1
General Pond Arrangement

The model was constructed and operated as a distorted Froude Law model, having exaggeration of both the vertical length scale and the velocity scale. Vertical distortion of 2.5 times the horizontal scale was necessary in order to provide a workable depth of flow in the model; the horizontal scale was limited by the amount of available laboratory space. A 4.8 times exaggeration of the scaled Froude velocity was required to satisfy the Reynolds requirement for turbulence. If laminar flow were present in the model, flow patterns would be incorrectly represented because of relatively too high shear on model boundaries.(1)

Prior to the selection of the velocity scale used during model testing, the three preliminary tests listed in Table 1 were performed in order to demonstrate that circulation patterns would be relatively insensitive to flow velocity. The only variable between the tests was the velocity scale; the physical model (geometry, length scales) was unaltered.

TABLE 1: PRELIMINARY TESTS

Test	Velocity Scale	Time Scale	Flow Regime
1	1:3.1	1:80	Laminar
2	1:2.08	1:120	Turbulent
3	1:1.4	1:180	Turbulent

Photographs 1 through 3 show the test results after a period of 18 hours circulation in the study area. Although the overall flow patterns were similar, certain differences were apparent. The most obvious difference was a marked reduction in lateral spread as flow velocities were increased from a laminar to a turbulent condition. Circulation patterns observed in the two turbulent flow tests were similar well into the central portion of the pond; the differences observed in this region were attributed largely to the effects of unsteady eddy separation along the main flow path.

On the basis of similarity of circulation patterns for the preliminary tests, it was concluded that operation of the model at the 1:2.08 velocity scale, 4.8 times the Froude scale velocity, would result in flow patterns characteristic of prototype behaviour. It was not suggested that the selected velocity scale would more accurately represent prototype behaviour than, say, the 1:1.4 scale; velocity scale selection was influenced by consideration of having a convenient time scale and length of test.

In summary, the scales defining the circulation model and its operation were:

Vertical length	1:100
Horizontal length	1:250
Velocity	1:2.08
Time	1:120
Discharge	1:52,000

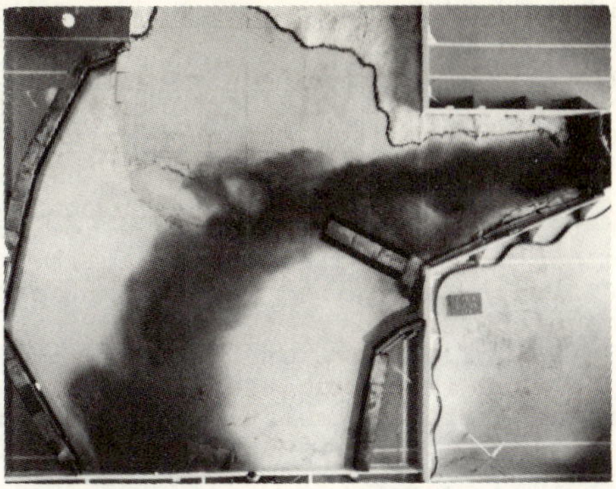

Photo 1; Test 1; 1:3.1 Velocity Scale; 18 hours.

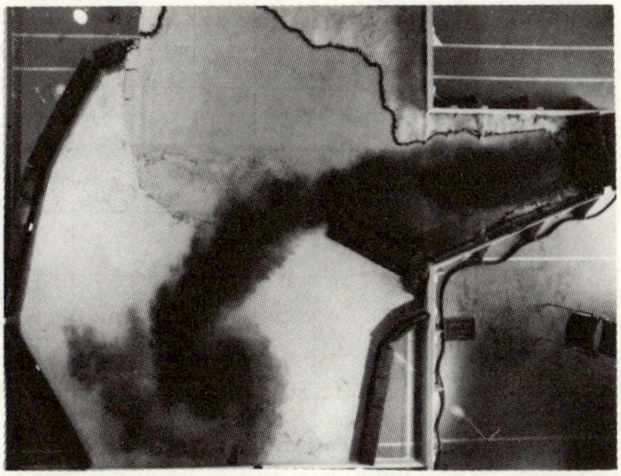

Photo 2; Test 2; 1:2.08 Velocity Scale; 18 hours.

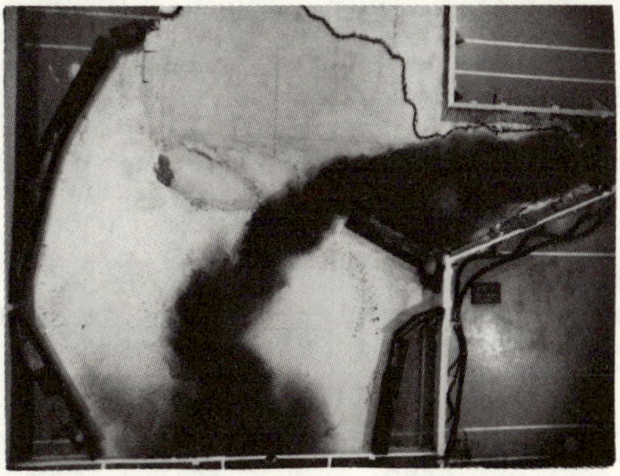

Photo 3; Test 3; 1:1.4 Velocity Scale; 18 hours.

Selection of wind effects to be examined in the model was made on the basis of long term wind frequency data for the area. A prevailing NNW wind was chosen for testing, as it appeared that this direction was most likely to produce adverse circulation patterns. Wind speeds of 2 m/s and 4 m/s were chosen as being representative of summer conditions.

Rather than attempt to simulate wind induced currents on the basis of a scaled wind velocity, the model wind was adjusted until a desired surface velocity was attained. The surface velocity of a wind driven current is approximately 2 to 3 percent of the mean wind velocity measured at a 10 metre reference elevation [2,3]. Model wind effects were therefore simulated by computing prototype currents on this basis and scaling surface velocities accordingly.

Two parameters were used to compare the relative effectiveness of alternative dyke designs: (i) the flow residence time, expressed as the time required for the flow front to pass through the study area; and (ii) pond utilization, expressed as the percentage of the model area occupied by the dyed inflow at the time a discernable flow front reached the outlet.

TESTING: SELECTION OF DYKE DESIGN

Seven dyke design alternatives were tested with conditions of water flow only. Prior to each test, the model was operated for a minimum of 4 weeks (prototype), to ensure steady state circulation patterns. Test procedure consisted of abruptly commencing a constant rate of dye injection into the recirculating flow, and photographing the flow pattern at one hour intervals.

The preliminary dyke design, proposed at the onset of the study, had been selected primarily by economic considerations; the length and alignment corresponded to that of an existing topographic ridge. However, as shown by Figure 2 and Photographs 4 and 5, this design was far from optimal, with only 35 percent of the pond area being utilized during a 13-hour period required for the flow front to short-circuit through the pond.

Subsequent model designs involved variations on dyke length and alignment. Also investigated were discontinuous dykes, which, although generally more effective, were subsequently deemed to be undesirable from the standpoint of field serviceability.

In tests of continuous dyke designs, it was consistently observed that the flow front was well defined, having a slow rate of lateral diffusion relative to the forward jet velocity. It was found that the maximum residence time attainable with a reasonable length of dyke was 21 hours, with a corresponding 50 percent utilization of surface area. These values resulted from the forced deflection of the flow jet towards the outer

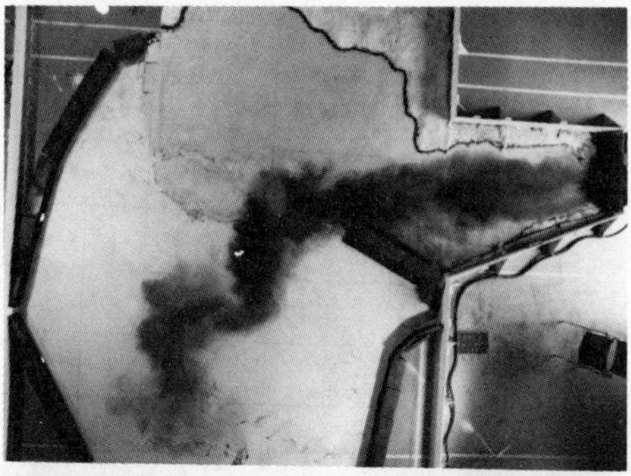

Photo 4; 12 hours

Photo 5; 24 hours

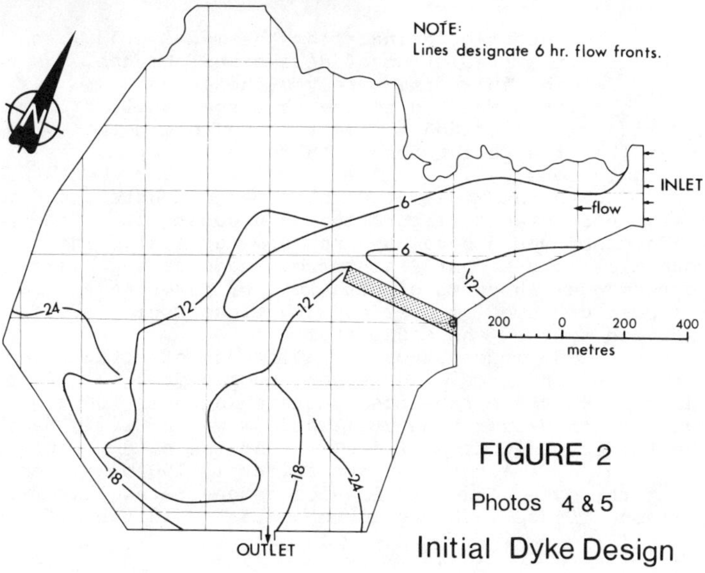

FIGURE 2

Photos 4 & 5

Initial Dyke Design

perimeter of the pond. Figure 3 and Photographs 6 and 7 show this circulation pattern for the final dyke design.

TESTING: EFFECTS OF WIND SHEAR ON CIRCULATION PATTERNS

Three tests were performed to investigate the effects of wind shear on circulation patterns for the final dyke design. Steady state circulation patterns were established prior to each test.

The first two tests were performed with constant NNW winds of 2 m/s and 4 m/s, respectively. In both cases, a dramatic increase in pond mixing was observed, greatly diffusing the flow front and thereby improving the circulation pattern. Currents generated by the 4 m/s wind were of sufficient magnitude that the main flow path was deflected across the centre of the pond. Figure 4 and Photographs 8 and 9 show this circulation pattern resulting from the 4 m/s wind.

A final test was performed to examine the rate of decay of currents induced by the 4 m/s wind; pond effectiveness could be impaired if the circulation pattern did not return to the non-wind condition after cessation of the wind. The flow patterns were observed to have returned to the windless condition within 40 hours of wind cessation; Figure 5 and Photographs 10 and 11 show the pattern of decay. The rapid decay of wind generated currents confirmed the selection of the dyke design in that long term pond circulation patterns were demonstrated to be not susceptible to temporary forces.

MODEL STUDY BENEFITS

Information gained from the model study, in particular the prediction of design performance, was not easily determinable from purely mathematical considerations. That is to say that the model, treated as a design aid, provided the most reasonable means by which to examine the performance of alternative dyke configurations.

Had the model study not been conducted, and the preliminary dyke configuration been constructed, the pond performance would have been far from optimal. The potential future cost rising from inadequate cooling performance (in terms of both loss of power revenue and the cost of remedial dyke construction) would have far exceeded that of the model study.

The model study was also cost effective when considered in terms of dyke construction costs. Given that the 'best' circulation pattern (as some function of dyke length versus pond utilization) was determined to be the forced deflection of flow around the pond perimeter, the final dyke length was approximately 150 m shorter than the first dyke alternative which demonstrated this flow pattern. Since each design to be tested in the model was

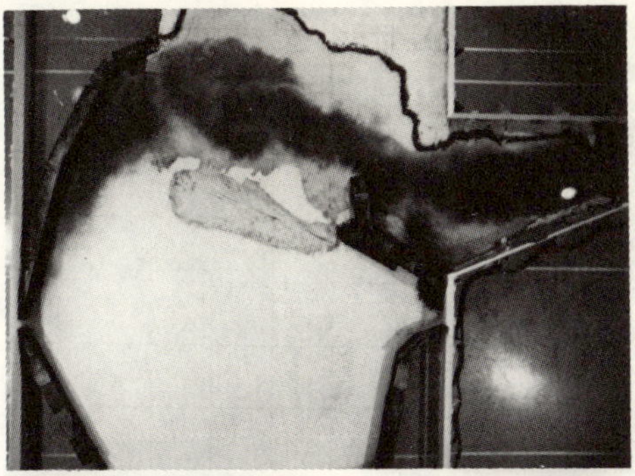

Photo 6; 12 hours

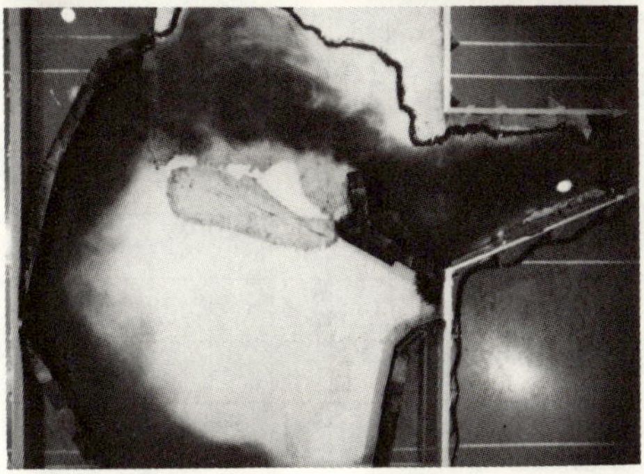

Photo 7; 24 hours

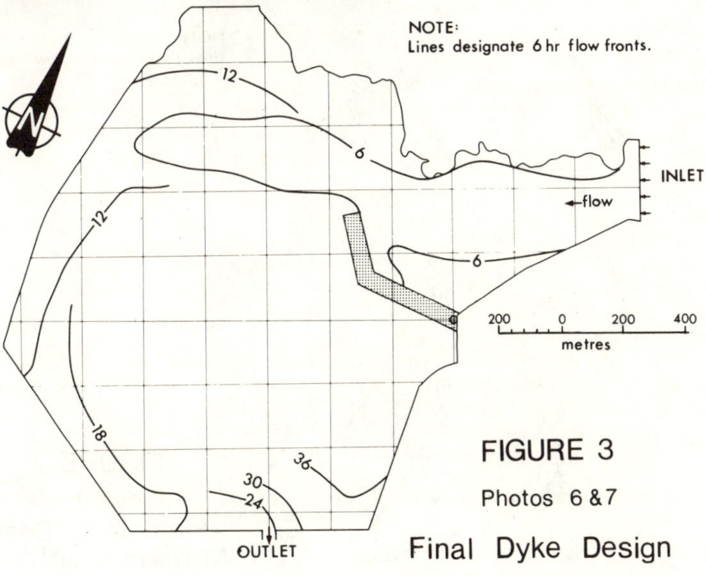

FIGURE 3

Photos 6 & 7

Final Dyke Design

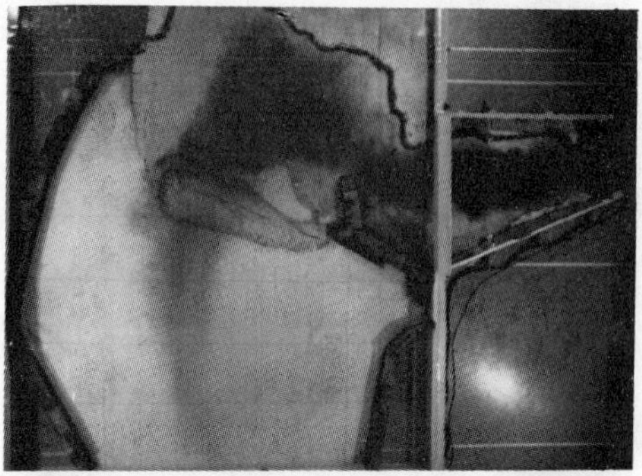

Photo 8; 12 hours

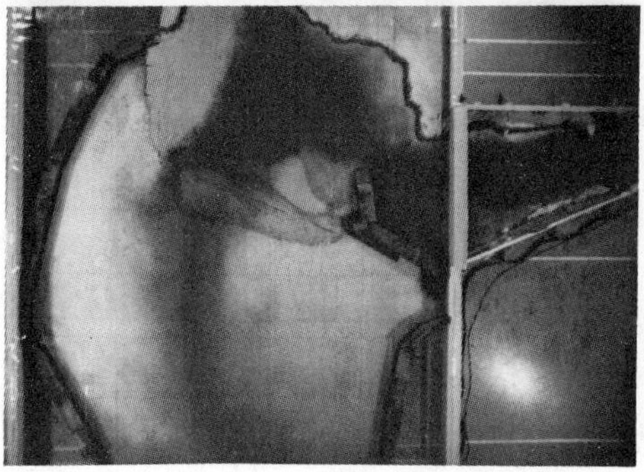

Photo 9; 24 hours

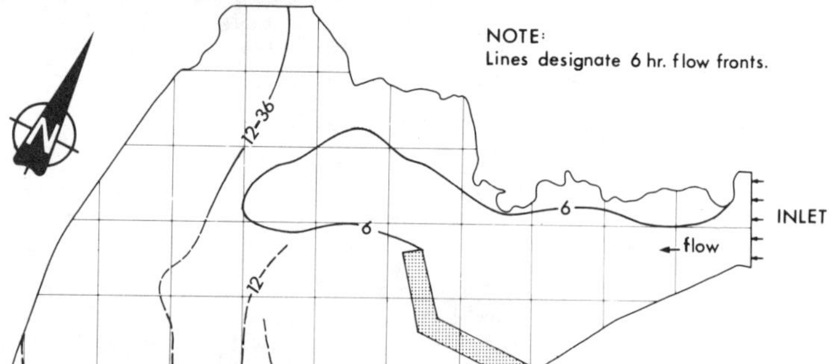

FIGURE 4
Photos 8 & 9
Final Dyke Design
With 4 m/s NNW Wind

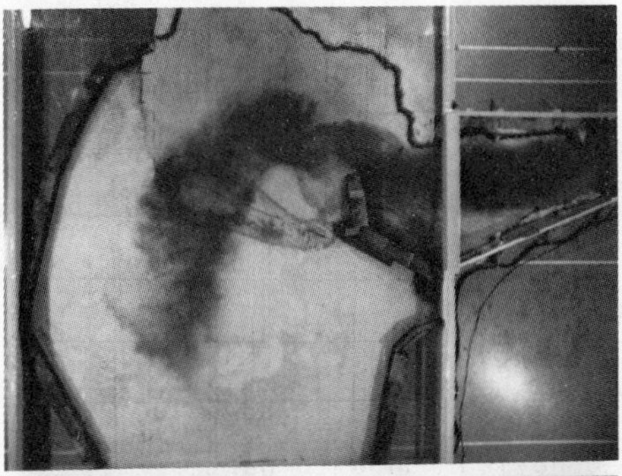

Photo 10; 12 hours

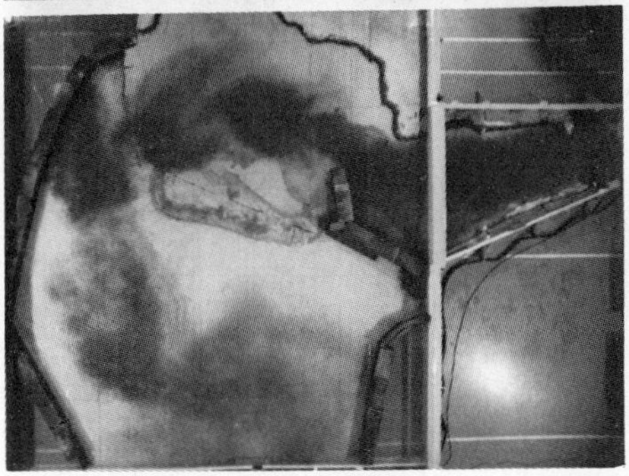

Photo 11; 24 hours

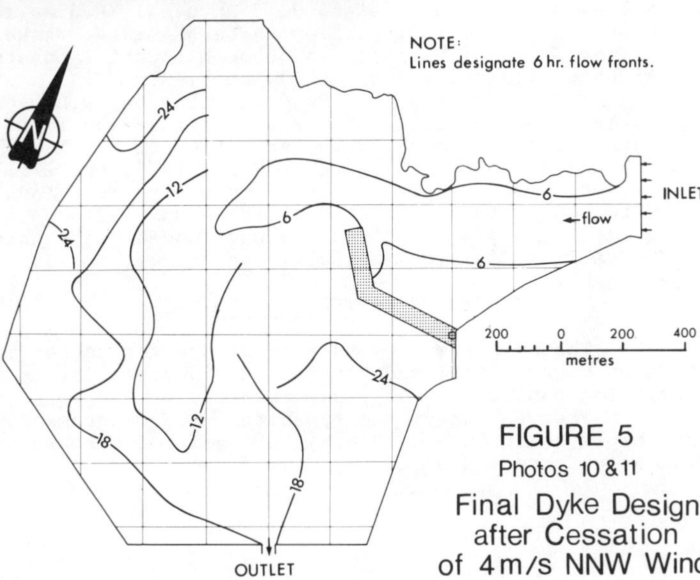

FIGURE 5
Photos 10 & 11
Final Dyke Design after Cessation of 4 m/s NNW Wind

selected through consultation with the project design engineers as being a viable alternative, it is not unrealistic to compare this construction cost savings against the cost of the model study. Roughly, the 150 metres of dyke represents a cost savings of approximtely 10 times the total model budget.

Although not as readily justifiable on a cost basis, two additional benefits of the model study are worthy of mention. Of primary importance, qualitative model information regarding circulation patterns made possible a subsequent numerical calculation of the thermal effectiveness of the cooling pond, conducted as a separate study. Secondly, observations of the model provided those involved with the study with a better general understanding of prototype behaviour and, in particular, the reaction of circulation patterns to wind activity.

CONCLUSION

The study presented herein demonstrates one of many instances for which a qualitative scale model can be highly cost effective. Not all hydraulic phenomenon lend themselves equally well to the sort of simplified analysis performed in this study, however the point of the matter is to draw attention to the role which a hydraulic design aid model can serve in the solution of practical engineering problems.

REFERENCES

1. Silberman, E., and Stefan, H., "Physical (Hydraulic) Modelling of Heat Dispersion in Large Lakes: A Review of the State of the Art", Argonne National Laboratory, Argonne, Illinois, Aug., 1960, pp. I-8.
2. Ali, K.H.M., Hedges, T.S. and Whittington, R.B., "A Scale Model Investigation of the Circulation in Reservoirs", Proceedings, Institution of Civil Engineers, Volume 65, Part 2, Mar., 1978, pp. 129-161.
3. Hecker, G.E., and Yale, G.A., "Modelling Wind Induced Water Currents", presented at the August 1973, Hydraulic Engineering and the Environment 21st Annual Hydraulics Division Specialty Conference, pp. 335-348.

ACKNOWLEDGEMENTS

The author wishes to acknowledge the cooperation of Calgary Power Ltd. and Montreal Engineering Co. Ltd. in permitting publication of study results.

Preparation of the typescript and diagrams for this paper has been assisted by the staff of Northwest Hydraulic Consultants Ltd.

THERMAL DISCHARGE INVESTIGATION

By Yung Y. Shen[1], M.ASCE and Richard S. McGinnis[2], M.ASCE

INTRODUCTION

Discharge modifications are proposed at Northern States Power Company's (NSP) Prairie Island Nuclear Generating Plant located on the Mississippi River approximately 40 miles (64 km) southeast of Minneapolis and St. Paul. Discussions with the Minnesota Pollution Control Agency (MPCA) regarding the National Pollutant Discharge Elimination System permit for the facility indicated several problems which led NSP to this proposal. The revised discharge configuration will eliminate recirculation of water to the plant intake and improve mixing in the main river channel. In conjunction with proposed intake modifications, this will also allow increased appropriation of water and thereby improve plant efficiency and offset the cost of modifications. A general plot plan showing the existing circulating water system and locations of the proposed discharge and intake structures is presented in Figure 1.

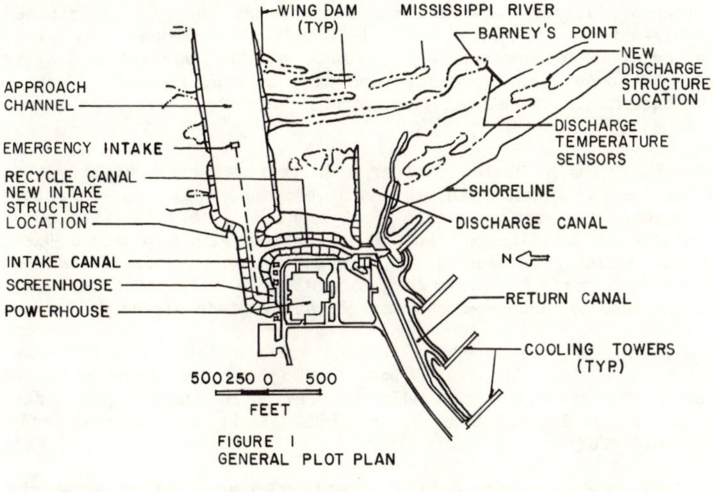

FIGURE 1
GENERAL PLOT PLAN

[1]Lead Hydraulic Engineer, Stone & Webster Engineering Corporation, Denver, Colorado

[2]Manager, Special Services, Northern States Power Company, Minneapolis, Minnesota

The proposed discharge modification will consist of four submerged pipes discharging into the Mississippi River's main channel approximately 2000 ft (610 m) south of the present discharge at velocities ranging from 8 to 10 ft/sec (2.4 to 3.1 m/sec) for all combinations of plant discharge flow (150 to 1390 cfs, 4.3 to 39.4 cms).

Prior to approval, regulatory agencies required a study of thermal effects under Section 316(a) of the Federal Water Pollution Control Act Amendments of 1972. This procedure required NSP to identify probabilities of various temperatures occuring in the river and the areas affected by those temperatures. Areas of concern extend downstream about 3500 ft (1.07 km) to Federal Lock and Dam No. 3 where complete mixing occurs. NSP decided to develop a mathematical model to predict areas of influence of certain isotherms for various combinations of river flow and plant discharge flow and thereby assess environmental impact.

HYDROTHERMAL MATHEMATIC MODELS

Because temperature distribution of heated discharge is governed by different heat transport mechanisms in the zone where jet diffusion predominates and in the area beyond this zone, two different models were used for these regions. In the zone close to the discharge point (near-field), free turbulence created by shearing action of the discharge with respect to the ambient water body causes jet diffusion of the heated water. Outside this near-field zone there is a considerably larger far-field region in which heat is distributed by buoyancy-driven currents and by diffusion caused by ambient turbulence. Furthermore, the process at the air-water interface, i.e., atmospheric heat exchange and wind stress, further affects heat distribution. Mathematical backgrounds for both models are described below.

NEAR-FIELD MODEL - Heated discharge is released horizontally into the ambient receiving water as submerged buoyant jets. However, because submergence over the pipes is equivalent to one pipe diameter, the thermal plume will reach the water surface within a short distance from the outlet. Consequently, analysis within the near-field zone is treated as a surface discharge jet instead of attempting to integrate the phenomena of submerged buoyant jet and surface discharge jet.

Various mathematical models of heated surface jets are available for prediction of two and three dimensional plume configurations. Jirka et al (1975) present a comprehensive review of thermal plume models. The model of Shirazi and Davis (PDS Model) is selected for the near-field study.

The basic partial differential equations discribing steady-state in the three-dimensional model consist of the equations of continuity, momentum and heat flow. These equations, shown in Cartesian coordinate system, are:

$$\frac{\partial u}{\partial x} + \frac{\partial v}{\partial y} + \frac{\partial w}{\partial z} = 0 \qquad (1)$$

$$\frac{\partial(u^2)}{\partial x} + \frac{\partial(uv)}{\partial y} + \frac{\partial(uw)}{\partial z} = \frac{g}{\rho a}\int_z^{\infty} \frac{\partial(\Delta\rho)}{\partial x}\, dz - \frac{1}{\rho a}\frac{\partial(Pd)}{\partial x} \tag{2}$$

$$\frac{\partial(uv)}{\partial x} + \frac{\partial(v^2)}{\partial y} + \frac{\partial(vw)}{\partial z} = \frac{g}{\rho a}\int_z^{\infty} \frac{\partial(\Delta\rho)}{\partial y}\, dz - \frac{1}{\rho a}\frac{\partial(Pd)}{\partial y} \tag{3}$$

$$\frac{\partial(uw)}{\partial x} + \frac{\partial(vw)}{\partial y} + \frac{\partial(w^2)}{\partial z} = \frac{-1}{\rho a}\frac{\partial(Pd)}{\partial z} \tag{4}$$

$$\frac{\partial(uT)}{\partial x} + \frac{\partial(vT)}{\partial y} + \frac{\partial(wT)}{\partial z} = 0 \tag{5}$$

$$\Delta\rho = \beta(T_a - T) \tag{6}$$

in which:

x, y, and z are Cartesian coordinates with x in direction of flow
u, v, and w are time-averaged velocities in x, y, and z
T_a is the ambient temperature
$\Delta\rho$ is the local density difference
ρa is the ambient density
g is the gravitational acceleration
β is the coefficient of thermal expansion
Pd is the dynamic pressure
T is the jet temperature

In order to reduce the above partial differential equations to ordinary differential equations, a Gaussian distribution is assumed on the lateral and vertical distribution of jet velocity and temperature. With assumed temperature and velocity profiles, the energy, volume, and momentum fluxes can be integrated across the plume at any cross-section.

Solutions by the PDS model are limited to a uniform and constant surface discharge of heated water from a rectangular channel into a large and deep water body that is either at rest or moving at a uniform and constant velocity. Effects of bottom and shoreline geometry on thermal plume are not included in the PDS model. Therefore, a subroutine to correct thermal plume configuration due to bottom and shoreline attachment was added to the program. Procedures describing these corrections are discussed in the following sections.

THERMAL PLUME CORRECTION DUE TO BOTTOM GEOMETRY - A heated plume discharged into a shallow receiving water is significantly affected by the bottom geometry. An obvious parameter to measure the relative water shallowness is the ratio of h_{max}/d, where h_{max} is the maximum jet depth and d is the local water depth. Using this parameter, experiments (Koester, 1974) show that for $h_{max}/d \leq 0.75$, the jet was typically unaffected by the presence of the bottom geometry. The dilution reduction factor, r_s, resulting from a best-fit of the experimental data is:

$$r_s = \left(\frac{0.75}{h_{max}/d}\right)^{0.75} \quad \text{for } h_{max}/d \geqslant 0.75 \tag{7}$$

Other laboratory and field data examined by Jirka et al (1975) show substantial agreement with Equation 7. Correction of plume centerline temperature due to bottom geometry, as suggested in the MIT Report (Jirka, et al, 1975) is:

$$\left(\frac{\Delta Tc}{\Delta To}\right)_{shallow} = \left(\frac{\Delta Tc}{\Delta To}\right)_{deep} \frac{\left[1 - \left(\frac{\Delta Ts}{\Delta To}\right)_{deep}\right]}{(r_s - 1)\left[1 - \left(\frac{\Delta Tc}{\Delta To}\right)_{deep}\right] + \left[1 - \left(\frac{\Delta Ts}{\Delta Tc}\right)_{deep}\right]} \tag{8}$$

Terms in Equation 8, $(\Delta Tc/\Delta To)_{deep}$, represents theorectical temperature prediction from PDS model. ΔTo is the temperature increase at the discharge point. ΔT_s is the temperature increase at the distance to which the near-field temperature predictions are valid.

THERMAL PLUME CORRECTION DUE TO SHORELINE ATTACHMENT - The presence of a downstream shoreline boundary may under certain circumstances, limit entrainment of ambient water into the jet from one side of the plume resulting in reentrainment of already mixed water into the jet.

Evaluation of available field and laboratory data suggested that some of the most important factors in controlling shoreline attachment are: (1) cross flow ratio, $R = Ua/Uo$, (2) shallowness factor, h_{max}/d, and (3) discharge angle θ. For a given discharge angle and a straight shoreline, a criterion for shoreline attachment as suggested in the MIT Report is (Jirka et al, 1975):

$$R\,(1 + \cos\theta) > 0.05\,(h_{max}/d)^{-3/2} \tag{9}$$

A method of image was used to correct for shoreline attachment which reflects the plume with respect to the shoreline. At any point (x, y), the temperature rise ΔT after the correction for shoreline attachment is computed as follows:

$$\Delta T = \Delta T_1 + \Delta T_2 - \frac{\Delta T_1 \times \Delta T_2}{\Delta T_o} \tag{10}$$

in which ΔT_1 and ΔT_2 are the temperature increases with shallowness correction at real and image plumes, respectively.

Centerline temperature rise and characteristic plume width and depth for the real and image plumes are not necessarily equal at any location (x, y). To reduce computational steps, only the area between the plume centerline trajectory and the shoreline has been included in the calculation of the shoreline correction. Shoreline attachment effects upon the area riverward of the centerline trajectory can generally be neglected, except in the case of extremely strong crossflows.

FAR-FIELD MODEL - An unsteady three-dimensional hydrothermal model developed by Yeh and Tsai (1976), was used for prediction of far-field temperature distribution. The base partial differential equation describing heat dispersion phenomenon is:

$$\frac{\partial(\Delta T)}{\partial x} + u\frac{\partial(\Delta T)}{\partial x} = Kx\frac{\partial^2(\Delta T)}{\partial x^2} + Ky\frac{\partial^2(\Delta T)}{\partial y^2} + Kz\frac{\partial^2(\Delta T)}{\partial z^2} - E(\Delta T) + S \qquad (11)$$

in which:

ΔT is the temperature rise
u is the velocity in the x direction
Kx, Ky, and Kz are eddy diffusivities in x, y, and z directions, respectively
S is the heat input
E is the surface heat exchange coefficient

Predictions of temperature in the Mississippi River, obtained from Equation 11 were compared with field surveys conducted in the Mississippi River near the Prairie Island Nuclear Generating Plant. Model calibration and its initial conditions were presented in a separate report (Stone & Webster, April 1978).

INTEGRATION OF ZONE MODELS

Maintaining continuity in predictions over the entire area of thermal discharge influence requires linkage of predictions in the near-field with those of the far-field zone. There is no clear-cut transition between these two zones. An intermediate region with mixed near- and far-field effects in which buoyant convection plays a dominant role is normally found between these two zones. No firm concepts have been advanced for this intermediate range.

The concept of heat conservation can be used as a rough approximation of the true physical situation to link the near-field and far-field models. Based on this concept, with an assumed source strength and discharge area and a given ambient velocity field, isothermal configurations predicted by the far-field model can be used to match those predicted by the near-field model at the transition zone. The following difficulties are encountered in trying to achieve this match:

1. Heat flux is proportional to the product of initial temperature rise and source discharge rate. If source discharge velocity is assumed to equal ambient river velocity, three variables remain to constitute the heat conservation; initial temperature rise, source width and source depth. Due to the number of cases studied, assumed values of the strength and dimension of the soure were relatively random and could not be consistent.

2. By assuming an initial temperature rise for the far-field model of 1.5 times that predicted by the near-field model at the transition point, source width is generally greater than 1,000 feet for the cases with large plant discharge (about 1,400 cfs). This is due to the limit of available river depth. This assumed factor of 1.5 in the initial temperature rise produces unrealistic source dimensions for the far-field model.

3. For most available far-field models, inital jet momentum is neglected. However, at the transition point, plume average velocity is still at least twice the ambient velocity. If far-field model predictions are linked directly with those of the near-field model, plume size within the intermediate zone is over-estimated. This is due to slow heat dissipation in the far-field.

Because of the above difficulties, a "best-fit" approach was used to link temperature predictions of the two models. In this method, for an assumed source width and depth and temperature rise for the far-field model, three profiles were used as a basis to compare the results of the two models. These are: (1) temperature profile with respect to depth along the plume centerline (Figure 2), (2) temperature profile with respect to plume width at the water surface (Figure 3), and (3) surface isothermal configuration (Figure 4). Temperature predictions by the near-field model at the transition point were used as a reference for comparisons. For each range of plant discharge, assumed source width and depth for the far-field model generally remain the same. Some deviations ($\pm$ 10 percent) resulted from variation in crossflow. Assumed temperature rises for virtual sources range from 15 to 25 percent higher than those predicted by the near-field model at the transition point.

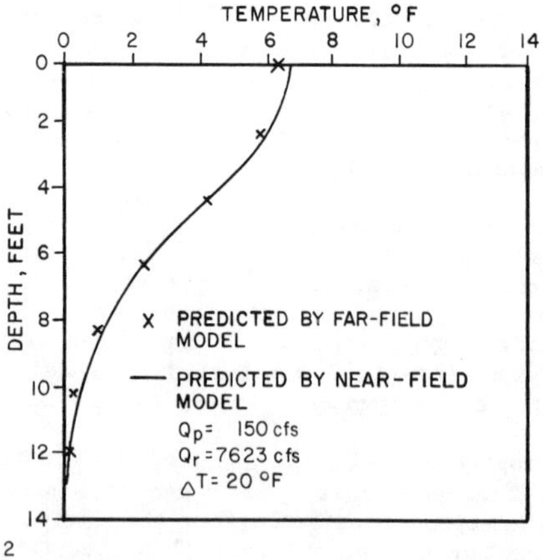

FIGURE 2
TEMPERATURE DISTRIBUTION VS. DEPTH AT CENTERLINE OF PLUME

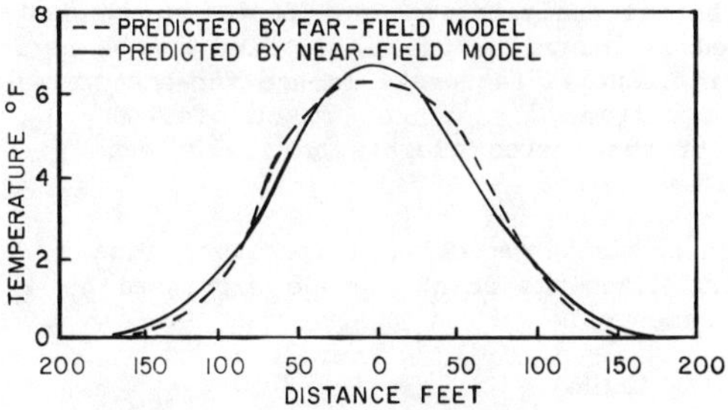

FIGURE 3
TEMPERATURE DISTRIBUTION VS. PLUME WIDTH AT WATER SURFACE

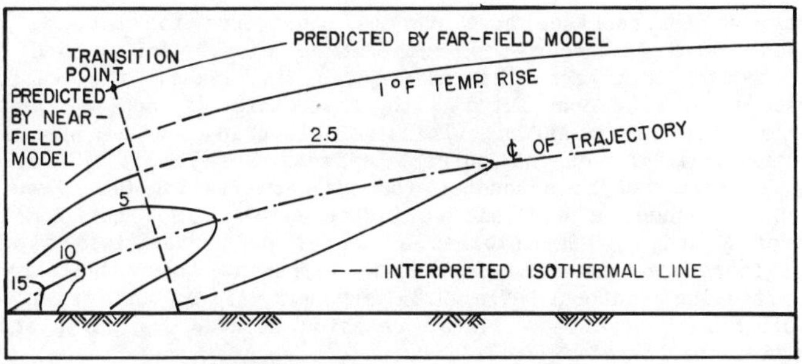

FIGURE 4
SURFACE ISOTHERMAL CONFIGURATIONS

CORRELATION STUDIES

NSP uses a simulation model to assess circulating water system performance. The combination of near- and far-field models discussed above is not readily adapted to a simulation technique because of the problem of linkage of the models and the computer time required. Therefore a need existed to develop a correlation describing specific elements such as plume centerline temperature at 1,000, 2,000, and 3,000 ft (305, 610, and 915 m) downstream from the discharge point and areas within various isotherms for use in the simulation model.

Heat discharge from the system into the Mississippi River is characterized by: (1) densimetric Froude number, Fo, (2) ratio of plume width to depth, A, (3) crossflow ratio, Ua/Uo, (4) initial discharge angle with crossflow, θ, (5) river bottom geometry, (6) shoreline attachment, and, (7) existing river flow turbulence. Effects of these factors on the temperature distributions can be presented as a series of curves. Since items four through six are fixed for the thermal analysis, their effects on the heated discharge can be considered as independent variables and can be neglected in the correlation of influential factors. In the far-field region, existing flow turbulence (Item 7) is a major factor in determining characteristics of the heated plume, and its effect is considered to be included in Item 3.

For a given plant discharge flow, temperature rise (Δ T) at some distance from the discharge point can be expressed as a function of the following parameters:

$$\Delta T/\Delta To = F(Fo, A, Ua/Uo) \qquad (12)$$

in which Δ To is initial temperature rise. The term Ua/Uo can be expressed as Qr/Qp in which Qr and Qp are the river flow and plant discharge flow respectively.

Factors governing surface area can be considered the same as those discussed above in regard to temperature rise. Asbury and Frigo (1971) showed that for thermal plumes in lakes, the ratio of temperature can be correlated with the ratio of surface area to discharge flow rate, As/Qp. This relationship can be applied to rivers as well if ambient current effects are small. Effects of ambient current can be accounted for with the factor Qr. Even this approach introduces a dimensional factor $AsQr^{0.85}/Qp$, but for given ranges of A and Fo, it combines a series of curves into a single curve. Therefore, it is selected to represent the correlationship among affecting factors. The 0.85 exponent of Qr was selected by determining the slope of a linear relation between log As/Qp and log Qr for given ratios of Δ T/Δ To.

DISCUSSION OF RESULTS

A total of 61 specific cases defined by NSP were evaluated to predict the heated plume distributions in the vicinity of the discharge area. An additional 37 cases were added for the correlation study.

Water surface isothermal configurations for two of the study conditions are shown in Figures 5 and 6. The distortion of plume temperature profile due to crossflow and shoreline attachment can be seen in these figures. Distortion can be significant in the case of strong crossflow (Figure 5) compared to weaker crossflow conditions (Figure 6). It was found that bottom geometry (or shallowness) affects only plumes with higher discharge flow rates (Qp $\geqslant$ 1,375 cfs, 39.0 cms). If the plume is affected by the river shallowness, high temperatures will persist due to increased bottom interference. For most cases studied, river flow rates were less than 10,000 cfs (283.4 cms). This results in a low main channel velocity ($\leqslant$ 0.5 fps, 0.15 mps) and slow heat dissipation in the far-field zone.

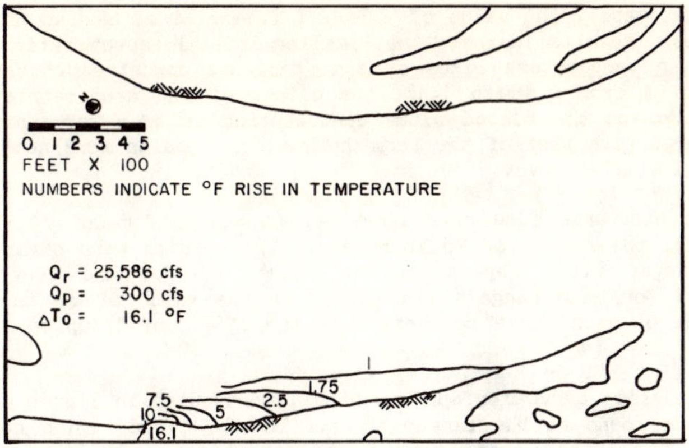

FIGURE 5
PREDICTED ISOTHERMAL CONFIGURATION IN THE MISSISSIPPI RIVER

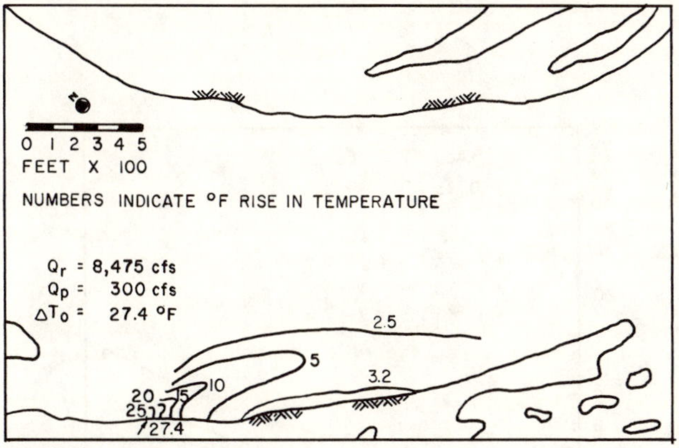

FIGURE 6
PREDICTED ISOTHERMAL CONFIGURATION IN THE MISSISSIPPI RIVER

Water quality standards established by MPCA require, in part, that temperature rise of the receiving water be limited at the edge of the mixing zone. Past administrative practices of the MPCA suggest that, as a guideline, mixing zones extend 1,000 ft (305 m) from the point of discharge.

The established correlation curves are presented in Figure 7 for the predictions of temperature rise at 1,000 ft (305 m) from the discharge point. It can be seen from this Figure that for a given value of Qr/Qp, as the value of A (the ratio of initial heated plume width to depth) decreases, the value of $\Delta To/\Delta T$ (the rate of heat dissipation) increases. Results indicate that smaller initial aspect ratios result in smaller temperature rises at some distance downstream. For given values of A from 0.45 to 1.18, the effect of the densimetric Froude number, Fo, on the heated plume distribution is of a secondary order as compared with that of the term Qr/Qp. The results were combined to develop a single curve.

As plant discharge flow rate increases (A = 1.75 for Qp = 1,375 cfs, 39.0 cms), the effect of Fo increases. The results were divided into three ranges with respect to the values of Fo, and curves were developed for each range (Figure 7). As the value of Fo increases, the slope of each curve reduces, i.e. the effect of Qr/Qp on $\Delta To/\Delta T$ lessens.

The correlation curve for surface area is presented in Figure 8. Some scatter is found at the larger areas. The far-field region comprises most of the larger area where large variations in plume areas are found only for small temperature variations.

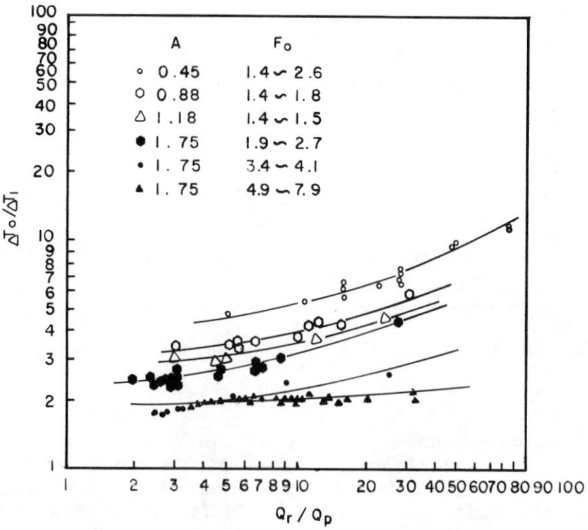

FIGURE 7

CORRELATION BETWEEN Q_r/Q_p AND $\Delta T_o/\Delta T_i$ AT 1000 FEET

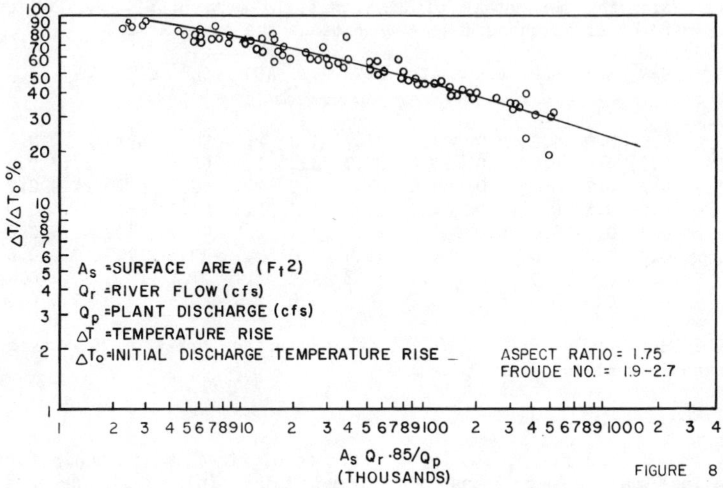

FIGURE 8
AREA OF SURFACE ISOTHERMS

SUMMARY

Based on equations developed in the correlation study described above, temperature at plume centerline at 1,000, 2,000, and 3,000 ft (305, 610, and 915 m) downstream from the point of discharge were calculated (Table 1) using NSP circulating water system simulation model. Likewise, areas within various isotherms were calculated (Table 2). Results such as presented in Tables 1 and 2 were given to the project biological consultant for assessment of environmental impact. Study results were presented to regulatory agencies and agreement has subsequently been reached on the proposed modes of operation. NSP plans to begin construction of the proposed modification in 1981 and operation in 1982.

Table 1 Percent of time calculated mixed ΔT is equalled or exceeded at 1,000 ft (305 m) downstream from the discharge along the plume centerline.

TEMP F°	JAN*	FEB*	MAR*	APR	MAY	JUN	JUL	AUG	SEP	OCT	NOV*	DEC*
13	16.5	12.9	1.8								2.7	16.6
12	66.2	70.4	98.5								9.9	55.6
11	98.9	95.8	58.8								18.9	85.1
10	100.0	97.1	70.2								29.4	93.8
9	100.0	99.0	78.5							0.2	46.6	98.7
8	100.0	100.0	87.2							1.5	75.4	99.8
7	100.0	100.0	94.9						0.1	6.5	88.6	99.8
6	100.0	100.0	96.6					0.0	1.0	20.6	92.0	100.0
5	100.0	100.0	98.4			0.2	0.0	0.8	6.4	47.7	98.5	100.0

* - Assumes open cycle operation (no cooling towers).

Table 2 Estimated percentage of time that the area within the 20° C (68.0° F) isotherm equals or exceeds the given areas.

AREA HECTARES	JAN	FEB	MAR	APR	MAY	JUN	JUL	AUG	SEP	OCT	NOV	DEC
0.001	0.0	0.0	1.0	0.0	93.6	27.7	1.0	2.1	56.1	43.3	27.9	0.2
0.005	0.0	0.0	0.8	0.0	81.2	27.7	1.0	2.1	56.1	43.1	27.7	0.2
0.01	0.0	0.0	0.6	0.0	65.8	27.7	1.0	2.1	56.0	41.8	24.4	0.2
0.05	0.0	0.0	0.0	0.0	24.5	26.6	1.0	2.1	51.7	25.4	9.5	0.0
0.1	0.0	0.0	0.0	0.0	13.3	25.0	1.0	2.1	47.4	18.2	2.5	0.0
0.5	0.0	0.0	0.0	0.0	4.9	16.1	1.0	2.1	33.8	7.5	0.0	0.0
1.0	0.0	0.0	0.0	0.0	3.3	12.8	1.0	2.1	27.3	3.9	0.0	0.0
2.5	0.0	0.0	0.0	0.0	0.7	6.7	1.0	2.0	20.3	1.0	0.0	0.0

ACKNOWLEDGMENT - The correlation study for surface area, by Mr. James G. Kuiken is gratefully acknowledged.

REFERENCES

Asbury, J. G. and Frigo, A. A., "A Phenomenologized Relationship for Predicting the Surface Areas of Thermal Plums in Lakes", Argonne National Laboratory, Report No. ANL/ES-5 (1971).

Jirka, H., Gerhard et al, "An Assessment of Techniques for Hydrothermal Prediction", Department of Civil Engineering, M.I.T., Technical Report No. 203, 1975.

Koester, G. E., "Experimental Study of Submerged Single-Port Thermal Discharge", M.S. Thesis, Department of Civil Engineering, M.I.T., 1974.

Shirazi, M. A. and Davis, L. R., "Workbook of Thermal Plume Prediction, Volume 2, Surface Discharge", Environmental Protection Technology Series, Report No. EPA-R2-72-0056, U.S. Environmental Protection Agency, Corvallis, Oregon, 1974.

Stone & Webster Engineering Corporation, "Thermal Discharge Analysis, Prairie Island Nuclear Generating Plant, Northern States Power Company, April 1978".

Stone & Webster Engineering Corporation, "Thermal Discharge Analysis, Alternate Discharge System, Prairie Island Nuclear Generatin Plant, Northern States Power Company, June 1978".

Stone & Webster Engineering Corporation, "Supplement to Report on Thermal Discharge Analysis, Alternate Discharge System, Prairie Island Nuclear Generating Plant, Northern States Power Company, January 1979".

Yeh, G. T. and Tsai, Y. J., "Analytical Three-Dimensional Transient Modelling of Effluent Discharge", Water Resource Research, Vol. 12, No. 3, P. 533-540, June 1976.

MODELING HOURLY WATER TEMPERATURE DYNAMICS

Russ T. Brown[1]

ABSTRACT

An hourly water temperature model of a 27 km reach of Guntersville Reservoir was schematized as a series of eight fully mixed segments with three adjacent embayment volumes. A one-dimensional unsteady flow model provided the exchanges between segments. Heat balance formulations similar to those for daily scale models were used. Simulations of four months were used to demonstrate the seasonal response of the model. Daily and hourly scale dynamics were compared with data from the downstream boundary. Results of sensitivity analyses for October and January indicated that the model was responsive to the heat exchange rate coefficients and that the initial estimates produced the best match with the data. The April and August simulations demonstrated that the model was limited by the fully mixed assumption when actual conditions were slightly stratified.

I. INTRODUCTION

The development of hourly scale water temperature models involves much more than changing the time step of the models from a day to an hour. The entire modeling system, which includes the individual heat exchange and mixing process formulations, the required meteorologic, hydrologic and inflowing temperature data, and the recorded data that are used to calibrate and validate the model, must be transformed to an hourly level of detail.

Several processes influence water temperatures at an hourly time scale. The diurnal heating caused by solar radiation and the convective nighttime mixing due to surface cooling are dominant processes that cannot be accurately described with a daily scale model. The intermittent stratification and heating of the surface temperatures that develop during the day cause the surface heat exchange rates to shift relative to those calculated from daily average temperatures. Wind-induced mixing is another hydrothermal process that often occurs at a time scale of less than a day. An hourly scale model is more appropriate for simulating these near surface temperature dynamics that are driven by diurnal variability in the meteorology. Hourly scale temperature modeling is required to accurately simulate water temperature effects of peaking power releases from dams. Tailwaters or downstream reservoir temperatures can be significantly affected by these variable flows. An application of hourly scale temperature modeling on Guntersville Reservoir in northeast Alabama is described.

[1]Civil Engineer, Tennessee Valley Authority, Norris, Tennessee

II. APPLICATION OF THE HOURLY TEMPERATURE MODEL

A. Study Reach

The area of interest is a 27 km reach of Guntersville Reservoir as shown in Figure 1. The upstream portion of Guntersville Reservoir is relatively narrow (500 m), approximately eight m deep and can be described as riverine with flows of 300-2000 m^3/sec and corresponding velocities of 0.1-0.5 m/sec. The associated turbulence is usually sufficient to prevent any persistent stratification from developing. Releases from the upstream dam, Nickajack, produce significant variations in both flow and surface elevation in this reach. Several large embayments within the reach complicate the model schematization. These areas are considered as storage zones in the flow model and exchange with the main reservoir volume as the reservoir surface elevation changes.

The main reservoir segments have a mean depth of three-four m, while the embayment segments are more shallow with mean depths of one-two m. With summertime river flows averaging 765 m^3/sec and wintertime flows averaging 1250 m^3/sec, travel times through the main channel portion of the model are about two days in the summer and one day in the winter. Flows are quite variable from day to day from 300 m^3/sec to 2000 m^3/sec, so the travel times vary from less than a day to over three days. Hourly flow variations will not affect the travel time through the reach, but are important in governing the heating effects from the power plant, the exchange with the embayments and the interactions between solar heating and vertical mixing.

B. Available Data

A survey and compilation of available hourly data is an important step in the development of an hourly scale temperature model. A model can only be adequately developed and validated if there exists sufficient meteorological, flow, and water temperature data for specifying the boundary conditions and calculating the heat exchange rates. Recorded hourly temperature data that span a suitable range of conditions to provide data for calibration-modification exercises and independent validation data must also be available.

Hourly meteorological data including drybulb and dewpoint temperatures, windspeed, and solar radiation are available at Widows Creek. Atmospheric radiation terms were estimated throughout the study. Solar radiation, air temperatures and windspeed data were estimated during periods of lost record to provide a complete hourly data base of meteorological and radiation variables. Details are discussed in Reference 1.

Hourly flow data were available at the upstream and downstream dams. The flow model calculated flows between each model segment for each hour along with volumes and surface areas. These hourly data describing the physical and hydraulic characteristics of the study reach were arranged into monthly files for input to the temperature model.

Two automatic water temperature monitors provide data that can be used to guide the formulation of the hourly model algorithms,

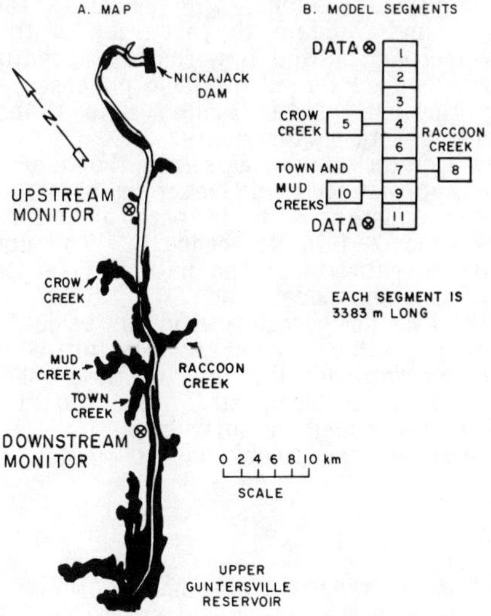

Figure 1 — Geometrical Schematization of Guntersville Reservoir

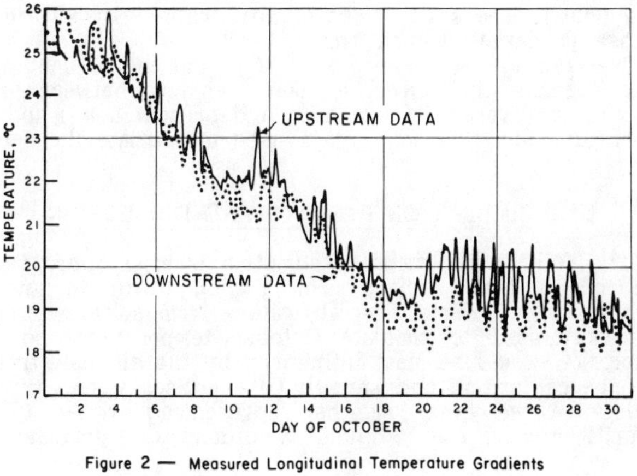

Figure 2 — Measured Longitudinal Temperature Gradients

specify boundary conditions, calibrate the modeled heat transfer and mixing process rates, and validate the accuracy of the model output under a range of meteorological and flow (mixing) conditions. Because the same data cannot be used for all of these purposes, decisions must be made regarding the use of the limited data to best meet these important but conflicting data requirements.

An hourly monitor at the upstream boundary of the study reach is located in approximately six-seven m of water and records temperatures at several depths. It is part of a network of water temperature stations described in Reference 2. The upstream monitor was used to provide an estimate of the natural river temperatures at the upstream boundary of the model reach.

The downstream temperature monitor varies in depth from approximately 0.5-1 m. The diurnal temperature dynamics at the depth of the sensor are recorded but no information about the vertical gradients at this location is obtained. These data were used to calibrate and modify the model formulations. It was also the only resource for validation of the model on this reach of Guntersville Reservoir.

C. Model Formulation

The eight main reservoir segments as well as the three embayment segments (Figure 1) are considered to be fully mixed at all times. The flow model calculates the advective exchange rates between the segments of the main reservoir which are simply the hourly flows at the boundaries of the segments. The exchanges between the embayments and the adjacent reservoir segments are computed from the change in volume of the embayments which depends on the change in water surface elevation at the adjacent reservoir segment. For simplicity, no longitudinal dispersion terms are included in the model.

The heat transfer formulations follow methods that are well known from daily scale temperature modeling (Reference 3). No changes have been made to any of the coefficients in transforming them to an hourly time scale. These initial rate estimates are used in the base case simulation of each month.

The average of the previous hour and the present hour flow is used to estimate the advective heat transfers between segments. The surface area (average of last hour and present hour) and the last hour temperature for each segment is used to estimate the net surface heat flux.

III. SIMULATION RESULTS AND DISCUSSION

The fully mixed model assumption is most appropriate when convective mixing from surface cooling along with turbulent river mixing minimizes any surface stratification from solar heating. The model was first used to simulate October temperatures so that the initial calibration would be least influenced by the diurnal stratification effects. Once calibration and sensitivity exercises were completed for October data, three other months representing winter (January); spring (April); and summer (August) conditions were simulated.

A. October Simulations

The upstream and downstream temperatures for October 1978 are shown in Figure 2. The average longitudinal temperature gradient was only -0.2°C (cooling). The entire reservoir reach was responding to the meteorology simultaneously in time, and uniformly in space. The uniformly high solar radiation produced a large diurnal variation in the drybulb air temperatures and a correspondingly large diurnal variation in the water temperatures recorded at both.

The initial base case simulation results for October are shown in Figure 3. The comparison between the model and the data should be made at all time scales of significant variability. In the present modeling study, the seasonal time scale is important; the calibration of the model for the four representative months will provide an indication of the ability of the model formulation to accurately predict temperatures for the seasonal range of meteorological conditions. The daily time scale is also important so that the variations due to storm fronts and other weather patterns can be accurately modeled. For October, the daily scale match was excellent. The third important time scale in the present study is the hourly or diurnal variations. A match at this time scale demonstrates that the model is accurately responding to the range of meteorological variation that characterizes the diurnal heating and cooling. For October, the match of the diurnal (hourly) time scale was also very impressive.

Although temperatures at the model inflow and the model outflow are almost identical throughout October, this does not suggest that the modeled heat exchange processes are unimportant. The diurnal heating-cooling effects cannot be simulated without accurate heat balance formulations. Neither will the outflow temperatures simultaneously follow the inflow temperatures unless the heat exchanges within the model are accurately simulated. Figure 4 provides a graphic illustration of the importance of the heat exchange formulations during October by showing the model results assuming no surface heat exchange. The diurnal variations were gone and the outflow temperatures were too high since they were the inflow temperatures lagged by the travel time through the reach (which is two-three days). The modeled heat exchange processes must be formulated correctly to maintain the proper temperatures throughout the modeled reach, even when no longitudinal temperature gradient is present.

The model validity was further tested with a series of sensitivity simulations for October in which one of the modeled heat exchange rates was increased or decreased. The evaporative cooling was modified to determine the effects on the predicted temperatures. Figure 5 shows the recorded downstream temperatures compared to the modeled temperatures for 2X and 0.5X evaporation. The model appears quite sensitive to the evaporative cooling estimates. The overestimated evaporation produced a net cooling of 1-3°C below the data. The simulated temperatures with the evaporation heat flux reduced to half of the base case were 1-2°C warmer than the data. This is fairly strong confirmation that the evaporative heat exchange formulation is both important for the overall heat balance and quite accurately estimated in the present model. In both these modifications to the evaporative heat fluxes, the diurnal component of the temperature dynamics was not significantly changed.

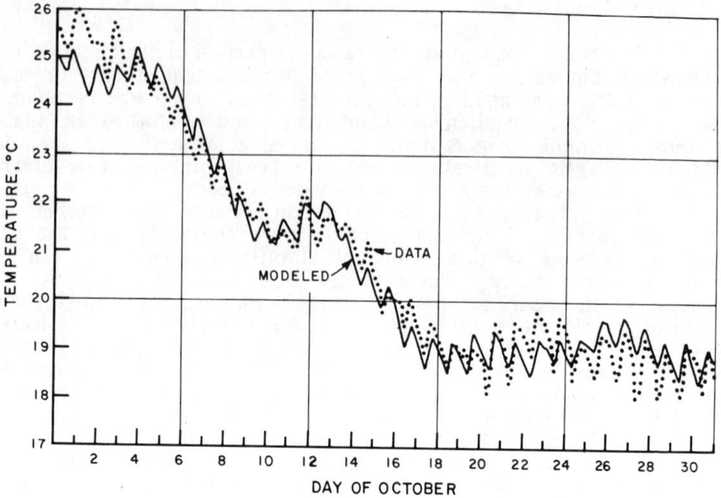

Figure 3 — Comparison of Downstream Data with Base Case Simulated Temperatures

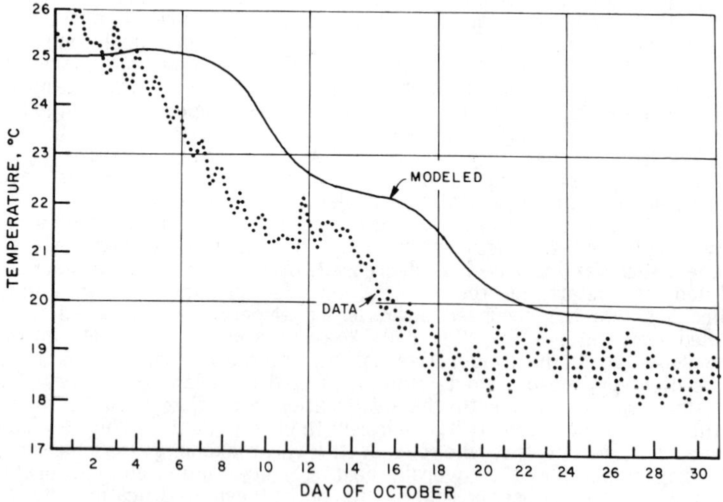

Figure 4 — Downstream Data Compared to Modeled Temperatures with No Surface Heat Exchange

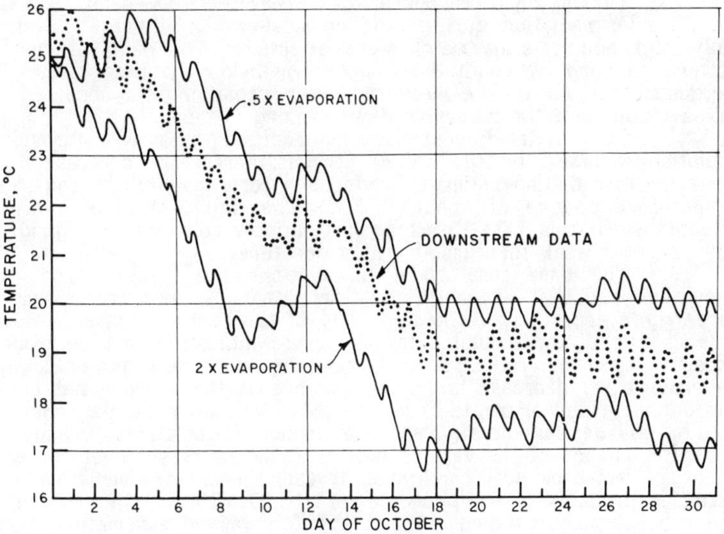

Figure 5 — Comparison of Downstream Data with Modeled Temperatures Using 2X and .5X Evaporation Heat Flux

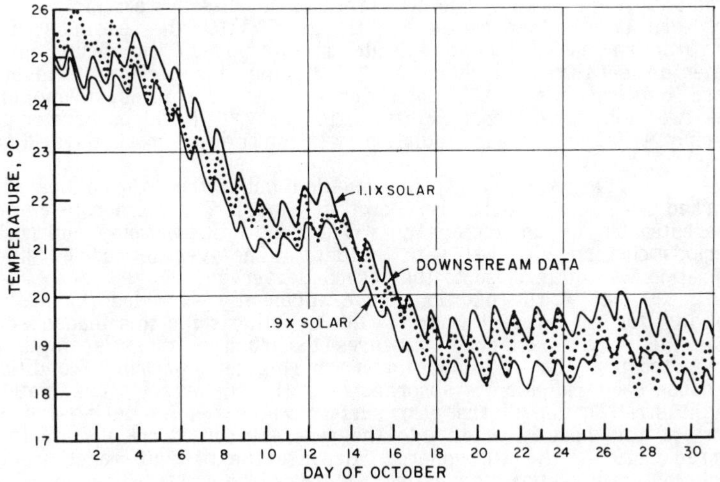

Figure 6 — Comparison of Downstream Data with Modeled Temperatures Using .9X and 1.1X Solar Input

The modeled response to increased (1.1X) and decreased (0.9X) solar radiation during October is shown in Figure 6. Both the daily and hourly scale match were affected. The daily temperature pattern changed by about 0.5°C above or below the base case. This suggests that all of the measured solar radiation was absorbed and mixed throughout the reservoir depth during October.

The modeled response to modified longwave atmospheric radiation was also tested. A 20 percent increase or decrease in the atmospheric radiation estimate produced a corresponding average water temperature change of about 1.5°C. The diurnal variation was not affected since this heat flux term is relatively constant throughout the day, being a weak function of air temperature.

The base case model formulations for October produced a particular net heat exchange that simulated the downstream temperatures quite accurately. Changing any of the terms produced a noticeable change in the simulated temperatures, indicating that the model is sensitive to each term. But an increase in one term might be counterbalanced by a decrease in another such that the average net balance throughout a day or month is maintained. Validation of the model must include a demonstration that the model formulations produce an accurate balance of the various heat transfer processes over the entire range of meteorological conditions at each important time scale. The model is probably more applicable for the October conditions than for the other months studied because the fully mixed assumption is more realistic. The range of conditions for which the simulations remain valid must be determined.

B. April Simulations

The temperature data at the upstream and downstream monitors for April 1978 showed that the longitudinal temperature gradients were small. The temperatures increased from 13°C to 18°C by April 17 and then cooled slightly to 17°C for the remainder of the month. The major cooling episode of April 19-20 is reflected in the water temperatures which cooled 2°C during this period. Windspeeds were between five and 15 mph during most afternoons, dropping to one-two mph during most nights. Wind effects should be greater than during October. Solar radiation was strong on most days during April.

The base case April simulation using the formulations which worked well for October is shown in Figure 7. Temperatures were overestimated by an average of 2°C although the warming and cooling periods corresponded well with the data. The average modeled diurnal variation was quite close to the recorded average.

The April conditions have apparently exceeded the limits of validity for the present model. One hypothesis for this inadequacy is that the fully mixed model captures too much of the solar input and overestimates the heating under springtime warming conditions. Because solar radiation is absorbed near the surface of turbid water, a slight stratification of the near surface water results during the day. Most of this heat is mixed into the reservoir but some of it is transferred back to the atmosphere. Because the present model does not include any vertical structure, these diurnal stratification effects cannot be accurately reproduced. One possible remedy would be to modify the modeling scheme to include a vertical temperature gradient.

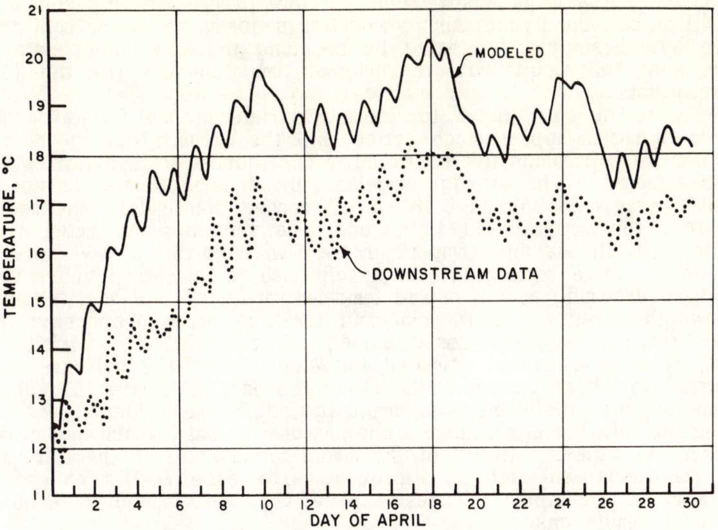

Figure 7 — Comparison of Downstream Data with Base Case Modeled Temperatures

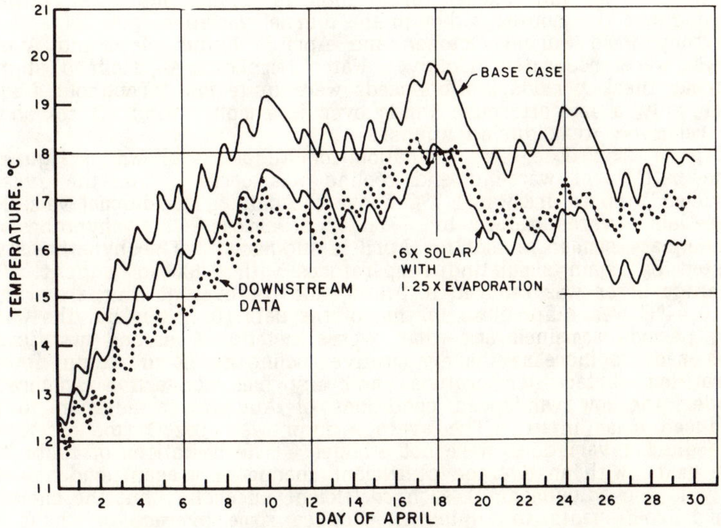

Figure 8 — Comparison of Downstream Data with Modeled Temperatures Using .6 x Solar, 1.25 X Evaporation and 1.25 Advection

This will be a major modification. Before proceeding with this model modification, the hypothesis of model inadequacy should be confirmed. The hypothesis can be tested by reducing the solar input data and comparing the results with the original simulation and with the downstream data.

The solar input for April was reduced to 0.6X. The daily scale dynamics appear much better, but the warming during the first half of April remains overestimated. The diurnal scale dynamics were also affected. The average modeled diurnal variation was reduced to 0.55°C (the data showed 0.78°C). A second modification was made to improve the match based on an additional hypothesis of model inadequacy. If the surface temperatures are warming due to the sun, then the evaporative cooling effects should also be increased above those modeled assuming a fully mixed temperature for the surface. Figure 8 shows the results of 0.6X solar and 1.25X evaporation and advection. The match is slightly closer during the first warming period, but temperatures are underestimated following the cooling episodes. The diurnal variations remain smaller than the data indicates. The limitations of this coefficient adjustment procedure are evident. By considering only average changes, the seasonal or daily match may become closer, but the sensitivity of the model formulations to the full range of conditions will not be improved. The same coefficient changes determined for April will result in a very poor match if used for October simulations.

C. August Simulations

The recorded temperatures during August 1978 were relatively stable, beginning at 28-29°C, cooling to 26-27°C by August 10 and slowly rising to 27-28°C by the end of the month.

The air temperatures indicate significant solar heating throughout the month, although the diurnal variations are not as large as they were during October and April. August 4-16 and August 29-30 were moderately cloudy. Water temperatures declined slightly during these periods. Windspeeds were quite low throughout August with only a few afternoon winds over five mph. Wind effects should be relatively weak during August.

The base case simulation for August is shown in Figure 9. The pattern of warming and cooling was correct, but the average modeled temperature was 1.7°C above the data. The diurnal variations were also overestimated, by a factor of almost 2X. A hypothesis of inadequacy similar to that for April is suggested. This hypothesis was tested by again simulating temperatures with 0.6X solar input. The average error was reduced to 0.6°C, and the average diurnal variation of 0.44°C was quite close to that of the data (0.42°C), but the warming periods remained somewhat overestimated. A second modification was made to increase the evaporative cooling by 2X to account for the stratified surface temperatures and any free convective evaporation under the low windspeed conditions of August, in addition to the reduced solar input. The average error was -0.2°C (too cool), and the diurnal variations were not affected. The limitations of calibrating the model with an average coefficient change were again evident. Not only is the adjustment of each coefficient uncertain, but the choice of which coefficients to adjust to produce some average match is not obvious.

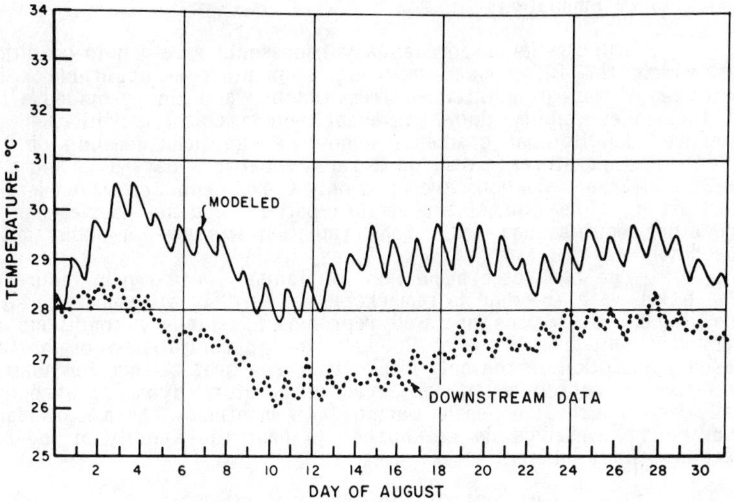

Figure 9 — Comparison of Downstream Data with Base Case Simulation

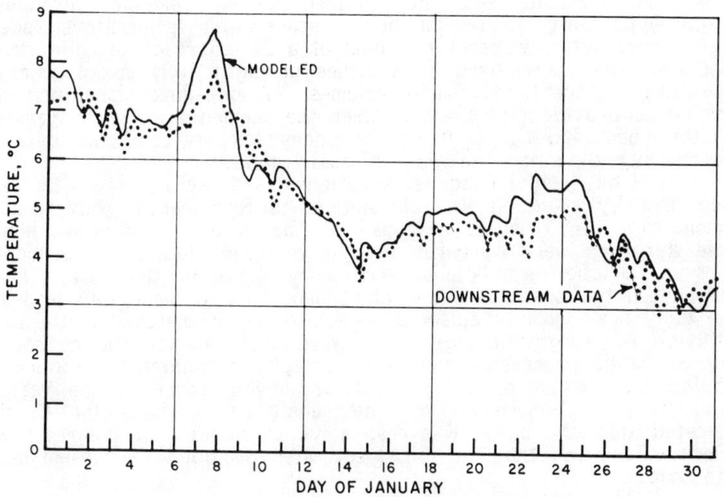

Figure 10 — Comparison of Downstream Data with Base Case Simulation

D. January Simulations

The simulation for January represents a return to conditions for which the fully mixed modeling assumption is applicable. The meteorology was quite different from October and this provided a test of the model validity under different meteorological conditions. The measured longitudinal gradients showed a significant cooling of 0.5-1.5°C. Temperatures cooled quite rapidly between days 9-13 and days 26-29. Diurnal variations averaged only 0.4°C, and were much less on most days. The diurnal temperature pattern was not as clear as for the other months simulated. Solar radiation was low for about half of the days.

The base case simulation for January is shown in Figure 10. The match with the data is remarkable. The daily dynamics as well as the diurnal fluctutions are well reproduced. January conditions are very different from those of October; the appropriateness of the fully mixed assumption is the only similarity. The heat balance formulations are able to reproduce the observed temperature dynamics at both a daily and hourly time scale during both months. The meteorological conditions themselves do not appear to limit the validity of the heat exchange formulations.

IV. SUMMARY AND CONCLUSIONS

The Tennessee Valley Authority (TVA) has collected an extensive series of hourly water temperature measurements in the vicinity of several power plants (Reference 2). Meteorological stations at several power plants throughout the Tennessee Valley provide hourly meteorological data that can be used to calculate heat exchange rate (Reference 1). Hourly flow data are available from the TVA dams. The availability of the required hourly scale data provides a unique opportunity to develop hourly scale water temperature models.

A water temperature model of a 27 km reach of Guntersville Reservoir was schematized as a series of eight fully mixed segments with three adjacent embayment volumes. A one-dimensional unsteady flow model provided the flows between the segments. Two temperature monitors provided the upstream boundary temperatures and data for calibration and testing the range of model validity.

The heat balance formulations used were similar to those used in daily scale models. Simulations of four months were used to demonstrate the seasonal response of the model. Daily and hourly scale dynamics were compared with downstream data for each month.

October simulations were very close to the data. Large diurnal fluctuations due to strong solar heating were well modeled. The daily scale cooling episodes were also closely matched to the data. Although the longitudinal gradients were small, it was shown that the heat exchange processes must be accurately formulated to produce the simultaneous response of the entire reservoir reach to the variable meteorology. The results of the sensitivity analysis for October indicated that the model was responsive to changes in the rate coefficient values and that the initial estimates provided the closest match with data.

The April and August simulations demonstrated that the present model was limited by the fully mixed assumption when actual conditions are slightly stratified. The initial simulations were consistently high by 1-2°C although the diurnal variability from day to day was well reproduced. It was hypothesized that the model inadequacy resulted from the fully mixed assumption. When the reservoir stratifies even slightly during the day, a significant amount of the solar energy does not mix into the reservoir volume, but is transferred back to the atmosphere. This hypothesis of model inadequacy was tested by uniformly decreasing the solar input to 0.6X the measured values and increasing the evaporative cooling. The daily scale temperature dynamics appeared much closer to the data, but warming periods continued to be overestimated and cooling periods were underestimated. The diurnal variations were also reduced below that observed.

The limitations of a coefficient adjustment procedure to account for the effects of hydrothermal processes that are not included in the model were evident. Coefficient adjustments may improve the average seasonal or daily match with data, but they cannot improve the sensitivity of the model formulations to the full range of meteorological conditions. The January simulations confirmed the basic accuracy of the heat exchange formulations, subject to the fully mixed assumption. October and January represent very different meteorological conditions yet water temperatures for both months could be very accurately simulated.

Since the major limitation of the hourly temperature model appears to be the fully mixed assumption, the model should be modified to include the vertical stratification effects. Several additional process formulations may be required to successfully simulate these near surface hourly temperature dynamics. The vertical absorption of solar radiation will become more important. Data describing the absorptivity of the water will be required. A formulation that describes the vertical mixing of slightly stratified reservoir segments as a function of the river flow may be required. A wind mixing formulation is also needed. Some preliminary work on formulating these effects at an hourly time scale has begun (Reference 4). Hourly water temperature measurements of the vertical temperature structure are required for the development and testing of an hourly temperature model that includes near surface stratification effects. The TVA has collected suitable data in the vicinity of several power plants. These data provide a unique opportunity to continue the development of hourly scale temperature models. These models can increase the understanding of natural water temperature dynamics and improve the analysis of thermal impacts.

REFERENCES

1. Brown, R. T., W. L. Reid III and L. W. Hsieh, "The Analysis of Meteorological Data for Water Temperature Modeling," Report No. WR28-1-540-106, TVA Water Systems Development Branch, Norris, Tennessee, 1980.

2. Brown, R. T., "The TVA Hourly Water Temperature Monitoring System Through December 1978," Report No. WM28-1-540-101, TVA WSDB, Norris, Tennessee, 1979.

3. "Heat and Mass Transfer Between a Water Surface and the Atmosphere," Lab Report 14, TVA Engineering Laboratory, Norris, Tennessee, 1972.

4. Brown, R. T., and L. W. Hsieh, "Modeling Hourly Water Temperature Dynamics," Report No. WR28-1-540-104, TVA WSDB, Norris, Tennessee, 1979.

5. Jobson, H. E., "Thermal Modeling of Flow in the San Diego Aqueduct, California, and Its Relation to Evaporation," USGS Prof. Paper 122, 1980.

6. Jobson, H. E., and T. N. Keefer, "Modeling Highly Transient Flow, Mass Heat Transport in the Chattanooga River Near Atlanta, Georgia," USGS Open File Report 79-270, 1979.

7. Brown, R. T., "Modeling Hourly Water Temperatures between Widows Creek and Bellefonte," Report No. WR28-1-540-108, TVA Water Systems Development Branch, Norris, Tennessee, 1980.

LOCALIZED SCOUR BELOW SUBMERGED VERTICAL GATES

By Hilmi Dogan Altinbilek,[1] M.ASCE and Yakup Basmaci[2]

ABSTRACT: The subject of this study is an investigation of the local scour under vertical gates by horizontal submerged jets on cohesionless bed. The analysis of localized scour is based upon the conservation of mass with scour hole as control volume. The differential volume and sediment transport rate terms of differential equation are analyzed separately using the experimental results. A total of 26 model tests were performed covering a range of Froude number from 1.6 to 24.8 and using three different bed materials. Measurements were made to determine the variation of scour hole geometry with time. Relations were developed for the terminal depth of scour and for the change of the scour depth with time.

INTRODUCTION

The prediction of the extent of the local scour and the prevention of damaging scour are important in the design of the hydraulic structures. The subject of this study is an investigation of the localized scour under vertical gates by horizontal submerged jets on cohesionless bed. The object is to establish similarity relations for localized scour.

THEORETICAL ANALYSIS

A method for analyzing localized scour is given by Carstens (1). Assuming that the control volume is the scour hole volume, the principle of the conservation of mass can be written as follows:

$$dV/dt = Q_{so} - Q_{si} \qquad (1)$$

in which dV/dt is the rate of change of the scour hole volume V, with respect to time t. Q_{so} and Q_{si} are the volume rates of sediment being carried out of and into scour hole respectively.

[1]Assoc. Prof., Dept. of Civil Eng., Middle East Techn. Univ., Ankara, Turkey. Presently, Civil Eng. Dept., King Abdulaziz University, Jeddah, Saudi Arabia.

[2]Asst. Prof., Faculty of Earth Sciences, King Abdulaziz University, Jeddah, Saudi Arabia.

In the case of localized scour under the vertical gates by horizontal jets, it is assumed that no sediment is transported into scour hole other than the sediment that would slide from the wides into the scour hole, Q_{si} is zero. Assuming Q_{so} and V are functions of scour hole geometry, Eq. 1 is separable and integrable as follows:

$$t = \int_0^S dV/Q_{so} \qquad (2)$$

The analysis of scour-hole development is based upon a separate determination of terms in Eq. 2 in which S is scour depth. One approach is to use an analysis of the forces on the surface grains in the formulation of sediment transport function. Model tests provide the values of the constants needed by development and also to provide an expression for scour hole volume.

A sediment-pickup function for the rate of transport per unit area out of a flat bed in absence of deposition was proposed by LeFeuvre, Altinbilek and Carstens (2). because of the fundamental similarity of all localized-scouring situations, the sediment-pickup function can be expected to be applicable in other situations. A general form of output sediment transport function, Q_{so}, can thus be formulated as follows:

$$Q_{so}/(uD_gB) = f \{ \text{scour-hole geometry, sediment-grain geometry, flow parameter, } S/b \} \left[C_D' N_s^2/(8.2 \cos \alpha) - \tan \theta - \tan \alpha \right] \qquad (3)$$

In Eq. 3, u is the velocity just above particles on the bed. D_g is the geometric mean diameter of sediment particles. B is the width of the channel. S is the depth of scour and b is the characteristic length of the obstruction along which sediment pickup occurs. C_D is the coefficient of drag on a particle. g is the magnitude of acceleration of gravity. θ is the angle of repose of bed material, and α is the angle of inclination of the bed from the horizontal. N_s which is equal to $u/((s-1)gD_g)^{1/2}$, is called the sediment number where s is the specific weight ratio. The sediment number is a form of Froude number.

For the case of localized scour under vertical gates by horizontal submerged jets, the characteristic length, b, can be chosen as the thickness of jet at the section of vena contracta. Initial jet velocity, U, is taken as reference velocity. Geometric variable, S/b, is necessary to establish the stage of scour-hole development. Sediment grain geometry parameter includes size, shape and gradation of the bed material. Since scour hole reaches an equilibrium condition due to reduction of jet velocity over the scour hole, flow parameter is important in the determination of terminal condition.

Using above considerations and assuming that all particles are picked up from a relatively flat area, Eq. 3 can be written as follows:

$$Q_{so} = fUBD_g \left(C_D' N_s^2/8.2 - \tan \theta \right)^{5/2} \qquad (4)$$

in which f is a functional coefficent to be determined experimentally for each localized scour situation. Eq. 4 expresses the rate of sediment transport out of the scour hole under vertical gates by horizontal submerged jets. The scour hole depth as a function of time is obtained by integration of the equation of continuity of sediment in the scour hole, Eq. 2.

EXPERIMENTS

Model tests were performed to study the development of the scour under vertical gates utilizing the existing glass flume at Hydraulics Laboratory of the Middle East Technical University of Ankara, Turkey. Measurements were made to determine the variation of scour hole geometry with time. Change in flow conditions and characteristics of bed material are also recorded for each run.

The test section of the flume is 19.67 ft (6 m) long and has glass walls in order to allow observations. At the beginning of the glass-walled test section, there was a vertical gate made of brass. The bottom of the flume upstream to the gate was concrete. Downstream of the gate there was a sediment bed of 7.9 in. (20 cm) depth. The upper plane of the sediment bed was at the level of the channel bottom. The sediment bed was extended 9.84 ft (3m) in the longitudinal direction.

During a run, horizontal and vertical dimensions of the scour hole can be measured visually by means of a grid system marked on the glass wall. Point gages were provided for upstream water elevation and tail water elevation measurements. Tail water elevation was controlled with an adjustable overflow gate at the downstream end of the flume.

Model tests were performed on flat bed. Development of scour hole was observed for different values of Froude numbers in submerged jet flows. Three experiments were performed for each case at varying tail water conditions to observe the effect of tail water depth on the development of scour.

In all experiments, the surface of the bed initially was level. The run is started by closing the fully open gate to desired opening over the established flow in channel. At the beginning of each run, the scour hole starts forming at a rapid rate. Bed material was transported as bed load. In this period, vertical dimensions of scour hole and the dune increased faster than the horizontal dimensions. After the slope of the upstream face of the dune exceeds the natural angle of repose of the sediment, the mechanism of transport changed into suspension. The location of the most active scour was at the point of impingment of the jet which was pendulating in the scour hole. The sand was entrained by the flow at the point of impingment and was lifted in suspension by counter clockwise eddy motion. A large part of the sediment in suspension was settled on the upstream face of the dune. The remaining part which passed the crest of the dune constitutes the sediment removed from the scour hole.

During the later periods of a run, very few number of particles were carried by the flow. There was no sediment in suspension. Configuration of the scour hole and the dune were stabilized. At that time it was assumed that the scour has reached to the state of equilibrium.

Visual recordings of dimensions of scour hole development and photographs taken during a run provide an accurate estimation of scour hole profile. Development of scour was recorded with an accuracy of $\pm$0.079 in. (2 mm).

Three different bed materials were used. Two of the bed material were quartz sand of 0.256 in. (6.5 mm) and 0.051 in. (1.3 mm) geometric mean diameter. The specific gravity of quartz sand was 2.6. The other bed material was tuff material whose median size was 0.104 in. (2.65 mm) and specific gravity was 1.30.

A total of 26 runs were performed. A run is terminated when the scour hole depth reached channel bottom or when rate of development of scour hole is extremely small. If the flow conditions were out of designed limits an additional run was performed. The tests covered a Froude number range of 1.6-24.8.

The experimental apparatus, the procedure and the scope of experiments were reported elsewhere in more detail (3,4).

ANALYSIS OF EXPERIMENTAL RESULTS

Geometry of Scour Hole: The nondimensional profile of the scour is shown in Fig. 1 and is similar to a vertical axis parabola meeting the plane of the gate. The scour hole profile remains geometrically similar as the development of scour hole progresses. Hence the scour hole geometry can be expressed by a single parameter, namely the scour depth, S. Profile of scour hole can be approximated to isosceles triangle having side slopes equal to the tangent of the angle of repose θ of the bed material. Scour hole volume can then be written as follows:

$$V = (2B/\tan\theta)\, S^2 \qquad (5)$$

In eq. 5, the term B is the width of the channel. Triangular scour hole profile assumption is in close agreement with the experimental data. The change of the scour hole volume, dV, can be obtained by differentiating Eq. 5 with respect to S.

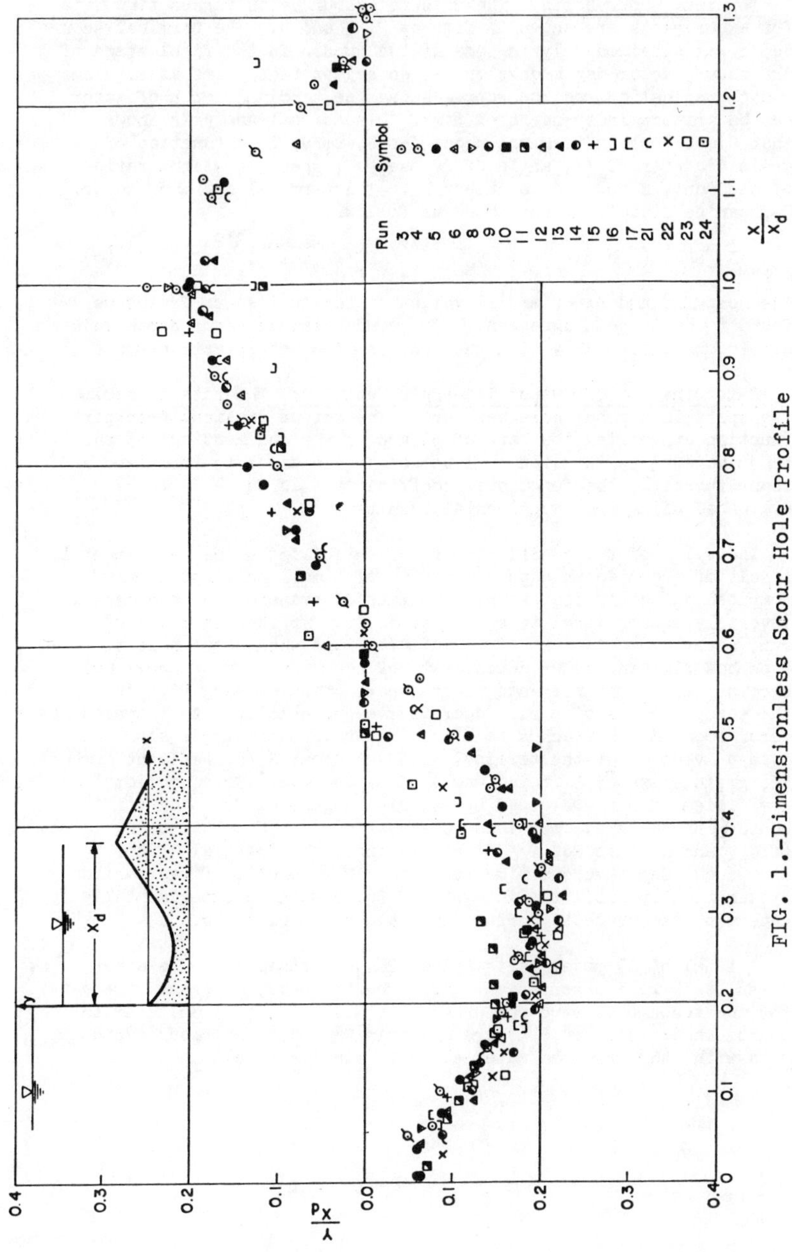

FIG. 1.-Dimensionless Scour Hole Profile

Terminal Scour Depth: The measured scour depth versus time data for experiments are shown in Figures 2, 3 and 4. The terminal scour depth was attained only in some of the runs. In the final stage of the scour, scour depth develops as an asymptote to time axis. Thus a mathematical expression representing the terminal depth of scour can be written independent of time. Dimensional analysis shows that dimensionless terminal scour depth, S_T/b, is a function of grain diameter, D_g/b; angle of repose, ; specific weight ratio of sediment, s and Froude number F_r. The terminal depth of scour, S_T, can be closely approximated as follows:

$$S_T/b = (b \tan\theta / D_g)^{1/2} (F_r / \sqrt{s-1})^{3/2} \quad (6)$$

The computed and experimental values of terminal scour depths were found to be in good agreement. Computed terminal scour depth values also compare favorably with the results of other investigators (5,6).

Determination of Output Transport Function: The rate of sediment transport into scour hole was zero. The output sediment transport function expressing the rate of pickup of the sediment out of the scour hole, Q_{so}, is given by Equation 4. In order to formulate a scour function, the functional coefficient f in Eq. 4 is to be evaluated using the experimental results.

The value of the coefficient f was calculated using experimental result of scour depth S as a function of time. An inspection of computed values indicates that the rate of transport of sediment is inversely proportional to scour depth S/b. At the beginning of a run, scour depth is zero; f is infinity, and when scour hole approaches limiting scour depth, rate of transport out of scour hole becomes zero. The net rate of transport which is very high during the early periods of a run, decreases exponentially. As f approaches zero, scour depth reaches to a terminal value. Analysis show that f is a function of the terminal depth of scour S_T/b, sediment number N_s, grain diameter D_g/b and angle of repose θ. The functional coefficient f was evaluated by assuming a general function for f conforming the above mentioned observations, then substituting values of Q_{so} and dV into Eq. 2 and integrating. The integral of Eq. 2 is a scour-depth versus time function. The function f was varied systematically until the integral of Eq. 2 matched experimentally determined scour-depth versus time data of model tests.

A total of 19 model tests out of 26 were analyzed. The other seven tests were observation runs. The scour-depth versus time data for the tests that were performed with 0.256 in. (6.5 mm) diameter quartz sand (runs 3.4,5,6) are given in Fig. 5. The scour depth data with other two bed materials give similar results.

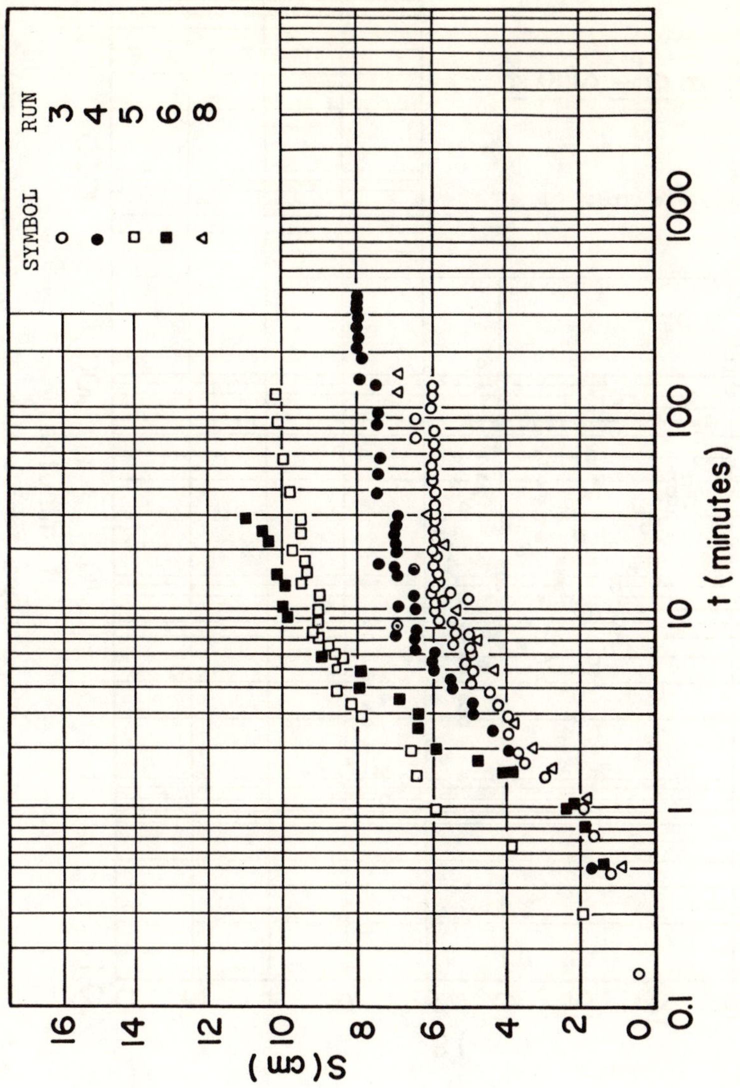

FIG. 2.-Scour Depth Versus Time, D_g=0.256 in.(6.5 mm)

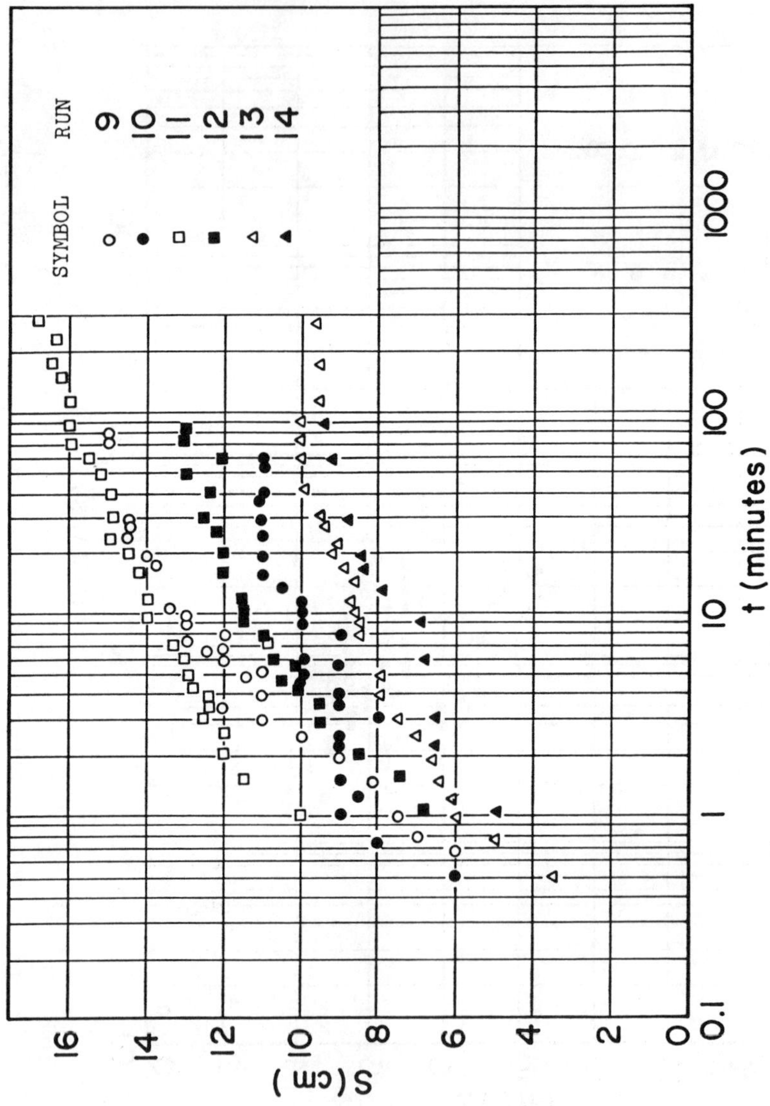

FIG. 3.-Scour Depth Versus Time, D_g=0.104 in.(2.65 mm)

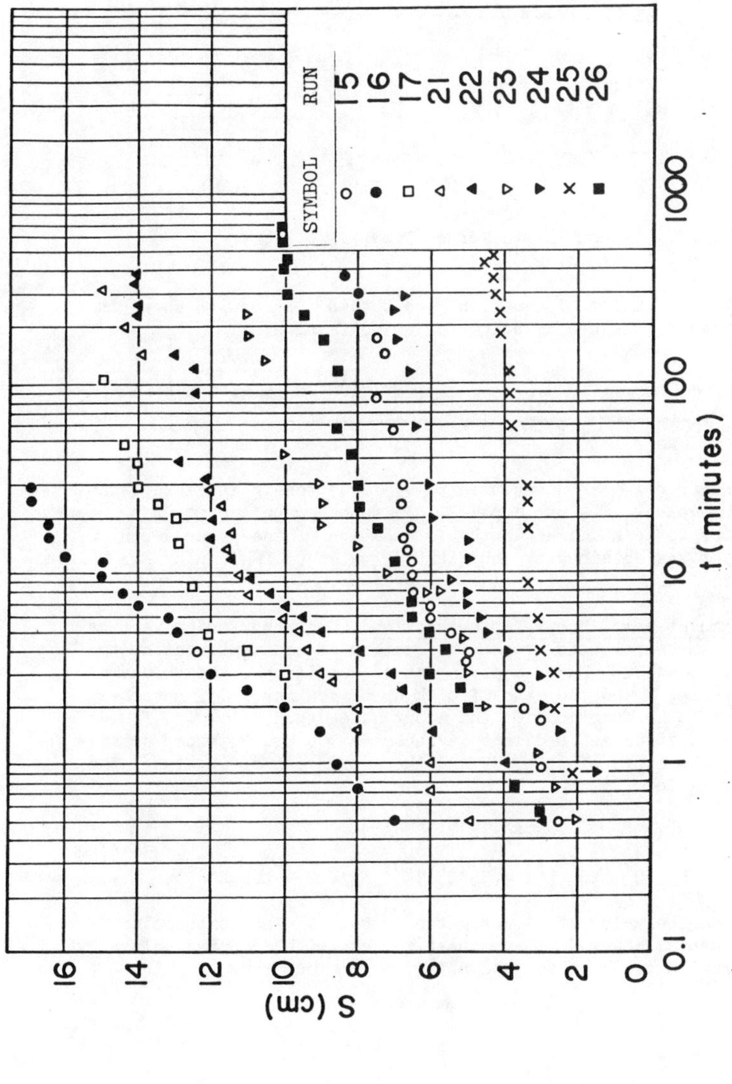

FIG. 4.-Scour Depth Versus Time, D_g=0.051 in.(1.3 mm)

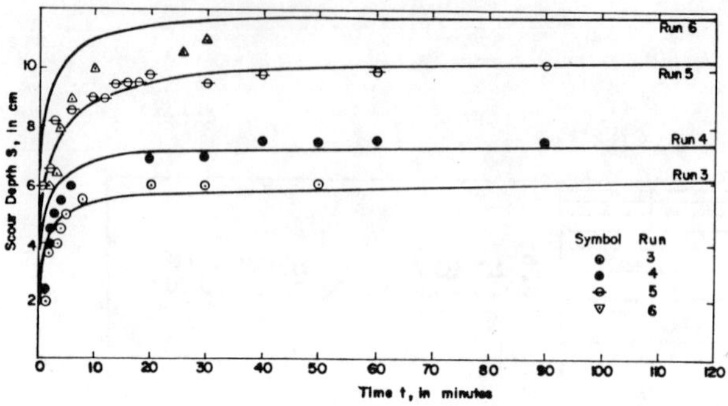

FIG. 5.-Scour Depth Versus Time, D_g=0.256 in.
(1 in.=2.54cm)

The integration of equation 2 to obtain the results shown in Fig. 5 were obtained by using following f value:

$$f = \frac{\tan\theta}{N_s^3 (D_g/b)^{1/2}} \left[\log \frac{\left(\frac{\tan\theta}{D_g/b}\right)^{1/2} \left(\frac{F_r}{\sqrt{s-1}}\right)^{3/2}}{(S/b)} \right]^2 \qquad (7)$$

A numerical method was used to evaluate the integral Eq. 2 at various values of scour depth S. Thus the output sediment transport function to be utilized in the formulation of the scour depth function was obtained by substituting value of f given by Eq. 7 into Eq. 4.

Formulation of Scour-Depth Function: The scour under vertical gates can be computed by the integration of sediment continuity equation, Eq. 2. Under a given set of conditions, the volume of scour hole, V and the rate of pick-up of sediment, Q_{so} have been assumed to be functions of the scour-hole-depth, flow and fluid characteristics and sediment variables. The Eq. 2, after substituting the expressions of dV and the expression of Q_{so} takes the following form:

$$t = \int_0^S \frac{4(D_g/b)^{0.5} N_s^3 \, S \, dS}{U D_g \tan^2\theta \, (\log(S_T/S))^2 (C_D' N_s^2/8.2 - \tan\theta)^{2.5}} \qquad (8)$$

in which the value of S_T was given by Eq. 6. The comparision of experimental scour-depth-versus-time data with computed values by integration of Eq. 8 are shown Fig. 5 for four runs.

The nondimensional form of Eq. 8 can be used for the preparation of nondimensional scour depth charts. The dimensionless scour depth versus time curve which is given in Fig. 6 is prepared for an angle of repose 38.6°, specific weight ratio of 2.60 and Froude number of 4. In computation of Froude number, the maximum jet velocity was used. The characteristic length b is taken equal to the thickness of jet at vena contracta.

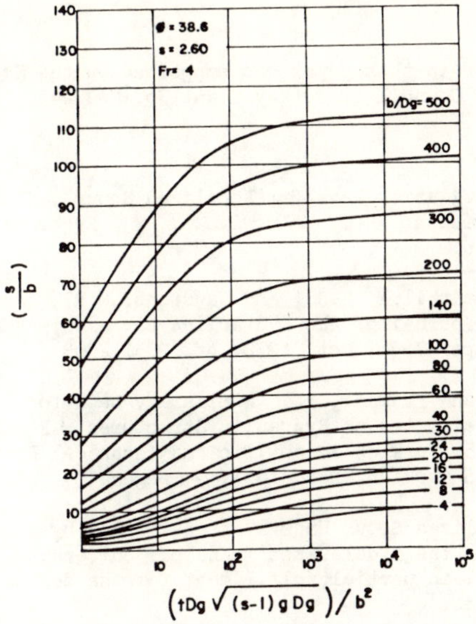

FIG. 6.-Nondimensional Scour Curves

CONCLUSIONS

A method has been presented for computing the terminal depth of localized scour and for determining the time variation of the scour depth under vertical gates in cohesionless bed under the action of horizontal submerged jets. The conclusions of this study are limited to homogeneous, uniform materials. The range of experiments covers Froude number, F_r, from 1.6 to 24.8 and sediment number, N_s, from 3.8 to 13.6. From this investigation following conclusions are drawn.

1. Geometry of scour hole remains similar during the development of scour hole and can be represented by a vertical axis parabola. The distance from the plane of the gate to the crest of the dune is approximately equal to five times the scour hole depth.

2. Terminal depth of scour is proportional to Froude number, mean grain diameter, specific weight ratio, gate opening and the angle of repose of the bed material. The relationship is given by Eq. 6.

3. The depth of scour hole as a function of time can be found by the integration of continuity equation. The general form of the relation is given in Eq. 8.

ACKNOWLEDGMENT

The work reported on in this paper was supported by the Scientific and Technical Research Council of Turkey Grant No. MAG-245.

APPENDIX.-REFERENCES

1. Carstens, M.R.,"Similarity Laws for Localized Scour", Journal of the Hydraulics Division, ASCE, Vol. 92, No. HY3, Proc. Paper 4818, May 1966, pp. 13-36.

2. LeFeuvre, A.R., Altinbilek, H.D., and Carstens, M.R., "Sediment Pickup Function", Journal of the Hydraulics Division, ASCE, vol. 96, no. HY10, Proc. Paper 7636, Oct. 1970, pp. 2051-2063.

3. Altinbilek, H.D., and Basmaci, Y., "Similarity Laws for Localized Scour Under Vertical Gates of Hydraulic Structures", Final Report, Project No. MAG-245, Turkish Scientific and Technical Research Council, Ankara, in Turkish, Dec. 1971, 94 pages.

4. Basmaci, Y., "Localized Scour Under Vertical Gates by Submerged Jets" thesis presented to the Middle East Technical University at Ankara, Turkey in Dec. 1971 in partial fulfillment for the degree of Master of Science, 63 pages.

5. Tarapore, Z.S., "Scour Below a Submerged Sluice Gate", thesis presented to the University of Minnesota at Minneapolis, Minn., in 1956 in partial fulfillment of the degree of Master of Science.

6. Laursen, E.M., "Observations on the Nature of Scour", Proc. of 5th Hydraulic Conf., State University of Iowa, Bull. 34, 1952, p. 179.

SAN FRANCISCO OCEAN OUTFALL PORT VALVE DEVELOPMENT

by

Loring E. Hanson[1], M. ASCE, Henry H.C. Lin[2], M. ASCE
Allen L. Davis[3], M. ASCE, John S. Readshaw[4]
and Robert T. Hudspeth[5], M. ASCE

INTRODUCTION

The City and County of San Francisco has a combined sewer system. During dry weather (May through November), sanitary wastewater flows vary from approximately 50 mgd to 150 mgd with an average flow of about 100 mgd. During wet weather, (December through April), combined wastewater peak flows exceed 1,500 mgd. San Francisco is implementing an extensive wastewater management program to upgrade treatment and storage facilities. All wastewater will ultimately be transported to the Southwest area of the City for discharge through the ocean outfall system. The maximum design capacity for the ocean outfall is 670 mgd.

The basic outfall configuration selected to satisfy discharge requirements with these flows is a three-pipe system. One 9-foot diameter pipe will discharge continuously all flows up to its capacity of 190 mgd at a distance of approximately 4½ miles from shore and a depth of 80 feet. Two 9-foot diameter pipes will discharge all flows in excess of 190 mgd up to their capacity of 480 mgd at a distance of approximately 2½ miles and a depth of 50 feet. Figure 1 shows the schematic layout of the SWOOP outfall system configurations. The wet weather pipes will discharge only about 5 percent of the time on an annual basis and will be dormant the balance of the time.

---

[1]Project Manager, CH2M HILL, INC., San Francisco, California

[2]Project Engineer, CH2M HILL, INC., San Francisco, California

[3]Manager, Hydraulics Division, CH2M HILL, INC., Corvallis, Oregon

[4]Project Engineer, Western Canada Hydraulics Laboratories, Port Coquitlam, B.C.

[5]Associate Professor, Civil Engineering Department, Oregon State University, Corvallis, Oregon

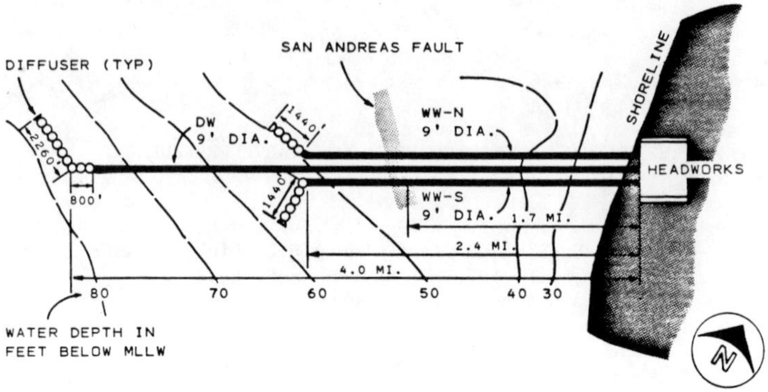

FIGURE 1
SCHEMATIC
670-MGD SYSTEM OUTFALL CONFIGURATION

The outfalls are to be installed in an area of relatively shallow water, with a loose, silty, sand bottom and high wave energy. The outfall pipes including the diffusers will be buried below the sea bottom to minimize possible damage from wave forces and ship anchors. The diffusers will be constructed with vertical risers terminating in turret-like multiple port riser caps extending above the seabed to discharge the wastewater.

Each wet weather outfall pipe has a diffuser length of 1,440 feet. The two wet weather diffusers each have 24 risers at 60-foot spacing. The riser pipe is 24 inches in diameter. Each riser cap has four discharge ports with a nominal port diameter of 8.1 inches. The design discharge capacity of each riser is 10 mgd (15.8 cfs).

Loss of capacity in the wet weather diffuser resulting from sediment intrusion and marine growths (biofouling) was considered to be a major problem. This would occur as a result of wave pumping of seawater into and out of the ports during periods of no discharge. After consideration of a number of alternatives for solving this problem, it was decided that the approach offering the most promise would be the installation of check valves on the discharge ports.

EVOLUTION OF THE CHECK VALVE CONCEPT

Any proposed check valves would need to be of simple durable construction, resistant to abrasion, and preferrably resistant to biofouling in order to function in the hostile ocean environment. They should also have a low head loss at design flows. In a review of standard manufactured check valves the only type that appeared to have potential was the rubber "duckbill" type. In an early test of these valves at Hydro Research Science in Santa Clara, California, it was determined that if these valves were stiff enough to resist back pressures the head loss during discharge was excessive.

Further consideration of alternatives resulted in the decision that the type of valve most nearly meeting the requirements would be a reinforced flexible rubber flap mounted over the exterior face of each port. The rubber would be "NOFOUL," a proprietary product of the B.F. Goodrich Co., which resists biofouling. In recognition of the vulnerability to damage of these flap valves, being exposed on the exterior of the riser caps, it was further decided that a poppet type check valve would be installed on the top of the riser pipe to serve as a back up in the event one or more of the flap valves were damaged. The lining of the cavity and the materials of the poppet valve are to be resistant to biofouling. The entire riser cap is removable to facilitate repairs. Figure 2 illustrates the configuration of a diffuser riser cap.

VALVE TEST PROGRAM

Because the success or failure of the entire outfall project would be critically dependent on the successful performance of the check valves, a test program was undertaken to verify that the valves would in fact perform as intended, and to determine hydraulic characteristics.

Preliminary hydraulic tests were conducted by CH2M HILL on a 1:12 scale model. Overall, the results of the preliminary tests were promising.[6] During the tests, the poppet valves appeared stable at all flows. It must be noted, however, that waves were not simulated in testing this preliminary model. The preliminary port valve test pointed out the need for a more comprehensive port valve development program.

The riser cap assembly model study was conducted in a two-phase program. During the first phase, two sets of tests were performed. Full scale and 1:4 scale models were tested under steady state conditions at Western Canada Hydraulic Laboratories (WCHL) in Vancouver, B.C. These initial tests established actual hydraulic losses in the prototype and verified the similarity of hydraulic relationships between the full-scale and 1:4 scale models. The 1:4 scale model was then tested in the wave tank at Oregon State University (OSU) to check performance under dynamic wave action.

The second phase was conducted on 1:4 scale model only to refine the design of the various system components and optimize system performance and reliability.

---

[6] CH2M HILL, INC., San Francisco Southwest Ocean Outfall Project Sanitary and Hydraulic Status Report, Supplement No. 1, (April 1978), and Supplement No. 2, (November 1978).

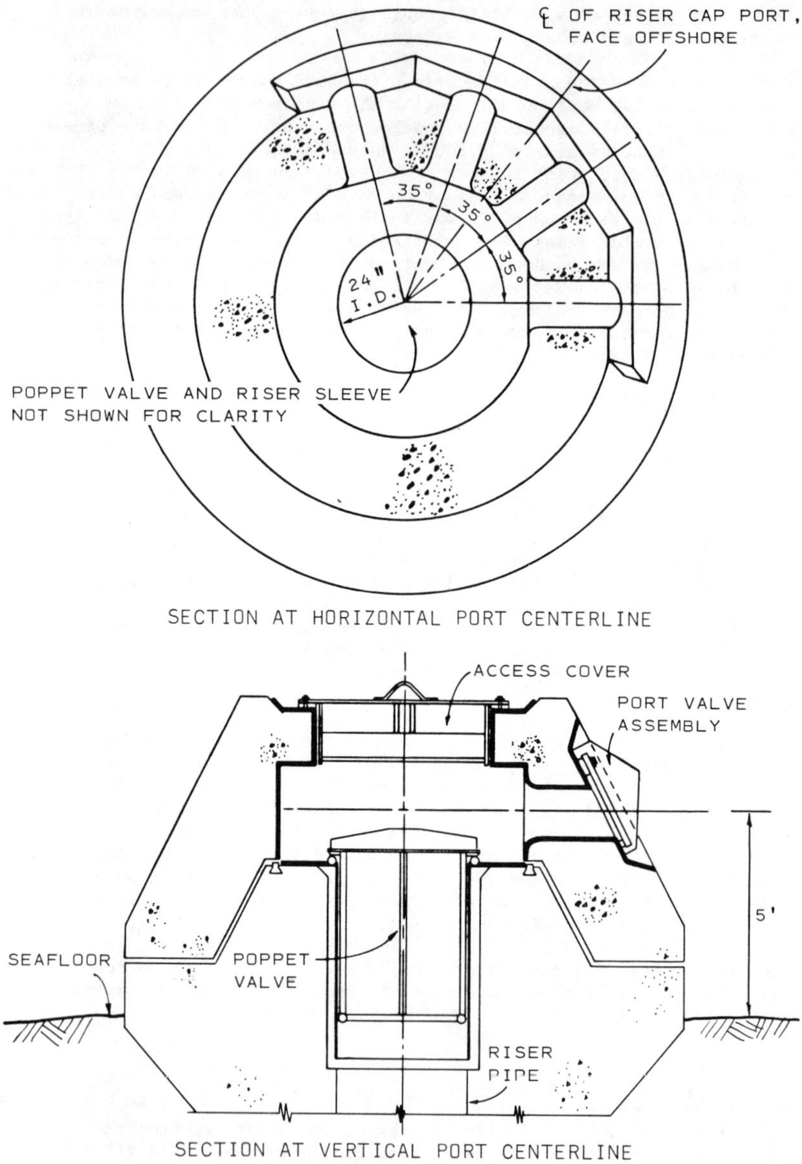

FIGURE 2
RISER CAP CONFIGURATION

MODEL SIMILITUDE LAW

Flow characterized by a free surface, such as in the gravity waves, is generally dominated by gravitational forces. Dimensional analysis indicated that dynamic similarity would be obtained by making the dimensionless Froude number the same in both the model and prototype. Flow in an enclosed system, such as in pipe flow, is generally controlled by fixed boundaries and dominated by viscosity. In that case, the Reynolds number of the model and prototype must be equal in order to achieve dynamic similarity in the system.

Hydraulic modeling of the valve system in a submerged condition with wave action, represents a situation simultaneously involving gravitational and viscous forces. For models with low Reynolds numbers, i.e., less than 10,000, similitude is attained only by satisfying Froude and Reynolds laws simultaneously.

If the same fluid is used in both model and prototype, similitude can only be achieved by constructing the model at full-scale. However, no existing hydraulic laboratory facilities have the capability to test the effect of wave action on the full size diffuser riser cap system.

Experience suggests that lengths, velocities, and forces can be scaled according to Froude Laws with good accuracy if the Reynolds number exceeds approximately 10,000. A length ratio of the model-to-prototype of 1:4 was selected to permit scale model testing in the wave tank at the wave research facility at OSU. The wave tank develops wave heights of 3.75 feet in 12 feet of water, which corresponds to a 15 feet wave height in 48 feet of water for the 1:4 scale riser cap assembly. Water from the local municipal system was used in the test, since the effects of the water density used in the model study was assumed to be negligible.

MODEL FACILITIES AND INSTRUMENTS

Both the full-scale model and 1:4 scale model were made according to the riser cap dimensions shown on CH2M HILL preliminary design drawings. They were also both constructed of clear plastic to aid observation of flow patterns and sediment intrusion.

The steady state testing was performed in a 60' long x 25' wide x 12' deep concrete tank at WCHL. The flow to the model was supplied by 75 hp and 60 hp centrifugal pumps. A plenum chamber and flow straightener were provided to ensure a uniform flow into the model. The poppet valve was fabricated with mild steel with corrected weight, and the flap valve was fabricated with Goodrich's "NOFOUL" rubber and reinforced with a steel plate.

The OSU wave tank has dimensions of 342' long x 12' wide x 15' deep. A hydraulically driven, hinged flap wave generator is located at the end of the flume to generate periodic or random waves. A constant headbox was constructed at the site of the wave tank to control the test flow into the model.

Piezometers were provided on the full-scale model to measure static pressure in the riser stems, the riser cap, and in the test tank. Linear Variable Differential Transformers (LVDT) were provided to measure the displacements of the poppet valve and one of the flap valves. An orifice plate was used to measure the flow into the model.

A Marsh-McBirney electromagnetic flowmeter was used to measure the flow from the constant headbox into the 1:4 scale model at OSU. Pressure transducers were used to measure pressures in the riser stem, riser cap, and in the wave flume. The wave water surface fluctuations in the tank were measured with a sonic water level profiler.

Data acquisition and recording were accomplished by using a mini-computer PDP-11E10 operating under the RT11 system. The collected data were then reduced and analyzed by using a Finite Fourier transform (FFT) algorithm.

The second phase testing was more qualitative than quantitative. A seven channel strip chart recorder was used to continuously record the analog signals representing wave height, ambient pressure adjacent to the diffuser port, pressure in the valve chamber, pressure in the riser stem, poppet valve displacement, discharge rate, and the water surface fluctuation in the headbox. An underwater TV camera and monitor enabled observation of model behavior during testing in the wave tank and video tapes were made of all tests.

In wave tank dynamic testing, experiments were performed on the 1:4 scale model for wave heights of 3, 7, and 15 feet, with riser discharges of 0, 2, 4, 8, 12, and 15.8 cfs, for wave incidence of 0°, 45°, and 90°. The angle of wave incidence was defined as the angle between the axis of symmetry of the discharge ports and the incident wave crests.

In phase two testing, various valve assemblies, flow, and wave conditions were simulated to test refinements in valve configuration. A total of 27 test runs were performed including one to determine the hyraulic characteristics of the valve system under static conditions without the poppet valve.

DISCHARGE COEFFICIENT DERIVATION

One of the objectives of the port valve hydraulic model studies was to determine discharge coefficient, C_D, for the diffuser system head loss calculations. The tests of the full-scale model and 1:4 scale model conducted in phase one study at steady state conditions concluded that the 1:4 and full-scale models have similar head loss characteristics and consequently a common discharge coefficient.

The test results indicated that the head loss was relatively insensitive to Reynolds number (for Reynolds numbers greater than 10,000), and the Froude modeling law was adequate to provide dynamic similitude for this study.

The total head loss across the prototype valve system was 2.69 feet at the design discharge of 15.8 cfs. The measured head loss was the difference in total head between a point just below the entrance to the riser cap and a point just outside the riser cap assembly.

Figure 3 shows the relationship of discharge and head loss on a single riser cap assembly. Hydraulically, a relationship between the discharge through the riser, Q, and the difference in total head, E, (between a point inside the diffuser pipe to a point outside the discharge port) was required to calculate the diffuser system performance.

Conceptually, E can be expressed as the sum of (1) entrance loss from diffuser pipe to riser, (2) friction loss in the riser, (3) elbow loss in the riser, (4) loss in the riser cap assembly, and (5) velocity head of the jet. The latter two components constitute the head loss, E_r, measured in the port valve tests, since the riser and diffuser section were not part of the tested configuration.

$$E = \frac{V_d^2}{2g} + (X_e + X_1 + f_r \frac{L_r}{D_r}) \frac{V_r^2}{2g} + E_r \qquad (1)$$

where V_d and V_r are the velocities in the diffuser and riser respectively; X_e is an entrance coefficient whose value is about 0.4 for a sharp edged entrance and 0.1 for a rounded entrance; X_1 is the elbow loss coefficient; f_r is the Darcy-Weisbach friction factor for the riser; L_r and D_r are the length and diameter of the riser. The quantity E_r, measured in the port valve testing, is the sum of (1) velocity head of the jet (2) head loss due to poppet valve, (3) head loss due to the flap valve, and (4) head loss in the riser cap chamber. Koh[7] has analyzed the test data and derived the following empirical relation:

$$E_r = \frac{V_p^2}{2g} + 0.7 + 0.21e^{-(\frac{Q}{6.6})^2} + 0.25 \frac{V_r^2}{2g} \qquad (2)$$

where V_p is the jet velocity.

Thus, defining the discharge coefficient, C_D, by

$$Q = C_D A_p \sqrt{2gE} \qquad (3)$$

in which A_p = Area of port

[7] Memorandum to CH2M HILL, INC., by Dr. Robert C.Y. Koh. May 18, 1979.

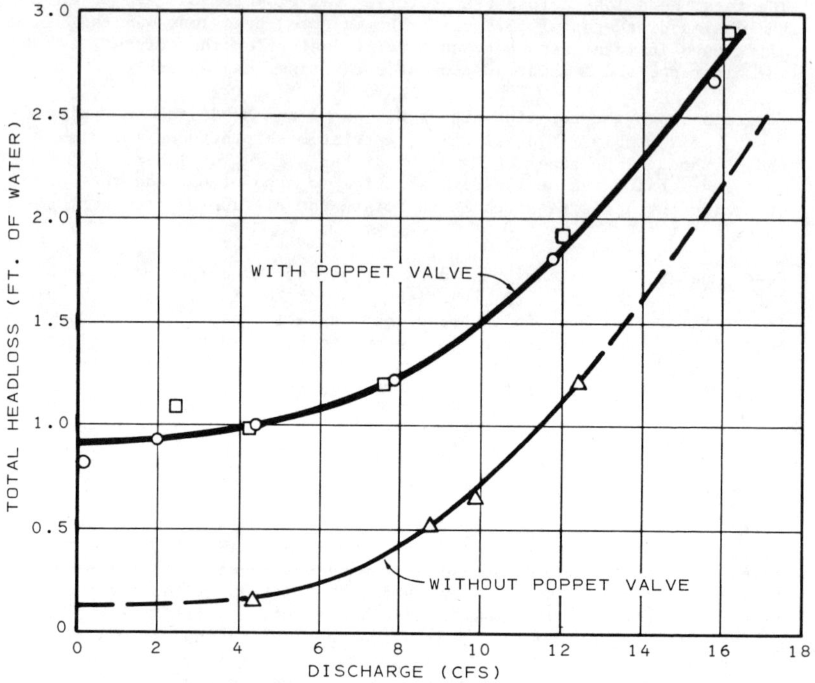

NOTE: ALL VALUES IN PROTOTYPE UNITS

KEY: ○ 1:1 SCALE MODEL, WITH POPPET VALVE
□ 1:4 SCALE MODEL, WITH POPPET VALVE
△ 1:4 SCALE MODEL, WITHOUT POPPET VALVE

FIGURE 3
DIFFUSER RISER CAP HEADLOSS

then,

$$C_D = \frac{1}{\sqrt{X}} \left(1 - \frac{V_d^2}{2gE} - \frac{0.7 + 0.21e^{-(Q/6.6)^2}}{E}\right)^{1/2} \tag{4}$$

where $X = (X_e + X_1 + f_r \frac{L_r}{D_r} + 0.25) (\frac{A_p}{A_r})^2 + 1$

A_r = Area of riser

Steady state tests were performed during phase two study to determine the hydraulic characteristics of the valve system with the poppet valve removed. The total head loss for the riser cap assembly without the poppet valve was reduced from 2.69 feet to 2.10 feet for 15.8 cfs discharge. The measured head loss data were superimposed on Figure 3 and compared with the head loss measured for the whole riser cap assembly. The comparison indicated that the head loss term of 0.7 feet as shown in equation (2) was closely associated with the poppet valve.

TEST RESULTS

1. In general, the wet weather riser cap assembly geometry tested is hydraulically satisfactory in the steady state. The appearance of a strong circulation and vortex inside the valve chamber, while it may be undesirable hydraulically, may be beneficial in scouring any sediments which may settle in the diffuser cap.

2. In phase one study, an unsatisfactory aspect of operation was that orbital velocities caused the flap valves to open 180 degrees. The flap valves were modified to provide stirrups and side shrouds on the face of the outlet ports. Second phase testing confirmed that the modified valve design was effective in correcting this problem.

3. Under no discharge conditions with waves the water surface level at the headbox was naturally pumped down to near the trough elevation of the waves. It is not clear to what extent this will happen in the prototype because the storage capacity between the multiple risers, the effect of the multiple diffuser riser caps with varying phase angle with respect to passing wave crests, and the momentum effects in the prototype riser pipe, have not been modeled.

4. Under a no-flow discharge condition, the poppet valve would seat itself when the headbox water surface level was pumped down naturally by waves. However, none of the valve configurations tested prevented the flap valves from opening under wave conditions, even when the poppet valve was removed from the valve system.

5. The flap valves effectively reduced the amounts of sediment entering the valve chamber; however, they could not completely prevent seawater or sediment from intruding into the riser cap.

6. The addition of two 1-inch diameter holes in the poppet valve reduced opening of the flap valves under no discharge conditions. Also, the addition of extra weight to the flap valves for a total prototype weight of 19.2 lbs reduced the amount of the flap valves opening under wave actions; however, it did not keep the valves in closed position. These two findings were incorporated into the final port valve system design.

ANTI-BIOFOULING MATERIALS SELECTION

The successful control of fouling organisms depends how well antifouling materials, (e.g., high copper content alloys, or organotin compounds) prevent attachment and growth of fouling organisms. The efficacy of these materials is known only in seawater. Exposure to sewage effluent on a continual or intermittent basis may alter the chemistry of these materials, and may result in different effectivenesses in preventing organism attachment and different corrosion rates.

An extensive coupon testing program was performed on different antifouling materials to determine the best antifouling material for both effectiveness and life span, and to create specifications for material description and thickness. The material coupon testing program concluded that copper-nickel alloys (90:10 copper-nickel and 70:30 copper-nickel) appeared most suitable for use in the outfall design.

To minimize the biofouling problem, the rubber in the flap valves and poppet valve seats will be made of an anti-biofouling compound, such as NOFOUL rubber manufactured by B.F. Goodrich Company.

CONCLUSIONS

The port valve model study provided much information regarding the valve hydraulic characteristics, the valve behavior under wave conditions, and the effectiveness of the valve system in limiting the extent of sand intrusion.

Both full-scale and 1:4 scale models were useful in verifying the valve system performance under steady state and dynamic wave actions. Although some refinements resulted from these tests, the valve and riser cap assembly performed essentially in accordance with design predictions.

Although extensive testing has been conducted in the model program it must be recognized that those tests cannot simulate the severe environment on the ocean bottom where the valves will be installed. Consideration is being given to further testing in the ocean to simulate as nearly as possible the actual operating conditions.

An extensive inspection program is recommended in the early stage of the outfall system in operation to observe the valve system effectiveness, and to refine the system components design, if necessary.

ACKNOWLEDGEMENT

This study was planned and administered by CH2M HILL, INC., a subcontractor of PBQ&D, INC., which is the prime design contractor on the Southwest Ocean Outfall Project (SWOOP) for the City and County of San Francisco. CH2M HILL, INC., is responsible for the hydraulic and sanitary design of the project. Among the many individuals who made substantial contributions to this work, were Roger W. Lindquist, Robert Beckley, Rod Hoffman, Kevin Clancy, and Larry Well of CH2M HILL, INC., and Dr. Norman H. Brooks, Dr. R.C.Y. Koh of California Institute of Technology, Pasadena, California.

Perforated Plates as Hydraulic Energy Dissipators

By

J. Paul Tullis,[1] M. ASCE, James J. Powers III,[2] A.M. ASCE,
Paul F. Shiers,[3] M. ASCE, and Whitney W. Hall,[4] A.M. ASCE

ABSTRACT

The hydraulic and cavitation characteristics of perforated plates (multi-hole orifice plates) were studied to assess their suitability as hydraulic energy dissipators in piping systems. Experimental data yielded curves for hydraulic resistance and for four cavitation levels: choking, long-term allowable, short-term allowable, and excessive. Cavitation intensity was gaged subjectively with aural and visual observations. The testing program results showed that perforated plates have significant advantages as hydraulic energy dissipators. In particular, no cavitation erosion damage was found during the test, and the perforated plates could be spaced closely together in series.

INTRODUCTION

Research conducted at the Utah Water Research Laboratory of Utah State University has provided some insight into the properties of perforated plates (multi-hole orifice plates) as hydraulic energy dissipators. Hydraulic energy dissipation devices are often used to provide the necessary resistance in a fluid system to yield desired flowrates. This occurs commonly in piping networks as a system design develops, is changed by operational modifications, or when a system is required to serve several different functions. In this case it had been determined that flow balancing devices were required in a cooling water system, operating as a siphon, at a steam-electric power station. The area in which the flow restrictions were to be located had only short piping lengths available, and pressures could be well below atmospheric. It was decided to investigate the suitability of perforated plates for the application. Extensive literature is available on hydraulic and cavitation research conducted on the more conventional energy dissipation devices such as throttled valves, single-hole orifices, and other specialized sudden enlargement type devices. However, due to a lack of similar literature, model test data were required to develop design criteria for both hydraulic resistance and cavitation characteristics of perforated plates in a form that could be generalized for design use within the range of pipe sizes and flow conditions in the prototype system.

---

1. Prof. of Civil Engineering, Utah State Univ., Logan, Utah and Prof. of Civil Engineering, Colorado State Univ., Ft. Collins, Colorado.
2. Engineer, Hydraulic Division, Stone & Webster Eng. Corp., Boston, MA.
3. Sr. Hyd. Eng., Hyd. Division, Stone & Webster Eng. Corp., Boston, MA.
4. Hyd. Eng., Hyd. Division, Stone & Webster Eng. Corp., Boston, MA.

A full scale model test program was conducted to assess the benefits and characteristics of a multi-hole concept.(8,9) A preliminary program of testing evaluated headloss and cavitation characteristics of perforated plates, and developed suitable spacing requirements for multiple plates in series. More extensive testing was then performed to collect a greater range of detailed data, after preliminary results showed that there were several benefits to the use of a multi-hole design.

BACKGROUND

Previous research work on cavitation and its effects on piping systems has been carried out for valves, single-bore orifices, and many specialized sudden enlargement designs.(1,2,3,4) After investigating the available literature and some of the existing applications of perforated plate orifices, it was determined that the multi-hole concept could be advantageous for use as an energy dissipation device.(7) In particular, it was the authors' judgment that perforated plate orifices would be more resistant to cavitation effects and could be spaced closer together in series than single-bore orifices. Placing single-bore orifices in series requires long runs of straight pipe which are not always available.(5) Failure to provide the proper spacing between orifices in series leads to uncertainty concerning the hydraulic performance of the orifices. Advantages in cavitation resistance and spacing requirements would therefore make perforated plates a preferable alternative to conventional single-bore orifices while retaining the benefits of relatively low initial and replacement cost, maintenance free operation, and elimination of the potential for inadvertant operational misalignments to which throttled valves are susceptible. There is, however, a lack of the experimentally proven design data for perforated plates that exists for the corresponding single-bore orifice. Therefore, a model test was conducted to provide the necessary information.

Earlier research on cavitation of valves and orifices has shown that intensity, and therefore damage potential, increases with pipe size for sudden enlargement type flow controls.(6) This size scale effect is a function of the strength of vortices generated by viscous shear in the flow separation zone downstream from the expansion. Larger pipe sizes allow the formation of larger vortices of greater angular momentum and strength. The rotation of a vortex results in an instantaneous localized pressure reduction at its center. If the pressure reduction is sufficient to reach vapor pressure, a vapor cavity will open, the size of which is dependent on the strength of the vortex. The collapse of these vapor cavities as the pressure recovers downstream of the expansion results in the objectionable noise, vibrations, and erosion damage associated with cavitation. Therefore, larger pipes tend to exhibit heavier cavitation due to the greater vortex strength generated during energy dissipation.

One advantage of a multi-hole design is in the reduced vortex size and strength generated in the separation zone as shown in Figure 1. Additionally, the interference of multiple jets causes vortices of opposite rotation to interact, greatly reducing the maximum vortex strength and leading to quick dissipation of angular momentum. This results in the implosion of smaller vapor cavities closer to the jets.

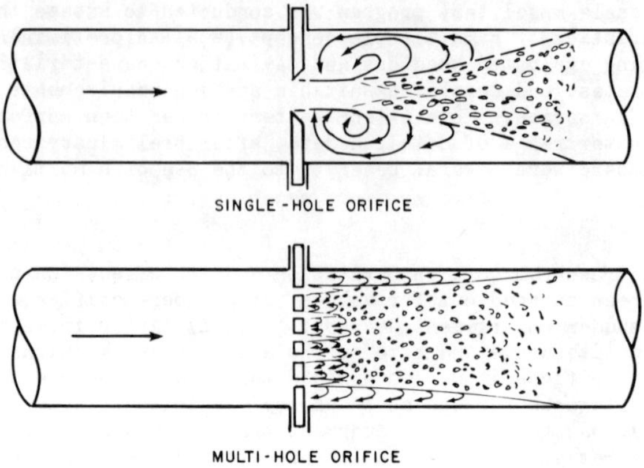

FIG. 1. - Comparison of Vortex Strengths for Single-hole and Multi-hole Orifice Plates

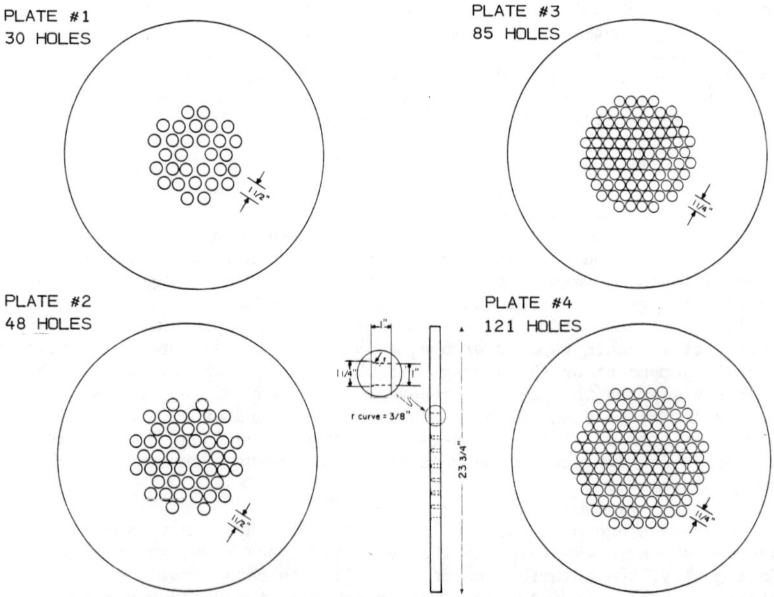

FIG. 2. - Typical 20-inch Perforated Plate Orifices

On the basis of this reasoning, a series of six 20-inch diameter perforated plate orifices were fabricated for detailed pressure drop and cavitation testing. Two smaller plates, 16-inch and 10-inch diameter, were also tested in order to assess size scaling characteristics. Typical 20-inch diameter plates, as shown in Figure 2, had geometrically similar hole patterns consisting of varying numbers of 1-inch diameter holes in triangular arrays at either 1¼ or 1½-inches center to center spacing. The two hole spacings were tested to check on hydraulic effects of spacing, and to permit designers to use the larger spacing where required to meet allowable plate stress requirements for higher pressure differentials. The plates used in this study were made of 1-inch thick aluminum plate . The plates tested were: 48,85, and 121 holes, 1¼-inches spacing, 20-inch diameter; 30,48, and 85 holes, 1½-inches spacing, 20-inches diameter; 33 holes, 1½-inches spacing, 16-inches diameter; and 19 holes, 1¼-inches spacing, 10-inches diameter.

The hole inlets were rounded at a radius of 3/8-inch to prevent hydraulic whistling. This annoying phenomenon can occur when the jet separates and reattaches to the orifice hole wall. In previous experimental work, sharp edged holes produced a noise sufficiently irritating to cause the particular design to be excluded from a sudden enlargement test program.(1)

The equations governing the hydraulic and cavitation performance of perforated plates are the discharge coefficient

$$C_d = \frac{V}{\sqrt{\dfrac{2(Pu-Pd)}{\rho} + V^2}} \qquad (1)$$

where V is the average pipe velocity, (Pu-Pd) is the measured recovered piezometric pressure drop across the orifice, ρ is the mass density of the fluid; and the cavitation index

$$\sigma = \frac{Pd-Pv}{Pu-Pd} \qquad (2)$$

where Pd is the recovered pressure 4 to 5 pipe diameters downstream of the orifice, Pv is the vapor pressure at the liquid temperature, and Pu is the pressure 1 pipe diameter upstream of the orifice.(2) For perforated plates in series, the combined discharge coefficient is expressed as

$$C_d' = \frac{1}{\sqrt{\dfrac{1}{C_{d_1}^2} + \dfrac{1}{C_{d_2}^2} + \dots + \dfrac{1}{C_{d_n}^2} - 1}} \qquad (3)$$

where (n) is the number of perforated plate orifices.

EXPERIMENTAL PROCEDURE

A 20-inch diameter pipe section was positioned in a test loop as shown in Figure 3. High flowrates at low pressures were supplied from a small reservoir, lower flowrates at high pressures were supplied by pumps. A highly polished soft aluminum plate was positioned flush with the pipe wall just downstream of the test orifice to detect cavitation erosion pitting. Previous work with single-hole orifices in a similar test loop arrangement has shown that damage due to cavitation can easily be detected on polished aluminum plates.(6)

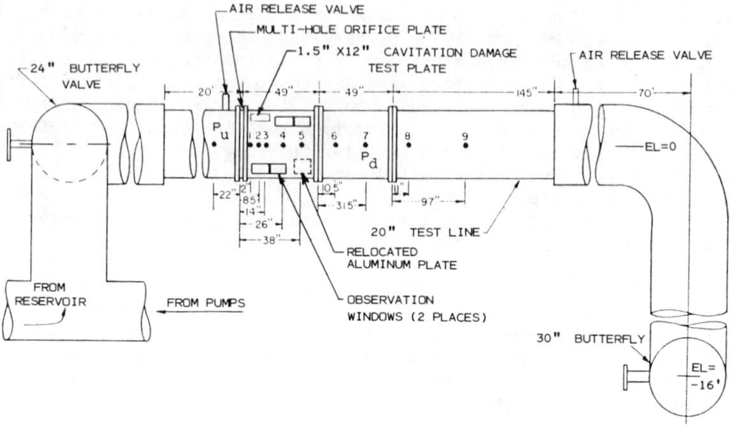

FIG. 3. - 20-inch Multi-hole Orifice Plate Test Line Configuration

Preliminary testing was conducted using a series of 9 pressure taps extending 10 pipe diameters downstream of the orifice to define the pressure recovery zone. Accelerometer readings were taken at each of the pressure taps to aid in locating the point of maximum vibration, which would correspond to the region of vapor cavity implosion when operating under cavitating conditions. This data was needed to fix the minimum allowable spacing between perforated plates in series. Based on the results of these pressure and vibration readings, several of which are plotted in Figure 4, it was determined that a minimum spacing of $2\frac{1}{2}$ pipe diameters for perforated plates in a series would provide adequate margin to prevent cavitation damage to the downstream plate. Flanges for two additional orifices were then installed in the test loop at this spacing.

In the course of the study, observation windows were added to the test section to view the first $2\frac{1}{2}$ pipe diameters downstream of the test orifice. The soft aluminum plate mounted in the pipe wall was relocated just upstream of the second test orifice position to check that area for pitting damage. Pressure transducers were mounted at three locations in the pressure recovery zone to measure the amplitude vs. frequency of the pressure fluctuations. Detailed development of hydraulic and cavitation performance characteristics was accomplished by using pressure readings taken at 1 pipe diameter upstream of the orifice, and at $\frac{1}{3}$ and $4\frac{1}{2}$ pipe diameters downstream of the orifice.

The procedure for testing both single test plates and multiple test plates in series was similar. The plates were first calibrated using volumetric tanks to determine C_d. The test conditions were then varied over a wide range of flows and pressures to evaluate the effects of cavitation. Cavitation intensity was gaged subjectively by aural and visual observations. When sufficient data was collected on a given test orifice

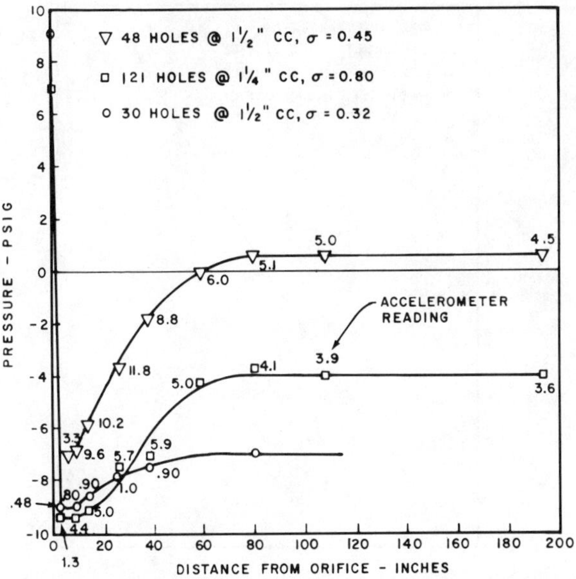

FIG. 4. - Pressure and Vibration Readings in Jet Recovery Zone

to determine the point of undesirable cavitation, the test loop was run at this condition for an hour, after which the pipe wall aluminum plate was removed and inspected for cavitation damage pitting.

Two test loops were also fabricated for testing the 16-inch and 10-inch diameter perforated plates. The plates were calibrated to determine C_d and tested to determine the various operating conditions as in the 20-inch diameter test loop. No inspections were made for cavitation damage in the smaller test loops since none had been found in the larger 20-inch diameter loop.

RESULTS

A curve of C_d vs. flow area ratio (Ao/Ap) is plotted in Figure 5 for the various perforated plates tested. The curve is uniform and compares favorably with single-hole orifice and nozzle data which are also plotted for comparison. The data for single-hole orifices have been expressed in terms of the reduced area of the jet at the vena-contracta in order to give comparable results.

The rounded hole inlets of the perforated plates preclude the formation of a vena-contracta and therefore the area of the holes equals the area of the jets, as is the case with a flow nozzle. This result suggests that perforated plates of similar geometry to those tested can be designed with any hole size provided the plate thickness is not great enough to introduce significant frictional losses. The two different hole spacings tested during the study had no significant effects on the

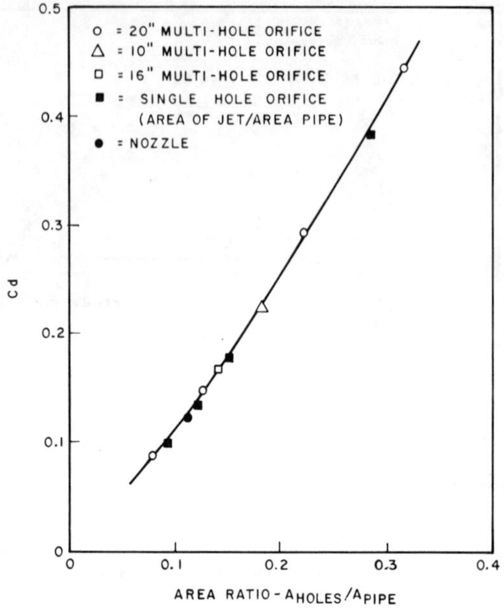

FIG. 5. - C_d for Perforated Plate Orifices

hydraulic characteristics of the plates. In design applications the factors influencing selection of hole size, hole spacing, and plate thickness mainly concern evaluating the size of solid particles in the fluid medium and meeting design code allowable stresses in the plate.

During the course of the experimental work it was determined that the discharge coefficient of the perforated plates would decrease below a certain cavitation index. Plots of the percent change in Cd vs. σ are given in Figure 6. It was found that the multi-hole orifice plate begins to choke well before the pressure at the pipe wall boundary just downstream of the orifice plate reaches vapor pressure. Apparently the interactions of the many small jets resulted in very localized high shear stresses and pressure reductions which were not felt at the pipe wall. Rouse has shown that the pressure distribution varies only a few percent across a single-hole orifice.(10) The results of this study suggest that this is not true for multiple-hole orifices, and further work is required in this regard.

Five operating limits were evaluated during the study and are plotted in Figure 7.

1. Onset of Choking, σ_{ch}. This limit represents the condition at which the discharge coefficient of the perforated plate begins to decrease due to cavitation.

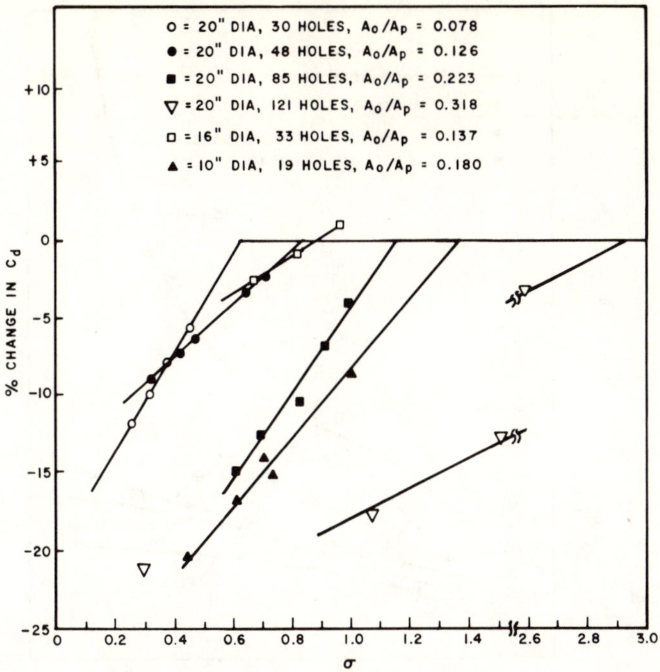

FIG. 6. - Change in C_d for Perforated Plate Orifices as a Function of Cavitation Index

2. Long-term operating limit, σ_{lt}. This limit was determined subjectively and identifies the lowest value of sigma for which the system could be operated continuously without excessive noise or vibration and with no cavitation erosion damage.

3. Short-term operating limit, σ_{st}. This limit was determined subjectively and identifies the flow conditions where the noise and vibration intensites are high enough that the system should not be operated at this condition for extended periods of time. The test program indicated that there is no cavitation erosion damage, but the level of noise and vibration was high enough to be of concern.

4. Excessive operating limit, σ_{ex}. This operating limit was determined subjectively and represents a condition where noise and vibration levels are excessive.

5. Incipient cavitation damage, σ_{id}. This represents the condition at which pitting first occurs on a soft aluminum strip placed on the pipe boundary. This limit was investigated but under no condition was it possible to find any cavitation erosion damage on the boundary of the pipe downstream from the orifice plate. Incipient damage values for single-hole orifices are plotted in Figure 7 for comparison.

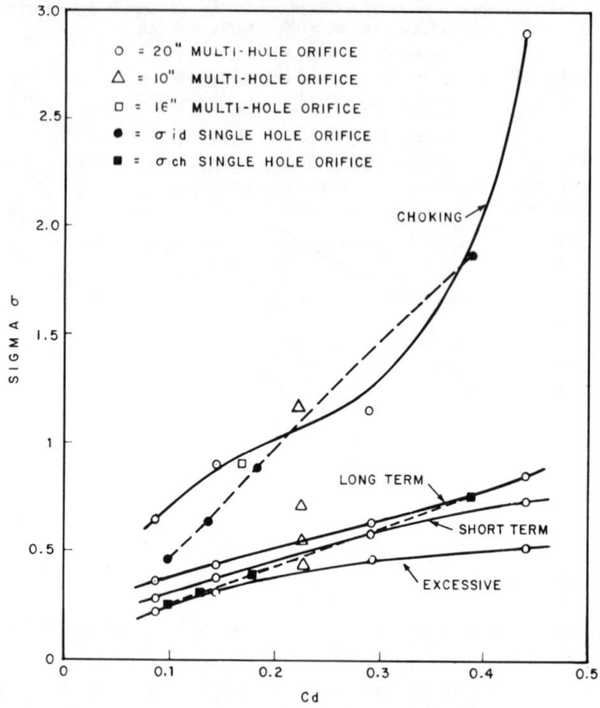

FIG. 7. - Cavitation Operating Limits for Perforated Plate Orifices

The cavitation level curves of Figure 7 are based on data collected on the 20-inch diameter multi-hole orifice plates. Comparison curves for single-hole orifices are also shown.(6) Although previous work in cavitation research has shown that cavitation intensity decreases with decreasing pipe sizes, this was not generally the case with the smaller 16-inch and 10-inch diameter test orifices. The 16-inch diameter test orifice performance in the various cavitation levels was comparable to the 20-inch diameter test orifice. The 10-inch diameter test orifice had significantly lower cavitation resistance than was predicted. This size scaling effect appears to result from the fact that the size and geometry of the perforated plate holes was not changed as the pipe diameter was decreased. This resulted in reducing the beneficial effects of many small jet interactions. If the pipe diameter were further reduced it would be reasonable to assume that the cavitation suppression performance of the fixed multi-hole geometry would further deteriorate until, in the limit, it would approach the performance of a single-hole orifice. Maintaining geometric similitude when reducing pipe diameter would probably be the more preferable means of designing perforated plates in pipes smaller than 20-inch diameter. Further experimental work is required in this regard. The mean downstream pressures were higher in the 10-inch diameter test loop than in the larger diameter test loops. This probably added a pressure scaling effect which increased cavitation intensity.

Observations of the flow field in the pressure recovery zone of the 20-inch diameter test loop revealed that large amounts of air were coming out of solution in the form of many tiny bubbles. During conditions of heavy cavitation the air bubbles were so profuse that it was impossible to see the test orifice through the observation windows. This deaeration effect probably serves as a major contributor to the cavitation resistance of perforated plates. The test program was conducted with subatmospheric mean pressures in the jet recovery zone. The cavitation intensity limits of Figure 7 should therefore be used with caution when significantly higher positive downstream pressures are expected in an application, since this could prevent deaeration from occurring.

Plots of pressure fluctuation amplitude vs. frequency were taken for the jet recovery zone in an attempt to identify any trends that would make the cavitation level evaluations less subjective. Unfortunately it was found that the pressure variations were random, and there appeared to be little correlation between cavitation intensity and frequency vs. amplitude that could aid in quantifying a given cavitation intensity.

It was found that the pressure recovery zone downstream of the perforated plates normally extended 4 to 5 pipe diameters, which corresponds to the distance required for hydraulic energy dissipation as the jets expand. This is approximately the distance required for expansion of a single-hole orifice jet.(11) The small jets issuing from the perforated plates apparently combine as they expand to form a unified jet. This effect was probably enhanced by the close spacing patterns of the orifice holes. In order to check the effects of spacing perforated plates in series $2\frac{1}{2}$ pipe diameters apart, two different 20-inch diameter plates were positioned in the test loop and operated for 12 hours in a condition when the upstream plate was slightly below the short-term operating limit. At the end of this test no cavitation erosion pitting was seen on the polished upstream face of the downstream orifice plate. Cavitation resistance in general was so pronounced that in order to produce pitting on the aluminum plate in the pipe wall it was necessary to plug all but the outer row of orifice holes, which were within 2-inches of the pipe wall, and operate under severe cavitation. This was the only circumstance in which damage was detected. As pictured in Figure 2, the orifice hole patterns were developed to concentrate the holes in the center of the pipe. Holes were not located closer than 2-inches to the pipe wall, which in this case was two hole diameters.

Three different combinations of three plates placed in series were calibrated to determine if the combined C_d of each set of plates would be affected by the hydraulic interference caused by their close spacing. The difference between the calculated and the measured combined C_d ranged from +4.7 to +5.2 percent. Details of the plate arrangements and resulting C_d variations are shown in Table 1.

Orifices	C_d for each plate	Combined C_d'		
		Calculated	Measured	% Diff.
Upstream: 48 hole @ 1.25" cc Middle: 48 hole @ 1.5" cc Downstream: 85 hole @ 1.5" cc	0.147 0.145 0.295	0.0979	0.103	5.2
Upstream: 48 hole @ 1.25"cc Middle: 85 hole @ 1.25" cc Downstream: 85 hole @ 1.5"cc	0.147 0.290 0.295	0.121	0.127	5.2
Upstream: 48 hole @ 1.25" cc Middle: 121 hole @ 1.25" cc Downstream: 85 hole @ 1.5" cc	0.147 0.444 0.295	0.127	0.133	4.7

Table 1. Evaluation of Pressure Drop Characteristics of Orifices in Series

CONCLUSION

Multi-hole orifice plates have significant advantages as hydraulic energy dissipators. Their cavitation properties are such that fewer plates are required to dissipate head than if single-hole orifices were used, while retaining the economic benefits of a simple design. Additionally, multi-hole orifices in series can be spaced closer together than single-bore orifices without significant effects to their hydraulic characteristics. This can represent a savings in material, piping, and space for applications where a fixed hydraulic resistance is required in a piping system. No cavitation erosion damage was observed at the pipe wall boundary or on downstream perforated plates for any of the conditions tested. This beneficial property gives the designer some margin such that if system conditions are misjudged, or change with time, cavitation damage is not likely to occur before noise and vibration become unacceptable.

APPENDIX I - REFERENCES

1. Tullis, J. P., "Cavitation Scale Effects for Valves," Journal of the Hydraulics Division, ASCE, Vol. 99, No. HY 7, Proc. Paper 9874, July 1973, pp. 1109-1128.

2. Tullis, J. P. and Govindarajan, R., "Cavitation and Size Scale Effects for Orifices," Journal of the Hydraulics Division, ASCE, Vol. 99, No. HY 3, Proc. Paper 9605, March 1973, pp. 417-430.

3. Russell, S. O. and Ball, J. W. "Sudden Enlargement Energy Dissipator for Mica Dam," Journal of the Hydraulics Division, ASCE, Vol. 93, No. HY 4, Proc. Paper 5337, July 1967, pp. 41-56.

4. Watson, W. W., "Evolution of Multijet Sleeve Valve," Journal of the Hydraulics Division, ASCE, Vol. 103, No. HY 6, June 1977, pp. 617-631.

5. "Fluid Meters - Their Theory and Application, " Report of ASME Research Committee on Fluid Meters, ASME, 6th Edition, 1971.

6. Ball, J. W. and Tullis, J. P. and Stripling, T., "Predicting Cavitation in Sudden Enlargements," Journal of the Hydraulics Division, ASCE, Vol. 101, No. HY 7, Proc. Paper No. 11442, July 1975, pp. 857-870.

7. Idel'Chik, I. E., "Handbook of Hydraulic Resistance, Coefficients of Local Resistance and of Friction," U.S. Department of Commerce National Technical Information Service Document No. AEC-TR-6630, 1960.

8. Tullis, J. P. "Cavitation of Perforated Orifice Plates - Phase A," Prepared for Stone & Webster Engineering Corporation, Hydraulics Program Report No. 26, December, 1979.

9. Tullis, J. P. , "Cavitation of Perforated Orifice Plates - Phase B," Prepared for Stone & Webster Engineering Corporation, Hydraulics Program Report No. 30, May, 1980.

10. Rouse, H. A. and Jedinsky, V. "Cavitation and Energy Dissipation of Conduit Expansions," Proceedings of the 11th Congress of International Association for Hydraulics Research, Leningrad, USSR, 1965, Vol. I, pp. 1-8.

11. Sami, S. "Velocity and Pressure Fields of a Diffusing Jet," Dissertation presented to Iowa State University in partial fulfillment of the requirements for the degree of Doctor of Philosophy, August, 1966.

APPENDIX II - NOTATION

The following symbols are used in this paper:

Cd = Dimensionless discharge coefficient (Equation 1).

C_d' = Dimensionless combined discharge coefficient for multiple perforated plates (Equation 3).

C_{dn} = Individual discharge coefficient for a perforated plate in series.

Ao = Flow area of perforated plate orifice, square feet.

Ap = Flow area of pipe, square feet.

g = Gravitational constant, in feet per second per second.

n = Number of perforated plate orifices.

Pd = Pressure 4 to 5 pipe diameters downstream from orifice measured beyond zone of flow establishment.

Pu = Gage pressure measured 1 pipe diameter upstream of orifice.

Pv = Gage vapor pressure of flowing liquid.

V = Average pipe velocity, in feet per second.

ρ = Mass density of liquid.

σ = Dimensionless cavitation index (Equation 2).

FLOOD FREQUENCY MODELING AND OUTLIERS

by Krishan P. Singh,[1] M.ASCE

ABSTRACT: The methodology recommended by the U.S. Water Resources Council for determining flood-flow frequency has been analyzed with special reference to detection of outliers and regionalization of skew values. The low-outlier criterion is shown to be too severe, resulting in rather very low values (algebraically) of skew. Statistics of storms producing floods perceived as high outliers and the associated antecedent soil moisture conditions are used for confirming such outliers. A methodology has been developed and applied to modify low and/or high outliers. The best statistical plotting position can be used for checking the fit of the observed and computed floods. Physiographic and flow-section characteristics can be used for defining hydrologically homogeneous areas. The regionalization of skew is shown to be more precise and meaningful if the outliers are modified before analyzing the flood series. Because of the interaction between the standard deviation and skew values, the regionalization of skew alone is open to question.

INTRODUCTION

The U.S. Water Resources Council published Bulletin No. 15 (8) in 1967 to promote a uniform, consistent, and accurate technique for determining flood-flow frequency. Extensions and updates of this bulletin have been published as Bulletin No. 17 (9) in 1976 and Bulletin No. 17A (10) in 1977. The recommended method checks for low outliers, incomplete record, historical floods, etc., and fits log-Pearson type III, LP3, distribution to observed annual flood peaks. A regional skew value is used from a generalized skew coefficient map (10) because of the errors inherent in computing skew, g, from small and medium size samples. With the exception of lower southern Illinois, the map indicates a generalized g of -0.4 for the State (Fig. 1). There is a wide variation in average skew values for the one-degree quadrangles, ranging from +0.25 to -1.55, and there is no systematic pattern or trend in variation of average g from one quadrangle to the other. There is considerable variation from the average within a quadrangle also.

The basic factors such as the storm, basin and stream, soil, and floodplain characteristics were investigated (7) for their contribution to variation in skew values from one sub-basin to another within the

[1] Principal Hydrologist, State Water Survey, Urbana, IL.

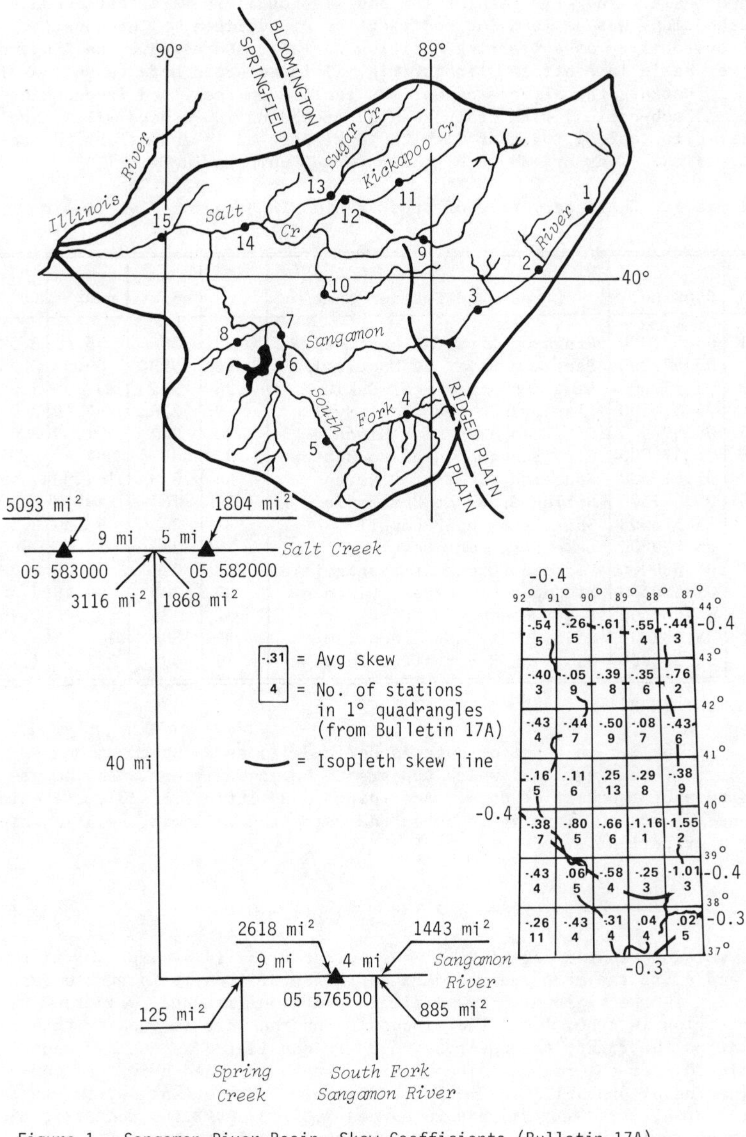

Figure 1. Sangamon River Basin, Skew Coefficients (Bulletin 17A), and Major Tributary Junctions

same basin. The criteria for low and high outliers were tested and a methodology was devised for modification of outliers. The question of suitability of a plotting position was also addressed. The Sangamon River basin in central Illinois (Fig. 1) is selected here as an example to illustrate the effects of various factors on the flood frequencies. The 15 sub-basins, with drainage area exceeding 100 square miles, are listed in Table 1 together with the available basin and stream characteristics. They are shown by their serial numbers in Fig. 1.

Table 1. Characteristics of Study Basins in Sangamon River Basin

No.	USGS No.	Stream and Station Name	n	A sq mi	L mi	S_l ft/mi
1	05 571000	Sangamon River at Mahomet	29	362	56.41	3.59
2	05 572000	Sangamon River at Monticello	68	550	80.04	2.75
3	05 572500	Sangamon River near Oakley	26	774	97.98	2.21
4	05 574500	Flat Branch near Taylorville	27	276	47.49	2.01
5	05 575500	S.F. Sangamon near Kincaid	33	562	51.07	2.01
6	05 576000	S.F. Sangamon near Rochester	27	867	84.82	1.32
7	05 576500	Sangamon River at Riverton	64	2618	164.83	1.48
8	05 577500	Spring Creek at Springfield	29	107	29.37	5.39
9	05 578500	Salt Creek near Rowell	34	335	53.80	2.59
10	05 579500	Lake Fork near Cornland	29	214	37.00	4.65
11	05 580000	Kickapoo Creek at Waynesville	29	227	36.08	6.23
12	05 580500	Kickapoo Creek near Lincoln	32	306	54.48	5.12
13	05 581500	Sugar Creek near Hartsburg	32	333	42.77	5.76
14	05 582000	Salt Creek near Greenview	35	1804	114.68	2.22
15	05 583000	Sangamon River at Oakford	59	5093	222.33	1.27

LOW OUTLIERS

An outlier in a set of data is defined (1) as an observation (or a subset of observations) which appears to be inconsistent with the remainder of that set of data. According to Bulletin 17A (10), the existence of low outliers is to be judged using the following relationship as a criterion:

$$\left|\frac{x_n - \bar{x}}{s}\right| > [2.5 + 1.2 \log (\frac{n}{10})]\ (1 - 0.4\ g_r) \qquad (1)$$

in which x_n is the logarithm of the lowest flood in a sample of size n, $\bar{x}$ and s are the mean and standard deviation of logarithms of floods, and g_r is the regional or generalized skew coefficient. When the criterion is true, i.e., the expression on the left is greater than that on the right, the observation x_n is considered an outlier and deleted from the record. The record is then analyzed using the combination of probabilities method. For n=20, 30, 40, and 50, the test statistic, $|(x_n - \bar{x})/s|$, has to exceed 3.32, 3.56, 3.74, and 3.87, respectively. The lowest flood in none of the 15 flood series tested was an outlier according to the criterion in Eq. 1.

The lowest 4 floods observed at each of the 15 study basins are given in Table 2. There are 5 basins (Nos. 2, 4, 6, 8, and 10) for which 1 or 2 lowest floods appear to be low outliers insofar as they do not follow the trend of the rest of the data and occurred during severe droughts. A methodology was developed to replace outliers (low, high, or both) with values conforming to the rest of the data. The analyst can define the low and/or high outlier(s) that need to be replaced or modified. The various steps in the process are:

1) Compute $\bar{x}$, s, and g from the observed flood series.
2) If no high outlier(s) is (are) to be replaced, go to step 7.
3) Compute recurrence interval, T, of the high outlier to be modified.
4) Calculate Δk with the sample g from the table with T and g as arguments; k is the LP 3 deviate.
5) Compute new value of outlier from the preceding x_2 which is not an outlier

$$\text{new } x_1 = x_2 + s\ \Delta k \tag{2}$$

6) Replace the old x_1 with the new x_1.
7) If no low outlier (s) is (are) to be modified, go to step 13.
8) Let new g = - (sample g).
9) Compute T of the low outlier, considering the floods are arranged in an ascending order of magnitude.
10) Calculate Δk with the new g from the table as in step 4.
11) Compute new or modified value of the outlier from

$$\text{new } x_n = x_{n-1} - s\ \Delta k \tag{3}$$

12) Replace the old x_n with the new x_n.
13) With the new data set, calculate new $\bar{x}$, s, and g.
14) If absolute value of the difference between the new and old $\bar{x}$ is $\leq$ 0.001, between new and old s is $\leq$ 0.002, and between new and old g is $\leq$ 0.005, the new values of $\bar{x}$, s, and g as well as the new data set is the final set. If not, label the new values of the parameters and data set as old values and repeat steps 2 through 14.

Usually 3 to 5 iterations were needed in replacing the outliers with values consistent with the rest of the data. Values of Δk were interpolated from the following table.

Values of Δk for given T and g

T \ g	-1.6	-1.2	-0.8	-0.4	0.0	0.4	0.8	1.2	1.6
100	0.0310	0.0701	0.1267	0.1957	0.2726	0.3541	0.4380	0.5231	0.6084
50	0.0502	0.0970	0.1579	0.2279	0.3031	0.3809	0.4599	0.5387	0.6167
40	0.0589	0.1084	0.1706	0.2407	0.3151	0.3915	0.4685	0.5449	0.6200
20	0.0992	0.1571	0.2281	0.2924	0.3633	0.4338	0.5028	0.5695	0.6331
10	0.1170	0.2439	0.3097	0.3761	0.4399	0.5003	0.5565	0.6079	0.6537

Table 2. Distribution Parameters and 100-Year Floods

No.	4 highest floods, cfs				4 lowest floods, cfs				s	g	Q_{100}
1	14600	11100	9100	9060	1020	1340	1580	1920	0.289	-0.098	18330
2	19000	16700	16600	14000	704	1370	1530	1670	0.293	-0.308	20470
					1110				*0.285*	*-0.143*	*21520*
3	16000	15300	13700	13200	2390	2660	3020	3020	0.253	0.473	26460
4	13000	9400	8620	7450	457	660	1770	1850	0.333	-0.822	12940
					1088	*1360*			*0.276*	*-0.032*	*15920*
5	21500	13700	11900	11100	749	1110	1470	1550	0.335	-0.108	23330
	17080								*0.329*	*-0.210*	*21170*
6	18100	14300	11000	11000	971	1230	2120	2150	0.317	-0.331	21740
					1210	*1570*			*0.302*	*-0.147*	*22670*
7	68700	41000	32900	30600	1040	1830	2540	2840	0.309	-1.257	36610
	43080				*1240*				*0.298*	*-1.353*	*33760*
8	6750	5100	4760	4700	217	225	639	640	0.377	-0.291	9110
					317	*479*			*0.338*	*0.187*	*10390*
9	24500	12400	10600	10300	829	1040	1090	1310	0.377	0.099	29930
	16260								*0.366*	*-0.033*	*25690*
10	6120	4900	4700	4570	152	548	680	1000	0.333	-0.961	6050
					432				*0.289*	*-0.096*	*8150*
11	15100	12800	8420	8300	1300	1330	1500	1530	0.286	0.250	19230
	14080	*11100*							*0.279*	*0.164*	*17700*
12	14800	13800	7970	7860	1400	1520	1600	1730	0.280	0.104	18970
	12230	*10070*							*0.267*	*-0.084*	*16000*
13	26800	10100	8500	8350	1620	1690	1820	1960	0.261	0.520	22660
	11850								*0.233*	*-0.191*	*14090*
14	41200	38100	26900	22000	3400	4020	4040	4880	0.278	-0.073	49120
	40910	*33760*							*0.275*	*-0.119*	*47150*
15	123000	46300	35800	44700	3480	3800	4630	5670	0.308	-0.520	78990
	51230								*0.295*	*-0.819*	*63150*

Figures in italics are the modified values for the outliers directly above them; and values of s, g, and Q_{100} in italics are for the flood series with outliers modified.

As an example, two low floods in the 27-year annual flood series for the Flat Branch near Taylorville (No. 4 in Tables 1 and 2) are selected as low outliers. The results from the computer output are given on the next page.

Sample: $\bar{x}$ = 3.5410 s = 0.3325 g = -0.8219

First iteration: 457 → 803.8 and 660 → 1172.3
$\bar{x}$ = 3.5593 s = 0.2906 g = -0.2556
Second iteration: 803.8 → 1008.3 and 1172.3 → 1309.4
$\bar{x}$ = 3.5647 s = 0.2795 g = -0.0845
Third iteration: 1008.3 → 1069.9 and 1309.4 → 1348.3
$\bar{x}$ = 3.5661 s = 0.2768 g = -0.0437
Fourth iteration: 1069.9 → 1084.8 and 1348.3 → 1357.5
$\bar{x}$ = 3.5665 s = 0.2761 g = -0.0344
Fifth iteration: 1084.8 → 1088.2 and 1357.5 → 1359.7
$\bar{x}$ = 3.5665 s = 0.2760 g = -0.0323

Therefore, the two very low floods (which appear to be inconsistent with the rest of the data) can be modified from 457 and 660 cfs to 1088 and 1360 cfs, respectively. The sample skew increases from -0.8219 with the outliers to -0.0323 when the outliers are modified. The standard deviation decreases from 0.3325 to 0.2760. The outliers and their modified values as well as the values of s, g, and Q_{100} with and without modification of outliers for basins 2, 4, 6, 8, and 10 are given in Table 2; Q_{100} denotes the 100-year flood estimate.

HIGH OUTLIERS

According to Bulletin 17A (10), "If a peak is considered a high outlier, a comparison with historical flood data and flood information at nearby sites should be made. With this information a plotting position is assigned to each outlier and it is then treated as a historic flood peak ... If it is not possible to assign a revised plotting position to the outlier, it is retained in the basic computations as is." Excluding basins 7 and 15 (in Tables 1 and 2) which have "major-junction" problems, one high outlier each is perceived for basins 5, 9, and 13. Storms producing these very high floods were analyzed together with the antecedent soil moisture conditions. These analyses yielded the following information.

Basin	Flood peak, cfs	Approx. storm T, years
5	21,500	50-75
9	24,500	100-150
13	26,800	about 100*

*very wet antecedent soil moisture condition when storm occurred.

Two statistical tests for defining high outliers were adopted from Barnett and Lewis (1) and they indicated one high outlier at 5% level for basins 9 and 13 only. The methodology developed for modifying the perceived outliers was applied to the outliers for basins 5, 9, and 13 as well as to 7 and 15 with major-junction problems, and 11, 12, and 14 with not too-high outliers. The outliers and their modified values as well as the values of s, g, and Q_{100} are given in Table 2. Modification of a high outlier decreases both s and g.

MAJOR-JUNCTION PROBLEMS

The Sangamon River basin has two junctions of major tributaries or rivers in the system: junction of the Sangamon River and Salt Creek, and junction of the Sangamon River and South Fork, as shown in Fig. 1. The relevant statistics for the first 5 top floods, in cfs, at Oakford (basin 15) and corresponding floods at Riverton (basin 7) and Greenview (basin 14) for the 35-year concurrent record 1942-1976 are given below. These floods occurred in April, May, or June.

Sangamon at Oakford			Salt Creek at Greenview			Sangamon at Riverton		
Rank	Peak	Date	Rank	Peak	Date	Rank	Peak	Date
1	123,400	5/20/43	1	41,200	5/19/43	1	68,700	5/19/43
2	45,800	4/25/73	5	21,000	4/25/73	4	27,000	4/24/73
3	44,700	4/26/44	4	22,000	4/25/44	2	30,600	4/25/44
4	42,900	6/25/74	2	38,100	6/24/74	not determined		
5	42,300	5/03/70	9	17,700	5/03/70	5	23,400*	6/18/70

*The annual peak was caused by a different storm event than at Oakford and Greenview. Thus, it is out of phase with Oakford and Greenview flood peaks.

For the first 20 top floods at Oakford, corresponding-year floods are out of phase for 5 years at Greenview and 4 years at Riverton. For the remaining 15 floods, these out-of-phase years are 9 and 10, respectively. Thus, it may be assumed that for 2-year and higher recurrence-interval floods, the chance of floods from the Sangamon River and Salt Creek being out of phase is about 1 in 4 or 5. A Q_{100} of about 100,000 cfs may be satisfactory for the Sangamon River near Oakford. The flood peaks from Greenview and Riverton both contribute to the high flood peaks at the Oakford gage. Because of this, the sample skew coefficient, g, becomes -0.520 at Oakford, compared with -1.257 at Riverton and -0.119 at Greenview. The low value at Riverton is caused by the junction of the Sangamon and South Fork Sangamon, 4 miles upstream; with drainage areas of 1443 and 885 square miles, respectively. For the 25-year concurrent record (1952-1976), the first 5 top floods at Riverton have 3 corresponding-year floods out of phase at Rochester and 4 at Oakley. This reduces the magnitude of top floods at Riverton, and hence the value of g.

Sangamon at Riverton			S.F. Sangamon, Rochester			Sangamon, Oakley		
Rank	Peak	Date	Rank	Peak	Date	Rank	Peak	Date
1	29,000	6/29/51	6	7,160*	7/02/51	14	4,500*	4/31/51
2	27,000	4/24/73	4	9,130*	4/25/73	6	7,480	4/23/73
3	23,400	6/18/70	5	11,100	6/18/70	12	5,500*	4/23/70
4	23,200	5/27/68	3	8,490*	5/29/68	4	13,200*	5/17/68
5	22,600	7/01/57	1	18,100	7/02/57	7	7,400*	4/30/57

*The peak is out of phase with that at Riverton.

It is evident that the value of *g* downstream of the junction of two major tributaries depends considerably on the concurrency of floods in the tributaries upstream of the junction.

PLOTTING POSITION

The results of statistical analyses are often compared by plotting the derived and observed floods on log-probability paper. The plotting position for the observed floods has been a mater of considerable controversy. A general formula for plotting positions (4) is

$$p = \frac{m - a}{n - a - b + 1} \tag{4}$$

in which m is the rank order of the flood values arranged in a descending order of magnitude, p is the probability of exceedence (recurrence interval is the reciprocal of p), and a and b depend on the distribution. For a symmetrical distribution, a equals b. Commonly used plotting position of

$$p = \frac{m}{n + 1} \tag{5}$$

is obtained by putting a=b=0 and is known as the Weibull plotting position. However, a value of about 0.38 (2,3) is considered the best for a sample size of 15 or larger. The distribution curves fitted to the modified and unmodified flood series for the Flat Branch near Taylorville are drawn in Fig. 2. The observed floods plotted with a=b=0.38 or the best plotting position can be differentiated easily on the graph from those with a=b=0 or the Weibull plotting position for 2 or 3 low and high floods. The points plotted with a=b=0.38 fit the LP3 curves better than with a=b=0.

OTHER RELEVANT FACTORS

Some other factors that affect the observed flood distributions and cause them to vary from one basin to another and from one sub-basin to another within a major basin are described below.

Physiographic Characteristics. The northeastern half of the Sangamon basin lies in the Bloomington Ridged Plain (5) which is characterized by low, broad moraine ridges with intervening wide stretches of relatively flat or gently undulatory ground moraine. The glacial deposits are relatively thick throughout the division and drainage development is generally in the initial stage. The valleys of principal streams have floodplains bordered by valley-train terraces. The southwestern half of the basin lies in the Springfield Plain which is distinguished mainly by its flatness and shallow entrenchment of drainage. Drainage systems are well developed and the Plain as a whole is in a late youthful stage of dissection. Most of the principal streams have low gradients and occupy broad alluviated and terraced valleys.

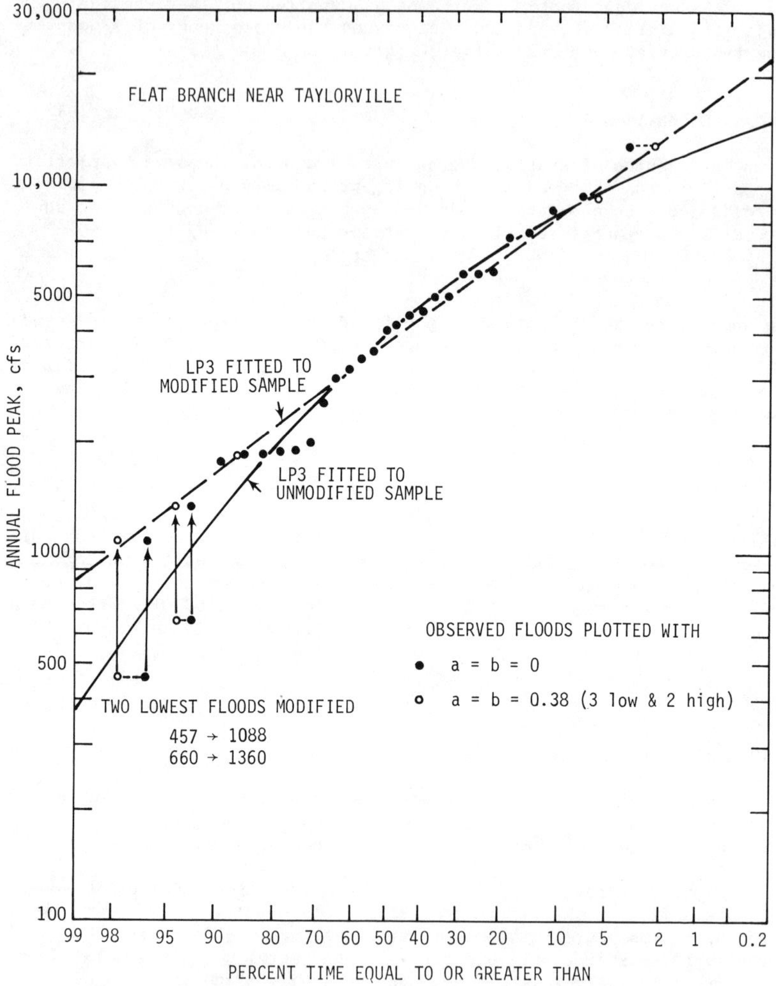

Figure 2. Effect of Modification of Very Low Floods on Curves Fitted by the LP3 Method: Flat Branch near Taylorville

Basin Characteristics. Regressions were run with main-channel length, L, and slope, S_l, as dependent and the drainage area, A, as the independent variables. The simple regression on log-transformed variables yields the following:

$$L = 2.134\ A^{0.545};\ r = 0.974 \text{ with log L and log A} \quad (6)$$

$$S_l = 39.32\ A^{-0.419};\ r = -0.790 \text{ with log } S_l \text{ and log A} \quad (7)$$

Slope-area equation indicates relatively higher observed values of S_l for basins 11, 12, 13, and 14, and relatively lower values for basins 4, 5, and 6 than those given by Eq. 7. These slope differences are attributable, in part, to the physiography of the two divisions. The L vs. A relation is highly significant but the S_l vs. A is also significant; the correlation coefficient r for 5% and 1% significance levels are 0.514 and 0.641, respectively.

Flow Section Characteristics. Bankfull discharge in a river is considered to correspond to about a 2-year or median flood. At higher flows, the water gradually covers the floodplain; the depth and width of this flooding depend on the floodplain geometry. The interaction between the fast-moving channel flow and the slow-moving floodplain flow at small depths of flooding leads to transport of the longitudinal momentum from the main channel to the floodplain (6). This reduces the flow velocity in the channel as well as the mean velocity for the composite section, and it is comparable to the introduction of an additional storage element into the system. The information on actual discharge measurements at various gaging stations was used in developing discharge versus velocity, water width, and flow-area curves. These curves indicate higher flow velocities for basins lying in the Bloomington Ridged Plain (e.g., for a 25-year flood a mean velocity of about 4 to 6 ft/sec compared to 2 to 4 ft/sec for basins in the Springfield Plain), a bankfull discharge close to the median flood compared to bankfull discharge of about one-half of the median flood or less for the Springfield Plain. These indications are in agreement with the physiographic contrasts.

REGIONALIZATION OF *g* AND *s*

Values of *g* and *s*, both without modification of any outliers and with modification of perceived outliers, are given in Table 2. Basins 7 and 15 have major-junction problems and basin 3 is affected by the regulation of Lake Decatur. Therefore, the statistics for *g* and *s* for the remaining 12 basins are given below.

Statistic	Mean	Std dev	Minimum	Maximum
g (observed series)	-0.168	0.419	-0.961	0.520
g (outliers modified)	-0.067	0.125	-0.210	0.187
s (observed series)	0.313	0.038	0.261	0.377
s (outliers modified)	0.294	0.036	0.233	0.366

It is evident that the modification of outliers reduces greatly the value of the standard deviation of the skew coefficient as well as its range.

The parameters g and s were correlated with the basin characteristics: drainage area, A, and main-channel length, L, and slope, S_l.

Unmodified flood series

$$g = -3.217 + 0.883 \log A + 1.510 \log S_l \quad (R = 0.596,\ S_e = 0.372) \tag{8}$$

$$s = 0.729 - 0.129 \log A - 0.163 \log S_l \quad (R = 0.803,\ S_e = 0.025) \tag{9}$$

Flood series with outliers modified where necessary

$$g = 0.600 - 0.259 \log A \quad (r = 0.649,\ S_e = 0.100) \tag{10}$$

$$s = 0.559 - 0.078 \log A - 0.124 \log S_l \quad (R = 0.589,\ S_e = 0.032) \tag{11}$$

R and r are multiple and simple correlation coefficients and S_e is the standard error of estimate. It is evident from the comparison of Eqs. 8 and 10 that much better regionalization of the skew coefficient can be achieved with suitable modification of outliers.

CONCLUSIONS

The following conclusions are drawn from this study.

1. The criterion for considering a low flood as a low outlier, given in the U.S. Water Resources Council Bulletin 17A, is too severe and needs to be modified.

2. Modification of low outliers decreases the standard deviation and increases significantly the skew coefficient. Hence, regionalization of one parameter alone is not justifiable.

3. A methodology was developed and computerized to modify the low and/or high outliers so that their modified values follow the trend of the rest of the data.

4. Analysis of the flood-producing storm and the antecedent soil moisture condition can be used to test if a very high flood is an outlier. Modification of a high outlier reduces both the standard deviation and the skew coefficient.

5. The statistically best plotting position is superior to the Weibull plotting position for testing the goodness of fit of the derived flood distribution curve and the observed floods.

6. The distribution parameters below the junction of two major tributaries are affected greatly by the degree of concurrence of flood peaks in time and flow magnitude. The standard deviation and the skew coefficient would be lower below the junction than if the tributary peaks had been more in phase.

7. The regionalization of skew should be attempted after obtaining skew values from the flood series, with the outliers modified if needed, for hydrologically homogeneous regions or subregions.

APPENDIX I —— REFERENCES

1. Barnett, V., and Lewis, T., *Outliers in Statistical Data,* John Wiley and Sons, New York, N.Y., 1978.

2. Blom, G., *Statistical Estimates and Transformed Beta Variables,* John Wiley and Sons, New York N.Y., 1958.

3. Cunnane, J., "Unbiased Plotting Positions--A Review," *Journal of Hydrology,* Vol. 37, 1978, pp. 205-222.

4. Harter, H.L., "Some Optimization Problems in Parameter Estimation," in *Optimizing Methods in Statistics,* edited by J.S. Rustagi, Academic Press, New York, N.Y., 1971, pp. 33-62.

5. Leighton, M. M., Ekblaw, G. E., and Horberg, L., *Physiographic Divisions of Illinois,* Illinois State Geological Survey Report of Investigation 129, Urbana, IL, 1948.

6. Rajaratnam, N., and Ahmadi, R. M., "Interaction Between Main Channel and Flood-Plain Flows," *Journal of the Hydraulics Division,* ASCE, Vol. 105, No. HY5, 1979, pp. 573-578.

7. Singh, K.P., *Regional and Sample Skew Values in Flood Frequency Analysis of Streams in Illinois,* Illinois State Water Survey Contract Report 223, Urbana, IL, 1980.

8. U.S. Water Resources Council, *A Uniform Technique for Determining Flood Flow Frequencies,* Bulletin 15, Washington, D.C., 1967.

9. U.S. Water Resources Council, *Guidelines for Determining Flood Flow Frequency,* Bulletin 17, Washington, D.C., 1976.

10. U.S. Water Resources Council, *Guidelines for Determining Flood Flow Frequency,* Bulletin 17A, Washington, D.C., 1977.

HYDRAULIC GEOMETRY COEFFICIENTS OF FLOODPLAINS

By

Nani G. Bhowmik,[1] M.ASCE

ABSTRACT

The "hydraulic geometry concept" is a method where the width, depth, area, and flow velocity in a stream are related to the stream's relative position within the basin. This methodology was found to be valid by various researchers for streams and rivers located in the humid areas of the world. The hydraulic geometry parameters for the streams are related to stream order and or the drainage area of the basin. The hydraulic geometry concept is thus a tool which enables the engineer to specify numerically the various hydraulic geometry parameters of the river at any location within the basin.

Research has been conducted to investigate and to postulate the validity of the hydraulic geometry concepts for the floodplains of the stream. Hydraulic geometry relationships for floodplains of eight river basins from Illinois and four river basins from outside of the state of Illinois have been developed. It was found that the concept of "hydraulic geometry" is also valid for the floodplains of the streams and rivers. Relationships between floodplain width, cross-sectional area, surface area, depth, sinuosity, incision and stream order have been developed. These hydraulic geometry parameters were also found to be related to the drainage area of the individual stream segments. The hydraulic geometry coefficients can easily be used in the mathematical modeling of the floodplains of the streams. The variations of the hydraulic geometry coefficients of the floodplains for streams outside of Illinois are similar in nature, but numerically they are different.

INTRODUCTION

Streams and rivers represent a dynamic system which has been working on the earth's crust from prehistoric time. The shape, the size, the extent, and the spread of the floodplains and stream cross sections are the end product of this dynamic system. The formation and development of the floodplain must have followed the laws of nature and consequently must have formed in a very systematic manner. This paper presents some of the results which to some extent confirm the hypothesis that the

[1]Hydraulic Engineer, Illinois State Water Survey, Post Box 232, Urbana, IL 61801

formation of the floodplain follows a definite pattern and that this formation and the existence of the present floodplains can be explained numerically.

BACKGROUND

Floods have been studied by numerous researchers and investigators. The development, the causative factors, and the propagation of flood waves have been studied for different geographic and geologic areas both theoretically and empirically. Various investigators have worked on the classification of floodplains based on different physical phenomena. Melton (1936) has classified the floodplain structure on the basis of stream activity.

The physical laws of mechanics and hydraulics were used by Horton (1945) in a comprehensive study of geomorphology in which he revealed a consistent pattern under which stream systems develop and to which they continually adjust. He showed that the number of streams, the length of streams, and the slope of the streams were all related consistently to stream order throughout any existing stream system. Later revisions of the Horton stream-ordering system were made by Strahler (1964). Because of the inherent flexibility and soundness of the Strahler system of stream ordering, it was used by Stall and Fok (1968) for Illinois streams, by Stall and Yang (1970) for 12 river basins in the United States, and by various other investigators. It provides a means of evaluating numerically the structure of a stream system, and has been used extensively to allow a place comparison for hydraulic characteristics at various locations within a stream system. The Horton-Strahler system of stream order and the associated laws of stream development have considerable hydrologic implications regarding floods, reservoir sedimentation, sedimentation in general, and many other factors.

A definite and clearcut correlation is needed between the stream ordering and the hydraulic geometry of the stream before the system of stream development can be defined. The important hydraulic characteristics of a stream channel are width, depth, and velocity of flow. Leopold and Maddock (1953) proposed the concept of hydraulic geometry relationships for stream systems in various physiographic locations. They showed that width, depth, and velocity of flow in a stream change along the course of the stream in a downstream direction, with a constant frequency of discharge.

The hydraulic geometry concept postulated by Leopold and Maddock (1953) and the Horton-Strahler (1964) stream-ordering system have been used by various investigators around the world to determine and to predict the hydraulic geometries of stream channels at various locations in the stream course. It appeared to the author that floodplains are also curved in a systematic way and that the development of floodplains seems to follow a pattern. It was suspected that the concept of hydraulic geometry and stream ordering found to be so immensely valuable for explaining the stream system should also be equally applicable for explaining the development and formation of floodplains of the streams. The detailed analyses of this research is given by Bhowmik and Stall (1979).

DATA ANALYSIS

Altogether, eight river basins in Illinois and four river basins outside of the state were selected for the present investigation. All these river basins are located in the humid areas of the United States. Hydraulic geometry relationships for the main channel of these rivers have already been developed by Stall and Fok (1968) for Illinois streams and by Stall and Yang (1970) for the streams outside Illinois.

After the selection of the river basins from within and outside Illinois, the topographic maps covering the drainage areas of each selected river were gathered. Next the Horton-Strahler stream orders were determined. According to Strahler (1964) visible unbranched streams shown on the topographic maps in blue were defined as the first order streams. When two first order streams join, a second order stream begins and so forth. For this investigation, only streams of third order or higher were used. The floodplain for each selected stream for each order was delineated on the corresponding topographic map. After delineating the floodplain for each order stream in each river basin, 10 to 60 cross section locations were marked on the map. The delineated floodplain on the map for each reach of the river was then planimetered and the down-valley length of the floodplain was determined. The ratio of the planimetered floodplain area in the plan view to the down-valley length gave the average width, W, of the floodplain for that specific reach of the river.

The average cross-sectional area of the floodplain, A, for each stream order in the river basin was computed by the following procedure. The distances from the edge of the floodplain of all the contour lines at each cross section line were measured. The difference in elevation between the contour line at the edge and in the floodplain determined the depth of the floodplain at that location. The cross-sectional area of the floodplain was then computed at each cross section and the average cross-sectional area for the entire reach was determined. From this average area, A, and the average width, W, determined from the planimetered area, the average depth of the floodplain, D, for each reach was computed.

HYDRAULIC GEOMETRY RELATIONSHIPS

The hydraulic geometry relationships for the eight river basins within Illinois and the four river basins outside the state have been developed.

The eight Illinois river basins for which the hydraulic geometry relationships have been developed are the Rock, Kankakee, Spoon, Sangamon, Kaskaskia, Big Muddy, Embarras, and Little Wabash Rivers.

The number of streams for each order for different river basins varied anywhere from 1 to 380. Normally there is only a river segment corresponding to the highest order stream in each river basin and about 73 to 380 river reaches for the third order streams. The number of segments for each stream order that were finally analyzed varied anywhere from 1 to 284. Then the hydraulic and geometric parameters were determined for each stream segment corresponding to each order stream for each river basin.

The basic equation that was found to be valid for streams is given by equation 1.

$$\ln Y = i + j F + k U \quad (1)$$

where Y = A, D, V, W, and Q; i, j, and k are coefficients; U is the stream order; V is the velocity; and F is the frequency of discharge. In the case of main streams, the value of F can vary anywhere from 0.1 to 0.9 with the best fitting value occurring when the value of F was equal to 0.1, that is the discharge for which the flow is exceeded only 10 percent of the time. However, in the case of floodplains, the floodplain-forming discharge is taken to be the 100-year discharge and as such the value of F is constant for all cases. Thus with a constant value of F, equation 1 can be transformed into equation 2 as follows.

$$\ln Y = n + m U \quad (2)$$

Here n and m are coefficients. Equation 2 is the basic equation that is postulated to be valid for the floodplains of different river basins.

The drainage area, D_A, was shown to be related to the stream order U for all streams investigated by Stall and Fok (1968) and Stall and Yang (1970). Thus stream order U in Eq. 2 can be replaced with drainage area, D_A.

Relationships between the average values of W, A, D, and Q_{100}, with stream order U are given in Figs. 1 through 4 for the Kaskaskia river basin. The frequency distribution or histograms for each parameter corresponding to third, fourth, and fifth order streams and the regression equations are also shown in these figures. Similar analyses with other parameters were also done.

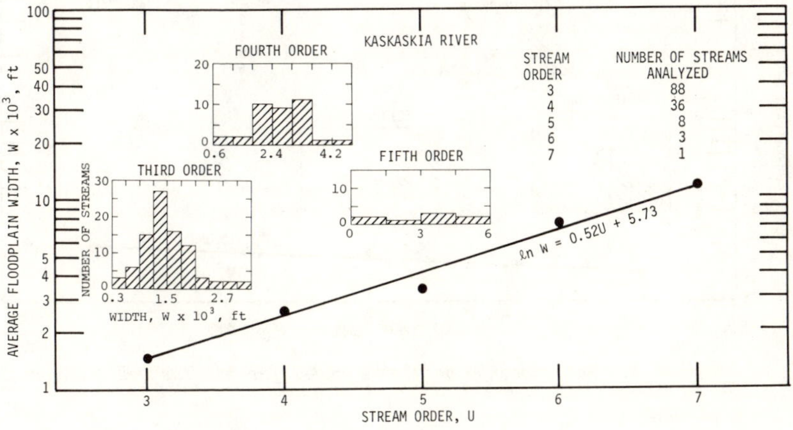

Figure 1. Relationship between W and U for the Kaskaskia River

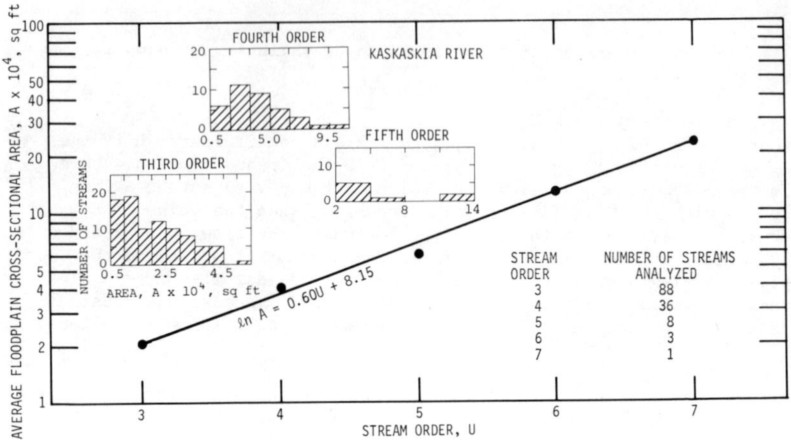

Figure 2. Relationship between A and U for the Kaskaskia River

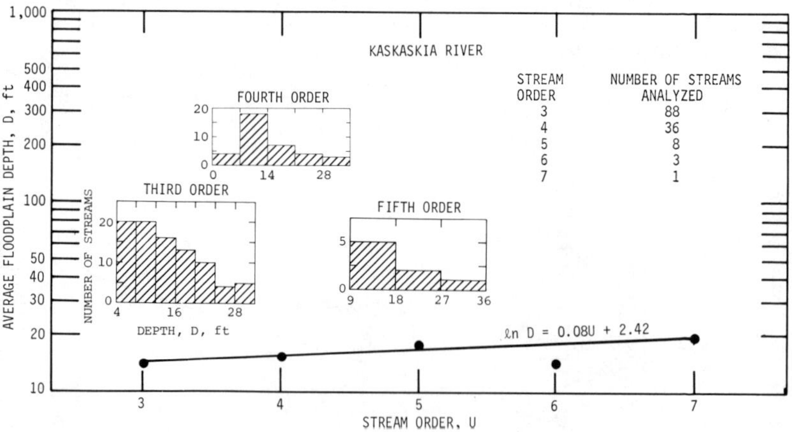

Figure 3. Relationship between D and U for the Kaskaskia River

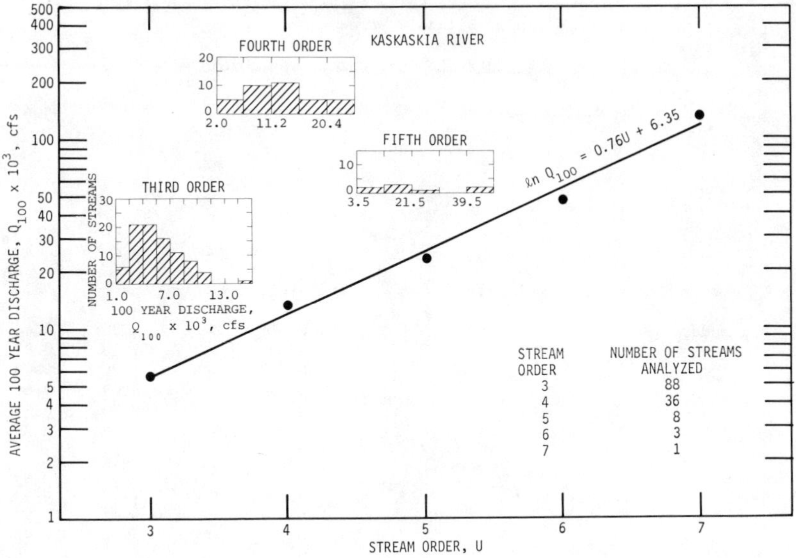

Figure 4. Relationship between Q_{100} and U for the Kaskaskia River

Similar analyses were also done for four river basins outside of Illinois. These river basins are: the White River in Indiana; the Big Sandy in Kentucky, West Virginia and Virginia; and the Susquehanna River in New York and Pennsylvania, and the Neches River in Texas.

Table 1 shows the hydraulic geometry coefficients of floodplains of the Illinois stream and table 2 shows similar information for the out of state streams.

The constant n shown in tables 1 and 2 is the intercept and the coefficient m gives the slope of the regression line. A standard statistical analysis was made to determine the different statistical parameters associated with the coefficient m and the constant n. These statistics are shown in tables 3 and 4.

The variance σ^2 indicates the spread of the variable values about a central value. Higher values of σ^2 will indicate a spread of the variable values over a wider margin. The skewness coefficient C_s is a descriptor of asymmetry of a distribution. The value of $C_s = 0$ indicates presence of a symmetrical distribution. The Kurtosis coefficient K measures the flatness or the peakedness of a distribution. A sharp peak in the distribution is associated with a value of K close to zero.

Table 1. Hydraulic Geometry Relationships of the Floodplains for Illinois Streams

(Stream order U is the independent variable)

$$\ln Y = n + mU$$

where:

$Y = W, A, D, Q_{100}, S_s, I,$ and D_A

n, m = coefficients

Name of stream	Maximum stream order	Hydraulic geometry parameter, Y	n	m
Big Muddy	7	W	5.70	0.67
		A	7.50	0.87
		D	1.80	0.20
		Q_{100}	6.09	0.66
		Ss	-0.53	0.20
		I	2.16	0.29
		D_A	-1.46	1.32
Kaskaskia	7	W	7.53	0.52
		A	8.15	0.60
		D	2.42	0.08
		Q_{100}	6.35	0.76
		Ss	-0.06	0.09
		I	3.12	0.09
		D_A	-0.42	1.27
Little Wabash	7	W	6.81	0.38
		A	8.02	0.67
		D	1.21	0.29
		Q_{100}	6.25	0.73
		Ss	-0.02	0.07
		I	2.09	0.26
		D_A	-0.73	0.31
Embarras	6	W	5.70	0.49
		A	7.60	0.54
		D	1.90	0.05
		Q_{100}	5.48	0.88
		Ss	-0.24	0.10
		I	3.26	0.02
		D_A	-1.60	1.54
Sangamon	7	W	5.03	0.67
		A	6.83	0.91
		D	1.80	0.24
		Q_{100}	5.90	0.79
		Ss	0.28	-0.02
		I	2.41	0.23
		D_A	-0.35	1.32

Table 1. Concluded

Name of stream	Maximum stream order	Hydraulic geometry parameter, Y	n	m
Spoon	6	W	5.88	0.43
		A	8.38	0.56
		D	2.50	0.13
		Q_{100}	5.75	0.79
		Ss	-0.27	0.12
		I	3.17	0.12
		D_A	-1.73	1.52
Kankakee	7	W	6.14	0.61
		A	7.57	0.75
		D	1.43	0.14
		Q_{100}	5.04	0.70
		Ss	0.20	0.01
		I	2.22	0.18
		D_A	-1.64	1.45
Rock	7	W	6.00	0.51
		A	8.40	0.63
		D	2.40	0.12
		Q_{100}	5.69	0.76
		Ss	0.21	0.01
		I	2.53	0.24
		D_A	-2.07	1.62

In table 3 for the coefficient m, the values of σ^2 for all the parameters are very small and the values of K - 3 are all negative. This indicates that the numerical values of the coefficient m are not too far away from a central value, and, at the same time, the distribution of m from various river basins has a sharp peak compared with a normal distribution. The skewness coefficient C_s indicates that the distribution of m for D_A is very close to a symmetrical distribution. On the other hand, the distributions of m for W, A, D, Q_{100}, and Ss are skewed to the right and for I it is skewed to the left.

These statistics for the coefficient m for the eight Illinois streams indicate that the variability of m among various river basins within the state is not very significant. Also, the numerical values of the coefficient m for all the parameters have a central tendency, i.e., a majority of them tend to cluster around a central value. This incidates that it may be possible to derive a mean value of m for the whole state of Illinois which can be used to determine various hydraulic and geometric parameters of floodplains within the state.

Table 3 also shows the standard statistics for the intercept n in Eq. 2. It appears that both the standard deviation σ and variance σ^2 are somewhat higher in magnitude compared with these values for the coefficient m. The distribution of n values are also skewed with the coefficient of skewness being both positive and negative, i.e., with long tails to the right and to the left, respectively. Excess Kurtosis coefficients of negative value indicate the presence of a distribution with a higher peak than the normal distribution. On the basis of the value of σ and

Table 2. Hydraulic Geometry Relationships of the Floodplains for Streams outside of Illinois

(Stream order U is the independent variable)

$$\ln Y = n + mU$$

Name of stream	Maximum stream order	Hydraulic geometry parameter, Y	n	m
White	6	W	4.56	0.76
		A	6.53	0.96
		D	1.97	0.20
		Q_{100}	5.03	1.14
		Ss	-0.13	0.07
		I	2.50	0.21
		D_A	-2.52	1.73
Big Sandy	8	W	5.28	0.27
		A	8.23	0.31
		D	2.95	0.04
		Q_{100}	5.33	0.82
		Ss	0.06	0.02
		I	3.55	0.09
		D_A	-1.43	1.23
Susquehana	7	W	5.71	0.42
		A	7.95	0.59
		D	2.24	0.17
		Q_{100}	3.61	1.20
		Ss	0.20	-0.02
		I	2.86	0.18
		D_A	-1.87	1.51
Neches	7	W	5.34	0.67
		A	7.68	0.73
		D	2.34	0.06
		Q_{100}	6.05	0.92
		Ss	-0.25	0.09
		I	3.07	0.04
		D_A	-1.23	1.50

σ^2 for n, it appears that a single intercept value (n) for all the streams in Illinois may not be feasible. The regression lines may be somewhat parallel to each other, but they all have different intercept values. This indicates that the rates of change of different parameters with stream order are approximately the same, but the regression equations have different base values.

Table 4 shows the standard statistics of m and n for all the river basins that were analyzed for the present investigation. The distributions of n corresponding to all the parameters are skewed ($C_s \neq 0$) and they have sharper peaks than the normal distribution, $(K - 3) < 0$. Thus it appears that lumping all the river basins from throughout the United States with the eight river basins from Illinois did not improve its predictability in obtaining a central value for the intercept n in Eq. 2.

Table 3. Standard Statistics of the Coefficient m and Constant n for Illinois Streams in the Equation $\ln Y = n + mU$

Variable	Mean	σ	σ^2	C_s	K-3
Statistics for m					
W	0.52	0.11	0.01	0.20	-4.66
A	0.67	0.14	0.02	0.52	-4.43
D	0.15	0.08	0.006	0.49	-4.29
Q_{100}	0.76	0.07	0.004	0.28	-3.88
Ss	0.06	0.08	0.006	0.26	-4.23
I	0.18	0.09	0.008	-0.47	-4.26
D_A	1.44	0.13	0.02	0.04	-4.91
Statistics for n					
W	6.16	0.74	0.54	0.37	-3.91
A	7.92	0.60	0.36	-0.26	-4.13
D	1.96	0.45	0.21	-0.27	-4.51
Q_{100}	5.82	0.43	0.18	-0.42	-4.16
Ss	-0.007	0.30	0.09	-0.33	-4.42
I	2.60	0.46	0.21	0.36	-4.77
D_A	-1.29	0.63	0.39	0.46	-4.61

σ = Standard deviation

σ^2 = Variance

C_s = Skewness coefficient

K-3 = Excess Kurtosis coefficient compared with a normal distribution

Table 4. Standard Statistics of the Coefficient m and Constant n for All Streams in the Equation

$$\ln Y = n + mU$$

Variable	*Mean*	σ	σ^2	C_s	K-3
Statistics for m					
W	0.52	0.14	0.02	0.01	-4.27
A	0.66	0.18	0.03	-0.01	-3.80
D	0.14	0.08	0.006	0.40	-4.13
Q_{100}	0.85	0.17	0.03	1.02	-3.35
Ss	0.05	0.07	0.005	0.39	-3.82
I	0.16	0.09	0.008	-0.27	-4.37
D_A	1.45	0.15	0.02	0.09	-4.30
Statistics for n					
W	5.87	0.79	0.62	0.42	-3.53
A	7.82	0.64	0.40	-0.51	-3.66
D	2.09	0.47	0.22	-0.16	-3.76
Q_{100}	5.54	0.75	0.56	-1.25	-2.05
Ss	-0.014	0.26	0.07	-0.28	-4.14
I	2.72	0.47	0.22	0.20	-4.51
D_A	-1.44	0.63	0.39	0.33	-3.86

APPLICATION OF THE HYDRAULIC GEOMETRY RELATIONSHIPS

The hydraulic geometry relationships and the generalized analyses that have been presented so far can be applied for actual field conditions in connection with hydraulic and hydrologic analyses of floodplains. Data from eight river basins in Illinois and four river basins from outside the state have been analyzed. Generalized relationships for the state of Illinois have also been developed. These analyses should form a broad base for the eventual application of the hydraulic geometry relationships. Some of the specific applications of the developed relationships are: (1) Determination of the Hydraulic Geometry Parameters of the Floodplains; (2) Estimation of Floodprone Areas; (3) Dispersion Studies in Open Channel; (4) Potential of Bank Erosion and Secondary Circulation, and (5) Low Flow Characteristics of Streams.

These are some of the applications of the hydraulic geometry relationships developed in the present research investigation. There are, of course, many other applications in which the hydraulic geometry relationships, the sinuosity, and the incision of the streams can be utilized. Geomorphologically, the river basins analyzed in the present investigation vary anywhere from mature to advanced stages of development. Thus the correlations developed should also be valid for a wide range of streams under different physiographic conditions within the country.

SUMMARY AND CONCLUSIONS

The hydraulic geometry of floodplains was investigated to determine the variability and similarity of the hydraulic geometry relationships for different river basins.

Data for eight river basins in Illinois and four river basins outside of the state were analyzed to determine the hydraulic geometry parameters of these basins. The conclusions are as follows.

1) The Horton-Strahler stream ordering system and the concept of hydraulic geometry that are valid for streams in the humid areas of the country are also applicable to the floodplains of the rivers.

2) The floodplains are carved in a systematic manner and the development of the floodplains follows a systematic pattern.

3) The floodplain hydraulic geometry parameters width, depth, and cross-sectional area are related to the stream order and they increase in the downstream direction. These parameters can also be related to the drainage areas of the individual streams.

4) The other parameters, including sinuosity, incision, 100-year discharge, and drainage area are also related with stream order and they increase in magnitude in the downstream direction.

5) The magnitudes of each hydraulic or geometric parameter within each river basin and for each stream order have frequency distributions varying anywhere from skewed to the left to symmetrical to skewed to the right. However, the frequency distribution of the hydraulic parameters for the mountainous stream for each order is generally nonskewed.

6) In Illinois, streams in the nonglaciated or infrequently glaciated areas are more sinuous than those in the glaciated areas.

ACKNOWLEDGMENT

The research described here was conducted by the author as part of his regular work at the Illinois State Water Survey. The work upon which this research is based was supported in part by funds provided by the U.S. Department of Interior as authorized under the Water Resources Act of 1964, Public Law 88-379, Project No. B-100-ILL.

REFERENCES CITED

Bhowmik, N.G., and Stall, J.B., "Hydraulic Geometry and Carrying Capacity of Floodplains," University of Illinois Water Resources Center Research Report No. 145, 1979, 147 p.

Horton, R.E., "Erosional Development of Streams and Their Drainage Basins -- Hydrophysical Approach to Quantitative Morphology," Geological Society of America Bulletin, Vol. 56 (3), 1945, pp. 275-370.

Leopold, L.B., and Maddock, T., "The Hydraulic Geometry of Stream Channels and Some Physiographic Implications," Professional Paper 252, United States Geological Survey, Washington D.C., 1953.

Melton, F.A., "An Empirical Classification of Flood-Plain Streams," The Geophysical Review, Vol. 26, 1936, pp. 593-609.

Stall, J.B., and Fok, Y.S., "Hydraulic Geometry of Illinois Streams," University of Illinois Water Resources Center Research Report, No. 15, 1968.

Stall, J.B., and Yang, C.T., "Hydraulic Geometry of 12 Selected Stream Systems of the United States," University of Illinois Water Resources Center Research Report, No. 32, 1970.

Strahler, A.N., "Quantitative Geomorphology of Drainage Basins and Channel Networks," in Handbook of Applied Hydrology, Sect. 4-II, V.T. Chow, ed., McGraw-Hill Book Company, New York, 1964.

A Model Structure for Stream Systems

by

John B. Stall,[1] Fellow, ASCE

Abstract

The hydraulic channel characteristics of natural streams are interrelated in a complex manner. In this paper a model structure is provided for these relationships. An existing natural stream system is evaluated using this structure. The fitting coefficients and exponents are given. Discharge, cross-sectional area, width, depth, and velocity of a stream at a particular location in the stream system are each related to the frequency of occurrence of the discharge, F, in percent of days per year, and the drainage area, A_d, in square miles. The resulting five hydraulic geometry equations represent the characteristic interrelationship of these factors throughout the stream system.

Background

The concept of hydraulic geometry was first published by Leopold and Maddock (1953). Here it was suggested that channel characteristics of natural streams are interrelated in a complex manner. The authors showed how the nature of a particular river system can be described quantitatively. They first described this interrelated system as the hydraulic geometry of the stream system. These authors gave three principal equations as follows:

$$W = a\ Q^b \qquad (1)$$

$$D = c\ Q^f \qquad (2)$$

$$V = k\ Q^m \qquad (3)$$

where W = width, D = mean depth, V = mean velocity, Q = discharge, and a, b, c, f, k, and m are numerical constants. Leopold and Maddock (1953) showed that these relations, even for stream systems in greatly different physiographic settings, appear to be consistent.

The Roanoke River System

Shown in Figure 1 is the Roanoke River System which was analyzed in this study. The basin covers 8410 sq mi and 37 stream gaging station records were available. Annual runoff is 13.2 inches. The length of the main stem of the river is 278 miles, total relief is 2470 feet; and average river slope is 8.9 feet per mile.

[1]Consulting Research Hydrologist, Urbana, IL

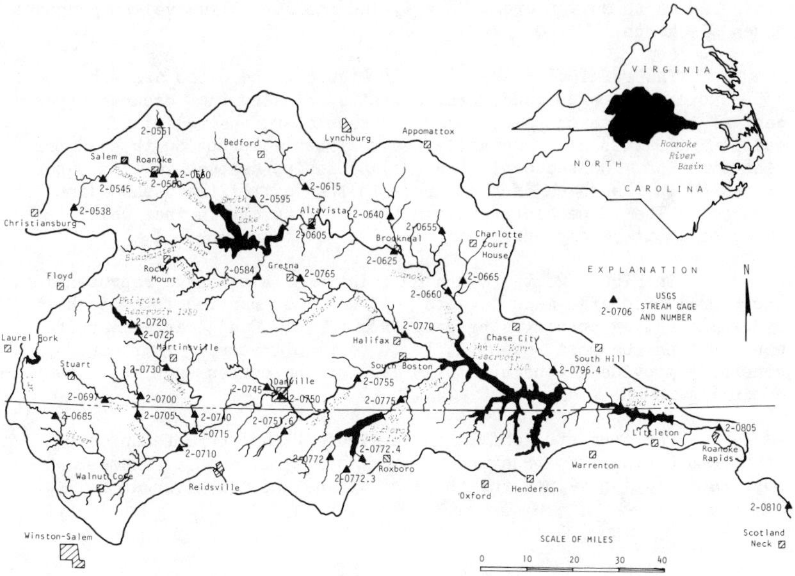

Figure 1. Roanoke River Basin

Field records of discharge measurements made by the U. S. Geological Survey were recorded on their Form 9-207. From 50 to 500 measurements were available at each station. Data available from each measurement were the width W, channel cross-sectional area A, average velocity V, and the discharge Q. The average depth D was computed from D = A/W with D being defined as the hydraulic depth.

Hydraulic Rating Curves

The first step in depicting graphically the channel conditions for a reach of stream represented by particular stream gaging stations was the plotting of station hydraulic rating curves. The four parameters A, W, D, and V were plotted on log-log paper versus Q. Figure 2 shows hydraulic rating curves for the gaging station on the Dan River at Danville, Virginia, in the Roanoke River Basin. The points represent actual measurements, straight lines were fitted to the points by eye to best represent the general relation. The scatter of points is due to the variation in local conditions, but the general pattern for the curves is evident. The curves are drawn in such a way as to be consistent. If a vertical section is taken at a given discharge, the values from the curves will satisfy two physical laws: the product of width and depth equals area, WD = A, and the area times velocity equals given discharge, AV = Q.

The vertical dashed line in Figure 2 indicates a discharge of 3850 cubic feet per second, cfs. This is the flow that occurs 10 percent of the days each year. As discussed later, the relationships developed in this study were limited to flows at or below this 10 percent duration. Consequently, the relationships derived in this paper are based on the shape of the curves to the left of the dashed line in Figure 2. The 10 percent duration flow of 3850 cfs is that which is exceeded about 36 days per year, and is less than bankfull.

In Figure 2, there are 89 points for each curve representing individual discharge measurements It was the practice to plot these hydraulic rating curves using data from 50 to 200 discharge measurements during the most recent 10 to 20 years of record. This would generally provide a consistent pattern for the curves such as illustrated in Figure 2.

The four straight lines in Figure 2 can be represented by equations having the structure of equations 1, 2, and 3. These represent <u>station</u> values for these relations and are given as follows:

$$W = 268.0\ Q^{0.07} \qquad (4)$$

$$D = 0.0636\ Q^{0.49} \qquad (5)$$

$$V = 0.0553\ Q^{0.45} \qquad (6)$$

$$A = 17.05\ Q^{0.56} \qquad (7)$$

Time Sampling

In order to compare hydraulic factors for various discharges at one location, a duration curve of daily discharges was derived for a representative period of years. The discharge, Q, which is exceeded 10% of the days per year is assigned a frequency of F = 0.10 and the discharge exceeded 90% of the days each year is assigned F = 0.90 for use in this model. Then, throughout the stream system, a steady-flow condition is assumed with the Q being that, at each gaging station, which has F = 0.10 or F = 0.20, etc including F = 0.90.

Place Sampling

Horton (1945) described a system of stream orders under which a stream develops; Strahler (1964) provided a slightly more flexible system of stream orders. He also showed how the term A_d = drainage area of a stream in sq mi is related to stream order. The consistent pattern of these relations supports our choice of A_d for this model to show how hydraulic geometry relations vary throughout a stream system.

Hydraulic Geometry Relations

The hydraulic reating curves as shown in Figure 2 for station 2-0750, on the Dan River at Danville, Virginia, in the Roanoke

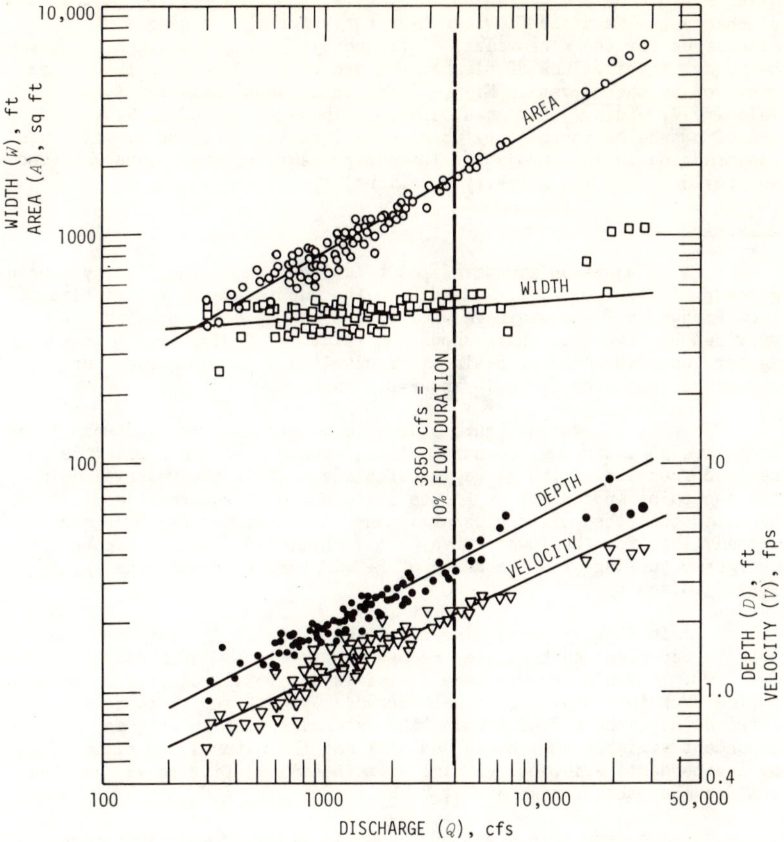

Figure 2. Hydraulic Rating Curves For Gage 2-0570, Dan River at Danville, Virginia, in the Roanoke River Basin

River Basin show the variation in A, V, W, and D as related to the discharge Q. From the flow duration curve for this gaging station, discharges for the nine values of frequency, F = 0.10 through 0.90, can be read. These values of discharge Q are read, along with the values provided by the curves in Figure 2 for cross-sectional area A, velocity V, width W, and mean depth D. These values of A, V, W, and D can of course be computed by using equations 4, 5, 6, and 7, which was the practice in this project. These represent the basic data analyzed to provide hydraulic geometry relations.

Equations

An equation was developed relating each of the five hydraulic geometry factors Q, A, V, W, and D to the two independent variables of flow frequency F and drainage area A_d. An example of this form is provided by Figure 3, which shows the relation of discharge Q to F and A_d for the Roanoke River Basin. For clarity, only two curves for values of frequency F = 0.10 and F=0.90 are shown.

Also shown in Figure 3 are the points used to fit these lines. There are 35 individual points defining each curve. Each point is derived from one of the 37 gaging stations used in the Roanoke Basin; the two points not shown in Figure 3 lie below 10 square miles. It can be noted that the points cluster around the curve for F = 0.10 more closely than for the lower curve. It is believed that the higher discharges occurring at a frequency of F=0.10 best represent the channel forming discharge.

In this project, nine curves were similarly fitted to the data to represent each of the frequencies from F = 0.10 through F = 0.90. All nine curves were fitted as one total multiple regression to the 37 points that define each curve. The fitting process was carried out mathematically using the natural logarithm of Q as the dependent variable and the values of F and the natural logarithm of A_d as independent variables. A similar mathematical fitting was carried out for the other parameters A, V, W, and D.

The result is a set of hydraulic geometry equations representing this entire Roanoke River System:

$$\ln Q = 0.47 - 2.35\ F + 1.05 \ln A_d \qquad (8)$$

$$\ln A = 0.54 - 1.46\ F + 0.92 \ln A_d \qquad (9)$$

$$\ln V = -0.08 - 0.88\ F + 0.13 \ln A_d \qquad (10)$$

$$\ln W = 1\ 52 - 0.34\ F + 0.54 \ln A_d \qquad (11)$$

$$\ln D = -0.98 - 1.13\ F + 0.38 \ln A_d \qquad (12)$$

where Q is in cfs, A_d in sq mi, A in sq ft, V in fps, W and D in ft. The designation ln means a logarithm to the base e.

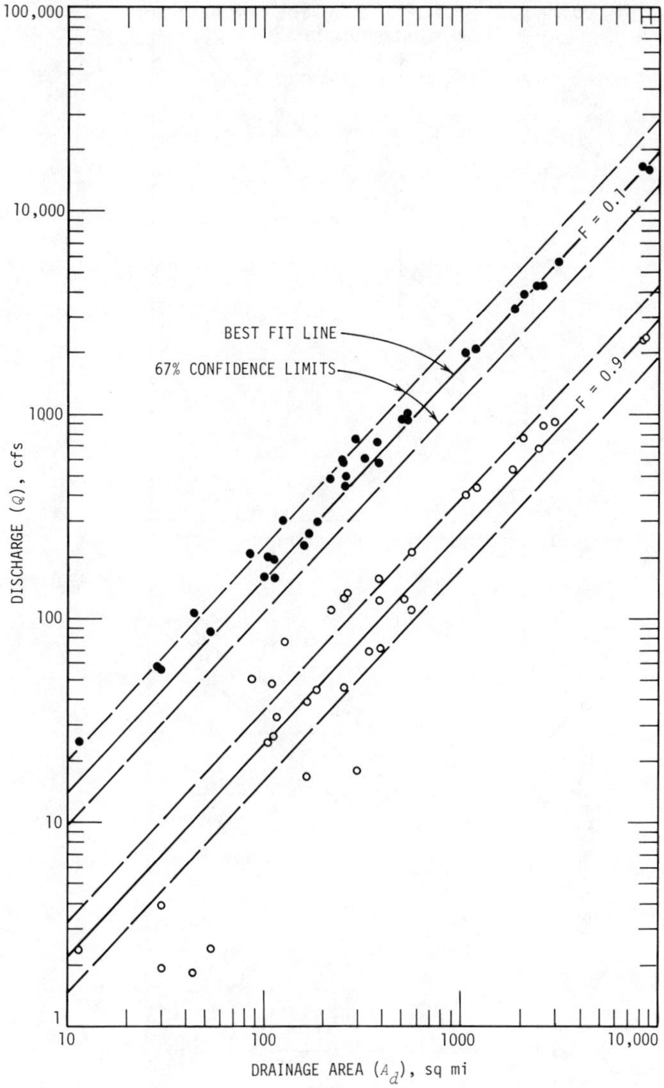

Figure 3. Discharge as Related to Drainage Area for Two Frequencies Showing the 67-percent Confidence Intervals, Roanoke River Basin

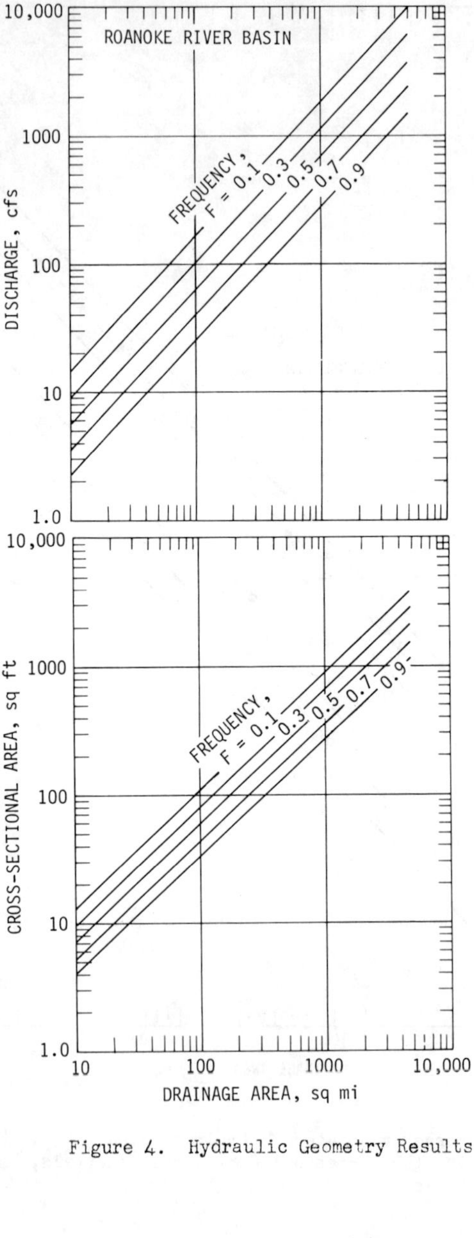

Figure 4. Hydraulic Geometry Results

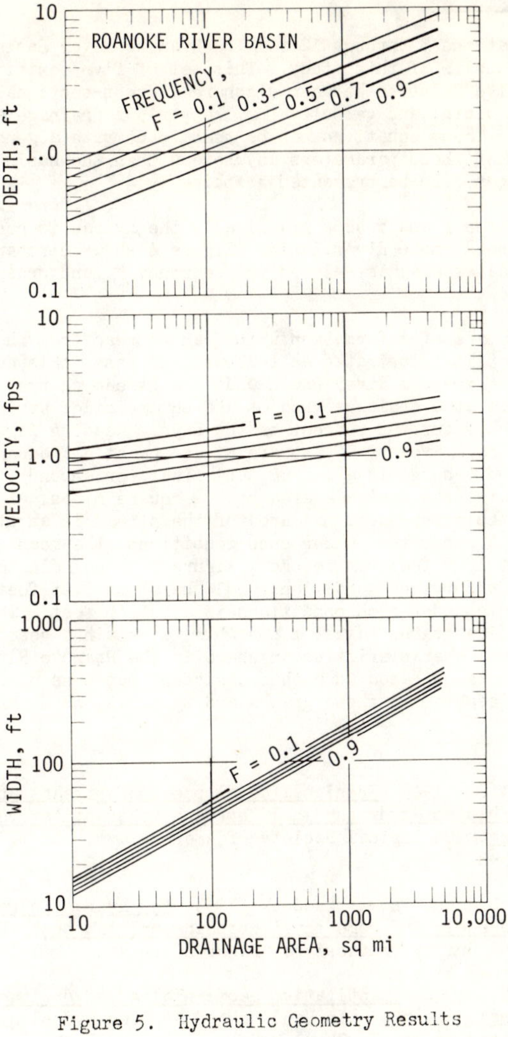

Figure 5. Hydraulic Geometry Results

Equations and Graphs

Equations 8 through 12 for the Roanoke River Basin exemplify the primary results of this study. This set of five equations shows the quantitative relation between discharge, cross-sectional area, stream depth, width, and velocity for the various drainage areas within the basin. These equations can be used to compute a generalized value for any of these parameters anywhere within the basin. Results of this type can also be presented graphically.

Figures 4 and 5 show graphically the hydraulic geometry results for the Roanoke River Basin. Figure 4 shows discharge Q, and cross-sectional area A, as related to frequency F, and drainage area A_d.

The type of information which can be read from the graphs in Figures 4 and 5 is illustrated as follows. Suppose it is desirable to know within the Roanoke River Basin what the stream channel characteristics might be at a drainage area of 100 square miles during the discharge which occurs 50 percent of the days. Reading Figure 4 shows that at a drainage area of 100 square miles and at a frequency of F = 0.50, the discharge is about 60 cubic feet per second. Figure 4 also shows that for a drainage area of 100 square miles, at a frequency of F = 0.50, the cross-sectional area of the stream is about 56 square feet. Figure 5 shows that under such conditions, the mean velocity would be about 1.09 feet per second. Figure 5 then indicates that the average depth under such conditions would be about 1.25 feet, and that the stream width under such conditions is about 46 feet. Thus, the use of these five graphs makes it possible to obtain a general estimate as to the stream characteristics anywhere in the Roanoke River Basin, and this estimate is based upon the consistent patterns of hydraulic geometry as presented in these graphs and by equations 8 through 12.

References

Horton, Robert E. 1945 Erosional development of streams and their drainage basins--hydrophysical approach to quantitative morphology. Geological Society of America, vol. 56 (3), pages 275-370

Leopold, Luna B., and Thomas Maddock 1953 The hydraulic geometry of stream channels and some physiographic implications. U. S. Geological Survey Professional Paper 252.

Strahler, A. N. 1964 Quantitative geomorphology of drainage basins and channel networks. In Handbook of Applied Hydrology, Section 4-11, edited by Ven T. Chow, McGraw Hill, New York.

SIMULATION MODEL FOR OPEN PIT MINE DEWATERING

By Catherine Kraeger-Rovey,[1] M. ASCE

INTRODUCTION

During 1979, a hydrologic computer model study was conducted at the request of Cyprus Mines Corporation, operator for the proposed Hansen Project, an open-pit uranium mine which is planned for development in south-central Colorado. The study was conducted by Wright Water Engineers, Inc. as part of a comprehensive hydrologic investigation. The purposes of the model study were (1) to support the results of Theis-type analyses which had been conducted to determine well locations and the schedule of pumping rates required for dewatering, and (2) to analyze the total hydrologic effect of the dewatering operation. In addition to the environmental assessment and mine dewatering plan, the results of this model study were used in the development of the post-mining reclamation plan, in the mine water supply and water management plan, and in the water rights augmentation plan for the project.

SITE DESCRIPTION

The proposed Hansen Mine site is located about thirty-five miles northwest of Canon City, Colorado in the Currant Creek Basin, which is tributary to the Arkansas River. The location of the project site is shown in Fig. 1. The mine pit will occupy a 7,000-foot long section of the valley of the Middle Fork of Tallahassee Creek, which drains into Currant Creek. The region in which the

---

[1]Geohydrologist, Denver Regional Manager, Ott Water Engineers, Inc., Denver, Colorado; formerly Director of Special Projects, Wright Water Engineers, Inc., Denver, Colorado.

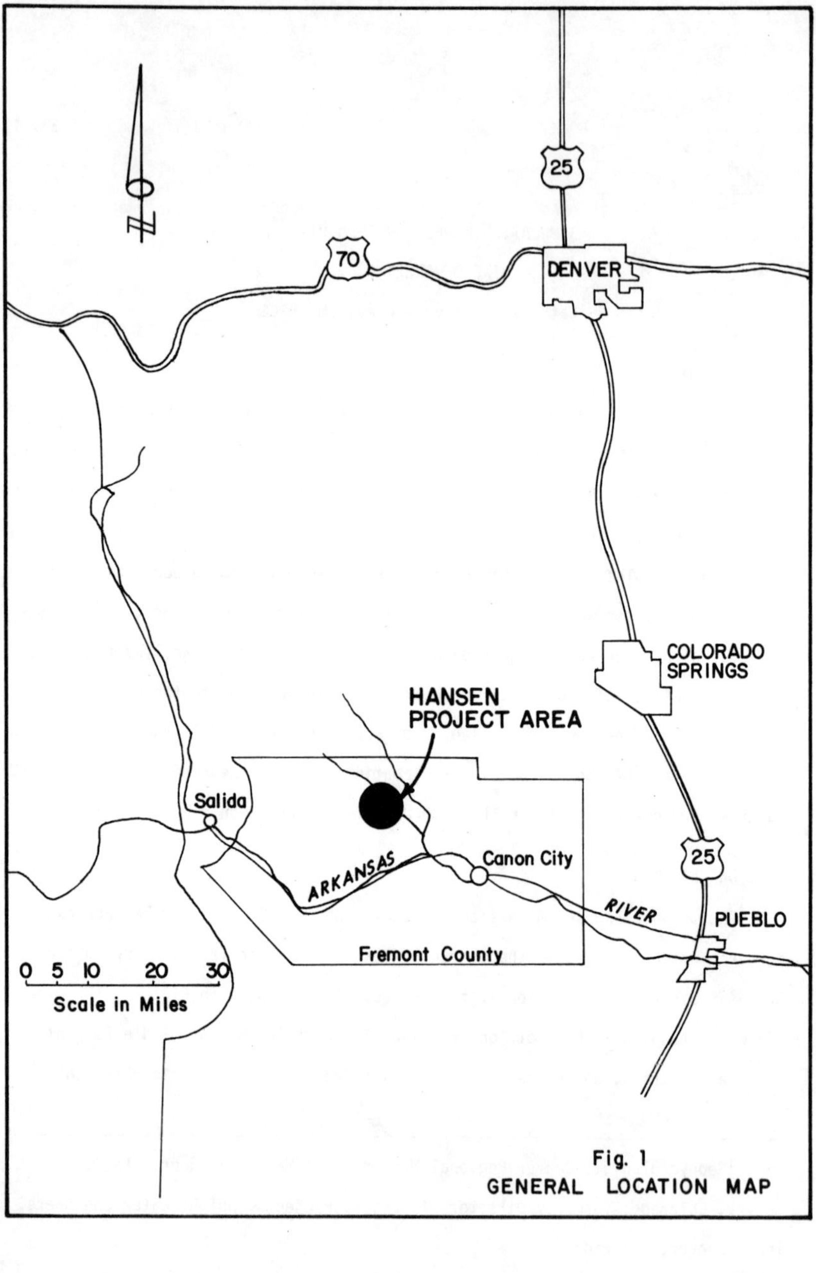

Fig. 1
GENERAL LOCATION MAP

Hansen Project is located is in the foothills of the Rocky Mountains, and the geology is complex.

The Hansen orebody is located in a layered system of water-bearing formations which occupies a paleovalley, bounded below and on three sides by relatively impervious Precambrian bedrock. The host rock of the uranium orebody is the Echo Park Formation, which consists principally of interbedded siltstones and sandstones. Paleochannels present in the Echo Park Formation contain significant fractions of relatively coarse material. The next formation above the Echo Park Formation that is of major hydrologic significance is the Tallahassee Creek Conglomerate, a volcanic formation. A bentonitic clay layer separates the Tallahassee Creek Conglomerate from the uppermost layer in the system. This layer retards vertical seepage, but does not prevent hydraulic connection between the lower and upper layers of the system. Above the clay layer are the Thirtynine Mile Andesite and other surficial deposits, including the stream alluvium of Middle Tallahassee Creek. The andesite is not continuous throughout the area, having been eroded away in some areas, primarily in the stream valley.

Permeabilities in the Hansen Mine site range from less than 1.0 to 170 gal/day/ft^2 (5×10^{-5} to 8×10^{-3} cm/sec). In general, the higher values in this range are associated with localized fracture zones near Precambrian outcrops, principally in the Tallahassee Creek Conglomerate. Storage coefficients for the materials in the layered system are typically low. A range of artesian storage coefficients of 0.001 to 0.005 and a water table storage coefficient of 0.1 were used in the model study. The alluvium of Middle Tallahassee Creek is of such limited extent that it is not considered of major hydrologic significance for the mine dewatering operation, even though its hydraulic conductivity is high relative to the underlying materials.

PRINCIPLES AND CAPABILITIES OF THE MODEL

The computer model used in the Hansen Project study is an adaptation of a three-dimensional, finite difference model developed by Rovey (1974) for simulating flow in a stream-aquifer system. The Dewater (DWTR) model is capable of simulating unsteady flow under confined and unconfined conditions, and accounts for gravity drainage and resaturation of the porous matrix due to water table fluctuations. The basic equation of the DWTR model is obtained by combining the continuity principle with Darcy's Law. The resulting flow equation, in three-dimensional finite difference form is:

$$\frac{\partial}{\partial x}\left[K_x \frac{\partial H}{\partial x} \Delta y \Delta z\right] \Delta x + \frac{\partial}{\partial y}\left[K_y \frac{\partial H}{\partial y} \Delta x \Delta z\right] \Delta y +$$

$$\frac{\partial}{\partial z}\left[K_z \frac{\partial H}{\partial z} \Delta x \Delta y\right] \Delta z = \phi \, \Delta x \Delta y \Delta z \, \frac{dS}{dH} \frac{\partial H}{\partial t} + Q ,$$

where

Δx, Δy, and Δz are model grid length, width, and depth, respectively,

K_x, K_y, and K_z are hydraulic conductivities in the longitudinal, lateral, and vertical directions of the model, respectively,

H is phreatic head,

ϕ is porosity,

S is storativity (storage coefficient),

Q is pumping rate.

The model from which DWTR was developed has the capability to simulate the stream-aquifer interface and to include the effects of precipitation, surface irrigation and evapotranspiration. However, these capabilities were not included in the DWTR model for its application on the Hansen Project. Streams in the area are small and flow intermittently. What little surface flow does exist will be routed around the mine area through a system of diversion channels, and therefore is expected to have little influence on mine dewatering. Detailed studies of the surface hydrology of the area were conducted for other phases of the project, and the results of these analyses were integrated with the results of the DWTR model study for the assessment of environmental impacts and post-mining reclamation needs.

MODEL SETUP FOR HANSEN PROJECT STUDY

To facilitate the use of the model, a three-dimensional grid system was developed for the entire thickness of the layered, water-bearing system, from the surface down to Precambrian bedrock in the Hansen Pit site and the area surrounding it for several miles on each side. The horizontal plane of the grid system is shown overlain on a map of the area in Fig. 2. The model extends several miles from the pit in each direction for two reasons: so that the constant head, low permeability boundary conditions assumed at the edge of the model do not significantly affect modeled flows and water levels near the pit, and so that the impact of dewatering on regional hydrology can be studied. A vertical cross-section of the model grid system is shown on Fig. 3. In the pit area, the bottom two grids represent the Echo Park Formation, which has generally higher permeabilities toward the bottom. The third grid from the bottom in the pit area represents the Tallahassee Creek Conglomerate formation. The fourth grid in the layer is thin by comparison to the thickness of the other grids and represents the bentonitic clay layer that separates the Tallahassee Creek Conglomerate from the Andesite. The uppermost grid primarily represents the Andesite and other surficial materials. The grid system is distorted in the vertical dimension to account for variations in the geology.

Prior to executing production runs with the model, a sensitivity analysis was conducted to determine the effects of input data variations on model results.

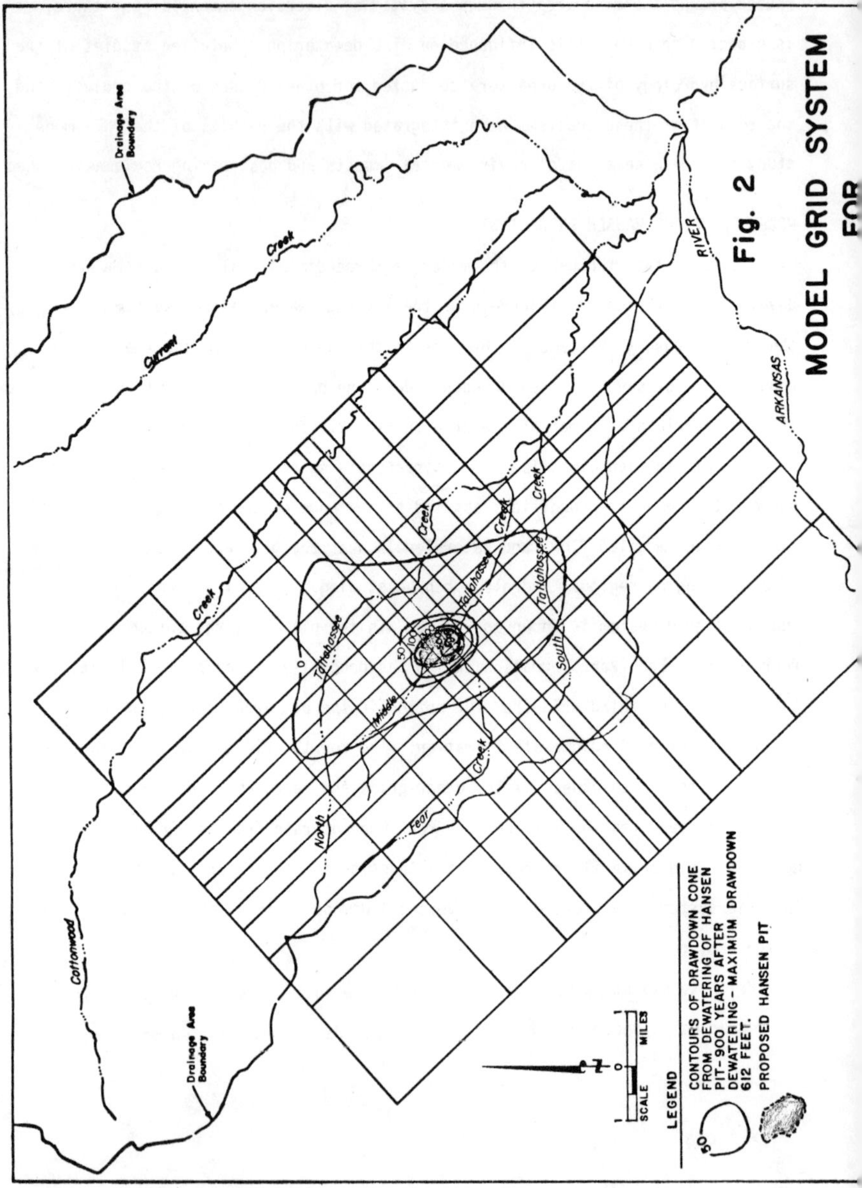

Fig. 2

MODEL GRID SYSTEM FOR

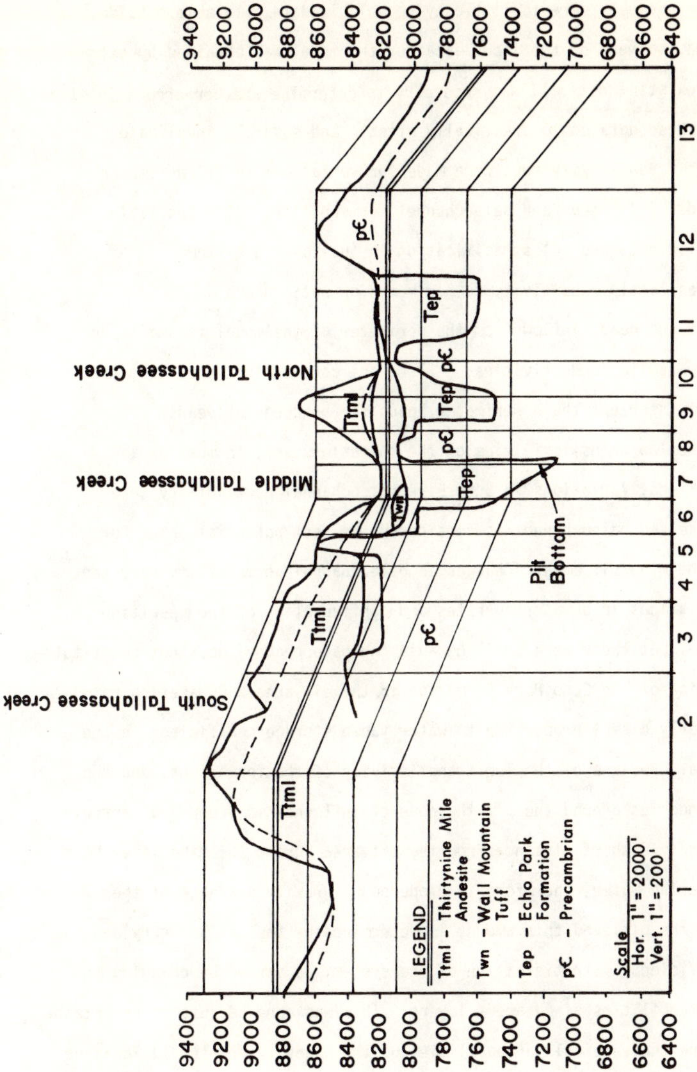

Fig. 3
VERTICAL CROSS-SECTION
OF THE MODEL GRID SYSTEM

This was done because there is little hydrogeologic data available outside the mine area, so much of the input data for the model was obtained by extrapolation of existing data. It was necessary to determine whether errors in extrapolation of these data could drastically affect, and possibly invalidate model results. Sensitivity runs were made for variations in (1) perimeter phreatic heads, (2) Echo Park paleochannel permeability, (3) water table storage coefficient, and (4) stratification in the layered system.

There was little sensitivity to perimeter phreatic heads. Regardless of the head, inflow from the perimeter of the model is small, due to the low hydraulic conductivities. Also, the cone of influence from the pit would not intercept these perimeter heads for hundreds of years, because of the low transmissivities of the formations and, in most areas, geologic barriers. Variations in Echo Park paleochannel permeability were tested because the paleochannel is considered the best potential path for inflows to the mine pit area. Changes in paleochannel permeability resulted in minor variations in dewatering rates and configuration of the resulting drawdown cone, but these were small because the paleochannel does not constitute a large portion of the Echo Park Formation in the pit area. Dewatering rates varied in nearly direct proportion to water table storage coefficient in the mine pit area. Because of the low transmissivity of the formations, and the geologic boundaries around the pit the cone of influence is steep and narrow, and travels only a short distance from the pit area during the mine life. Most of the dewatering water, therefore is produced by gravity drainage of the materials in the pit, and this amount is determined by the pit geometry and the storage coefficient. Stratification of the system was varied by changing the relative permeabilities of adjacent layers. This was done because of uncertainties regarding the vertical permeabilities of the Andesite and the Tallahassee Creek Congomerate. Variations in stratification resulted in altered pumping rates as a function of time and significantly different total dewatering volumes, but had little effect on the configuration of the cone of influence.

MODEL OPERATION

Mine pit dewatering is simulated in the model by specifying as boundary conditions the phreatic surface elevation in the mine area which is necessary for the scheduled mining to take place at each time step during the model run. Once the head is specified in the grids in the mine area, the groundwater flow equation is solved for the entire system of grids to determine the effect of lowering the phreatic surface in the mine pit on all the grids in the model system. At each time step, the model also computes the pumping rate from each pit grid necessary to maintain the specified head. During the run new boundary conditions are input as mining progresses. To determine the recovery of the phreatic surface at the end of the mining, the model is run for a number of years after the end of mining without the boundary conditions of specified heads in the pit, and with specified reclamation conditions.

RESULTS

Mine Dewatering. Pumping rates required to maintain dewatered conditions throughout the life of the mine were predicted for the range of hydrologic conditions that may exist once dewatering begins. The field tests that have been conducted thus far have all been under fully saturated, artesian conditions. It will not be possible to observe the behavior of the system under water table conditions with drainage until dewatering begins. There is some uncertainty regarding whether there will be good hydrologic communication between the layers once dewatering begins, and how much influence fracturing will have on flows in the Tallahassee Creek Conglomerate. Both conditions are tested by varying the hydraulic conductivity of the Tallahassee Creek Conglomerate.

To simulate poor vertical communication and little influence of fracturing, a run was made with a low hydraulic conductivity in the Tallahassee Creek Conglomerate. The conductivity in the mine pit area was set still lower, to a value of 0.05 gal/day/ft^2 (2.4 x 10^{-6} cm/sec) to account for the influence of stratification on vertical seepage.

Results from the run made with conditions of poor vertical communication and low influence of fracturing indicated dewatering rates ranging from approximately 2400 gal/min (0.15 m^3/sec) to less than 100 gal/min (0.006 m^3/sec). The mean dewatering rate for the entire mining period was approximately 650 gal/min (0.04 m^3/sec). High flow rates corresponded to times when excavation into new areas or into deeper formations took place. Actual pumping rates are not expected to reach the extremes predicted by the model; the mining plan is introduced as a series of discrete steps to the model, whereas in reality mining will be a continuous process.

To model better vertical communication and more influence of fracturing, a run was made with the conductivity values in the Tallahassee Creek Conglomerate in the pit area raised one order of magnitude from the values used in the previous run. Dewatering rates predicted in this run ranged from 8800 gal/min (0.56 m^3/sec) to less than 2000 gal/min (0.13 m^3/sec), with the average for the mining period being approximately 3400 gal/min (0.21 m^3/sec). As with the earlier run, high flows corresponded to excavation into previously untapped formation. The results of the second run are considered less realistic than the results of the first run. Even if good communication exists between the layers in the Hansen Pit, existing hydrogeologic information indicates that the surrounding formations do not have sufficient hydraulic conductivities to sustain such high rates of flow into the pit for the life of the mine.

Results of the run simulating low conductivity in the Tallahassee Creek Conglomerate and poor vertical communication between the layered formations in the mine pit area agreed well with results of analytical studies conducted for the development of a dewatering system. These results serve to support the validity of assumptions and simplifications made to facilitate the use of analytic techniques.

Regional Hydrologic Impacts. The runs which were made to determine mine pit dewatering rates were extended for 900 years after decommissioning of the mine, to predict long-term regional hydrologic effects of the dewatering operation. Because the principal limit on the extent of the drawdown cone is structural rather than hydraulic, there was little difference in the drawdown cone configuration predicted by the two runs. The cone of influence predicted by the model for 900 years after the end of mining is shown in Fig. 3. The maximum distance the cone of influence is expected to extend away from the mine pit is about three miles to the northwest, along the Echo Park paleochannel. From the end of mining to 900 years later, there is little change in the slope of the drawdown cone. The reasons for this are (1) almost no recharge available from the perimeters of the affected area, and (2) low transmissivity of the indigenous material. As was stated previously, the model only accounts for subsurface flows, and does not account for the effects of evaporation, precipitation recharge, and surface inflow. The model's predictions of regional effects are therefore conservative. Surface water inflow and recharge are expected to account for most of the anticipated water table recovery in the pit, which is to be partially reclaimed as a lake.

Impacts on area streams are expected to be limited to those areas inside the "zero" drawdown contour shown on Fig. 2. Increased seepage losses from these streams are anticipated as a result of the lowered water table. The total amount of flow lost will not be significant to the regional hydrology, because the streams are small and intermittent. The identification of this impact zone has made it possible to anticipate the effects on water rights in the area, and to make provisions for augmentation.

CONCLUSIONS

The hydrologic computer model study which was conducted to determine mine dewatering rates and regional hydrologic impacts substantiated the results of earlier analytic studies conducted to develop a mine dewatering system. The anticipated average dewatering rate, based on model results is about 650 gal/min (0.04 m^3/sec).

The model also produced a conservative estimate of the extent of the zone of influence for the mine dewatering drawdown cone. The maximum distance the cone is expected to extend in any direction is about three miles (five kilometers) to the northwest. As a result of the drawdown in this zone, some seepage loss from the small, intermittent streams in the area is expected.

APPENDIX-REFERENCE

Rovey, Catherine Kraeger, "Numerical Model of Flow in a Stream-Aquifer System" Ph.D. Dissertation, Colorado State University 1974, Hydrology Paper No. 74, August 1975.

Aquifer Tests at San Lorenzo Dam, El Salvador

Jing-Yea Yang,[1] and Sanford I. Strausberg[2]

ABSTRACT

Aquifer tests were run in December 1977 and January 1978 in alluvial materials underlying the right bank of the Rio Lempa in El Salvador, Central America. The purpose of the testing was to determine aquifer characteristics of alluvium of the San Lorenzo Dam site. Quantitative values of permeability are used for design of the foundation, which includes seepage control facilities and drainage features of the dam. Two aquifers and two aquitards comprise the alluvial section of the right bank of the San Lorenzo Dam site. They have limited areal extents and are shaped like strips parallel to the river because of abutments against bedrock walls. Various methods were used to analyze aquifer test data from drawdown and recovery readings for the determination of horizontal and vertical permeabilities in aquifers and aquitards. Hydrogeologic units are under semi-confined conditions and the major recharge is from the underlying bedrock. Groundwater normally discharged into the Rio Lempa. Alluvial hydrogeologic units respond elastically to river level fluctuations and well pumpages. Permeability values indicate that no-open work gravel conditions (taken at >10 cm/sec) are identified in the alluvium. This conclusion is reached after assumption that all aquifers are one order of magnitude thinner than determined from soil and geophysical logs.

INTRODUCTION

Aquifer tests were run in December 1977 and January 1978 in alluvial materials underlying the right bank of the Rio Lempa. Locations of test pumping and observation wells are shown on Exhibit 1. The purpose of the testing was to determine aquifer characteristics of alluvium of the San Lorenzo Dam site. Quantitative values of permeability are used for design of the foundation which includes seepage control facilities and drainage features of the dam.

Hydrogeologic data derived during the testing period were also used to define the pre-construction hydrogeologic regime of the alluvial valley for definition of groundwater conditions encountered during excavation and construction.

---

[1]*Argonne National Laboratory, Argonne, Illinois.*

[2]*Leggette, Brashears & Graham, Incorporated, Wilton, Connecticut.*

This work was conducted when both authors were with Harza Engineering Company, Chicago, Illinois, who was the consulting engineer for El Salvador's Comision Ejecutiva Hidroelectrica del Rio Lempa (CEL).

SAN LORENZO PROJECT LOCATION OF PUMPING AND OBSERVATION WELLS USED DURING AQUIFER TESTING PROGRAM DEC.1977 AND JAN.1977

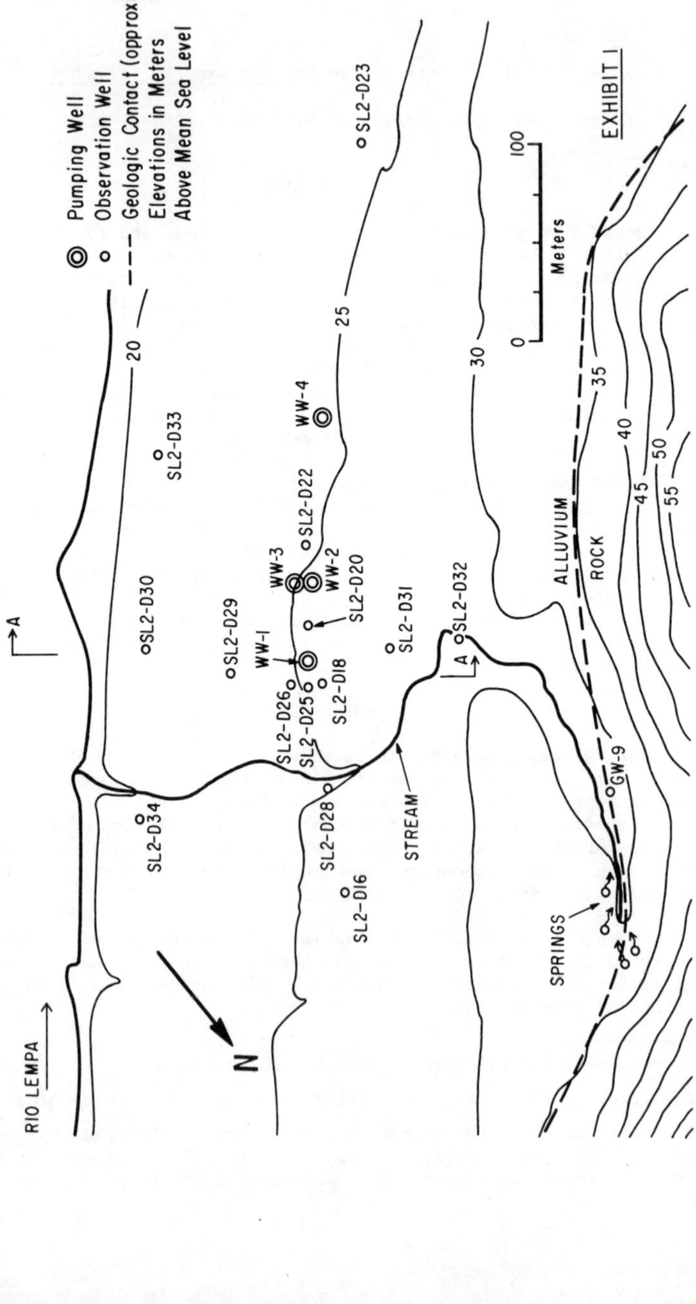

GENERAL HYDROGEOLOGIC CONDITIONS

Hydrogeologic Units--The alluvial valley is underlain by Quaternary alluvium, which in turn overlies extrusive volcanic rocks of Tertiary or Quaternary age.

Thicknesses of aquifer zones were determined from soil borings and borehole geophysical logs. Two aquifer zones are identified.

The Upper Aquifer is capped by an aquitard (semi-confining fine grained alluvial layer) that extends from the ground surface to the top of the Upper Aquifer. This surficial zone contains some relatively permeable lenses consisting of silty sand and is herein called the Upper Aquitard.

The top of the Upper Aquifer is about 15 to 30 meters below ground level (b.g.l.). The general thickness of this unit is about 15 to 22 meters. Fine grained alluvial lenses are found throughout the Upper Aquifer.

Below the Upper Aquifer is another aquitard, herein called the Lower Aquitard, generally about 4 to 14 meters thick. This unit consists of numerous intercalated lenses of sand, silt and gravel and is relatively less permeable than the aquifer units above and below.

The Lower Aquifer underlies the Lower Aquitard and extends to bedrock. At least one areally extensive fine grained layer, about 2 to 3 meters thick, and one lenticular fine grained layer are embedded in the Lower Aquifer along with numerous fine grained alluvial lenses of varying thicknesses and areal extents. The top of the Lower Aquifer is about 43 to 57 meters b.g.l. The thickness of the Lower Aquifer is greatest under the right bank of the river, where it is about 60 meters thick. However, under the portion of the alluvial plain closest to the valley wall (vicinity of test pumping wells) the thickness generally ranges between 15 and 40 meters.

Exhibit 2 shows the conceptual cross sectional configuration of the alluvial hydrogeologic units on the right bank of the Rio Lempa. Under the central part of the river (not shown on Exhibit 2), the aquitards appear to be thin or absent and consequently a single aquifer may be considered for flow analyses in this region. Table 1 gives screen settings and effective aquifer thicknesses at pumping and observation wells. Effective aquifer thickness was estimated as the thickness of the coarse grained section of the aquifer open to the well screen. Fine grained lenses were not included in effective thickness estimates.

Non-Pumping Groundwater Levels--Exhibit 2 shows the relationships of static potentiometric levels from different hydrogeologic units during the aquifer testing period.

Highest potentiometric levels were in the bedrock as evidenced by flowing conditions in SL2-D16 and SL2-D32 when drill casing extended into that material. Potentiometric levels at both these sites were about 2 meters above ground level in December 1977. Additional evidence

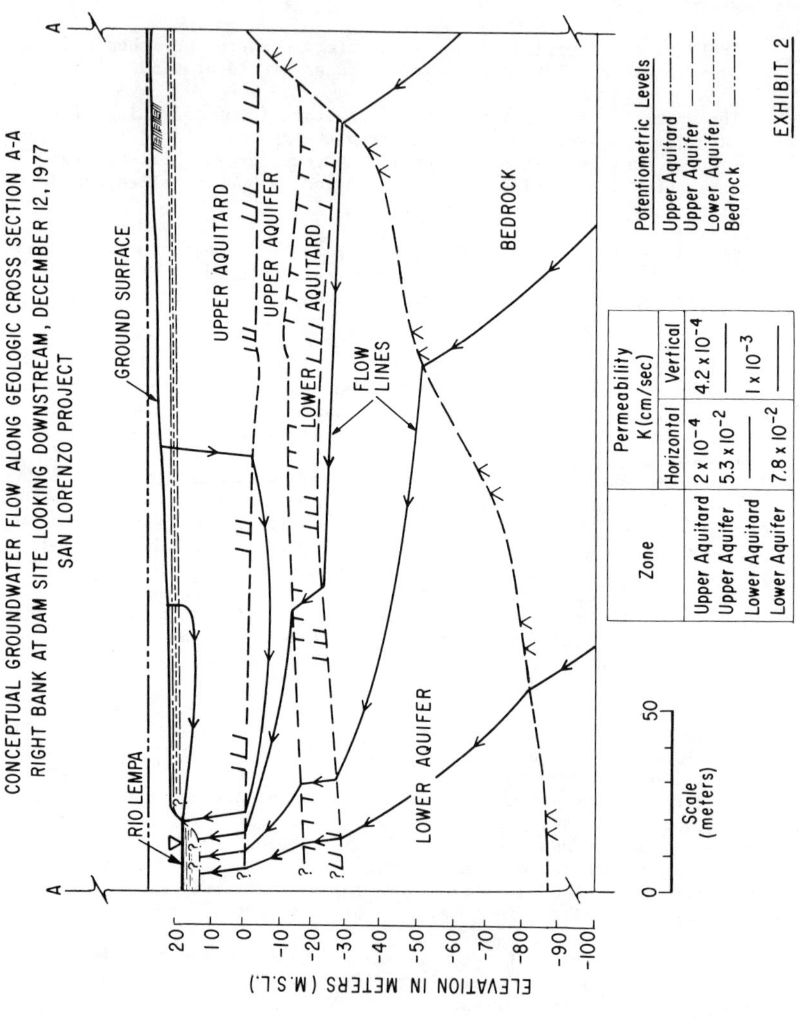

Zone	Permeability K (cm/sec) Horizontal	Vertical
Upper Aquitard	2×10^{-4}	4.2×10^{-4}
Upper Aquifer	5.3×10^{-2}	——
Lower Aquitard	——	1×10^{-3}
Lower Aquifer	7.8×10^{-2}	——

Table 1

General Data of Pumping and Observation Wells

Observation Well No.	Ground Elev. (M above MSL)	Screened Interval (MSL datum in M)	Estimated Effective Thickness of Aquifer (M)[6]	Thickness of Upper Aquitard (M)
SL2-D16[4]	25.12	-	12.8	-
SL2-D18[5]	25.32	-30.68 to -42.68	16.2	-
SL2-D20[3]	24.82	-37.68 to -43.68	7.9	-
SL2-D22[3]	24.01	-29.99 to -35.99	18.3	-
SL2-D23[3]	24.16	-23.84 to -29.84	22.3	-
SL2-D25[2]	24.73	-4.73 to -10.73	17.4	12.5
SL2-D26[1]	24.45	18.45 to 12.45	-	12.5
SL2-D28[3]	24.50	-23.5 to -29.5	15.9	-
SL2-D29[3]	22.67	-30.83 to -36.83	17.1	-
SL2-D30[3]	21.15	-32.85 to -38.85	28.7	-
SL2-D31[5]	26.61	0.61 to -5.39	13.0	-
SL2-D33[3]	21.02	-27.48 to -33.48	5.8	-
SL2-D34[3]	22.20	-29.8 to -35.8	30.5	-
Well #1[3]	24.86	-21.92 to -36.92	17.7	-
Well #2[2]	24.14	-3.61 to -12.97	18.3	11.6
Well #3[3]	24.17	-26.08 to -38.60	18.0	-
Well #4[2]	24.52	-3.59 to -12.98	14.3	12.2

[1]Screened in Upper Aquitard

[2]Screened in Upper Aquifer

[3]Screened in Lower Aquifer

[4]Open to Bedrock

[5]Multiaquifer well because of screen position and/or construction

[6]Coarse grained portion of aquifer open to well screen

of high bedrock potentiometric levels was seen at GW-9 where flow was measured at about 3 to 3.9 liters per second during the testing period and from springs emerging from bedrock near GW-9 as shown on Exhibit 1.

The Lower Aquifer had potentiometric levels about 18.5 to 19 meters above mean sea level (MSL) (about 6 meters b.g.l. at the site of pumping wells). Hydraulic gradient averaged 0.002 m/m toward the river (Exhibit 3).

The general configuration of the potentiometric level of the Upper Aquifer was judged to be similar to that of the Lower Aquifer based on water level measurements made in the three wells (2, 4 and SL2-D25) open to the former. However, the hydraulic gradient of the Upper Aquifer may be steeper than that of the Lower Aquifer. The static potentiometric level of the Upper Aquifer was about 0.5 meters lower than that from the Lower Aquifer at pumping well sites.

Water levels from relatively permeable zones in the Upper Aquitard were higher than those from the two aquifer zones but were lower than those from the bedrock. Water level in SL2-D26, screened in the Upper Aquitard, was 4.5 meters b.g.l. or 1.8 meters above that from SL2-D25 screened in the Upper Aquifer, at noon on January 11, 1978 (non-pumping conditions).

River level was lower than all static groundwater levels during December 1977 and January 1978. Groundwater levels fluctuated in elastic response to river level changes, caused mainly by regulation of flow from upstream power generation.

Pre-construction Groundwater Movement--Groundwater in the aquifers was recharged mainly from the bedrock as shown in Exhibit 2. Some recharge to the aquifers may have come from horizontal underflow through upstream alluvial sands and gravels. However, the latter was considered to be minor compared to recharge from the bedrock as indicated by the potentiometric map of the Lower Aquifer (Exhibit 3). This map indicates that horizontal component of groundwater flow was from the bedrock wall to the river.

Precipitation falling on the alluvial valley and infiltration from the spring fed streams and irrigation ditches crossing the valley also recharged the Upper Aquitard system. Part of this recharge leaked downward into the Upper Aquifer in the alluvial plain. The remainder of the recharge to the Upper Aquitard system discharged laterally into the river bank after reaching the zone of saturation.

The above conditions were valid only during the dry season. At times of high river level during the wet season, river water recharged aquifers and aquitards with temporary bank storage.

As shown in Exhibit 2, groundwater from the bedrock flowed upward, from the bedrock to the Lower Aquifer, through the aquitard separating the two aquifers, into the Upper Aquifer after which it was discharged laterally toward the river. Under the river, groundwater normally

POTENTIOMETRIC LEVEL OF LOWER AQUIFER, NOON, JAN. 11, 1978 SAN LORENZO PROJECT, EL SALVADOR

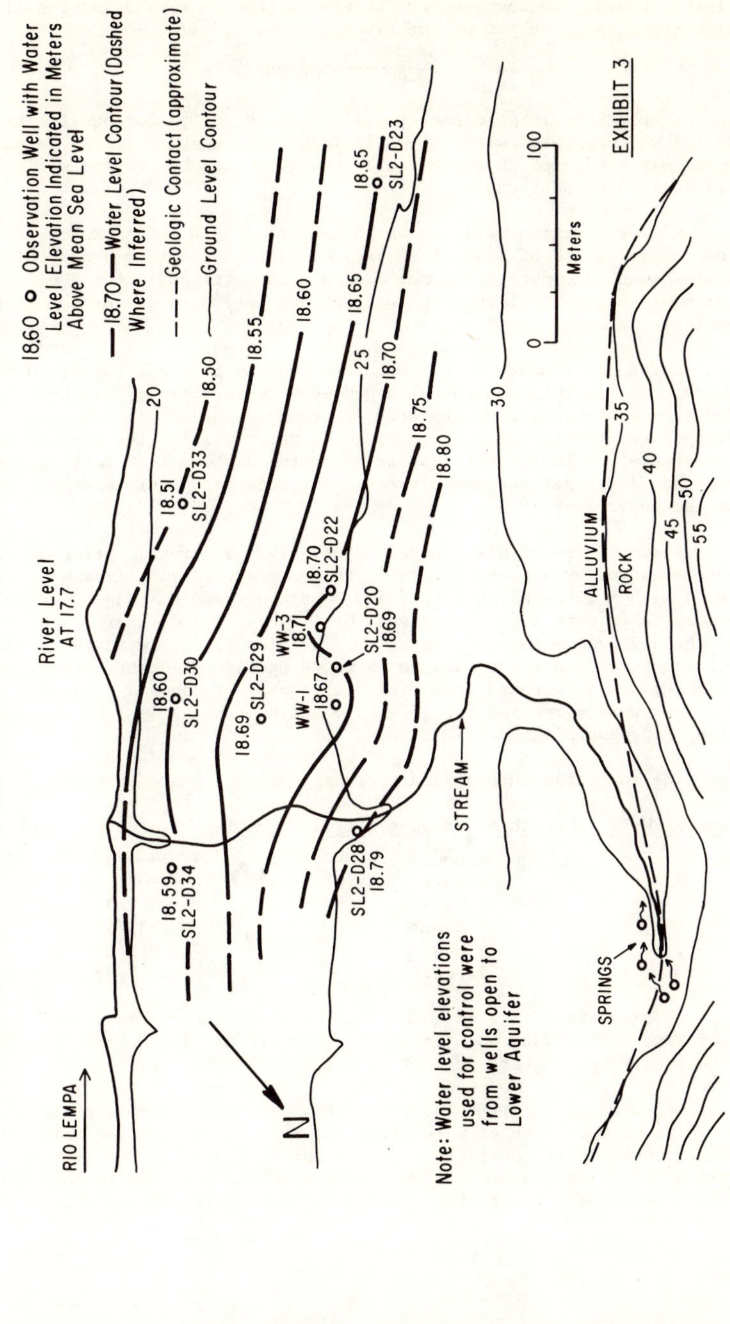

EXHIBIT 3

flowed upward from the aquifer underlying the river, and was finally discharged into the bed of the river.

AQUIFER TESTS

Scope--Four large diameter wells were installed for aquifer testing. Well 1 was completed with 30 cm (12-inch) casing and Wells 2, 3 and 4 were completed with 20 cm (8-inch) casing. Most observation wells had 4.5 cm (1 3/4-inch) casing.

The observation wells shown in Exhibit 1 were judged to be reliable for determination of aquifer characteristics of distinct hydrogeologic units based on construction methods, screen settings, water level measurements and water level fluctuations and were used for hydrogeological analyses.

Water level measurements were made mainly by steel tapes, to the nearest millimeter. Wells being pumped were measured by electric tape. Two automatic water level recorders were used in January 1978 for continuous records in non-pumping large diameter wells. These records were compared with river levels obtained for the January 1978 testing period from an additional automatic water level recorder previously installed by the government.

Temperatures of air, pumped water, river water and spring water were measured by hand held mercury thermometer. Spring flow was measured with a five-gallon bucket and a stop watch. Pumping of wells was accomplished with a diesel driven turbine pump and discharge was measured by the orifice method. Pumpages for individual tests were kept constant with no more than 4% variations in rates throughout tests. Recovery measurements were taken following shutdowns. Drawdown and recovery readings were corrected for effects of river fluctuations by extrapolation of pre-pumping trends.

The following pumping and recovery tests were performed:

Pumping Well	Length of Pumping Period	Pumping Rate m^3 per day/USgpm
2	2 hours	1908/350
4	2 days	1908/350
1	5 days	2028/372
1	5 hours	2987/548
2	5 hours	2028/372
3	5 hours	2142/393

A "slug" test (instantaneous injection of a known volume of water) performed in SL2-D26, open to the Upper Aquitard, was utilized to determine aquifer parameters of that unit (1, 2, 3, 7).

Groundwater Temperature, Spring Flow and Water Levels Regarding Recharge--Temperature of water from all four large diameter wells remained essentially constant at 32.5°C during pumping. Temperature of spring water, as measured at GW-9 also remained essentially constant at 32.5°C throughout the testing period. River water generally was 29°C to 30°C. Therefore, temperature data indicated that groundwater was not being recharged by river water during the testing period and that hydraulic connection existed from the bedrock to alluvial aquifers.

Spring flow measurements showed that pumpage did not reduce spring flow rates, indicating that recharge from bedrock exceeded the quantity discharged by pumping. This was also obvious from water level measurements in observation wells compared to corresponding river levels. The respective heights of groundwater levels in Wells 2 and 3 remained constant above corresponding river levels during the five day pumping period when Well 1 was discharging about 2028 m^3/day (0.54 million gallons per day). More than 10,000 m^3 (2.5 million gallons) was recharged from the bedrock in that pumping period and the cones of depression were stabilized. Water level data from the other pumping tests indicated the same relationship. Even when Well 1 was pumped at 2987 m^3/day (548 gpm) for 5 hours, recharge from the bedrock to the alluvial aquifer system effected by pumping exceeded the discharge due to pumping.

Confinement--Confinement is indicated by elevations of static water levels in all wells and groundwater level response to river level fluctuations, although hydrogeologic analyses show that complete confinement does not exist. Exhibit 2 illustrates the leaky nature of the system and also that static water levels of individual hydrogeologic units are higher than the tops of the respective units with the exception of the Upper Aquitard. However, permeable lenses in that unit are below its potentiometric level, indicating that confinement also exists in the Upper Aquitard.

Several methods have been used to analyze aquifer test data from drawdown and recovery readings for the determination of aquifer coefficients (4, 5, 6, 8, 9).

The coefficients of storage of the Upper Aquitard is in the order of 10^{-4} based on slug test analyses of SL2-D26. The coefficients of storage of the Upper Aquifer averages 7.3×10^{-4} and that of the Lower Aquifer averages 7.5×10^{-4}. The averages of all values from time-drawdown analyses and those from all distance-drawdown analyses were given equal weight for estimation of each overall individual value. This procedure was also followed for averaging permeability.

Storage coefficients for the Lower Aquitard and the bedrock could not be determined from test results, but were assumed to be in the same order of magnitude as those of the alluvial aquifers.

Semi-confined conditions are indicated by storage coefficients in the range of 5×10^{-4} to 5×10^{-3}. Therefore, semi-confinement is indicated for the aquifers and probably the aquitards and bedrock. Vertical leakage was identified by the stabilization of drawdowns and recoveries within 10 to 30 minutes after respective startups or shutdowns.

Permeability--Water level measurements made in pumping and observation wells during the drawdown and recovery phases of aquifer tests were the basis for computation of transmissivity, utilizing derivatives of the Theis Non-Equilibrium Equation. Permeability was determined from the formula:

$$K = \frac{T}{m} \times 1.1574 \times 10^{-3}$$

where K is permeability in cm/sec,
T is transmissivity in m^2/day, and
m is effective aquifer thickness in meters.

Horizontal permeability of the Upper Aquitard, computed from the slug test data, is estimated at 2×10^{-4} cm/sec. Vertical permeability of the Upper Aquitard was computed by leaky aquifer formulas and averaged 4.2×10^{-4} cm/sec.

Horizontal permeability of the Upper Aquifer averaged 5.3×10^{-2} cm/sec. Vertical permeability was not determined but is probably less than horizontal permeability because of alluvial stratification.

Vertical permeability of the Lower Aquitard, estimated by the Neuman-Witherspoon variation of the Theis Non-Equilibrium Equation (6), is estimated at 1.0×10^{-3} cm/sec. Horizontal permeability was not determined but is probably more than vertical permeability of the aquitard and is undoubtedly less than horizontal permeability of the aquifers.

Overall horizontal permeability of the Lower Aquifer averaged 7.8×10^{-2} cm/sec and vertical permeability is considered to be less because of stratification of numerous fine grained lenses in the unit.

A highly permeable lens in the Lower Aquifer was identified in the general vicinity of Wells 1, SL2-D29 and SL2-D30 as seen in some early time graphs of the latter two wells. The presence of this lens was also intimated by the flattening of non-pumping hydraulic gradient between Wells 1 and SL2-D29 (Exhibit 3). The horizontal permeability of this lens was estimated as 3.6×10^{-1} cm/sec and was considered to be the highest permeability of any sizable mass of material in the alluvial section identified to date. However, analysis of all available data indicated that this lens was localized in a narrow strip elongated normal to the river.

Permeability of the bedrock was not determined but must be relatively substantial considering the amount of upward recharge to the alluvial system. However, permeability of the bedrock is undoubtedly lower than alluvial aquifers.

Boundaries--Negative (discharging) boundaries result from the abrupt termination of the alluvial valley at the walls consisting of volcanic rocks having a lower permeability than alluvial aquifers. However, the pumpage was not sufficient to negate the recharge effects on the drawdown and recovery patterns of most wells. Consequently, the discharging effects of negative boundaries were largely masked except at Wells SL-D23, SL2-D29, and SL2-D30 screened in the Lower Aquifer. Exact locations of negative boundaries were not determined although the distance to a negative boundary from Well SL2-D23 was calculated at 180 meters, which may reflect the bedrock wall east of the well. Negative boundaries were also indicated by potentiometric contours under pumping conditions.

The major positive (recharge) boundary would be the Rio Lempa if pumpage was great enough to lower cones of depression below river level. However, because this was not accomplished, induced infiltration could not occur and the positive boundary was not identified.

CONCLUSIONS

Two aquifers and two aquitards comprise the alluvial section on the right bank of the San Lorenzo Dam site. They have limited areal extents and are shaped like strips parallel to the river because of abutments against bedrock walls. Hydrogeologic units are under semi-confined conditions and the major recharge is from the underlying bedrock. Groundwater normally discharged into the Rio Lempa before construction of the dam. Alluvial hydrogeologic units respond elastically to river level fluctuations and well pumpages.

Hydrogeologic parameters of the alluvial units are averaged below:

Unit	Coefficient of Storage (Dimensionless)	Horizontal Permeability cm/sec	Vertical Permeability cm/sec
Upper Aquitard	1×10^{-4}	2×10^{-4}	4.2×10^{-4}
Upper Aquifer	7.3×10^{-4}	5.3×10^{-2}	not determined
Lower Aquitard	not determined	not determined	1×10^{-3}
Lower Aquifer	7.5×10^{-4}	7.8×10^{-2}*	not determined

*Localized lens in the vicinity of Wells 1, SL2-D29 and SL2-D30 has horizontal permeability of 3.6×10^{-1} cm/sec. This value is not included in averaging in the above table.

Permeability values indicate that no-open work gravel conditions (taken at >10 cm/sec) are identified in the alluvium. This conclusion is reached after assumption that all aquifers are one order of magnitude thinner than determined from geologic and geophysical logs.

REFERENCES

1. Cooper, H.H., Jr., Bredehoeft, J.D., and Papadopulos, I.S., Response of a finite diameter well to an instantaneous charge of water, Water Resources Research 3 (1), 263-269, 1967.

2. Ferris, J.G., and Knowles, D.B., The slug test for estimating transmissibility, U.S. Geol. Surv. Ground Water Note 26, 1-7, 1954.

3. Ferris, J.G., Knowles, D.B., Brown, R.H., and Stallman, R.W., Theory of aquifer tests, U.S. Geol. Surv. Water Supply Pap. 1536-E, 104-105, 1962.

4. Kruseman, G.P., and N.A. deRidder, Analysis and Evaluation of Pumping Test Data, International Institute for Land Reclamation and Improvement, Wageningen, The Netherlands, 1970.

5. Lohman, S.W., Ground-Water Hydraulics, U.S. Geological Survey Professional Paper No. 708, U.S. Gov't Printing Office, Washington, D.C., 1972.

6. Neuman, S.P., and Witherspoon, P.A., Field determination of the hydraulic properties of leaky multiple aquifer system. Water Resources Research 8 (5), 1284-1298, 1972.

7. Papadopulos, I.S., Bredehoeft, J.D., and Cooper, H.H., Jr., On the analysis of "slug test" data. Water Resources Research 9 (4), 1987-9, 1973.

8. Stallman, R.W., Effects of water table conditions on water level changes near pumping wells. Water Resources Research 1 (2), 295-312, 1965.

9. Walton, W.C., Selected Analytical Methods for Well and Aquifer Evaluation, Bulletin 49, Illinois State Water Survey, Urbana, 1962.

SHORT THERMAL RIVER DIFFUSERS

By A.D. Parr[1], A.M. ASCE, J.G. Melville[2], A.M. ASCE, and J.J. Dlubac[3]

Abstract

Heated water from steam-electric generating plants is often discharged into rivers through multiport diffusers. A diffuser that does not span the entire river width can act as a jet pump and induce ambient flow into the diffuser plume from a portion of the river width greater than the diffuser length. This effect increases as river discharge decreases and must, therefore, be considered when designing diffuser outfall systems to perform at low flow conditions. A mathematical model is developed to determine the induced flow and to predict the dilution in the nearfield region of a short river diffuser plume. The results are presented in the form of dimensionless graphs that give dilution as a function of relevant ambient and effluent parameters. The model is also compared with field measurements for a prototype diffuser system.

Introduction

Submerged multiport diffusers are used to discharge thermal effluents into rivers when rapid mixing is desired. Ideally, the transverse effluent and ambient discharge distributions would be proportional to insure uniform mixing across the channel. Typically, however, economic or regulatory restrictions preclude the use of diffusers that span the entire river width and short diffusers that discharge in only part of the river are used. At low river flow conditions, the large momentum input of a unidirectional diffuser acts as a jet pump and draws ambient water into the discharge plume from wide portions of the upstream width. This effect can greatly alter the discharge distribution of the river near the diffuser and affect mixing performance.

Adams (1) and Lee, Jirka, and Harleman (5) analyzed the nearfield temperature and velocity fields downstream from a unidirectional diffuser in shallow, flowing water of infinite lateral extent. Their results are quite useful for offshore diffusers but do not apply directly to river diffusers where the supply of ambient water for entrainment is limited. This paper considers the case of a short diffuser in a shallow, laterally confined, flowing waterbody. A control volume analysis is used to formulate a set of equations to determine the induced flow in the nearfield region of the diffuser plume. The equations are solved

[1]Asst. Prof., Dept.of Civ. Engrg., Auburn Univ., Auburn, Alabama.

[2]Assoc. Prof., Dept. of Civ. Engrg., Auburn Univ., Auburn, Alabama.

[3]Res. Asst., Inst. of Hydr. Res., Univ. of Iowa, Iowa City, Iowa.

for the induced flow and the dilution at the downstream end of the nearfield region is determined. The results are presented in a set of dimensionless graphs that can be used to design diffusers to meet environmental standards for river environments. The mathematical model is also compared with field measurements in the nearfield region of the Quad Cities diffuser.

Mathematical Model

A multiport diffuser is essentially a large manifold pipe located near the bottom of a water body from which a series of discharge ports emanate. A definition sketch of a river diffuser of length L_d is shown in Figure 1. The diffuser has N discharge ports, each with cross-sectional area, a_o. Water is discharged at a constant temperature, T_o, and velocity, u_o, into an idealized rectangular cross section of depth, H_a, and width, W. The ambient temperature and velocity of the river are assumed to be constant and equal to T_a and u_a, respectively.

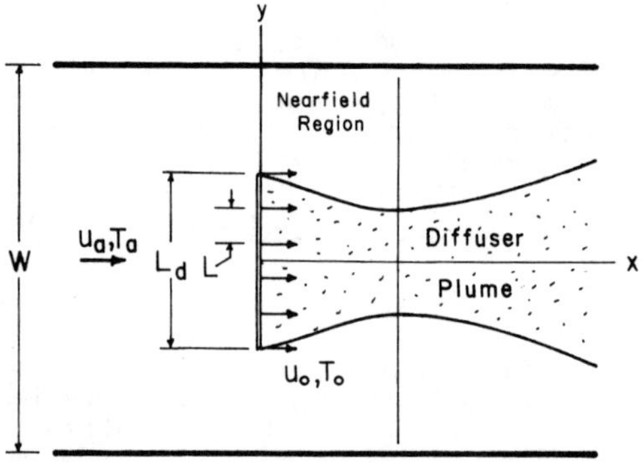

Figure 1. Definition Sketch of Short River Diffuser

The individual heated jets for a typical diffuser rapidly coalesce so that the flowfield is virtually two-dimensional within about 5 port spacings downstream form the efflux section (2,8). Furthermore, complete vertical mixing occurs within about 5 ambient depths for diffusers in shallow water (4,5). The nearfield portion of the diffuser plume, wherein flow is dominated by momentum, was determined by Lee, et al. (5) to extend about 0.5 L_d downstream from the diffuser. Since the flow in this region can be considered inviscid, a control volume analysis, similar to the approach of Adams (1), will be used to develop the equations for the flow in the nearfield region of a short river diffuser plume. A detailed sketch of the nearfield region is shown in Figure 2. The width of the plume at the end of the nearfield region is σL_d, where σ is called the contraction coefficient. The induced and the affected ambient velocities at Section 4 are denoted by u_n and u_b, respectively. The river, effluent and induced discharges are Q, Q_o, and Q_n, respectively.

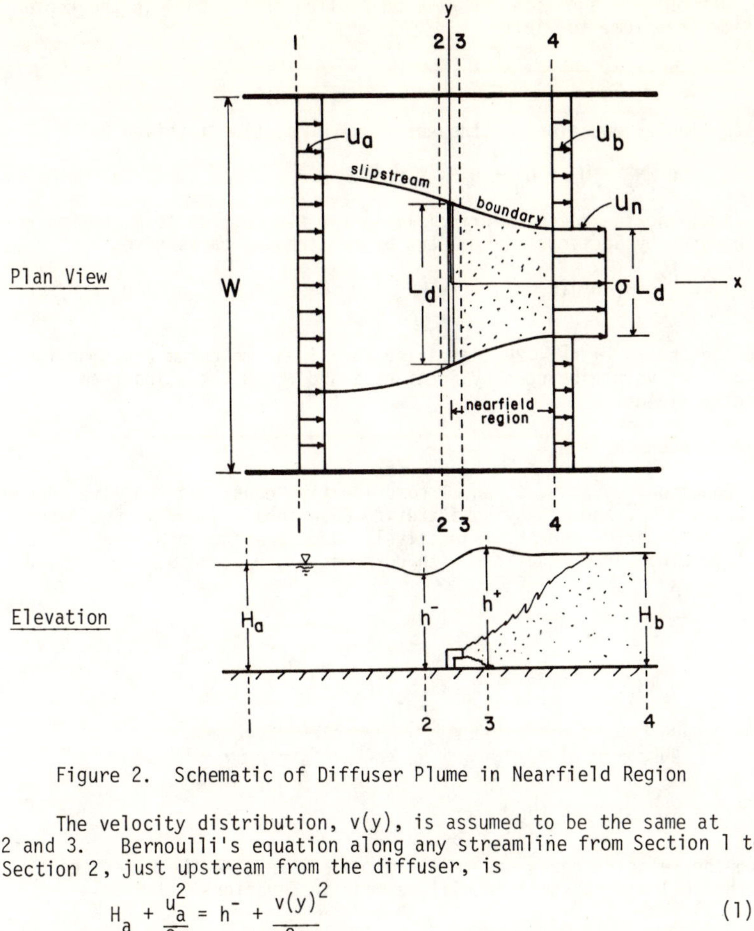

Figure 2. Schematic of Diffuser Plume in Nearfield Region

The velocity distribution, v(y), is assumed to be the same at 2 and 3. Bernoulli's equation along any streamline from Section 1 to Section 2, just upstream from the diffuser, is

$$H_a + \frac{u_a^2}{2g} = h^- + \frac{v(y)^2}{2g} \tag{1}$$

Applying Bernoulli's equation along the same streamline from 3 to 4 gives

$$\frac{u_n^2}{2g} + H_b = h^+ + \frac{v(y)^2}{2g} \tag{2}$$

Equations 1 and 2 can be combined to give the following expression for the rise in the water surface across the diffuser

$$\Delta h = h^+ - h^- = \frac{u_n^2 - u_a^2}{2g} + H_b - H_a \tag{3}$$

Thus, as noted by Adams (1), Δh is constant along the diffuser although h^+ and h^- are dependent on y as shown in Equations 1 and 2.

Bernoulli's equation can also be applied from 1 to 4 in the region outside the plume to yield

$$H_a + \frac{u_a^2}{2g} = H_b + \frac{u_b^2}{2g} \quad (4)$$

The continuity equation for the entire river section is given by

$$u_a W H_a + Q_o = Q_n + u_b H_b (W - \sigma L_d) \quad (5)$$

Applying the momentum principle in the x-direction to a control volume bounded by Sections 1 and 2 and by the channel banks gives

$$\frac{gW}{2}(H_a - H_b)(H_a + H_b) = -u_a^2 W H_a + u_b^2 (W - \sigma L_d) H_b - Q_o u_o + u_n^2 \sigma L_d H_b \quad (6)$$

Assuming that $h^+ + h^- = 2H_a$, applying the linear momentum equation for the control volume bounded by Sections 2 and 3 and the slipstream boundary yields

$$g L_d \Delta h H_a = Q_o u_o \quad (7)$$

Equations 3, 4, 5, 6, and 7 comprise five equations and five unknowns: Δh, u_n, u_b, H_b, and σ. By substituting Equations 4 and 7 in Equations 3, 5, and 6 for H_b and Δh, respectively, and by assuming $H_a + H_b = 2H_a$, the following three equations with unknowns u_b, u_n, and σ, are derived

$$\frac{Q_o u_o}{g L_d H_a} = \frac{u_n^2 - u_b^2}{2g} \quad (8)$$

$$u_a W H_a + Q_o = Q_n + u_b (W - \sigma L_d)\left(H_a - \frac{u_a^2 - u_b^2}{2g}\right) \quad (9)$$

$$g W H_a \left(\frac{u_b^2 - u_a^2}{2g}\right) = -u_a^2 W H_a + u_b^2 (W - \sigma L_d)\left(H_a + \frac{u_a^2 - u_b^2}{2g}\right) - Q_o u_o + u_n^2 \sigma L_d \left(H_a + \frac{u_a^2 - u_b^2}{2g}\right) \quad (10)$$

Since the velocity heads are small in comparison with the water depth, H_a, the following assumption will be made in Equations 9 and 10

$$H_a + \frac{u_a^2 - u_b^2}{2g} = H_a \quad (11)$$

Equations 8, 9, and 10 can be expressed in the following dimensionless forms

$$\frac{2}{A} = v_n^2 - v_b^2 \quad (12)$$

$$v_a A \ell + 1 = v_b A \ell + \sigma A (v_n - v_b) \quad (13)$$

$$\frac{v_a^2 - v_b^2}{2} = \frac{\sigma}{\ell}(v_n^2 - v_b^2) - \frac{1}{A \ell} \quad (14)$$

in which $A = H_aL_d/Na_o$ = area ratio; $v_n = u_n/u_o$ = induced velocity ratio; $v_b = u_b/u_o$ = affected ambient velocity ratio; $v_a = u_a/u_o$ = ambient velocity ratio; and $\ell = W/L_d$ = confinement ratio.

Equations 12 and 14 can be used to obtain the following equations for v_n and σ as functions of v_b

$$v_n = \sqrt{\frac{2}{A} + v_b^2} \tag{15}$$

$$\sigma = \frac{A\ell}{4}(v_a^2 - v_b^2) + \frac{1}{2} \tag{16}$$

which can be substituted into Equation 13 to give the following equation in one unknown, v_b,

$$v_bA\ell + A\left(\frac{A\ell}{4}(v_a^2-v_b^2) + \frac{1}{2}\right)\left(\sqrt{\frac{2}{A} + v_b^2} - v_b\right) - v_aA\ell - 1 = 0 \tag{17}$$

This equation can be solved numerically for v_b. The induced velocity ratio, v_n, and the contraction coefficient, σ, can then be calculated from Equations 14 and 15.

Applications

The parameter of most importance is usually dilution, which is defined by the equation

$$S = \frac{Q_n}{Q_o} \simeq \frac{u_n\sigma L_dH_a}{u_oNa_o} = \sigma Av_n \tag{18}$$

or in terms of v_b

$$S \simeq \sigma A\sqrt{\frac{2}{A} + v_b^2} \tag{19}$$

Graphs of S versus Q/Q_o for various confinement ratios are shown in Figure 3 for v_a = 1/10, 1/15, 1/20, and 1/30. These curves can be used to determine the minimum length of diffuser needed to achieve a desired nearfield dilution for specified ambient and effluent conditions. The limiting condition of perfect mixing, $S = Q/Q_o+1$, is represented by the lines for $W/L_d = 1$ in Figure 3. These lines indicate the condition when the diffuser entrains all of the river flow as shown in Figure 4. The regions above the perfect mixing lines have no physical meaning.

The mathematical model was derived for a diffuser centered in a rectangular channel. The application of this model for other diffuser configurations will now be considered. Figure 5 shows a diffuser of length L_d attached to one side of a rectangular channel of width, W, and depth, H. Also shown in the figure is the mirror image of the diffuser-channel section. It is apparent from symmetry that the dilution and the induced velocity for the shore-attached case will be the same as for the centered diffuser shown in Figure 1, given the same ambient and effluent velocities and port geometries. It follows, therefore, that the model can also be used to predict nearfield dilution and induced

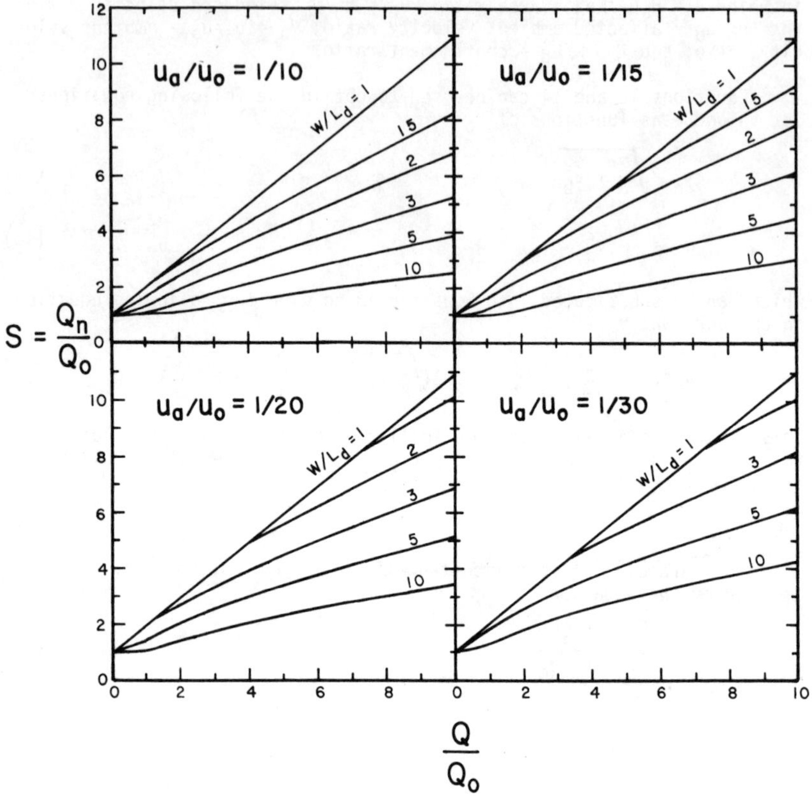

Figure 3. Dilution, S, as a Function of Q/Q_o, u_a/u_o, and W/L_d.

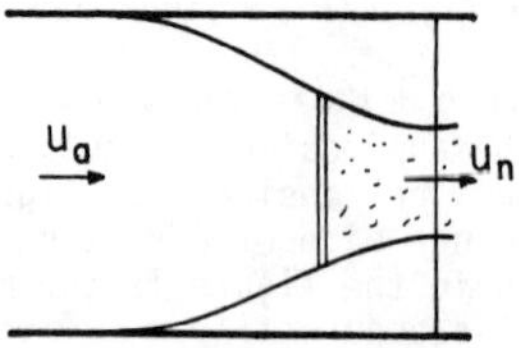

Figure 4. Sketch of Diffuser for Total Nearfield Entrainment of Ambient River Flow.

flow for arbitrary diffuser locations in a channel. The distance that the nearfield region extends downstream from the diffuser would be unknown, however, ranging from 0.5 L_d for the centered diffuser to L_d for the shore-attached diffuser.

Since natural rivers are seldom rectangular, certain assumptions must also be made in order to apply the mathematical model developed herein to nonrectangular channels. By assuming that the degree of diffuser confinement is mainly due to the availability of ambient mixing water rather than to the actual location of the lateral boundaries, an equivalent rectangular channel can be determined for a natural river by applying the following criteria

(i) The depth and ambient velocity for the equivalent channel equal the average depth and ambient velocity in the region of the river cross section in which the diffuser discharges

(ii) The ambient discharge for the equivalent channel equals the river discharge

(iii) The width of the equivalent channel is given by the equation

$$W_{eq} = \frac{Q}{\overline{u}_a \overline{H}} \tag{20}$$

in which $\overline{u}_a$ = average ambient velocity upstream from the diffuser; and $\overline{H}$ = average river depth upstream from the diffuser.

This approach will be illustrated by example for the Quad Cities diffuser system.

A schematic of the Quad Cities Nuclear Power Station diffuser system, the normalized transverse distributions of river and effluent discharge, and the depth profile are shown in Figure 6. At the diffuser the river width is about 2,200 ft and the average depth in the main channel is about 25 ft (3). The diffuser-pipe system actually consists of two 16-ft diameter manifold pipes. During some periods, however, only the one pipe shown in Figure 6 is used. This pipe is equipped with five small ports of 1.8-ft diameter and twenty large ports of 2.7-ft diameter. All of the ports are directed in the downstream direction at a 20° angle with the horizontal. About 90% of the total effluent discharge passes through the large ports in a section of the river 394 ft wide. The rest of the heated water is discharged through the small ports. For simplification, it will be assumed that all of the effluent discharge passes through the portion of the diffuser pipe equipped with the large ports.

From Figure 6, the average value of the normalized ambient unit discharge upstream from the large ports is about 1.95. Therefore

$$q = 1.95\,\overline{q}^{W} = 1.95\left(\frac{1}{W}\ q\ dy'\right) = 1.95\,\frac{Q}{W} \tag{21}$$

Substituting $q = \overline{u}_a\overline{H}$ into Equation 21 and rearranging gives

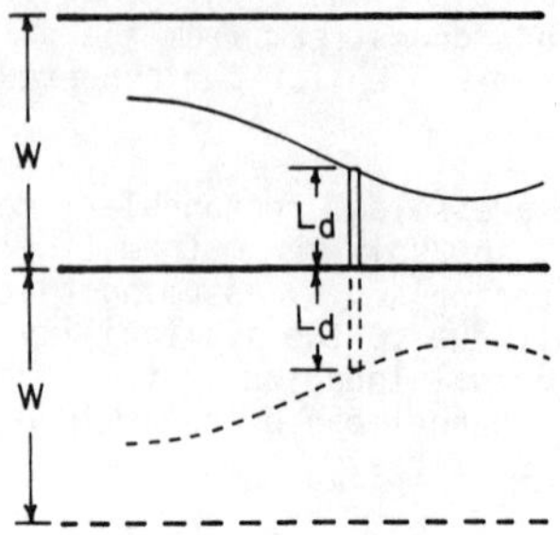

Figure 5. Sketch of Shore-Attached Short Diffuser

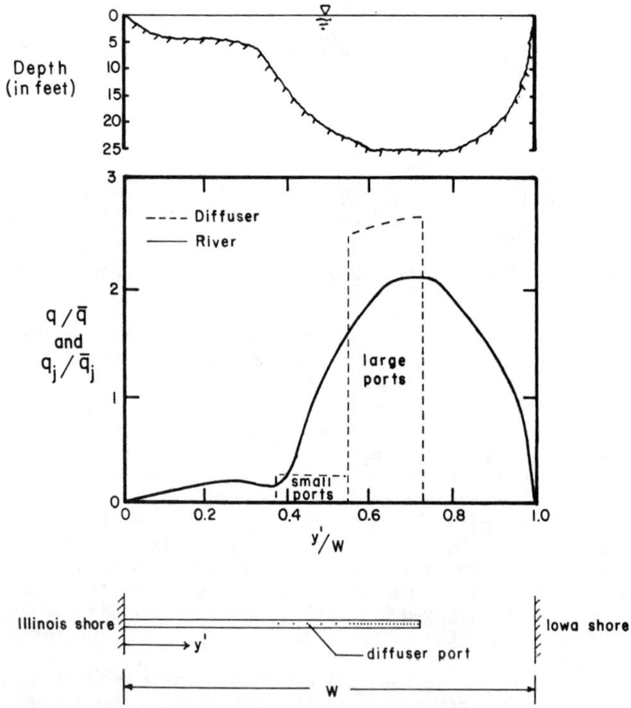

Figure 6. Schematic of Diffuser, Normalized Transverse Distributions of River and Diffuser Discharge, and Depth Profile for the Quad Cities Installation (One-Pipe Mode).

$$\bar{u}_a = 1.95 \frac{Q}{W\bar{H}} \tag{22}$$

which, when combined with Equation 21 yields

$$W_{eq} = \frac{W}{1.95} = \frac{2{,}200 \text{ ft}}{1.95} = 1{,}130 \text{ ft} \tag{23}$$

for the Mississippi River at the Quad Cities site.

A series of field measurements described by Parr and Sayre (7,8) and Sayre, et al. (9) were performed at the Quad Cities site for the one-pipe discharge mode of operation. Temperatures were measured upstream and downstream from the discharge section of the river in order to determine the excess temperature distribution at cross sections 150 and 500 ft downstream from the diffuser. Background data for the surveys is given in Table 1. The parameters in Columns 5 to 8 were used together with Equations 17 and 19 to determine the theoretical dilution, S, in Column 9. The theoretical excess temperature at the end of the nearfield zone was determined from the equation

$$\Delta T = \frac{(T_o - T_a)}{S} \tag{24}$$

where $(T_o - T_a)$ is the excess temperature of the effluent. Values of the theoretical excess temperature are shown in Column 10 of Table 1. The maximum depth-averaged excess temperature measured downstream from the large diffuser ports is shown in Column 11.

The agreement between the theoretical and the measured excess temperature, as shown in Columns 10 and 11 of Table 1, is reasonably good. The measured excess temperature values at x = 500 ft are consistently lower than the theoretical values. This should be expected since, from theory, the nearfield zone ends approximately 197 to 394 ft downstream from the diffuser ($0.5L_d$ to L_d).

Summary and Conclusions

A mathematical model was developed to predict the induced flow and dilution in the nearfield region of a short river diffuser. The results were presented in the form of dimensionless graphs relating dilution to relevant ambient and effluent parameters. The model was compared with prototype diffuser field measurements.

The model predicted the dilutions measured in the field reasonably well. Therefore, the dilution graphs of Figure 3 should be useful in designing diffusers for river environments where nearfield effluent standards must be met.

Survey Number 1	Distance D.S. x (ft) 2	Ambient River Discharge Q (cfs) 3	Effluent Excess Temp. (T_o-T_a) (^oF) 4	Relative Discharge Q/Q_o 5	Relative Velocity u_a/u_o* 6	Confinement Ratio W_{eq}/L_d 7	Area Ratio $\frac{HL_d}{Na_o}$ 8	Predicted Dilution S 9	Predicted Excess Temp. ΔT (^oF) 10	Measured Depth-Averaged Excess Temp. $(\bar{T}^d-T_a)_{max}$ (^oF) 11
1	150	26,700	40.1	46.0	.208	2.87	86	15.6	2.55	2.73
2	150	37,600	45.5	54.5	.245	2.87	86	17.1	2.65	1.97
3	500	22,200	11.7	19.1	.0863	2.87	86	9.83	1.19	1.15
4	500	37,100	25.2	43.6	.198	2.87	86	15.2	1.66	1.40
5	500	32,000	32.0	36.4	.163	2.87	86	13.6	2.35	2.17
6	500	27,000	37.3	41.5	.187	2.87	86	14.7	2.54	2.07
7	500	25,300	13.7	14.5	.0652	2.87	86	8.82	1.55	1.75
8	500	11,900	15.3	7.08	.0320	2.87	86	7.35	2.08	1.87
9	500	24,800	21.0	17.2	.0776	2.87	86	9.41	2.23	2.17
10	500	18,200	27.9	26.8	.120	2.87	86	11.5	2.43	2.17

* $u_o = Q_o$/(total area of small and large ports); $u_a = Q/28{,}200$ in ft/sec from Eq. 22

Table 1. Background Data and Predicted and Measured Temperature Dilution for Field Surveys at the Quad Cities Nuclear Power Station.

References

1. Adams, E. E., "Submerged Multiport Diffusers in Shallow Water With Current," thesis presented to the Massachusetts Institute of Technology, at Cambridge, Mass., in 1972, in partial fulfillment of the requirements for the degree of Master of Science.

2. Cederwall, K., "Buoyant Slot Jets into Stagnant or Flowing Environments," W. M. Keck Laboratory, Report No. KH-R-25, California Institute of Technology, Pasadena, California, April, 1971.

3. Jain, S. C., et al., "Model Studies and Design of Thermal Outfall Structures Quad-Cities Nuclear Plant," IIHR Report No. 135, Iowa Institute of Hydraulic Research, University of Iowa, Iowa City, Iowa, September, 1971.

4. Jirka, G. H., Abraham, G., and Harleman, D. R. F., "An Assessment of Techniques for Hydrothermal Prediction," R. M. Parsons Laboratory Report No. 203, Massachusetts Institute of Technology, Boston, Massachusetts, July, 1975.

5. Lee, J. H. W., Jirka, G. H., and Harleman, D. R. F., "Modelling of Unidirectional Thermal Diffusers in Shallow Water," Technical Report No. 228, R. M. Parsons Laboratory for Water Resources and Hydrodynamics, Massachusetts Institutes of Technology, Cambridge, Massachusetts, June, 1977.

6. Parr, A. D., and Sayre, W. W., "Prototype and Model Studies of the Diffuser-Pipe System for Discharging Condenser Cooling Water at the Quad-Cities Nuclear Power Station," IIHR Report No. 204, Iowa Institute of Hydraulic Research, University of Iowa, Iowa City, Iowa,June, 1977.

7. Parr, A. D., and Sayre, W. W., "Power Plant Diffuser-Pipe System," Journal of the Environmental Engineering Division, ASCE, Vol. 105, No. EE6, Proc. Paper 15048, December, 1979, pp. 1075-1093.

8. Parr, A. D., and Sayre, W. W., "Multiple Jets in Shallow Flowing Receiving Waters," Journal of the Hydraulics Division, ASCE, Vol. 105, No. HY11, Proc. Paper 14991, November, 1979, pp. 1357-1374.

9. Sayre, W. W.,"Quarterly Reports on River Temperature Surveys at the Quad-Cities Nuclear Power Station - July-Sept. 1976, Apr.-June 1977, July-Sept 1977, Jan.-Mar. 1978," Iowa Institute of Hydraulic Research, University of Iowa, Iowa City, Iowa.

EVALUATION OF COMPLIANCE CRITERIA FOR RIVER PLUMES

By

Paily P. Paily, M. ASCE*

ABSTRACT

Steam power plants which utilize once-through cooling, where fresh water is withdrawn from a natural river or lake and heated water is returned to the same after condenser circulation, are required to meet certain mixing zone standards for thermal plumes. Compliance with regulatory standards during plant operation is often verified from field measurements at specified time intervals. This paper describes how field data from thermal plume surveys can be used to develop mixing zone models which are useful in establishing the necessary operating criteria that will assure compliance of the standards during plant operation over the specified range of river flows.

INTRODUCTION

Operation of steam electric power plants with open-cycle cooling of plant condensors, where the heated water is directly discharged into a natural waterbody, results in the formation of a thermal plume within the receiving water in regions adjacent to the discharge outfall. The nearfield regions of the thermal plume, where the temperature gradient is the largest, is known as the mixing zone. Regulatory agencies of various state governments have established temperature criteria, as part of water quality standards, specifying the allowable sizes of mixing zones for various waterbodies, in an attempt to protect the biological populations in natural waters from temperature-induced hazards. During the operational licensing stage of a power plant, the allowable maximum sizes of mixing zone for various operating conditions are specified, and these are incorporated into the station technical specifications. For a river, mixing zone sizes corresponding to two different flow conditions, one a seasonal mean flow, and the other a seasonal worst-case flow, which usually equals the seven-day, ten-year flow for the season, are normally specified. During the operation of the plant these specifications are required to be complied with, and proof of such compliance is verified from field data collected during monthly thermal plume surveys.

Even though field measurements can be designed and planned to obtain data for various combinations of the different variables involved during plant operation, such as plant output, discharge flow rate,

*Associate Professor, School of Civil Engineering, Georgia Institute of Technology, Atlanta, Georgia 30332

temperature rise of discharge, river flow, meteorological conditions, etc., it is often difficult to obtain data at conditions matching those stipulated in the technical specifications. This is especially true of the requirements for the worst-case flow condition, since the chances are remote for having a seven-day, ten-year flow in the river during the initial periods of plant operation. However, once sufficient field data have been gathered, such data can be analyzed to develop reliable mathematical models for the mixing zone. If properly formulated, such models can then be used to evaluate the compliance criteria related to the plant operation for the entire range of river flows specified by the standards, thus eliminating the need for further field measurements, with subsequent savings of several thousands of dollars every year.

MIXING ZONE STANDARDS

Mixing zone is the immediate area of discharge within the receiving water where the heated effluent rapidly diffuses into the ambient flow, creating large temperature gradients within the thermal plume. The boundary limits of the mixing zone are established in terms of a specified linear distance, volume, or surface area such that outside the edge of the mixing zone temperature excesses due to the heated effluent are always below a specified limit. An examination of the mixing zone standards of various government agencies of states bordering the Missouri and Mississippi Rivers (3,7) indicates that, in most cases, the allowable temperature rise at the edge of the mixing zone is 5°F (2.78^{o}C) for rivers and 3°F (1.87^{o}C) for lakes and impoundments. The standards also state certain numerical limits, which are mainly intended as guidelines, applicable to mixing zone size, but often the size and location of mixing zone for any thermal discharge are determined on a case-by-case basis. The major consideration is that the mixing zone should be kept as small as possible, and should not be of a size or shape that could cause or contribute to the impairment of water uses. Important factors that are considered in determining the size of a mixing zone include:

(1) Size of the receiving water, volume of discharge, stream bank configuration, velocity ratio between the effluent and the stream flow, and other hydrologic, and physiographic characteristics;

(2) Present and anticipated future water use and water quality of the receiving body of water;

(3) Dilution ratio, defined as the ratio between the volume of effluent discharge to the seven-day, ten-year low flow of the receiving stream; and

(4) Required zone-of-passage to allow an adequate passageway for the movement or drift of aquatic life at all times.

The criterion related to the required zone-of-passage within the receiving water may be the most critical one for determining the size of a mixing zone. In rivers, lakes and reservoirs, continuous water routes of volume, area and quality are necessary to allow passage of free-swimming and drifting organisms with no significant effects produced on their populations. Because of this requirement, when there are several mixing zones close together, they should all be on the same side of the river so that a continuous passageway can be maintained. Also, overlapping of mixing zones should be avoided or minimized to prevent adverse synergistic effects. The standards also stipulate that the mixing zones should not intersect spawning and nursery areas of biological populations and their migratory routes, as well as municipal water intakes and river mouths.

For discharge of power plant thermal effluents into rivers, the size of the mixing zone is usually specified in terms of a fractional river width, and either a surface area or downstream distance in the streamwise direction, that a certain temperature-rise (usually 5°F) isotherm, which defines the edge of the mixing zone, should not exceed. For example, the mixing zone standards for the Quad-Cities Nuclear Power Station which discharges into the Mississippi River near Cordova, Illinois, as specified by the Iowa Water Quality Commission and the Illinois Pollution Control Board, stipulate the following (8): The mixing zone may not contain more than 25% of the cross-sectional area or volume of flow at any river cross section, and water temperature outside the mixing zone may not exceed the local ambient river temperature by more than 5°F (2.78°C). The Illinois regulations state, in addition, that no mixing zone shall exceed the area of a circle with 600-ft (180-m) radius, which approximately equals an area of 26 acres (10.5 ha).

ANALYSIS OF FIELD DATA

When the operation of a thermal power plant is regulated by prespecified mixing zone standards, compliance with such standards is demonstrated from thermal plume surveys. During these surveys, temperature measurements are made within the river so that the three-dimensional structure of the thermal plume can be mapped, and the actual areal extent of the mixing zone can be delineated. Knowing the river discharge, river width, and a representative river cross-section for the region occupied by the thermal plume, it is then possible to show if the size of the measured mixing zone is within the limits of regulatory criteria, or not.

Even though field measurements and associated data presentation are the best way of establishing compliance, application of this procedure for the entire range of conditions specified by the standards may not be possible from practical and economic considerations. Hence it is necessary to analyze and interpret the available field data to derive useful conclusions. Since the requirements of the standards are that the length, width and surface area within a certain temperature-rise isotherm should be within specified limits, if appropriate relationships can be established for these plume characteristics in terms of operating conditions, such relationships, or models, will be valuable in establishing compliance criteria for all conditions.

From the thermal plume maps prepared from field data for various plume surveys, the lengths, widths, and surface areas for several temperature-rise isotherms can be obtained using graphical techniques. These data can be correlated with the operating conditions to develop mixing zone models for plume center-line distance, surface area, and plume width as discussed below in relation to a case study.

CASE STUDY: NUCLEAR PLANT ON THE MISSOURI RIVER

Background: The Cooper Nuclear Station of the Nebraska Public Power District (NPPD) is located at Brownville, Nebraska, on the Missouri River. This station has a rated full-load capacity of 800 MW, and is designed to use a maximum cooling water flow rate of 1,450 cfs with a condenser temperature rise of 18°F in summer, and 744 cfs with a temperature rise of 36°F in winter. The heated effluent from the station is discharged into the Missouri River through a 120 ft -wide channel oriented approximately parallel to the river flow. The Environmental Technical Specifications for the station, established by the U.S. Atomic Energy Commission before the station went into operation in 1974, stipulate that the thermal mixing zone within the Missouri River is not to exceed a width of one-third of the river width, nor exceed more than 7500 ft downstream (5). The edge of the mixing zone is defined by the 5°F temperature-rise isotherm during summer months (May through October) and by the 10°F temperature-rise isotherm during winter months (November through April). Based on an average river width of 790 feet at the station site, the prescribed width and downstream extent of the mixing zone provide an allowable surface area of approximately 45 acres for the mixing zone (6).

The average summer and winter flows of the Missouri River at the station location are 35,000 cfs, and 22,000 cfs, respectively. The corresponding seven-day, ten-year flows are 31,000 cfs, and 8,000 cfs. It is required that the mixing zone size be kept below the allowable limits for both average and low flows in summer and winter.

Thermal plume surveys for the Cooper Station were conducted on a monthly basis from May 1974 through December 1978 by the consulting firm of Nalco Environmental Sciences, Northbrook, Illinois, and the data are available from several annual reports submitted by NPPD to the U.S. Nuclear Regulatory Commission. A summary of the measured data for summer plumes is presented in Table 1, and a similar summary for winter plumes is presented in Table 2. All the summer surveys were conducted at river flows above the summer average value, while a few surveys in in winter were made at flows close to the winter average value. None of the surveys were conducted at river flows approaching the specified seven-day, ten-year low flows.

Mixing Zone Models: Since compliance of the mixing zone standards at low flows could not be demonstrated from field data, it is necessary to develop appropriate mixing zone models for the Cooper Nuclear Station. Examination of data in Tables 1 and 2 indicates that the densimetric Froude number, which is a major similarity parameter for plume mixing, varied only within a small range so that the dependence of center-line temperature rise at any downstream location within the plume on the densimetric Froude number was only minor, if any. However, a clear dependence of the temperature rise on the flow ratio, which is defined as the ratio between the ambient river discharge (Q_R) and the thermal effluent discharge (Q_o) is observed. This is illustrated in Figure 1, which shows the measured surface isotherms on two occasions,

TABLE 1.
Summary of data from summer thermal plume surveys in the Missouri River near the Cooper Nuclear Station.

Sampling Date	Operating Load (%)	Initial ΔT (°F)	River Flow (cfs)	Discharge Flow (cfs)	Densi- metric Froude Number	5°F Isotherm Centerline Distance (ft)	5°F Isotherm Surface Area (acres)	5°F Isotherm Maximum Width (%)
23 May 1974	22	7.8	48200	725	2.83	400	0.8	14
27 June 1974	46	16.4	36600	725	2.00	450	1.2	15
30 July 1974	20	7.4	37100	762	3.15	550	1.0	15
22 August 1974	70	16.7	36100	1088	2.83	500	1.7	18
27 September 1974	52	8.6	37100	1352	6.06	300	0.9	15
17 October 1974	70	16.7	37100	1116	3.59	950	2.8	15
30 May 1975	40	13.4	45400	762	2.44	500	1.2	18
30 June 1975	35	10.1	57600	880	2.87	750	1.5	14
28 July 1975(#1)	78	15.9	64600	1234	2.89	160	0.3	14
28 July 1975(#2)	80	13.6	64600	1470	4.06	120	0.4	10
31 July 1975	85	14.6	62600	1470	3.89	200	0.3	11
1 August 1975	88	15.0	63400	1470	3.85	460	0.7	11
18 September 1975	46	10.5	66600	1088	4.03	350	0.6	14
28 May 1976	70	13.7	46800	1088	3.41	208	0.7	18
17 June 1976	70	19.1	42300	725	1.55	273	1.1	23
15 July 1976	77	19.1	40000	1088	2.50	465	2.0	42
19 August 1976	73	18.0	41100	1088	2.68	610	1.7	23
14 September 1976	69	13.0	04200	1088	3.42	1078	2.2	26
11 May 1977	93	19.6	37700	1089	2.84	974	2.4	28
28 June 1977	95	16.0	36600	1450	3.72	1948	6.8	32
26 July 1977	90	15.7	40400	1450	3.72	753	1.5	23
18 August 1977	91	18.5	43000	1089	2.69	974	2.1	20
14 September 1977	84	17.5	36500	1089	2.90	935	2.0	24
27 October 19t7	76	16.6	38200	1089	3.62	1175	2.7	27
8 May 1978	73	14.7	50000	1089	4.20	1092	2.2	8
27 June 1978	92	16.1	55000	1450	3.76	1011	2.2	24
25 July 1978	94	15.8	40100	1450	3.86	648	1.8	26
30 August 1978	82	13.9	58800	1450	4.10	773	1.7	10
20 September 1978	70	12.4	67300	1450	5.31	752	1.6	16
11 October 1978	84	14.9	57800	1450	4.35	661	1.1	12
8 November 1978	86	18.7	59400	1089	3.52	520	1.6	20

TABLE 2.

Summary of data from winter thermal plume surveys in the Missouri River near the Cooper Nuclear Station.

Sampling Date	Operating Load (%)	Initial ΔT (°F)	River Flow (cfs)	Discharge Flow (cfs)	Densimetric Froude Number	10°F Isotherm Centerline Distance (ft)	10°F Isotherm Surface Area (Acres)	10°F Isotherm Maximum Width (%)
23 January 1975	100	24.6	21000	1013	5.56	1500	2.3	23
28 Janury 1975	100	36.9	23400	675	2.37	1300	4.3	21
20 March 1975	100	30.6	38800	814	3.05	550	0.7	20
30 April 1975	47	16.1	61000	727	1.93			10
5 December 1975	92	24.1	57300	951	4.87	300	0.3	8
4 February 1976	88	35.8	30600	613	2.67	706	1.2	24
8 April 1976	80	26.1	45600	764	2.33			
23 November 1976	90	25.6	40800	876	4.41	597	1.3	26
3 March 1977	100	25.6	23900	974	4.06	1546	3.1	28
30 November 1977	89	31.9	26400	695	2.67	942	2.1	26
6 December 1977	89	33.3	22800	666	2.51	424	1.2	30
31 March 1978	72	28.4	52000	631	1.61	151	0.8	27
20 December 1978	97	34.4	31800	706	1.69	843	1.4	3

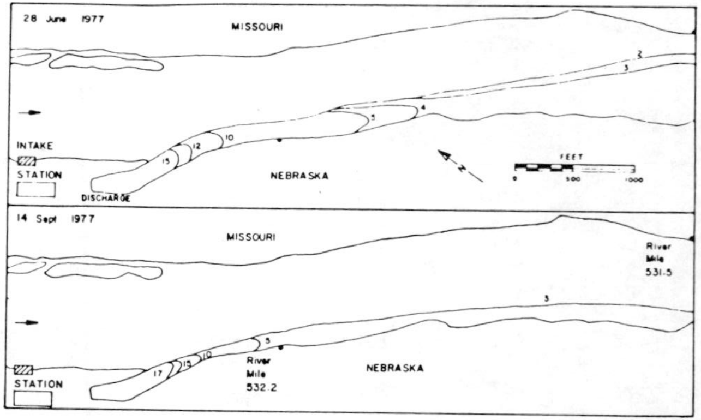

FIGURE 1

Measured Surface Isotherms in the Missouri River near the Cooper Nuclear Station on two Occasions

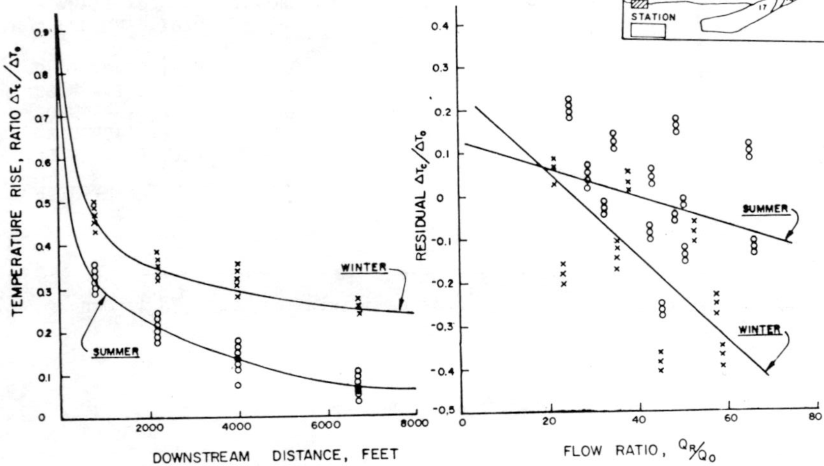

FIGURE 2

Exponential fit of measured center-line temperature rise ratio with downstream distance, and linear fit of residual values with flow ratio both for high station operating loads, and near average river flow conditions. At a lower value of flow ratio, Q_R/Q_o, equal to 25 on 28 June 1977, the isotherms are seen to extend farther downstream, compared to those with a flow ratio of 34 on 14 September 1977. Based on this, a functional relationship for the normalized center-line temperature rise can be written as

$$\frac{\Delta T_c}{\Delta T_o} = f_1 \left(\frac{Q_R}{Q_o}, \frac{X}{b_o}\right) \qquad (1)$$

in which

ΔT_c = plume center-line temperature rise;

ΔT_o = effluent temperature rise at the source;

Q_R = ambient river discharge;

Q_o = heated effluent discharge;

X = downstream distance from the source along the plume center-line trajectory to the temperature-rise isotherm ΔT_c; and

b_o = width of the effluent discharge canal.

For the Cooper Station plumes, the functional relationship in Eq. (1) is evaluated considering that there is a primary exponential relationship between $\Delta T_c/\Delta T_o$ and X/b_o, and a secondary linear relationship between $\Delta T_c/\Delta T_o$ and Q_R/Q_o, as illustrated in Figure 2. The resulting models for plume center-line temperature rise are

$$\frac{\Delta T_c}{\Delta T_o} = 0.13 - 0.0033\left(\frac{Q_R}{Q_o}\right) + 0.38 \exp\left(-0.03\frac{X}{b_o}\right) \qquad (2)$$

for summer plumes, and

$$\frac{\Delta T_c}{\Delta T_o} = 0.248 - 0.0097\left(\frac{Q_R}{Q_o}\right) + 0.396 \exp\left(-0.00816 \frac{X}{b_o}\right) \qquad (3)$$

for winter plumes. In these relations, temperatures are in ^oF, and other variables are in foot units.

For the plume surface area model, a power-law relationship between area and temperature rise is often assumed (1,2). Lau and Dick (4) has shown, from analysis of Great Lakes plume data, that plume area is also proportional to square of Q_o. In a river environment the surface area of plume should also vary inversely with the river flow, as higher dilution of heated effluent with the river flow leads to a lesser surface area of the plume. Hence a functional relationship for plume surface area will be

$$A\frac{Q_R}{Q_o^2}\sqrt{g'h_o} = f_2\left(\frac{\Delta T}{\Delta T_o}\right) \qquad (4)$$

in which

A = area enclosed by isotherm of temperature rise ΔT;

h = depth of heated effluent in discharge canal;

g' = $(\Delta\rho/\rho)g$ = apparent acceleration of gravity;

$\Delta\rho$ = $\rho_a - \rho$ = density difference between ambient water density, ρ_a, and heated water density, ρ.

Assuming a power-law relationship as shown in Figure 3 for the function in Eq. (4), the area models for Cooper Station plumes are

$$A\frac{Q_R}{Q_o^2}\sqrt{g'h_o} = 0.00848\left(\frac{\Delta T}{\Delta T_o}\right)^{-1.432} \tag{5}$$

for summer plumes, and

$$A\frac{Q_R}{Q_o^2}\sqrt{g'h_o} = 0.00384\left(\frac{\Delta T}{\Delta T_o}\right)^{-2.725} \tag{6}$$

for winter plumes. In these relations A is in acres, and the other variables are in foot units, with temperatures in ^oF.

The maximum plume width, the third plume characteristic required to define the mixing zone size, is mainly a function of plume shape for a plume of certain downstream length and surface area. The Cooper Station plumes were always shore-attached, as shown in Figure 1, with the maximum center-line temperatures occuring along the shore. By comparing the average value of the ratio of surface area to downstream distance with the measured plume width, a model for maximum width, W, is obtained as

$$W = 2.0\ A/X \tag{7}$$

where all variables are in foot units.

Using the above models for center-line temperature rise, surface area, and maximum plume width, the size of the mixing zone at the low-flow situation can now be determined. For full-load operation of the plant, the computed values are:

	Summer Low Flow (31,000 cfs)	Winter Low Flow (8,000 cfs)
Mixing Zone Length	2,220 ft.	15,930 ft.
Surface Area	5.43 acres	16.69 acres
Maximum Width	8.1% of river width	24.6% of river width

These results indicate that the Cooper Station plumes are in compliance with the mixing zone standards for the summer low flows, but will violate the allowable downstream distance criterion of 7,500 ft during winter low flows. Evidently, compliance can be achieved during winter low-flow periods only by reducing the station load, which will then proportionately reduce the heat rejected into the river. The questions then are: What is the minimum winter river flow above which full-load operation of the station is possible? Below this minimum river flow, what reduction in station load is needed to satisfy compliance at a certain river flow? The mixing zone model for center-line temperature applicable to winter periods, Eq. (3), can be used to seek solutions to these questions.

Simultaneous considerations of the full-load operating condition (Q_o=744 cfs; ΔT_o = 36°F) along with the distance-limiting mixing zone criterion ($\Delta T_c \leq 10^oF$ at X = 7,500 ft.) allow the minimum river flow to be determined as 16,300 cfs. This river flow can be considered as a "cut-off point" below which it will be necessary to operate the station at a partial load. The necessary compliance criteria for station oper-

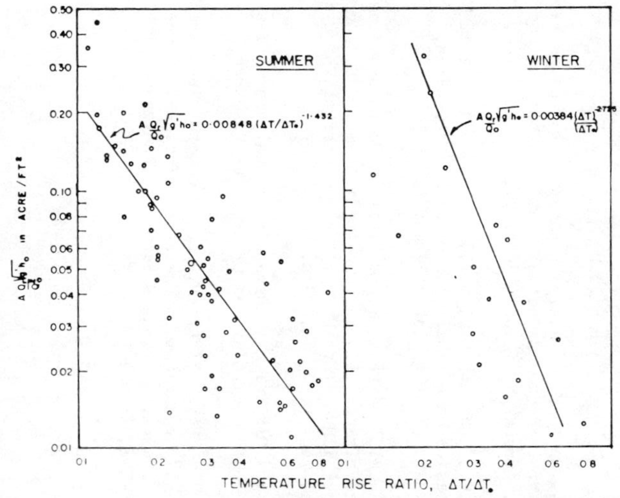

FIGURE 3

Relationship between plume area parameter and temperature rise ratio of isotherms

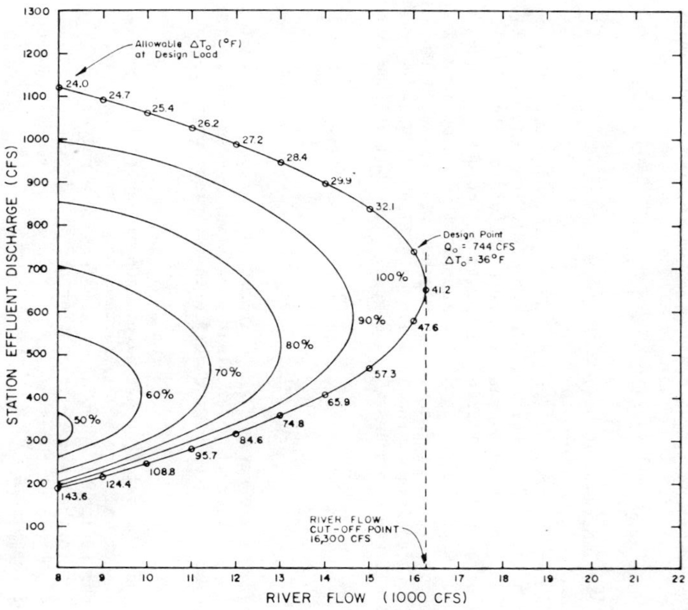

FIGURE 4

Estimated operating criteria for Cooper Nuclear Station at low flows ation at low river flows can now be evaluated from the two constraining requirements:

$$Q_o \, \Delta T_o \leq (744)\ (36); \tag{8}$$

and

$$\frac{10}{\Delta T_o} = 0.4858 - 0.0097 \frac{Q_R}{Q_o} \tag{9}$$

Equation (8) is directly obtained from the design load for the station, while Eq. (9) is obtained from Eq. (3) for the condition that $\Delta T_c = 10^oF$ at X = 7,500 ft. Equations (8) and (9) can be solved simultaneously to determine the percent load at which the station should be operated to maintain compliance with the standards and the results can be summarized as presented in Figure 4. For example, at a river flow of 8,000 cfs, it will be necessary to operate at 70% load if the effluent discharge rate is 700 cfs. The allowable condenser temperature rise then will be about 26.7^{o}F.

CONCLUSION

Plume data from field surveys can be analyzed to develop reliable mixing zone models for thermal power plant operation. These models are useful in developing proper compliance criteria for plant operation at low river flows.

Acceptance and approval of the type of data presented in Figure 4 by a regulatory agency which routinely reviews and verifies the station compliance with the applicable standards, can lead to a waiver allowing the cancellation of further plume surveys for the station. This will result in savings of several thousands of dollars every year in the operating cost of the plant.

REFERENCES

1. Asbury, J.G., and Frigo, A.A., "A Phonomenological Relationship for the Surface Areas of Thermal Plumes in Lakes," Report No. ANL/ ES-5, Argonne National Laboratory, Argonne, IL., 1971.

2. Elliott, R.V., and Harkness, D.G., "A Phonomenological Model for the Prediction of Thermal Plumes in Large Lakes," Proceed. 15th Conf. of Great Lakes Research, 1972.

3. Giaquinta, A.R., and Keng, T.C., "Thermal Regimes of the Mississippi and Missouri Rivers Downstream from the Southern Iowa Border," IIHR Report No. 211, Iowa Institute of Hydraulic Research, Iowa City, IA, 1978.

4. Lau, Y.L., and Dick, T.M., "Dimensional Analysis of Thermal Plumes," Seventeenth Int. Conf. on Great Lakes Research, 1974.

5. Nebraska Public Power District, "Environmental Technical Specifications," Appendix B to Operating License No. DPR-46 for the Cooper Nuclear Station, U.S. Atomic Energy Commission Docket No. 50-398, 1974.

6. Paily, P.P., "Thermal Plume Surveys in the Missouri River Near the Cooper Nuclear Station," Nalco Environmental Sciences Report to Nebraska Public Power District, 1976.

7. Paily, P.P., Su, T.Y., Giaquinta, A.R., and Kennedy, J.F., "Thermal Regimes of the Upper Mississippi and Missouri Rivers," IIHR Report No. 182, Iowa Institute of Hydraulic Research, Iowa City, IA, 1976.

8. Parr, A.D., and Sayre, W.W., "Power Plant Diffuser-Pipe System," ASCE, Jour. of Environmental Eng. Div., December 1979.

MATHEMATICAL MODEL FOR STAGED DIFFUSER DISCHARGE

Dominique N. Brocard, A.M.ASCE, Charles H. Beauchamp, A.M.ASCE, and Shih-Kuan Hsu

ABSTRACT

Staged diffusers are very effective in mixing cooling water or other effluents with ambient water, particularly when crossflows can occur in both directions. The offshore momentum also contributes to reducing re-entrainment after reversals of the crossflow direction.

This paper presents an integral model to predict the velocity and temperature distributions induced by staged diffusers in a crossflow. The model is restricted to cases where the plume occupies the full depth over the diffuser (shallow water conditions), but this condition is met for most applications. Beyond the diffuser, the plume usually stratifies and this region is also covered by the model.

The empirical coefficients used in the model are well defined for the plume region beyond the diffuser. The counterpart coefficients for the diffuser region are established here based on experimental data from several sources. The overall model results are compared with measurements in terms of surface temperature rise isotherms which are generally the most utilized products of thermal models. Overall, the predictions are very good over a range of diffuser designs and ambient crossflows.

INTRODUCTION

A number of recent hydrothermal model studies have shown that staged diffusers outperform other types of diffusers, such as alternating or tee diffusers, for discharges at sites where longshore currents can be in either direction (3), (6), (7), (8). Staged diffusers are characterized by an orientation of nozzles at a small angle, γ, on both sides of the diffuser axis, which is typically pointed in the offshore direction (Figure 1). Staged diffusers impart a net offshore momentum to the discharge (in contrast to alternating diffusers) thereby directing the heat away from the biologically productive nearshore, and reducing re-entrainment of previously discharged water after reversals of the longshore current direction. Also, in contrast with tee diffusers, the nearfield dilution performance of staged diffusers improves with increasing longshore current speed.

Alden Research Laboratory, Worcester Polytechnic Institute, Holden, Massachusetts, 01520.

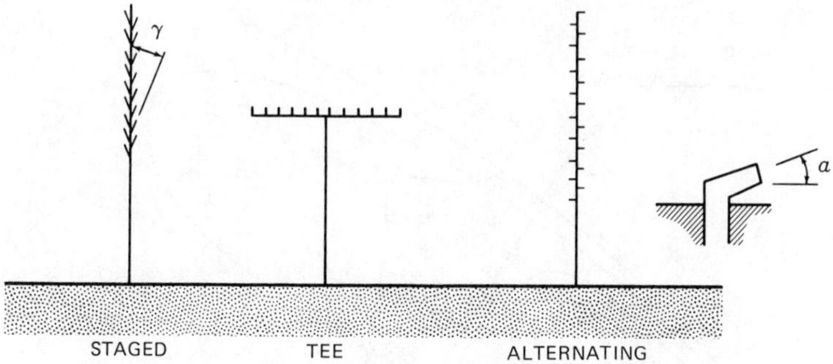

FIGURE 1 DIFFUSER CONFIGURATION

The nearfield characteristics of staged diffuser plumes in shallow and otherwise stagnant receiving water were calculated by Almquist and Stolzenbach (2) using an integral approach and, in effect, simulating the diffuser as a slot discharge of momentum and heat. (In this context, the term shallow water is applied to describe conditions where, over the diffuser, the plume occupies the full depth (Figure 2), as is often the case at economically acceptable distances from shore.) For most applications, however, thermal predictions are also needed beyond the diffuser nearfield and in the presence of cross-currents. This paper describes a mathematical model for the entire plume, including the effects of cross-currents. The emphasis is on the verification of the model with physical model data and on the selection of the empirical coefficients used in the model.

MODEL FORMULATION

In the integral approach, the mass, momentum, and energy equations are integrated over sections of the plume, becoming ordinary differential equations along the plume centerline. Local values of the velocity excess over ambient and temperature rise are obtained through assumed lateral and vertical similarity profiles. The integral approach has been extensively used and refined for surface buoyant jet predictions (5); to simulate the staged diffuser, mass, momentum, and heat source terms are added on the right-hand side of the integrated mass, momentum, and energy conservation equations. The complete set of equations, which have been derived elsewhere without the diffuser source terms (1), (5), becomes:

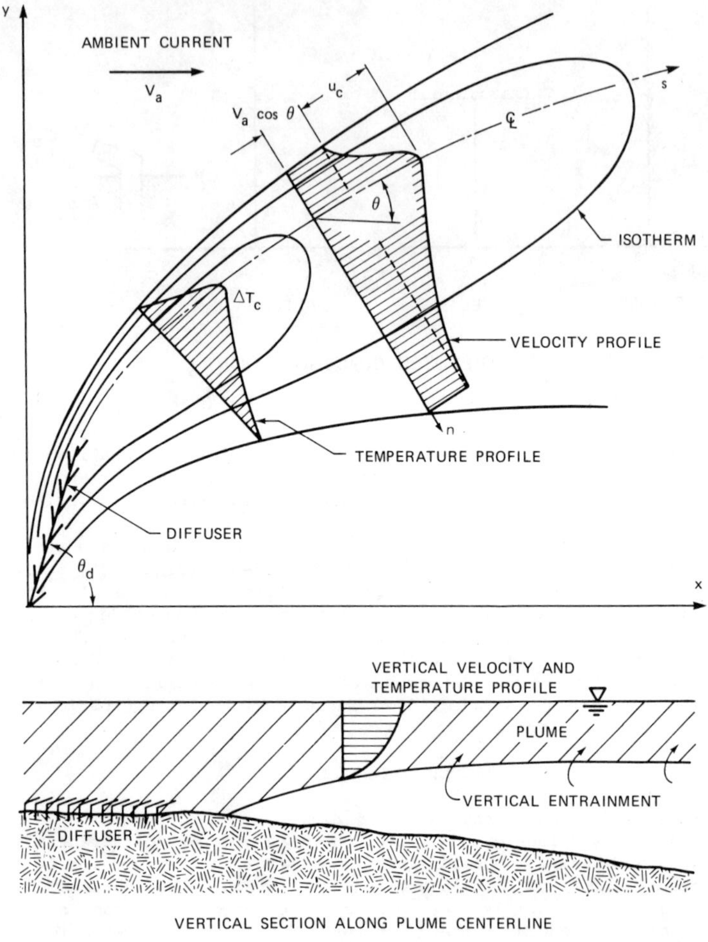

FIGURE 2 SCHEMATIZATION OF DIFFUSER PLUME

$$\frac{dQ}{ds} = 2\ \alpha\ h\ u_c + \alpha_c\ h\ V_a\ \sin\theta + 2\ \alpha\ u_c\ b\ I_8 + \left[\frac{Q_o}{L}\right]_{s<L} \tag{1}$$

$$\frac{dM}{ds} = \frac{dQ}{ds}\ Va\ \cos\theta - \frac{d}{ds}[2\beta\ g\ \Delta T_c\ h^2\ b\ I_5] - \frac{f}{4}\ b\ I_7\ u_c\sqrt{I_7^{\ 2}\ u_c^{\ 2} + (V_a\ \sin\theta)^2}$$

$$+ \left[\frac{M_o}{L}\cos(\theta_d - \theta)\right]_{s<L} \tag{2}$$

$$\frac{d\theta}{ds} = -\frac{1}{M}\left[\frac{dQ}{ds}\ V_a\ \sin\theta + (\frac{f}{4}\ b + \frac{C_D}{2}\ h)\ V_a\ \sin\theta\ \sqrt{I_7^{\ 2}\ u_c^{\ 2} + (V_a\ \sin\theta)^2}\right.$$

$$\left. + \left[\frac{M_o}{L}\sin(\theta_d - \theta)\right]_{s<L}\right] \tag{3}$$

$$\frac{dQE}{ds} = -2\ k\ b\ \Delta T_c\ I_6 + \left[\frac{QE_o}{L}\right]_{s<L} \tag{4}$$

$$\frac{db}{ds} = k_j\left[\frac{u_c}{u_c + 2\ V_a\ \cos\theta}\right] + \frac{k_b}{\sqrt{F^2 - 1}} \tag{5}$$

$$\frac{dx}{ds} = \cos\theta \tag{6}$$

$$\frac{dy}{ds} = -\sin\theta \tag{7}$$

in which

- s = absissa along plume centerline, from start of diffuser
- x, y = coordinates of plume centerline
- Q, M, QE = volume, momentum, and heat fluxes through a section of the plume
- h, b = plume depth and width length scales
- V_a = ambient current speed
- θ, θ_d = angles of plume centerline and diffuser axis
- $u_c, \Delta T_c$ = centerline excess velocity and temperature rise
- α, α_c = jet and cross-current entrainment coefficients
- k_j, k_b = jet and buoyant spreading coefficients
- f, C_D = friction and drag coefficients
- Q_o, M_o, QE_o = diffuser discharge volume, momentum and heat fluxes. If the discharge velocity is u_o, $M_o = u_o\ Q_o\ \cos\alpha\ \cos\gamma$
- L = diffuser length
- I_5, I_7, I_8 = profile integrals (see appendix)

In the shallow water regime, the plume occupies the full water depth over the diffuser ($s < L$). This regime is obtained if the overall plume momentum dominates over buoyancy forces which tend to produce stratification. The relative magnitude of momentum and buoyancy forces is given by the plume densimetric Froude number, $F_H = u_c/\sqrt{\beta \Delta T_c g H}$, and Almquist and Stolzenbach (2) set a lower limit of $F_H = 2.5$ for the existence of shallow water conditions. This value may be high, as (unpublished) data from a hydraulic model study (6) show a staged diffuser plume occupying the full water depth, with only a small temperature difference between the surface and bottom, at a densimetric Froude number of about 1.0.

When the plume occupies the full depth, the vertical entrainment (third term on the right-hand side of Eq. 1) is set equal to zero and the spreading equation, Eq. (5) is replaced by $h = H$. Beyond the diffuser, full depth conditions are maintained until the calculated plume depth, h, is less than the total water depth. The plume is then assumed to stratify and vertical entrainment is initiated.

The similarity profile used for excess velocity and temperature rise are laterally gaussian and vertically parabolic:

$$u(n, z) = u_c \left[1 - A_u \left(\frac{z}{h}\right)^2\right] \exp - \frac{n^2}{b^2} \tag{8}$$

$$\Delta T(n, z) = T_c \left[1 - A_T \left(\frac{z}{h}\right)^2\right] \exp - \frac{n^2}{\lambda^2 b^2} \tag{9}$$

in which

n, z = local lateral and vertical coordinates

λ = Schmitt number

A_u, A_T = vertical profile adjustment parameters variable between 0 (uniform profile) and 1 (fully parabolic profile)

For stratified plumes, the vertical profiles are assumed parabolic ($A_u = A_T = 1$). When the plume occupies the full water depth, more uniform profiles (truncated parabolic) are used ($A_u, A_T < 1$). Over the diffuser, measurement show that the temperature profiles are practically uniform ($A = 0$). The velocity profiles, however, may be expected to depend on the nozzle angle and plume densimetric Froude number, F_H. For values of F_H on the order of 12 to 15, measured velocity profiles in a model study (7) can be approximated by Eq. (8) with $A_u = 0.4$. The transition between full depth and stratified conditions was formulated in terms of the ratio, h/H, of the plume depth to the total water depth. Fully stratified conditions ($A_T = A_u = 1$) were assumed for $h/H < 0.8$ and a linear transition was postulated to full depth values ($A_T = 0$, $A_u = 0.4$) at $h/H = 1$. This transition is obviously simplified and a somewhat arbitrary; it however provides a practical model for an observed phenomenon.

SIMPLIFIED NEARFIELD SOLUTION

By neglecting friction, longitudinal pressure gradient and surface heat loss, a closed form solution can be obtained over the diffuser ($s < L$). The system of governing equations becomes

$$\frac{dQ}{ds} = 2\,\alpha\,H\,u_c + \alpha_c\,H\,V_a\,\sin\theta + \frac{Q_o}{L} \tag{10}$$

$$\frac{dM}{ds} = \frac{dQ}{ds}\,V_a\,\cos\theta + \frac{M_o}{L}\cos(\theta_d - \theta) \tag{11}$$

$$\frac{d\theta}{ds} = -\frac{1}{M}\left[\frac{d\theta}{ds}\,V_a\,\sin\theta + \frac{M_o}{L}\sin(\theta_d - \theta)\right] \tag{12}$$

$$\frac{dQE}{ds} = \frac{QE_o}{L} \tag{13}$$

If the current is normal to the diffuser ($\theta_d = 90°$) and is small enough that the deflection of the plume is negligible ($\theta \simeq 90°$), the above equations can be solved to give

$$u_c = \frac{M_o/Q_o}{\left(\frac{\alpha_c I_2}{2\,I_1}K_c + \frac{I_2}{2\,I_1}\right) + \sqrt{\left(\frac{\alpha_c I_2}{2\,I_1}K_c + \frac{I_2}{2\,I_1}\right)^2 + \frac{2\alpha\,I_2}{I_1}K^2}} \tag{14}$$

$$\Delta T_c = \frac{QE_o/Q_o}{\left(\frac{\alpha_c I_4}{2\,I_1}K_c + \frac{I_4}{2\,I_1}\right) + \sqrt{\left(\frac{\alpha_c I_4}{2\,I_1}K_c + \frac{I_4}{2\,I_1}\right)^2 + \frac{2\alpha\,I_4^2}{I_1\,I_2}K^2}} \tag{15}$$

in which

$$K = \frac{\sqrt{M_o}\,HL}{Q_o}\,, \qquad K_c = \frac{V_a\,HL}{Q_o} \tag{16}$$

and I_1, I_2, and I_4 are similarity profile integrals which are given in the appendix. This simplified solution shows that, as regards nearfield performance, the diffuser is characterized by two parameters: K and K_c. K describes the diffuser in the absence of current and K_c accounts for the effect of cross-current.

When there is no cross-current ($K_C = 0$), Eq. (15) becomes, after substitution of the appropriate numerical values for the profile integrals and recognizing that $QE_o/Q_o = \Delta T_o$ (discharge temperature rise):

$$\Delta T_c(0) = \frac{\Delta T_o}{0.39 + \sqrt{1.89\ \alpha\ K}} \tag{17}$$

This result is similar to that obtained by Almquist and Stolzenbach (2) who, however, also neglected the discharge flowrate of the diffuser (last term in Eq. 10). The effect of this assumption is smaller than the scatter of the experimental data which Almquist and Stolzenbach used to verify their expression. Therefore, their value for the numerical coefficient multiplying K in Eq. (17) can be used here, leading to $\alpha = 0.072$.

A feature of this simplified nearfield solution is that the temperature and velocity excess remain constant along the diffuser. The effect of bottom friction is to produce a slow increase of temperature along the diffuser, as shown on Figure 3. This increase is too low to be observed experimentally and does not justify the use of a larger value for the entrainment coefficient α. The values of the friction coefficient, f, listed in Figure 3 are for bottom friction. When the plume is stratified, the friction is interfacial and the coefficient was halved.

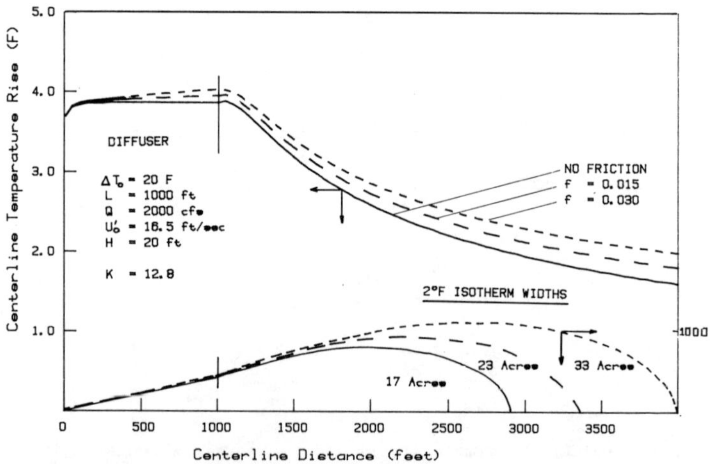

FIGURE 3 EFFECT OF FRICTION ON PLUME PREDICTIONS

Although the longitudinal centerline temperature distributions for the different values of friction appear little different, the plots of 2°F isotherms show significant size variation. This factor may be very relevant for hydraulic model studies in which the friction coefficient is often larger than prototype due to scale effects.

The simplified nearfield solution also allows the determination of the cross-current entrainment coefficient α_c. Comparison of Eq. (15) with measured data leads to $\alpha_c = 1.0$ (4). This implies that all the cross-flows intercepted by the diffuser becomes entrained in the plume, in addition to the turbulent jet entrainment.

MODEL VERIFICATION

Computed surface temperature rise isotherms with the parameters determined above were compared to measurements from two physical hydraulic model studies (3), (7). These studies covered a wide range of diffuser designs and cross-flows as characterized by K and K_c. Figures 4 to 7 show a limited sample of isotherm comparisons for two diffusers with K = 17.8 and 9.6 and with currents leading to K_c = 0 to 9.4. In general, the comparison is very good.

This model can be used to design staged diffusers with prescribed performance characteristics. Since the required computer time is relatively short, the model can be used to optimize staged diffuser designs relative to cost or other factors. Physical hydraulic model studies remain needed, however, when changes of current direction must be analyzed.

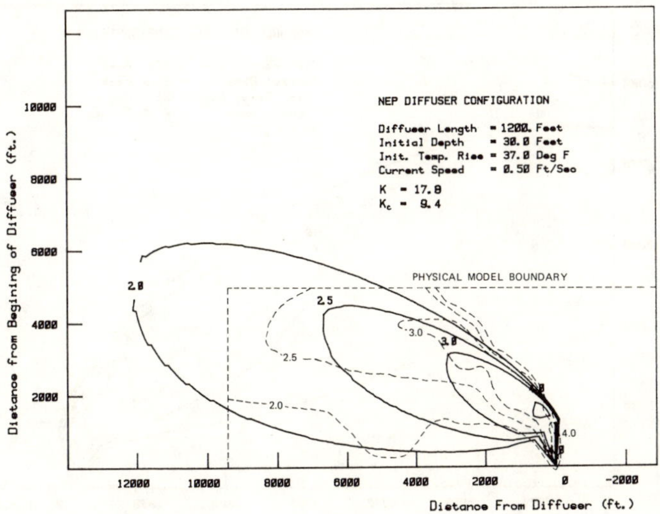

FIGURE 4 COMPARISON OF MEASURED AND CALCULATED SURFACE ISOTHERMS

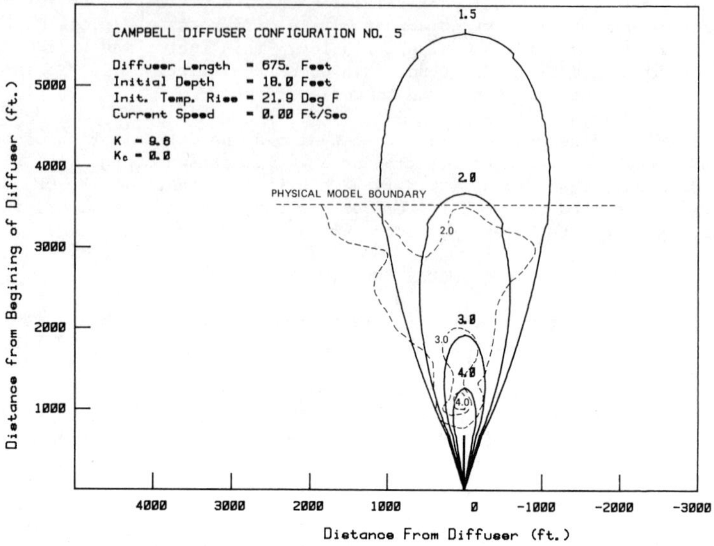

FIGURE 5 COMPARISON OF MEASURED AND CALCULATED SURFACE ISOTHERMS

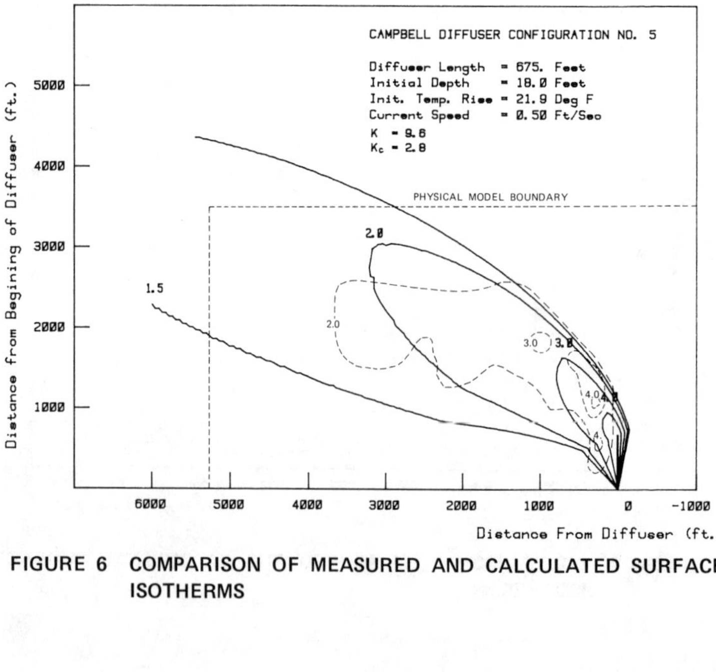

FIGURE 6 COMPARISON OF MEASURED AND CALCULATED SURFACE ISOTHERMS

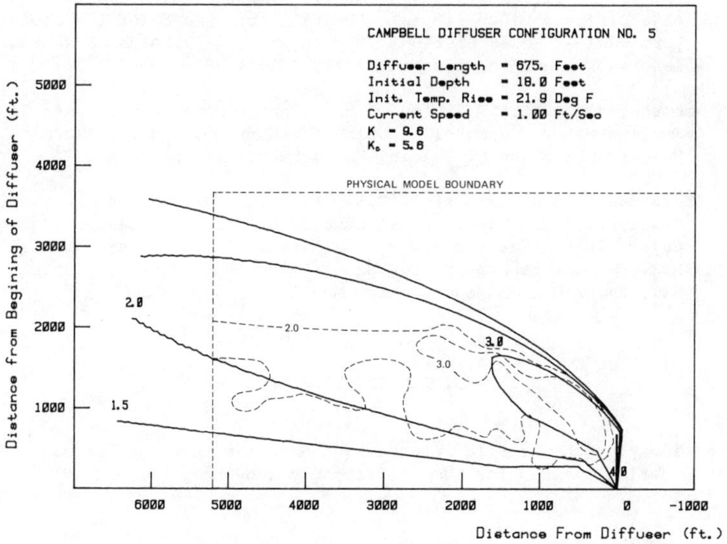

FIGURE 7 COMPARISON OF MEASURED AND CALCULATED SURFACE ISOTHERMS

REFERENCES

1. Alden Research Laboratory, "Surface Buoyant Jet Integral Model," Summary Report in Preparation.

2. Almquist, C.W., and Stolzenbach, K.D., "Multiport Staged Diffusers," Journal of the Hydraulics Division, ASCE, Vol. 106, No. HY2, February 1980.

3. Brocard, D.N., "Hydrothermal Studies of Staged Diffuser Discharge in the Coastal Environment, Charlestown Site," Report No. 137-77, Alden Research Laboratory, Holden, Massachusetts, September 1977.

4. Brocard, D.N., "Multiport Staged Diffusers - Discussion," to appear in Journal of Hydraulics Division, ASCE.

5. Dunn, W.E., Policastro, A.J., and Paddock, R.A., "Surface Thermal Plumes: Evaluation of Mathematical Models for the Near and Complete Field," Report No. ANL/WR-75-3, Argonne National Laboratory, Argonne, Illinois, May 1975.

6. Kirby, Jr., J.T., and Brocard, D.N., "Hydrothermal Model Studies of a Diffuser Discharge in a Coastal Environment for the Jamesport Nuclear Power Station," Report No. 127-79, Alden Research Laboratory, Holden, Massachusetts, July 1979.

7. Roberge, J.C., "Optimization of an Offshore Momentum Diffuser In Shallow Water - Campbell Station," Report No. 104-76, Alden Research Laboratory, Holden, Massachusetts, September 1976.

8. Stolzenbach, K.D., Almquist, C.W., Adams, E.E., and Freudberg, S.A., "Analytical and Experimental Studies of Discharge Designs for the Cayuga Station at the Somerset Alternate Site," Report No. 211, R.M. Parsons Laboratory for Water Resources and Hydrodynamics, MIT, Cambridge, Massachusetts, May 1976.

APPENDIX - SIMILARITY PROFILE INTEGRALS

Expressions for the profile integrals used in the model as well as their numerical values for the profiles given by Eqs. (8) and (9) are as follows.

$$I_1 = \iint \frac{u}{u_c}\, d(\frac{n}{b})\, d(\frac{z}{h}) = 0.846\ (1 - \frac{A_u}{3})$$

$$I_2 = \iint (\frac{u}{u_c})^2\, d(\frac{n}{b})\, d(\frac{z}{h}) = 0.624\ (1 - \frac{2}{3} A_u - \frac{1}{5} A_u^{\ 2})$$

$$I_4 = \iint (\frac{u}{u_c})\, (\frac{\Delta T}{\Delta T_c})\, d(\frac{n}{b})\, d(\frac{z}{h}) = 0.665\ (1 - \frac{A_u + A_T}{3} + \frac{A_u\, A_T}{5})$$

$$I_5 = \iint d(\frac{n}{b})\, d(\frac{z}{h}) \int_{z/h}^{1} \frac{\Delta T}{\Delta T_c}\, d(\frac{z}{h}) = 0.0784\ (6 - 3\ A_T)$$

$$I_7 = \int \frac{u}{u_c}\, d(\frac{n}{b}) = 0.846 \quad ; \quad I_8 = \int \frac{u}{u_c}\, \exp(-\ 5\ Ri)\, d(\frac{n}{b})$$

in which the integrals are on the plume cross-section or width.

INACCURACIES IN THE CHARACTERISTICS METHOD

E. Benjamin Wylie,[1] M. ASCE

ABSTRACT

The specified time intervals method of characteristics requires interpolations when analyzing nonlinear wave propagation, such as occurs in open channel flow. Inaccuracies have been noted: continuity imbalances during transient simulations, and an inability to maintain steady nonuniform flow conditions during repeated forward time steps. An alternative treatment is presented which utilizes time-line rather than distance-line interpolations. The result is a reliable simulation procedure for dynamic nonlinear wave propagation events.

INTRODUCTION

Wave propagation in open channel flow has been modeled by the method of characteristics by many investigators (1,2,4,7,9,10). In applications in which the complete dynamic wave model is needed to describe the physical phenomena, utilization of the characteristics grid method is recognized as the numerical method by which other methods may be judged for accuracy. Examples in which the complete dynamic wave model is needed (8) include high frequency flood waves in mild channels and transients generated by gate operations in irrigation canals. In the general case, implementation of the characteristics grid method is computationally unappealing, as the development of the spatial-time grid becomes unwieldy when the celerity and mean velocity vary during the transient. To circumvent this problem the regularly-spaced rectangular grid of the method of specified time intervals is used (6,12). Unfortunately an interpolation procedure is necessary so that calculations can proceed along characteristic lines. Interpolations may be troublesome in the method depending upon the manner of implementation and on the type of problem.

Although it is not specifically referred to in the open literature, some practitioners have run into complications using this approach. In many situations their numerical simulation performs beautifully, but in others it will neither hold steady state nor converge to steady state. Furthermore if gross mass conservation is checked over the

---

[1]Professor of Civil Engineering, University of Michigan, Ann Arbor, MI 48109.

entire system the user may find that in some cases the method is generating fluid within the system while in others fluid is being lost during the transient. Any one of these conditions is rather disconcerting.

This paper addresses these issues, sheds some light on the apparent randomness of their appearance, and offers an alternative treatment to circumvent the problems. A time-line interpolation procedure, which adheres to the Courant condition, is presented (12) as an improved simulation method for nonlinear wave propagation problems. This study is limited to subcritical flow although the concepts may be visualized more broadly.

EQUATIONS

The Saint-Venant equations for a prismatic channel without lateral flow are readily available in the literature (3,10):

$$A \frac{\partial V}{\partial x} + VT \frac{\partial y}{\partial x} + T \frac{\partial y}{\partial t} = 0 \qquad (1)$$

$$g \frac{\partial y}{\partial x} + V \frac{\partial V}{\partial x} + \frac{\partial V}{\partial t} + g(S - S_o) = 0 \qquad (2)$$

in which y = depth of water, V = the average velocity, A = cross-sectional water area, T = top width at the water surface, S = slope of energy gradeline, S_o = slope of bottom, g = acceleration due to gravity, x = horizontal distance, and t = time.

Equations (1) and (2) may be transformed using the method of characteristics (10,12) to the following form:

$$\pm \frac{g}{c} \frac{dy}{dt} + \frac{dV}{dt} + g(S - S_o) = 0 \qquad (3)$$

$$\frac{dx}{dt} = V \pm c \qquad (4)$$

in which $c = g\sqrt{A/T}$. The positive (negative)-valued compatibility equation, Eq. (3), is valid along the positive or forward (negative or backward)-characteristic line, Eq. (4).

The integration of Eq. (3) along the characteristic line requires an approximation since the celerity and friction slope are functions of the dependent variables. In this treatment, with the recognition that $dy = dA/T$, the positive-valued equations are written, for integration from R to P, Fig. 1, as follows:

$$\int_R^P dV + \int_R^P \sqrt{g/T} \frac{dA}{\sqrt{A}} + \int_R^P g(S - S_o)dt = 0 \qquad (5)$$

$$\int_R^P dx = \int_R^P (V + c)dt \qquad (6)$$

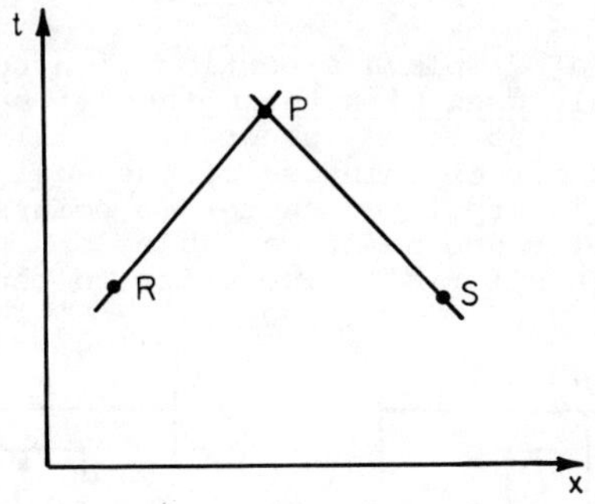

Figure 1. Characteristics in xt plane.

Integration yields

$$V_P - V_R + 4\sqrt{g}\left(\frac{\sqrt{A_P} - \sqrt{A_R}}{\sqrt{T_P} + \sqrt{T_R}}\right) + g\,\frac{(S_P + S_R - 2S_o)}{2}\,(t_P - t_R) = 0 \tag{7}$$

$$x_P - x_R = \frac{(V_P + c_P + V_R + c_R)}{2}\,(t_P - t_R) \tag{8}$$

In the evaluation of the second term in Eq. (5) exact integration is used for the area variation while the trapezoidal rule is used for the top-width variation. This is felt to be a reasonable approximation since, in all but the narrowest channels, the top width is a weaker function of water depth than the area. The trapezoidal rule is used in the last term, and in Eq. (6). With a similar treatment of the negative-valued equations the following set of four finite difference equations results.

$$F_P = V_P - V_R + 4\sqrt{g}\,\frac{\sqrt{A_P} - \sqrt{A_R}}{\sqrt{T_P} + \sqrt{T_R}} + \frac{g}{2}\,(S_P + S_R - 2S_o)\Delta T_P = 0 \tag{9}$$

$$\Delta T_P = \frac{2(x_P - x_R)}{V_P + c_P + V_R + c_R} \tag{10}$$

$$F_M = V_P - V_S - 4\sqrt{g}\,\frac{\sqrt{A_P} - \sqrt{A_S}}{\sqrt{T_P} + \sqrt{T_S}} + \frac{g}{2}\,(S_P + S_S - 2S_O)\Delta T_M = 0 \tag{11}$$

$$\Delta T_M = \frac{2(x_P - x_S)}{V_P - c_P + V_S - c_S} \tag{12}$$

RECTANGULAR GRID

Figure 2(a) displays a portion of a rectangular grid that is commonly used with the method of specified time intervals. A first order integration (12) of Eqs. (3) and (4) leads to a direct solution of the variables y and V at P in terms of interpolated values at points R' and S'. The values of the dependent variables are known at grid intersection points for all time prior to the current time

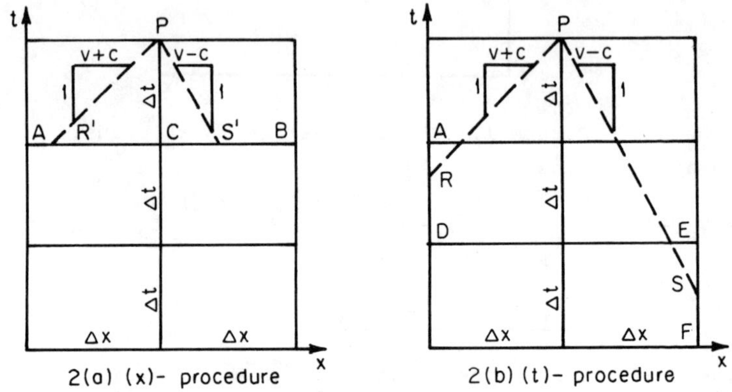

Figure 2. Rectangular Grid.

from initial conditions or from earlier calculations. Although there are variations, the usual method is to interpolate along the distance line ACB to find conditions at points R' and S' for use as known values in the integrated finite difference equations. A linear interpolation (12) between points A and C (C and B) is often used to find the values at R'(S').

A more precise solution procedure could use Eqs. (9)-(12) with the subscripts R and S in the equations referring to R' and S' in Fig. 2(a). This implies an iterative solution since the positions of R' and S' are functions of the solution at point P. The Newton-Raphson method could be used to solve the nonlinear equations. Although this method is numerically more accurate it does not remove the problems introduced by the interpolation procedure described above. Since interpolation errors are the root of the problems being addressed herein this type of interpolation in the x-direction at a particular instant will be referred to as the (x)-procedure.

As an alternative to the (x)-procedure, the characteristic lines that reach backwards in time from point P could be made to extend to the intersection with vertical time lines as shown in Fig. 2(b).

Values of the variables at points R and S could then be determined by interpolating in the time direction at specific channel sections (11). Thus a linear interpolation between points A and D (E and F) could provide the variables at points R(S) for use in Eqs. (9) to (12). This method is adopted herein and is referred to as procedure (t), for interpolations in the t-direction at a particular section. In many examples this procedure leads to dramatically improved results, as will be observed in the examples that follow.

DISCUSSION

A continuity balance over the entire system during the life of the transient provides a useful means of assessing the accuracy of the results of a numerical calculation. Although correct results are not assured in procedures that produce a good continuity balance, it is positive that incorrect results are generated by methods that do not satisfy mass conservation. If ν_i, ν_f represent the volumes stored in the channel at initial and final times, respectively, and ν_{in}, ν_{out} represent the volumetric inflow and outflow during the time T, then the continuity balance C_B may be evaluated at any time T after initiation of the numerical procedure.

$$C_B = \nu_{in} - \nu_{out} + \nu_i - \nu_f \tag{13}$$

A nondimensionalized parameter called the "percentage quantity surplus" (9) may be defined by dividing by a stored volume such as ν_i.

$$C_B^* = 100\ C_B/\nu_i \tag{14}$$

In some cases a dimensionless parameter normalized by the average through-flow volume may be more appropriate.

Procedure (x) is known to produce unacceptable continuity imbalances in some situations. A visualization of the problem may be obtained by examination of Fig. 3 which shows the profile for steady nonuniform flow in the physical plane. Linear interpolation for this case would yield depths and velocities at points R' and S' that would be larger than the actual values. The numerical solution would yield values at point P that would be too large, merely compounding the problem. If this form of interpolation does not produce numbers that match a steady nonuniform profile it follows that an iterative calculation in time will not converge to a stable nonuniform profile, nor will it maintain a steady condition in the channel while stepping forward in time. Not only will continuity not be satisfied but, with certain boundary conditions, the solution will ultimately drift to a totally unacceptable water surface condition. Various features minimize this scenerio, while others maximize the problem. Improvements are realized by minimizing interpolations and utilizing a finer

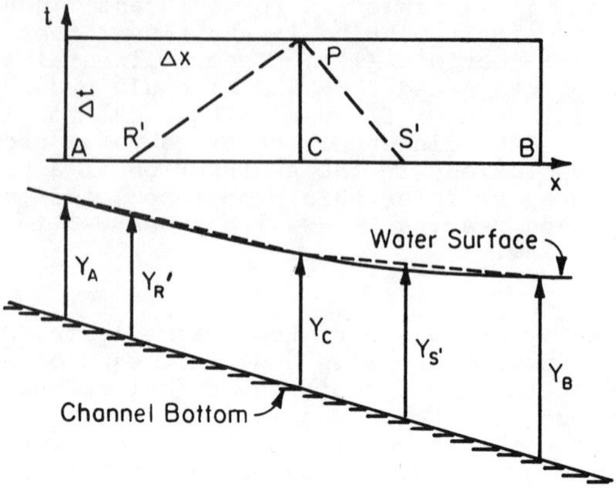

Figure 3. Linear interpolation in nonuniform flow.

spatial discretization. Of course there is no error in modeling steady uniform flow, so problems in which the profile is close to this condition are not nearly as significantly influenced by interpolation inaccuracies as others. It should be noted that some nonuniform profile shapes permit a loss of volume from the system while others generate mass during the numerical method.

APPLICATIONS

Numerical results are presented to illustrate the features of procedures (x) and (t). The example involves simulation of a flood wave in a trapezoidal channel: $B = 10$ m, $S_o = 0.0002$, Manning's $n = 0.022$, side slopes 0.5 horizontal to 1 vertical, length = 5000 m. Uniform flow, at a depth of 3.1 m and $Q_o = 37.92\ m^3/s$, is prescribed as the initial condition. A constant elevation reservoir is located at the downstream end of the channel, and an inflow hydrograph Q, is defined at the upstream end.

$$Q = Q_o + 30(t/20 \exp(1 - t/20))^{2.5} \qquad t \leq 20 \text{ min}$$

$$Q = Q_f + 30(t/20 \exp(1 - t/20))^{2.5} + (Q_o - Q_f)(t - 85)^2/4225$$

$$20 < t < 85 \text{ min}$$

$$Q = Q_f \qquad t \geq 85 \text{ min}$$

Two imposed hydrographs are examined, one with the final flow $Q_f = 48\ m^3/s$, and the second with the final flow $Q_f = 12\ m^3/s$. After passage of the stream rise each case results in steady nonuniform flow in the channel; the first is an M2 profile and the second an M1 profile. The inflow hydrograph is shown along with the computed downstream response in Figs. 4 and 6.

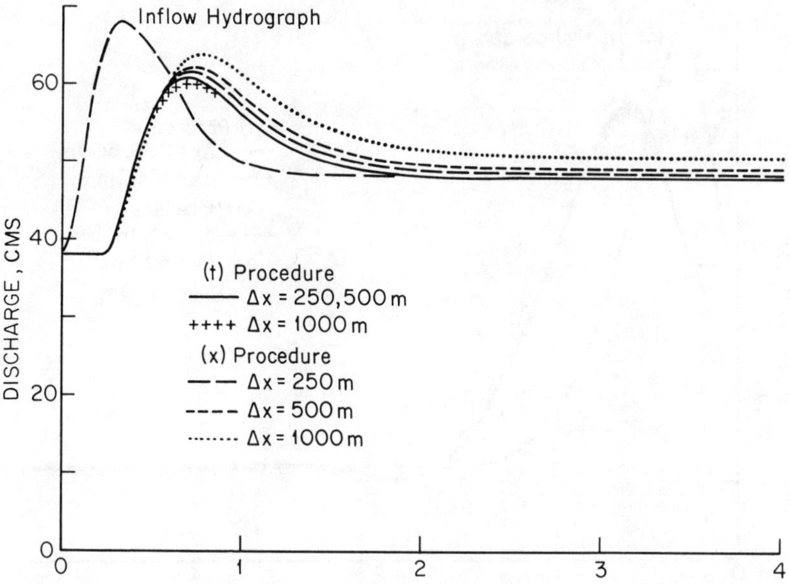

Figure 4. Discharge hydrograph, Q_f = 48 m^3/s.

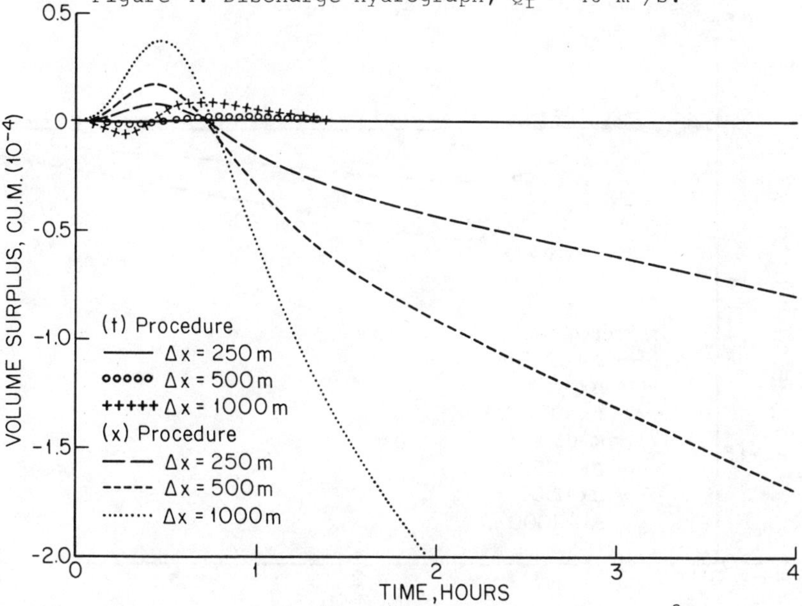

Figure 5. Volume surplus vs. time, Q_f = 48 m^3/s.

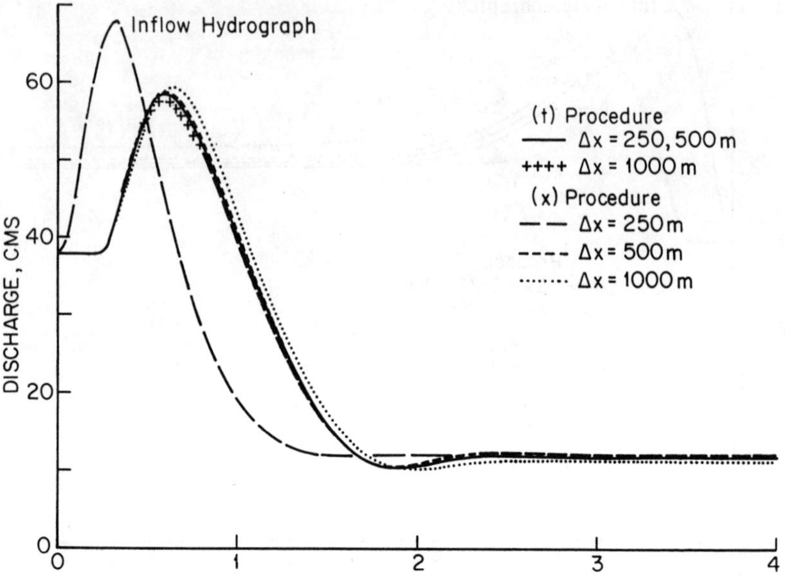

Figure 6. Discharge hydrograph, Q_f = 12 m^3/s.

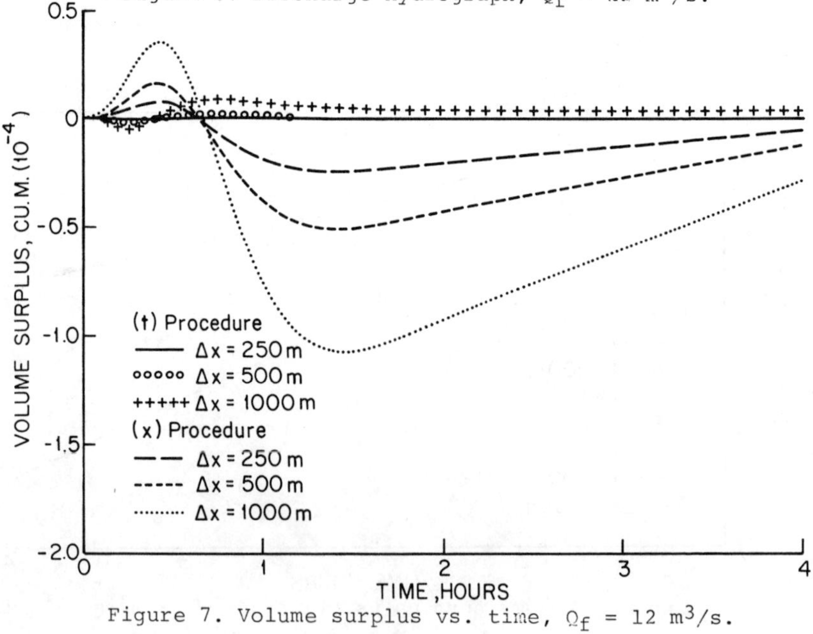

Figure 7. Volume surplus vs. time, Q_f = 12 m^3/s.

Each case was analyzed with 3 different levels of discretization, Δx = 1000 m, 500 m, and 250 m. Figures 4 and 6 show that the time-line interpolation procedure produced essentially the same results at each discretization level, while the distance-line results approach the time-line method when the shorter reach lengths were used. Four hours after initiation of the transient each numerical experiment showed essentially steady state in the channel. Table I shows the downstream computed outflows at 4 hours for each case for the distance-line method (x). The outflow equaled the inflow for each case with the time-line method (t).

TABLE I

Downstream Flows at 4 Hours, (x)-procedure.

Δx m	$Q = 48\ m^3/s$	$Q_f = 12\ m^3/s$
1000	50.61	11.11
500	49.08	11.58
250	48.49	11.79

Figure 5 shows the continuity imbalance, C_B, as it varies with time for the case of $Q_f = 48\ m^3/s$. The three levels of discretization for each method are indicated. As a percent of the total initial stored volume the maximum volume surpluses for each of the distance intervals of 1000 m, 500 m, and 250 m in the (x)-procedure are 2.09%, 1.09%, and 0.71%, respectively. The corresponding values in the (t)-procedure are 0.51%, 0.11%, and 0.03%. A minimum volume surplus is never established in the (x)-procedure since fluid volume is being lost continuously during steady state simulation at 48 m^3/s. The rate is shown in Table II. Figure 7 and Table II provide the volume surplus information for the case of $Q_f = 12\ m^3/s$.

The rate of gain of volume during steady state simulation, expressed as a percentage of volume stored at steady uniform flow, for each numerical experiment with the distance-line procedure is provided in Table II. The (t)-procedure held steady state in each case.

TABLE II

Rate of Gain of Volume at Steady State in %/hr, (x)-procedure.

Δx m	$Q = 48\ m^3/s$	$37.92\ m^3/s$	$12\ m^3/s$
1000	-5.25	0.	1.79
500	-2.17	0.	0.48
250	-1.03	0.	0.42

It has been pointed out in the past (4,12) that the distance-line procedure improves considerably in accuracy with finer discretization. However it is clear in these examples that the time-line procedure with the coarsest discretization produces results superior to the (x)-procedure with the finest discretization. The (t)-procedure utilized in this study used an iterative solution, while the (x)-procedure did not. The (x)-procedure followed the conventional first-order evaluation (12). This difference in approach no doubt contributed to some of the discrepancies in results, but not significantly. The iterative technique utilized with the (t)-procedure was 3.5 to 4 times more expensive in computer time than the direct (x)-procedure.

It should be noted that all potential numerical studies would not provide the striking discrepancies that were presented in this example. The differences generated by the distance-line procedure are less pronounced when flow conditions are near uniform flow, as already pointed out. The type of boundary conditions imposed on the problem also influence the results.

CONCLUSIONS

An alternative formulation of the specified-time intervals method of characteristics has been presented which involves interpolations along time-lines rather than the more conventional distance-line interpolations. Two variations of an example of flood-wave propagation in a prismatic channel illustrate the merit of the procedure. The method holds steady nonuniform flow, and preserves a continuity balance over the entire channel during iterations in the time domain. Each of these concerns has been known to be troublesome with the more traditional distance-line interpolation method.

ACKNOWLEDGEMENT

The research described herein was supported in part by the National Science Foundation Agreement NO. 2931-2 to The University of Michigan.

APPENDIX I - REFERENCES

1. Baltzer, R. A., and Lai, C., "Computer Simulation of Unsteady Flows in Waterways," Jour. of the Hydraulics Div., ASCE, Vol. 94, No. HY4, July, 1968, pp. 1083-1117.

2. Isaacson, E., Stoker, J., and Froesch, A., "Numerical Solutions of Flow in Rivers," Jour. of the Hydraulics Div., ASCE, Vol. 84, No. HY5, Oct., 1958, pp. 1810-1818.

3. Henderson, F. M., OPEN CHANNEL FLOW, Macmillan Co., New York, N.Y., 1966.

4. Jolly, J. P. and Yevjevich, V., "Simulations Accuracies of Gradually Varied Flow," Jour. of Hydraulics Div., ASCE, Vol. 100, No. HY7, July, 1974, pp. 1011-1030.

5. Liggett, J. A., and Woolhiser, D. A., "Difference Solutions of the Shallow Water Equations," Jour. of the Engineering Mechanics Div., ASCE, Vol. 93, No. EM4, Aug., 1967, pp. 39-71.

6. Lister, M., "The Numerical Solutions of Hyperbolic Partial Differential Euqations by the Method of Characteristics," Mathematical Methods for Digital Computers, Ralston and Wilf, Eds., Wiley and Sons, 1960.

7. Mahmood, K. and Yevjevich, V., Eds., UNSTEADY FLOW IN OPEN CHANNELS, Vols. I and II, Water Resources Publications, Fort Collins, Colo., 1975.

8. Ponce, V. M., and Simons, D. V., "Shallow Wave Propagation in Open Channel Flow," Jour. of the Hydraulics Div., ASCE, Vol. 103, No. HY12, Dec., 1977, pp. 1461-1476.

9. Sivaloganathan, K., "Flood Routing by Characteristic Methods," Jour. of the Hydraulics Div., ASCE, Vol. 104, No. HY7, July, 1978, pp. 1075-1091.

10. Strelkoff, Theodor, "Numerical Solutions of Saint-Venant Equations," Jour. of the Hydraulics Div., ASCE, Vol. 96, No. HY1, Jan., 1970, pp. 223-252.

11. Wylie, E. B., "Free Air in Liquid Transient Flow," Third International Conference on Pressure Surges, BHRA, Canterbury, England, Mar., 1980.

12. Wylie, E. B. and Streeter, V. L., FLUID TRANSIENTS, McGraw Hill Book Co., New York, N.Y. Chapter 16, 1978.

APPENDIX II - NOTATION

The following symbols are used in this paper:

A = area of cross-section below water surface;

c = $\sqrt{g\ A/T}$, wave celerity;

C_B = surplus continuity imbalance;

g = acceleration of gravity;

n = Manning's roughness coefficient;

P = point in xt plane at which conditions are unknown;

Q = discharge;

Q_o, Q_f = initial and final discharges respectively;

R, S = points in xt plane at which conditions are known;

S = slope of energy grade line;

S_o = slope of channel bottom;

T = width of water surface;

t = time

V = average velocity at a cross-section;

x = distance along channel;

y = depth of water at a cross-section;

Δx = distance interval.

TECHNIQUES AND EXPERIENCES IN THE UTILIZATION OF UNSTEADY OPEN-CHANNEL FLOW MODELS

by Chintu Lai [1/], M. ASCE, Robert A. Baltzer [1/], M. ASCE, and Raymond W. Schaffranek [1/]

ABSTRACT

Five attributes which are essential to a model intended for operational simulation of unsteady flow in open channels are set forth. The capability to simulate a wide range of unsteady flows and the flexibility to permit schematization of a diversity of complex prototype conditions are attributes which depend largely on selection of a sufficiently comprehensive set of equations with which to describe prototype flow conditions in open channels. Model stability, numerical reliability, accuracy, and operational efficiency are largely dependent upon the numerical computation scheme selected and upon its skillful interpretation and transformation into a sound, replete computer program. The functional utility of a model depends upon its ability to interact with an operational, user-oriented, data-handling system designed to facilitate modeling.

Experience with models underscores the wisdom of relying upon a comprehensive set of equations to depict open-channel flow. Such a set of equations is presented and the significance of the various terms is indicated. Examples illustrating the computation of tidal flow, flood flow, and regulated flow (which typically represent rapidly varied reciprocating flow, gradually varied unidirectional flow, and a broad spectrum of transient flows, respectively) are given to demonstrate the broad applicability of selected comprehensive models. From an operational viewpoint the critical, user-oriented capabilities of a functional flow model are identified in terms of attributes which provide flexibility in both input of the required prototype data and output of computed results.

INTRODUCTION

The advent of the modern-day computer has opened a new chapter in the history of hydraulics. Not only has the computer afforded new and powerful means by which to solve traditional problems, it has also kindled an awareness of new or previously ignored hydraulic problems. Moreover, it has stimulated creative rethinking with respect to certain basic concepts and accepted understandings of flow phenomena. As a result, a new discipline, computational hydraulics, has emerged, through which many old and new hydraulic problems, previously beyond the reach of the hydraulic engineer, have now become tractable.

[1/] Research Hydrologist, National Center, U.S. Geological Survey, Reston, Virginia

Among the first of the hydraulic problems to attract the attention of scientists and engineers using computer techniques was computation of various unsteady flows, particularly unsteady open-channel flow. This field of study offers important and challenging problems, as well as numerous immediate opportunities for practical applications. Research and investigation of such problems have been actively pursued by the U.S. Geological Survey for more than two decades. These studies into the various aspects of unsteady, open-channel flow have resulted not only in a series of technical publications, but also in a set of powerful, efficient, problem-oriented, computer flow-simulation models. They have also resulted in a body of techniques and experiences which impart direction to current and future research and development activities in flow modeling.

Emphasis in unsteady, open-channel, flow-model development over the past two decades has been focused principally upon the development and testing of various mathematical-numerical simulation schemes. Three such schemes were presented by two of the writers in a paper delivered at the Fifteenth Hydraulic Speciality Conference, ASCE (1). Development and improvement of these simulation schemes and others have been achieved both by Survey personnel and by others, as is evident in the recent literature (2,3). However, while the choice of a sound mathematical-numerical technique is of great importance and a necessary consideration in simulation model development, it is not a sufficient consideration. Lest one forget, simulation models are not an end in themselves. Their intended purpose is to depict hydraulic phenomena and to provide information, insight, and data to a model user and ultimately to a decision maker. Other elements in model development must be considered in order to achieve models which are both usable and useful.

The intent of this paper is to set forth various elements that experience has shown to be crucial in the design, development, and utilization of unsteady, open-channel, flow simulation models. The discussion is limited principally to subcritical flow conditions and particular emphasis is placed upon the flexibility and wide range of utility achieved with models based upon the complete set of dynamic, open-channel flow equations. Consideration is also given to the impact of the type, format, and availability of prototype boundary-value data on the design and utilization of simulation models and to the impact of rapid advances in computer systems. Examples illustrating the wide applicability and utilization of the two types of flow models designed by the writers are presented.

ESSENCE OF A USABLE FLOW MODEL

If one were asked to set forth those attributes which are most essential in a model designed for operational simulation of unsteady flow in open channels, the list would include the following:

1) the ability to simulate, with minimal distortion, the wide range of unsteady flows--flood flows, tidal flows, and regulated flows--encountered in prototype channels;

2) the adaptability to permit schematization of a diversity of complex prototype conditions; for example, variable channel

conveyance and cross-sectional properties, channel overbank storage, lateral inflows, branching channels and networks of channels, to name but a few;

3) the ability to generate accurate results repeatably by means of a stable, convergent, and numerically reliable computational scheme;

4) the ability to provide a high degree of computational efficiency whether used for short-term special studies or long-term routine operations; and

5) the ability to facilitate functional, user-oriented modeling by interacting with an operational data-handling system.

Other considerations of importance are the ease with which the model may be implemented and maintained on a computer system or adapted to operate in the present-day, rapidly changing, computer environment. Such considerations depend in large measure on the organizational structure and modularity inherent in the basic program design, on the programing language used, and on the scope and depth of program documentation provided. These attributes and considerations, then, are the essence of a usable flow model. The degree to which they exist differentiates between those models that are operationally and economically viable and those that are not.

MODEL FORMULATION CONCEPTS

The completeness of the set of partial-differential equations used to represent flow in open channels largely determines whether or not the model has the ability to simulate a wide range of unsteady flows and whether it can be adapted to represent various complex channel configurations. Likewise, the choice of a numerical computation scheme and the skillful implementation of this scheme as a replete computer program largely determines whether or not the model will be stable, convergent, accurate, and computationally efficient. Thus, the choice of both the equation set and the computational scheme must be exercised with understanding and insight.

Mathematical Aspects

Although unsteady open-channel flow is usually represented by the equations of continuity and motion, various choices do exist with respect to the dependent variables used and to the inclusion of specific terms. These choices must be based upon factors such as the purpose of the simulation, the type of flow, the complexity of the channel-system, the required accuracy of the simulation, and the nature and availability of field-data. Although some designers and users of models may argue otherwise, the writers favor using comprehensive and sophisticated numerical models employing the full dynamic equations to be set forth subsequently. The reasons for this choice are many: 1) most natural waterways are far more complex than can be depicted by a highly idealized flow model of inherently limited capability; 2) schematization to represent a particular waterway is usually more difficult when using an over-idealized flow model and sometimes results in irrational or inconsistent parametric coefficients; 3) calibration of an over-idealized model is often difficult, expensive to achieve, and subsequent model verification is frequently inconclusive; 4) the level of support needed to implement, maintain, and document several models, each of limited scope, is considerably more cumbersome and costly than that required for a single comprehensive model; 5) the difference in computer costs, when using up-to-date computing systems to execute models of different completeness, is relatively small--and decreasing--in comparison with the total costs associated with undertaking an entire modeling effort; and 6) the cost of collecting, processing, and making prototype boundary-value data available for modeling is no cheaper for models of limited scope than for a comprehensive model.

For illustration, a comprehensive set of one-dimensional flow equations is shown in the following form,

$$\overset{(1)}{\frac{\partial Z}{\partial t}} + \overset{(2)}{u\frac{\partial Z}{\partial x}} + \overset{(3)}{\frac{A}{B}\frac{\partial u}{\partial x}} + \overset{(4)}{\frac{u}{B}\frac{\partial A}{\partial x}\Big|_H} + \overset{(5)}{uS_o} - \overset{(6)}{\frac{q}{B}} = 0 \qquad (1)$$

$$\overset{(7)}{\frac{\partial u}{\partial t}} + \overset{(8)}{u\frac{\partial u}{\partial x}} + \overset{(9)}{g\frac{\partial Z}{\partial x}} + \overset{(10)}{\frac{g}{C^2R}u|u|} + \overset{(11)}{\frac{qu}{A}} - \overset{(12)}{\xi\frac{V_w^2}{H}\cos\theta} = 0, \qquad (2)$$

in which longitudinal distance, x, and time, t, together with water-surface elevation (stage), Z, and flow velocity, u, are used as two independent and two dependent variables, respectively. Other parameters, coefficients, and constants are: A = cross-sectional area, B = top width of cross section, H = depth of flow, S_o = channel-bottom slope, q = lateral inflow per unit length, g = gravitational acceleration, C = Chézy's coefficient, R = hydraulic radius, ξ = wind-stress coefficient, V_w = wind speed, and θ = angle of wind direction with the positive x-axis. These equations represent mass and momentum conservation in a channel without overbank storage and in which the dependent variables and parameters are cross-sectional-averaged quantities. They can, of course, be written for an incremental reach-length having both conveyance and storage cross sections. Moreover, the Boussinesq coefficient, used to compensate for nonuniform velocity distribution in the conveyance cross section, may be incorporated easily.

In a very wide channel or estuary both the Coriolis force and the wind force can become perceptible in the transverse direction, y. Under such circumstances representation of the prototype flow by a set of one-dimensional equations may be inappropriate. In order to better appraise these circumstances the following associated force-balance equation in the transverse direction can be written:

$$\overset{(13)}{g\frac{\partial Z}{\partial y}} + \overset{(14)}{2\,\omega_z u} - \overset{(15)}{\xi\,\frac{V_w^2}{H}\sin\theta} = 0. \qquad (3)$$

Here the first term, (13), represents the force due to the transverse water-surface gradient (super elevation) which counteracts the Coriolis force, (14), and the transverse wind force, (15). Careful examination of the significance of each term in Eq. 3 under prototype conditions can aid in determining whether or not an areal, two-dimensional set of equations (and hence, two-dimensional flow model) may be appropriate.

Analysis of the terms in the one-dimensional equations affords insight into their useful application. By definition a transverse component of flow velocity does not exist for one-dimensional flow, and hence, a Coriolis force in the longitudinal direction, being a function of the transverse velocity, does not exist. Term ⑫ becomes significant when a strong, constant wind persists along a relatively straight channel. The effect of the lateral inflow, represented by terms ⑥ and ⑪, is sometimes underestimated. It can, however, play a predominant role under certain conditions. If the wind stress term, ⑫, and the lateral inflow terms, ⑥ and ⑪, are neglected and if the channel is considered to be approximately prismatic, that is, term ④ is negligible, Eqs. 1 and 2, with the remaining terms, ①, ②, ③, ⑤ and ⑦, ⑧, ⑨, ⑩, respectively, represent a commonly encountered form of the unsteady, open-channel flow equations.

If term ⑦ in Eq. 2 is omitted the resulting equation consisting of terms ⑧, ⑨, and ⑩ is applicable for steady, non-uniform flow. Remembering that $Z = H + Z_b$, in which Z_b is the channel bottom elevation, and that term ⑩ may be written as gS_f in which S_f is the energy gradient, one can express the steady, non-uniform flow equation as the familiar backwater-curve equation. For uniform flow, term ⑧ is neglected and the two remaining terms, ⑨ and ⑩, form the Chézy equation. Likewise, in the continuity equation, the removal of the transient term ①, leaves only terms ②, ③, and ⑤, which form the steady-flow, continuity equation, $\partial Q/\partial x = 0$. For uniform flow, it reduces to the form $\partial H/\partial x = 0$.

If one prefers, flow equations, in which discharge, Q, and water level, Z, are the dependent variables, may be written as

$$B \frac{\partial Z}{\partial t} + \frac{\partial Q}{\partial x} - q = 0 \tag{4}$$

$$\frac{\partial Q}{\partial t} + \frac{Q}{A}\frac{\partial Q}{\partial x} + Q\frac{\partial (Q/A)}{\partial x} + gA\frac{\partial Z}{\partial x} + \frac{gk}{AR^{4/3}} Q|Q| - \xi B V_w^2 \cos\theta = 0 \tag{5}$$

in which frictional resistance is expressed in Manning-like form by

$$k = \left(\frac{\eta}{1.49}\right)^2 \quad \text{(or in the metric system, } k = \eta^2\text{).}$$

In this expression, η is the unsteady flow complement of Manning's "n". Equations 4 and 5 are comparable to Eqs. 1 and 2 in every respect and may be similarly analyzed as to the meaning and significance of each term.

The above discussion reemphasizes the fact that many of the traditional and commonly accepted means for analyzing open-channel problems are premised on rather simplified versions of the fundamental flow equations. There is nothing wrong with such approaches provided, of course, that one does not lose sight of their inherent limitations and attempt to use them where they are inappropriate. The above discussion also makes evident that use of a comprehensive set of the dynamic flow equations will afford more realistic representation of a broader range of open-channel flow, than would an abbreviated set.

Numerical Aspects

Many numerical methods have been employed by hydraulic engineers for simulation of unsteady open-channel flow. New methods and better schemes are continually being sought. Some of the methods used by the writers have been reported previously (1). These and other methods used by the U.S. Geological Survey are identified elsewhere (3). For practical application, for broad utilization, and for implementation into operational modeling systems, the writers find the finite-difference approach (in which the rectangular x-t grid is employed) most useful. They particularly rely on the four-point, weighted implicit scheme (8) and on the specified-time-interval, characteristics scheme (4) for simulating a wide range of varied flow conditions. The finite-difference approach is particularly well suited to handling time-dependent, boundary values and other similar time-series data. Because most of the field, data-collection stations operated by the U.S. Geological Survey are designed to automatically record time-series of discrete data at fixed time intervals, these data are readily suitable for model utilization. They are processed and filed in a data base without prior consideration of the numerical method to be employed.

Stability analyses, error criteria, and convergence criteria are important factors in assessing and selecting numerical schemes, and in utilizing flow models. Of particular concern are the differing wave propagation characteristics of numerical schemes. Despite the desirability of such information, relatively little has been done in this area of numerical analysis, particularly with regard to nonlinear systems. For the practical utilization of models, the criteria developed for linear cases, where they exist, have been borrowed as (local) guides for nonlinear computation. This practice may be both complemented and tempered by the more practical aspects of the modeling effort: realistic schematization of the prototype; proper selection of initial and boundary-value data; rational model calibration and adjustment; selected parameter sensitivity analyses; and controlled model verification.

APPLICABILITY OF FLOW MODELS

A flow model based on the comprehensive, dynamic flow equations described above and fulfilling various requirements identified previously can be utilized for a variety of unsteady flow investigations. These include tidal flow, flood flow, and regulated flow, which typically represent rapidly varying reciprocating flow, gradually varying unidirectional flow, and a broad spectrum of regulated transient flows, respectively.

Illustrated in Fig. 1 is a simulation of the tide-affected flow in the lower Hudson River. Tidal influence in this river is present as far upstream as Green Island Dam near Albany, New York, a distance of 80 miles (129 km) from the mouth. For this illustration, the flow was computed by the implicit method for a reach only 21.4 miles (34.4 km) in length which extends from the Rondout Creek tributary downstream to Clinton Point located downstream of Poughkeepsie. Computed discharge hydrographs at four locations including the upstream and downstream ends of the reach are plotted in this figure. The discharge hydrographs

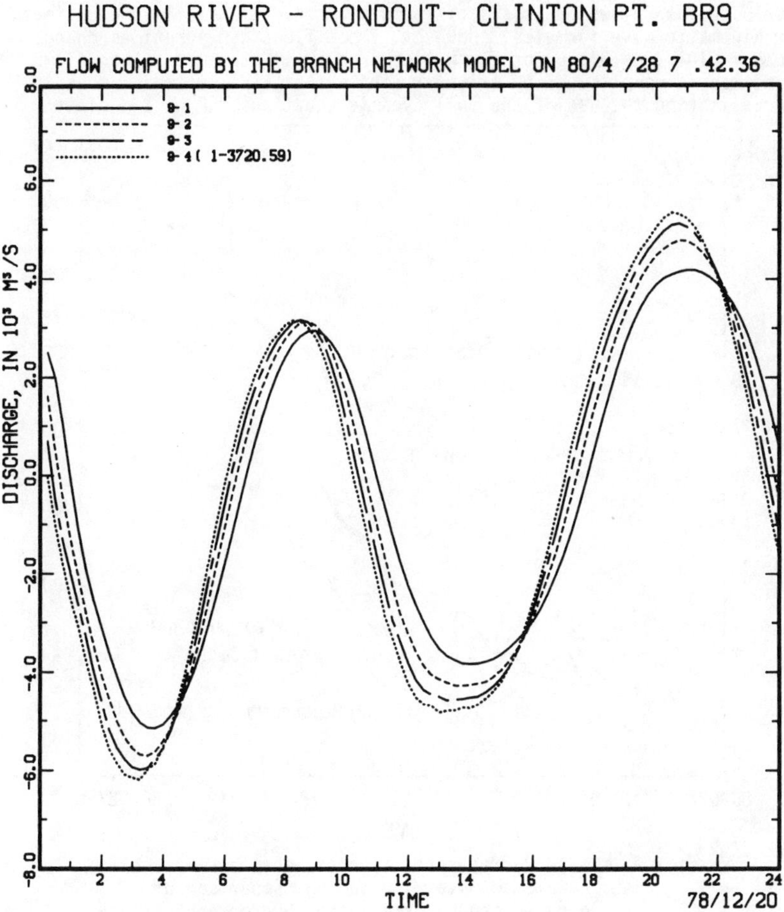

Fig. 1.--Computed Discharges at Four Locations on the Hudson River Between Rondout Creek and Clinton Point.

identified as 9-1 and 9-4 are for the Rondout Creek tributary and Clinton Point locations, respectively. The tidal effect in this reach is very strong. The positive (downstream) and the negative (upstream) flow hydrographs are similar in both magnitude and appearance. The manner in which the discharge varies along the length of the reach can also be ascertained from the illustration.

Because flood flows vary gradually, many hydraulic engineers still evaluate such flows by empirical formulas, conventional flood equations, or kinematic wave formulas. However, flood flows can sometimes change very rapidly, as occur for flash floods on small streams. Flood flows can also be complicated by tributary inflows or tidal motions, thus necessitating the use of the full dynamic equations. Fig. 2 depicts

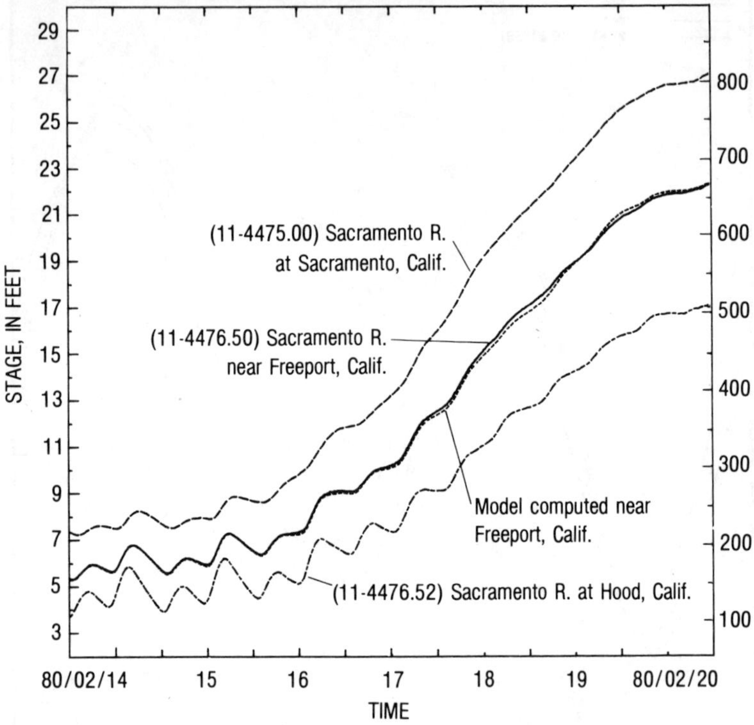

Fig. 2.--Stage Hydrographs at Three Locations on the Sacramento River, Including Comparison of Computed and Recorded Stage Hydrographs at Freeport.

sharp rises of water-surface elevations due to a flood observed at three water-level gaging stations (Sacramento, Freeport, and Hood) located along a reach of the Sacramento River in California. The unsteady flow in this reach was computed using a comprehensive, method-of-characteristics model. The model-computed stage hydrograph is also compared in Fig. 2 with the hydrograph recorded at Freeport. It is interesting to observe that the tidal effect, evident at low flow, is quickly suppressed as the flood stages rise. The flood stages continued to rise after February 20, 1980 (although not shown in the figure), and would have peaked early February 22, 1980, had not a controlled diversion been imposed. (Another example of flood flow, in this instance flash-flood flow on a small stream, Pheasant Branch near Middleton, Wisconsin, is given in Fig. 4 and discussed subsequently.)

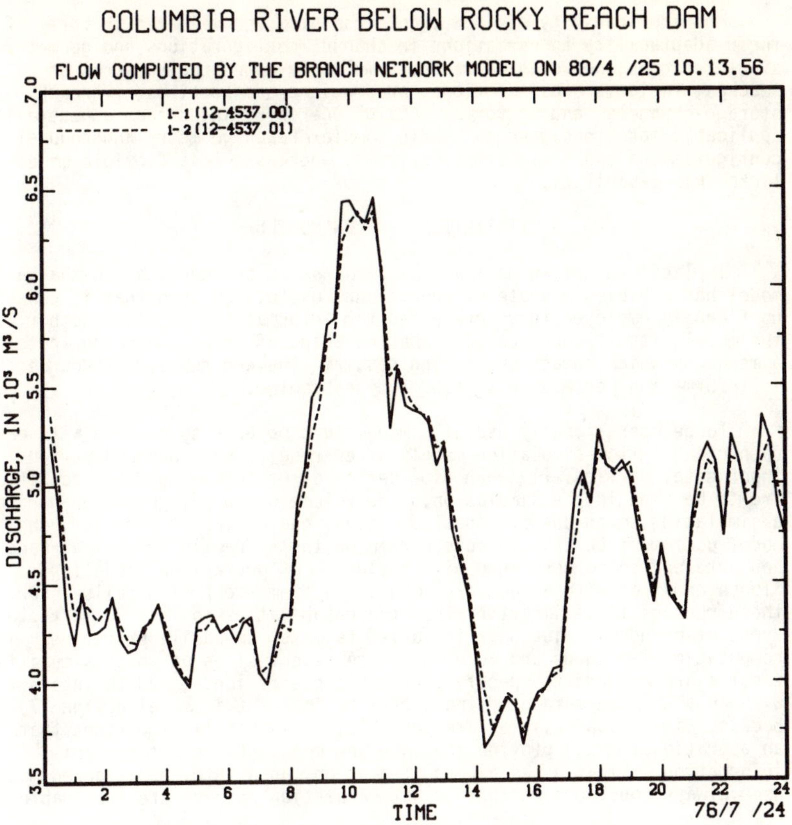

Fig. 3.--Computed Discharges for the Columbia River below Rocky Reach Dam.

Fig. 3 offers an example of the third type of unsteady flow, regulated flow in a reach of the Columbia River below Rocky Reach Dam near Wenatchee, Washington. The simulated discharge hydrographs for two field stations, 1.91 miles (3.07 km) apart, exhibit a distinct flow pattern different from either of the above flow types. The implicit method was used in the flow computation.

All three identified types of unsteady flow can be simulated effectively by any comprehensive model that is formulated on the full set of dynamic flow equations. If one anticipates that changes in flow type would take place in a particular reach during a simulation time period, such as was shown to have occurred in Fig. 2, then it is important to insure that the model to be used has the capability to handle the different flow types.

The applicability of flow models can also be evaluated in terms of their adaptability to variations in channel configurations and geometry--single reaches, networks of reaches, multiply-connected networks of reaches, variable cross sections, prismatic cross sections, overbank storage channels, and so forth. Again, one is reminded that a generally applicable model designed to handle complex reach geometry and channel configurations can treat simple systems, whereas a less flexible model lacks this capability.

UTILIZATION OF FLOW MODELS

Implicit in the words *model utilization* is the connotation that a model has achieved a state of operational usefulness such that it can be confidently employed to generate desired information. Stated another way, the model itself has ceased to be the point of focus, becoming instead a means by which to achieve an end result. The end result, of course, is information pertaining to a prototype problem.

To be operationally useful a model must be able to be used with ease. Open-channel flow-simulation models often require numerous and varied input data, particularly when time-series of prototype boundary conditions are to be used in the simulation. The volume of output generated during a simulation, even one of modest duration, can be vast. Moreover, the model user must be able to concentrate on the hydraulic problem at hand and not need to be concerned with a clutter of operational modeling "loose ends" or with a host of computer, systems-control details. For these reasons it is important that the capabilities to include multiple types of boundary-value data in varied formats, to easily prescribe computational methods and accuracies, to select sites at which simulation results are needed, to specify parametric coefficients, and to interact with an automated data-handling system be inherent in model design (7). Because simulation failures are possible, it is vitally important that an operational model provide adequate and meaningful error analysis information. Finally, as a word of caution, an astute, experienced model user always considers whether or not simulation results are reasonable.

Unsteady flow models can be used to investigate a variety of practical field problems. Flow models are being used to provide comprehensive flow information in support of complex interdisciplinary riverine and estuarine investigations, to appraise project design alternatives, and to support environmental impact appraisals. Selected examples of practical use of flow models by the writers can be cited. An investigation and analysis of the variations in discharge due to changes in flow-resistance was conducted using a model of Three Mile Slough near Rio Vista, California (4,5). Use of unsteady flow models in time-of-travel, wave crest movement, and injected particle motion studies have been demonstrated in separate investigations of the Connecticut River near Hartford, Connecticut, and of the Sacramento River near Sacramento, California (6). These examples illustrate flow-model utilization for short-term, special purpose studies.

However, in order for a model to be considered truly usable it must also accommodate long-term operational utilization. The writers' flow models are (or have been) in operational use on reaches of the Columbia River, Connecticut River, Delaware River, Sacramento River, and Pheasant Branch, a small stream near Middleton, Wisconsin. Their use at several other sites is being evaluated. The range of flow conditions illustrated by these examples varies from large-scale estuarine flows to small-scale riverine flows, thus emphasizing the flexibility required of a usable model.

The need to supply a wide range of riverine and estuarine flow information emphasizes perhaps the most critical user-oriented capabilities of a functional model, namely the provision for flexibility in both input of the required prototype data and output of computed flow results. The following three examples were produced by a flow model which possesses these desirable attributes. In each of these simulations, the required input boundary-value data were supplied to the model via an integrated data-handling routine which retrieved the data from a direct-access-type computer data base. Without such an automated mechanism undertaking these simulations would have been cumbersome, time consuming, and costly.

Figure 4 illustrates the use of a model to simulate flood flow in an extremely short reach. The reach, Pheasant Branch near Middleton,

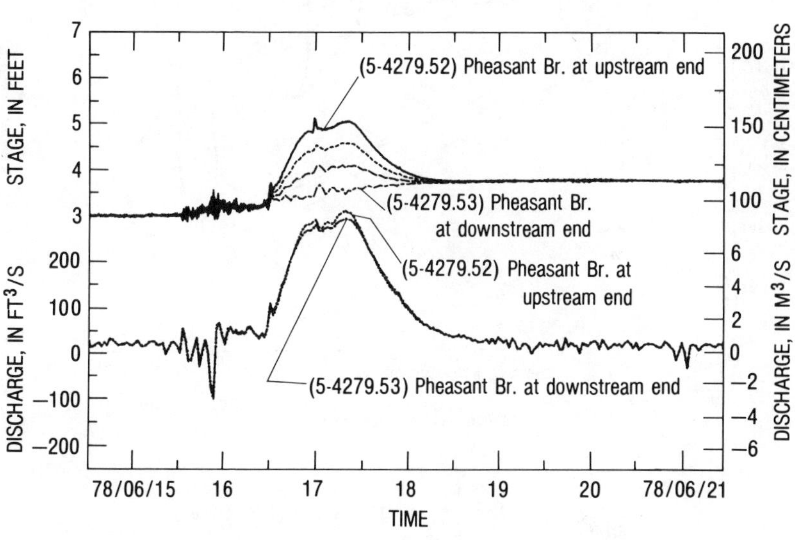

Fig. 4.--Computed Stage Hydrographs and Computed Discharge Hydrographs for Pheasant Branch, Near Middleton, Wisconsin.

Wisconsin, begins in marshland and ends 900 feet (274 meters) downstream at Lake Mendota. Backwater effects from this lake and storm-generated, transient flows necessitate the use of unsteady flow analysis. On June 17, 1978, a flood occurred which produced a sharp rise of stage and discharge. A simulation spanning the week including the flood was conducted using stage hydrographs recorded at the ends of the reach as boundary conditions. Model-generated stage hydrographs at four equidistant points, including the ends of the reach, and discharge hydrographs at the upstream and downstream ends, are portrayed in figure 4. It can be observed from these simulation results that such a small reach is extremely responsive and sensitive to changing climatological conditions.

Depicted in figure 5 are stage-discharge curves (loop-rating curves) for a 35 mile (56 km) reach of the Connecticut River. As can be seen, the

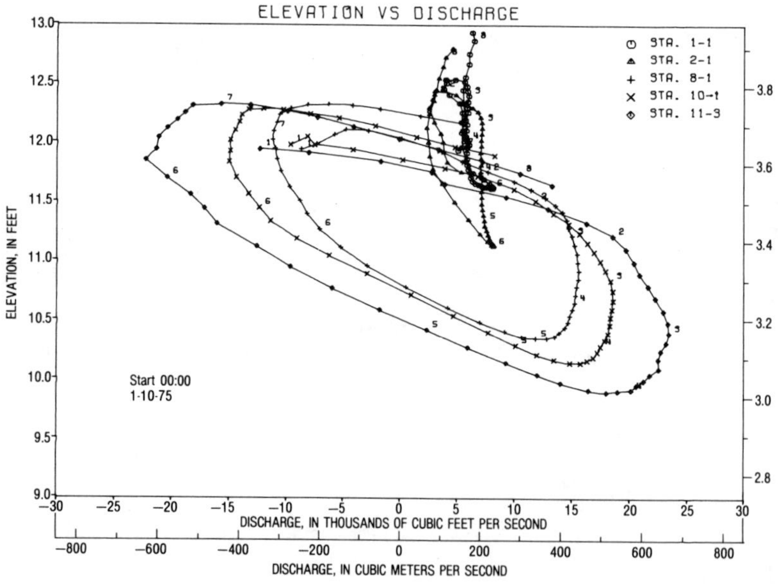

Fig. 5.--Curves of Computed Water-Surface Elevation Versus Discharge at Five Locations for the Lower Connecticut River.

simulation-generated loops expand as one views them in downstream order (from station 1-1 to station 11-3) as a result of the tidal effect becoming more influential. The long axis of the tidal loop extends from the lower-right to upper-left of the figure, as contrasted with that of the familiar riverine flood loop (lower-left to upper-right). This difference is due to the fact that the tide wave is propagating upstream from the ocean whereas the upland flood wave propagates downstream.

Of particular interest in this figure are several curve segments that remain nearly vertical. These indicate that although the stage is changing markedly ($\partial Z/\partial t$ has a large magnitude) the discharge is remaining essentially constant ($\partial Q/\partial t$ is nearly zero). The interplay between tidal and upland flow may be readily perceived throughout the length of this reach from the figure.

Table 1 gives a list of alternating flow-volumes occurring daily for the Hudson River at Poughkeepsie for the period October 17 to October 31, 1967. As can be seen, the tidal motion at this point is sufficiently strong to generate very large volumes of reverse flow. Tidal volume exchanges such as are presented in this table can be of value in appraising the flushing characteristics of rivers and estuaries.

OCTOBER 1967 FLOW IN MILLIONS OF CUBIC FEET AT POUGHKEEPSIE, N.Y.

DAY	FLOW	FLOW	FLOW	FLOW	FLOW
17	-347	949	-637		
18	-135	791	-918	805	-612
19	-178	1322	-459	1105	-213
20	-268	712	-814	669	-307
21	-379	717	-725	929	-158
22	-458	675	-827	764	-62
23	-380	417	-867	676	0
24	-490	525	-761	859	
25	38	-608	621	-791	609
26	282	-2	559	-717	557
27	387	-579	451	-675	431
28	605	-484	716	-775	215
29	701	-1402	447	-950	
30	-77	64	-1652	564	-975
31	249	-1520	830	-875	

Table 1.--Model-Produced Table of Accumulated Flow Volumes at Poughkeepsie, New York, for October 17-31, 1967.

These examples serve to illustrate three ways in which model results can be compiled and depicted to facilitate flow analyses. Numerous other output analysis techniques are available. Their ready utilization, however, is dependent upon a tractable, user-oriented simulation capacity to produce the basic flow information.

SUMMARY

The advent of computer methods has stimulated the study of various unsteady flows, particularly unsteady open-channel flow. While initial research emphasis was focused principally on the development and testing of various mathematical-numerical simulation schemes, attention is now being directed toward research and development leading to easily used, operational models designed to provide reliable information, insight, and flow data to a model user and ultimately to a decision maker. The principal attributes which are the essence of a usable flow model are identified: 1) the ability to simulate a wide range of flow conditions;

2) the adaptability to permit schematization of channels having diverse prototype geometry; 3) the ability to generate stable, accurate results repeatably; 4) the ability to provide a high degree of computational efficiency; and 5) the ability to facilitate functional, user-oriented modeling.

Experience with model design and implementation affirms the wisdom of relying on a comprehensive set of one-dimensional flow equations to achieve the desired attributes. Analysis of the terms in the equation set presented affords insight into their useful application. For practical application, for broad utilization, and for implementation into operational models, the writers find the finite-difference approach using the rectangular x-t grid most useful. They rely on the four-point, weighted implicit scheme and on a specified-time-interval, charactertistic scheme for simulating a wide range of varied flow conditions. Examples of model-generated tidal flows, flood flows, and regulated transient flows for widely differing conditions affirm the versatility of the models. For models to be used easily it is important that the capabilities to accept multiple types of boundary-value data in varied formats, to easily prescribe computational methods and accuracies, to select and vary parametric coefficients, and to interact with an automated data-handling system be inherent in the model design. In summary, utilization of flow models increases rapidly when models are designed to be easily used.

REFERENCES

1. Baltzer, R. A., and Lai, C., Computer simulation of unsteady flows in waterways: Journal of the Hydraulics Division, ASCE, v. 94, no. HY4, 1968, pp. 1083-1117.

2. Basco, D. R., Chairman, Sources of computer programs in hydraulics, (Task Committee Report): Journal of the Hydraulics Division, ASCE, v. 106, no. HY5, 1980, pp. 915-922.

3. Jennings, M. E., and Yotsukura, N., Status of surface-water modeling in the U.S. Geological Survey: Geological Survey Circular 908, 1979, 17 p.

4. Lai, C., Computation of transient flows in rivers and estuaries by the multiple-reach method of characteristics: Geological Survey Research, Geological Survey Professional Paper 575-D, 1967, pp. D273-D280.

5. Lai, C., Numerical simulation of wave-crest movement in rivers and estuaries: Use of Analog and Digital Computers in Hydrology; IASH/AIHS-UNESCO, Publication No. 81, AIHS, v. II, 1968, pp. 699-713.

6. Lai, C., A computer simulation study of travel times of injected particles and tide-wave in well-mixed estuaries: XIII Congress of IAHR, v. 3, Kyoto, Japan, 1969, pp. 123-130.

7. Lai, C., Schaffranek, R. W., and Baltzer, R. A., An operational system for implementing simulation models, A case study: Seminar on Computational Hydraulics, Proceedings, ASCE Hydraulics Division, College Park, Md., 1978, pp. 415-454.

8. Schaffranek, R. W., Baltzer, R. A., and Goldberg, D. E., A model for simulation of flow in singular and interconnected channels: Chapter 3, Book 7, Automated Data Processing and Computations, TWRI, U.S. Geological Survey, 1980 (in press).

FLOW ROUTING IN TAILWATER STREAMS

by Michael G. Ferrick[1], A.M. ASCE

ABSTRACT

A diffusion wave flow routing model was developed for application to routing in tailwaters downstream of hydroelectric dams which are operated for peak power production. Flow depths in these streams typically approach zero after relatively short periods of zero flow release at the dam. The modeling approach taken for this problem class is explained and justified. The performance of the numerical model is then studied and tested using laboratory dam break and field tailwater data sets.

INTRODUCTION

In the Tennessee Valley, tributary dams have created more than 300 miles of regulated streams. These streams or tailwaters generally differ from unregulated streams chemically, physically, and biologically. Accurate knowledge of unsteady flow in these riverine tailwater reaches is an important component in a number of environmental studies being conducted by the Tennessee Valley Authority (TVA). Studies of temperature and water quality and the effects of minimum flow requirements on tailwater fish populations ("instream flow" investigations) are current examples.

Flow in unstratified or weakly stratified reservoirs and in rivers having a significant base flow is generally modeled using the one-dimensional St. Venant equations of unsteady flow in open channels. The TVA utilizes several flow routing models (2, 4) which numerically solve these equations. Standard numerical techniques used to solve the dynamic St. Venant equations fail, however, at a point where the flow depth approaches zero. This condition can occur frequently in tailwaters where flow is controlled by a dam which is operated to produce power during times of peak demand.

The flow regime of tailwater streams, as defined in this paper, is very different from that found in natural rivers. Hydropower peaking operations can result in large flow changes which can occur in a short period of time. The typical duration of a given flow release is on the order of a few hours, and can be less. Tailwater streams in the Tennessee Valley have lengths of up to 50 miles and subsidence of short period waves can be significant in this distance.

[1]Civil Engineer, TVA Water Systems Development Branch, Norris, TN

MODEL SELECTION

The model selected for flow routing in tailwaters belongs to the class of simplified models which have been termed diffusion wave models. The conservation of momentum equation is simplified by neglecting the local and convective acceleration terms. The abbreviated set of equations can be solved without difficulty when the flow depth becomes small, as long as the continuum assumption is not violated.

Simplified routing methods have apparently not been applied to the class of flow problems found in tailwaters. The approach taken, therefore, requires justification. The linear analysis of Ponce et al. (8, 9) can be used to gain insight into the applicability of simplified models to a given routing problem. Attenuation and propagation characteristics of the diffusion wave model were checked against those of the dynamic model for a range of physical and flow parameters characteristic of tailwaters. Errors resulting from the omission of the acceleration terms were generally small.

Henderson (6) presents an order of magnitude analysis of the terms in the momentum equation for a kinematic wave. Comparing the water surface slope and bottom slope terms, he concludes, for a wide rectangular channel that

$$\frac{\partial y/\partial x}{S_o} \; \alpha \; \frac{\partial \tilde{q}/\partial t}{\tilde{q}^{2/3} \; S_o^{\;5/3}} \qquad (1)$$

where $\tilde{q}(t)$ is a given hydrograph of discharge per unit width. (Definitions of all notation used in the text are given in Appendix II.) In natural streams with a significant bottom slope, the water surface slope term may be neglected. In tailwaters, $\partial\tilde{q}/\partial t$ can be very large near the wave front. The water surface slope term can therefore be expected to play an important role. Wave attenuation is another role of the surface slope term when the reach of interest is long compared to the duration of the flow release.

Henderson compares the water slope to each acceleration term for the same set of conditions and concludes

$$\frac{(\partial y/\partial x)}{(\text{acceleration term})} = 0\;(Fr^2) \qquad (2)$$

where Fr is the Froude number of the flow. The square of the Froude number is of the order 0.1 for tailwater flows of interest in this investigation. Again, errors resulting from discarding the acceleration terms appear small.

THEORETICAL BASIS OF THE MODEL

Much of the theoretical basis of the model follows the work of Koussis (7). Weinmann et al. (10) relate the Koussis formulation to other simplified routing techniques. For convenience, the development will be summarized here. The conservation of mass equation can be written as

$$\frac{\partial Q}{\partial x} + \left[\frac{dA(y)}{dQ} \right]_x \frac{\partial Q}{\partial t} = q \qquad (3)$$

where

$$c = \left[\frac{dQ}{dA} \right]_x = \frac{dx}{dt}$$

With the simplifications discussed, and by neglecting the momentum contribution of the local inflow, the momentum equation is

$$\frac{\partial y}{\partial x} - S_o + S_f = 0 \qquad (4)$$

Using Equation 4 and Manning's equation, an expression for the stream rating curve can be stated as

$$V = \frac{1.486}{n} R^{2/3} \left(S_o - \frac{\partial y}{\partial x}\right)^{\frac{1}{2}} \qquad (5)$$

From the standpoint of computational ease, it would be an improvement if the spatial derivative in Equation 5 could be replaced with a quantity calculated at a point. Arguments from kinematic wave theory allow Equation 5 to be rewritten in a form of the "Jones Formula" as

$$V = \frac{1.486}{n} R^{2/3} \left(S_o + \frac{1}{c^2 B} \frac{\partial Q}{\partial t}\right)^{\frac{1}{2}} \qquad (6)$$

Henderson (5) addresses the adequacy of this approximation. The solution of Equations 3 and 6 forms the basis of the model used for flow routing in tailwaters.

Using the following expressions

$$\frac{\partial Q}{\partial x} \cong \frac{1}{\Delta x} \left(Q_{j+1}(t) - Q_j(t)\right) \qquad (7)$$

$$\frac{\partial Q}{\partial t} \cong \Theta \frac{\partial Q_j}{\partial t} + (1 - \Theta) \frac{\partial Q_{j+1}}{\partial t} \qquad (8)$$

Equation 1 can be rewritten as a first order linear ordinary differential equation

$$\dot{Q}_{j+1}(t) + a\, Q_{j+1}(t) = a\, Q_j(t) + a\, \Delta x q(t) - \frac{\Theta}{(1-\Theta)} \dot{Q}_j(t) \qquad (9)$$

where

$$a = \frac{\langle \bar{c} \rangle}{\Delta x(1-\Theta)} \qquad (10)$$

is assumed constant for a given reach in a time step, Δt. Assuming $Q_j(t)$ varies linearly over a time step, the solution of Equation 9 is,

$$Q_{j+1}((n+1)\Delta t) = C_1 Q_j((n+1)\Delta t) + C_2 Q_j(n\Delta t) + C_3 Q_{j+1}(n\Delta t) + C_4 \quad (11)$$

where

$$C_1 = 1 - \frac{\Delta x}{<\bar{c}>\Delta t}(1-C_3)$$

$$C_2 = \frac{\Delta x}{<\bar{c}>\Delta t}(1-C_3) - C_3$$

$$C_3 = \exp(-a\Delta t)$$

$$C_4 = [q((n+1)\Delta t) - q(n\Delta t)\exp(-a\Delta t)]\Delta x$$

The function, Θ, used to weight the time derivatives can be expressed as

$$\Theta = \tfrac{1}{2}(1-<\bar{Q}>/(\bar{B}S_o<\bar{c}>\Delta x)) \quad (12)$$

with the restriction that

$$0 \leqq \Theta < 0.5$$

As evidenced by Equations 11 and 12, successful tailwater routing depends upon the determination of wave celerity, c. Continuity considerations reveal that the wave celerity of sharp-fronted tailwater releases is not equivalent to that of a typical flood wave. In the extreme case of a surge moving down a dry channel, the wave celerity must be equal to the velocity of flow immediately behind the wave front. The monoclinal rising wave is a translatory wave of stable form (6). To determine an expression for wave celerity, the tailwater releases can be modeled as monoclinal rising waves during passage through a river reach, Δx. Assuming a wide rectangular channel and using the Chezy equation to describe the friction slope, the expression for wave celerity is then

$$c = \left(\frac{1 - (y_1/y_2)^{3/2}}{1 - (y_1/y_2)}\right) V \quad (13)$$

where y_1 is the depth ahead of the wave, and y_2 is the depth behind the wave. For the case of slowly rising floods, Equation 13 yields the familiar result, c = 1.5V. If y_1 is greater than or equal to y_2, the wave celerity can be modeled as 1.5V.

Application of the model to a tailwater requires that the user specify the reach length, Δx. The time step taken by the model is free to vary with flow conditions in the channel. The reach having the largest average value of celerity is used to compute the time step via the Courant condition

$$\Delta t = \frac{\Delta x}{<\bar{c}>_{max}} \quad (14)$$

Weinmann et al. indicate from flood routing experience with the Koussis formulation that optimal computational accuracy was achieved at this time step.

The final form of the law of conservation of mass, Equation 11, must be solved to advance the routing to the next time step. This equation is nonlinear and hence was solved iteratively. The equation describing the looped rating curve affects the computation of wave celerity and the conversion of discharge to stage directly and the computation of discharge indirectly through c. The weight function, Θ, can also vary in time and space with the flow, and must be continually updated during the iteration. Because the nonlinearity is not strong and a good initial estimate of the solution is available from the previous time step, the solution converges rapidly.

Experience with the model applied to tailwater flows has revealed that non-physical results generally precede a sudden flow increase or decrease. This "noise" can be suppressed by holding the stage or discharge about to change "non-physically" at its value from the previous time step. Numerical experiments were conducted to verify that this suppression device did not adversely affect the solution.

Model Testing

Small differences in the tailwater flow regime can often significantly alter an environmental assessment. An emphasis was placed, therefore, upon the accuracy of the flow model predictions. Extensive model testing was necessary to ensure that the desired accuracy could be achieved and to quantify the importance of various sources of error. A testing program was designed which included comparisons of model output with data obtained from laboratory and field studies.

Case Study I - WES Flume

A first step in the testing sequence was to utilize data obtained in the laboratory, eliminating many of the unknowns inherent in a field investigation. Tailwater flows are comparable on a smaller scale to those resulting from a dam failure. For this reason it was decided to test the model against laboratory studies of dam failures.

Dam break studies were conducted at the Corps of Engineers' Waterways Experiment Station in a 400-foot long flume. The results of a series of tests and details of the testing procedure are reported in Reference 3. A schematic diagram of the flume and downstream points of measurement are given in Figure 1. Two experimental cases were selected for comparison with the model. The first case (Test 8.1) involves the instantaneous removal of a partial height (0.3 feet) of the dam over the entire channel width. For this test the channel downstream of the dam was dry initially. The second test case (Test 1.1[20]) involved the instantaneous removal of the entire dam. A large base flow of about 38 percent of the peak flow existed in the channel prior and subsequent to the simulated failure. Both peak flows and the base flow of the second case were supercritical. Channel width, slope and roughness were measured and available for model input. The slope of the channel, 0.005, is steep, but comparable to some tailwaters in the TVA region.

<u>Dry Channel Study</u>--A five-second averaging of the experimentally determined hydrograph at the failed dam is presented in Figure

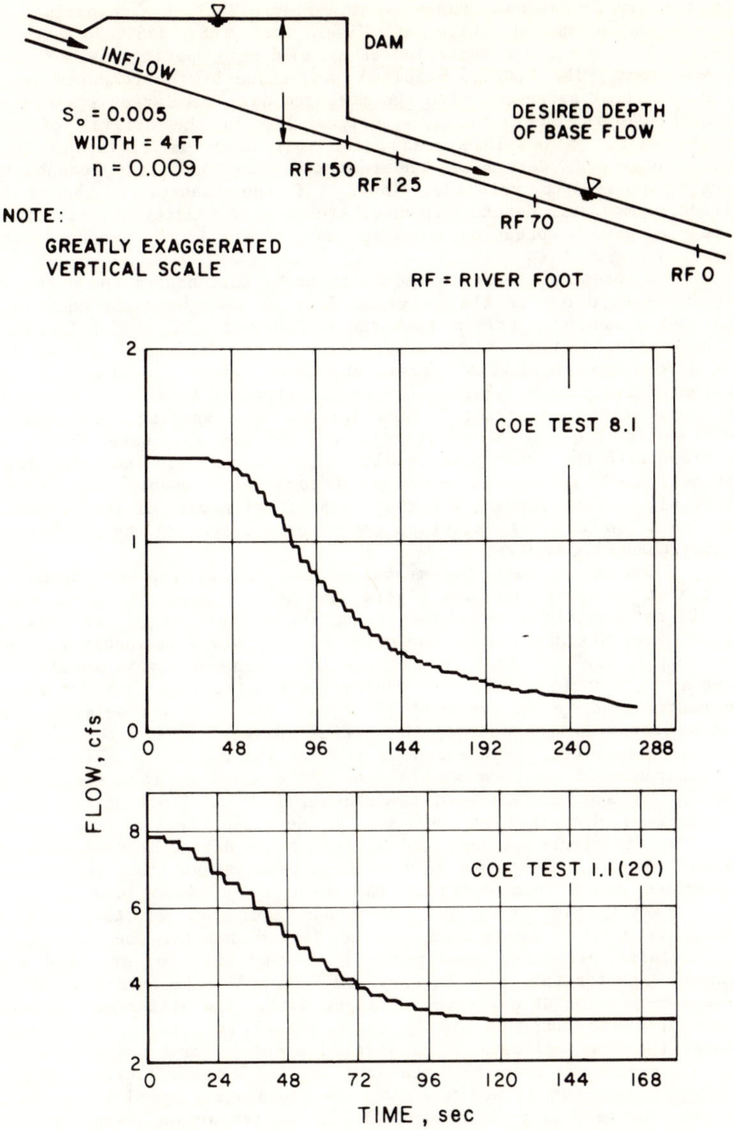

FIG. 1.—Schematic of COE Flume, Release Hydrographs at Failed Dam, RF150

1 for the dry downstream channel case. Figure 2 gives a comparison of measured and computed stage at River Foot (RF) 125, 70 and 0, respectively. For the computation an assumed mesh spacing, Δx, of five feet was used. The timing, magnitude and shape of the computed waves are in excellent agreement with the measured data. A larger scale view of these comparisons reveals a slight lag in the arrival of the computed wave. A possible reason for this error in timing is that channel roughness, determined under steady flow conditions, cannot be directly applied to unsteady flows. If the channel roughness is modified from 0.009, determined under steady flow conditions, to 0.008, the modeled wave propagation and shape move closer to the measured data (Figure 3).

A potential source of model error is that associated with the approximate solution of the original form of the governing equations (Equations 3 and 4). Errors that can be reduced with grid refinement are those attributable to truncation error of the difference scheme and those errors associted with approximations necessary for an analytical solution of Equation 9. Figure 4 presents expanded time scale views of model simulations comparing 5- and 10-foot mesh spacing. The coarse grid results lag the wave travel and diffuse the wave front by comparison with the finer grid results. This comparison indicates that additional mesh refinement would tend to improve model predictions. Another significant finding was that lack of refinement of the computational mesh can be an important source of error, even in the absence of changing channel geometry.

Consistent with the measured data and with observations of wave travel in steep, uniform rivers, the modeled wave travel for this case did not exhibit noticeable attenuation of the peak. The ability to model wave attenuation is the principle advantage of models retaining the water surface slope term over simpler kinematic wave models. A question which could be asked is whether or not the water surface slope term contributed to an improved solution. The term was deleted, and the case rerun. As expected, the rating curve which had been looped became single-valued. The speed of propagation of the wave front was the difference in the two simulations. The water surface slope term increased the apparent slope of the channel near the front of the wave. The effect of dropping the term was to decrease the celerity of the wave. The difference between the arrival times of the modeled wave at RF 0 was about 9.5 seconds. Even without wave attenuation, neglecting the surface slope term degraded the accuracy of model simulations.

Wet Channel Study--A five-second averaging of the experimentally determined hydrograph at the failed dam for the base flow study is also presented in Figure 1. A comparison of measured and computed stage for this case is shown in Figure 5. The numerical model is again responding strongly to the input data. The difference between the measured and computed steady state stages is due to a round-off error of the measured data. The modeled speed of wave propagation is again in agreement with the measured data. The coefficient of the velocity in Equation 13 which yields the propagation speed of the wave front is substantially increased (1.33) in comparison with the dry channel case (1.0). Comparison of Figures 2 and 5 reveals that the speed of wave propagation for this wet channel flow is greater by a factor of two than that of the dry channel case.

In an overall sense, the results obtained with this simplified model compare favorably with those of Chen (1) for this same case.

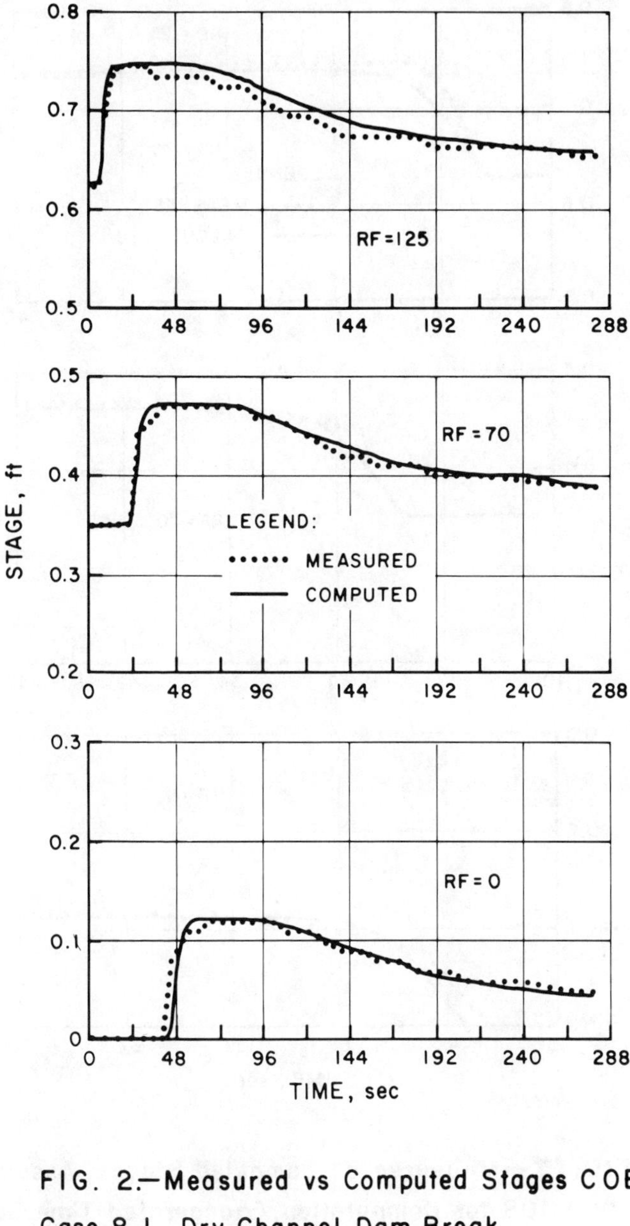

FIG. 2.—Measured vs Computed Stages COE Case 8.1, Dry Channel Dam Break

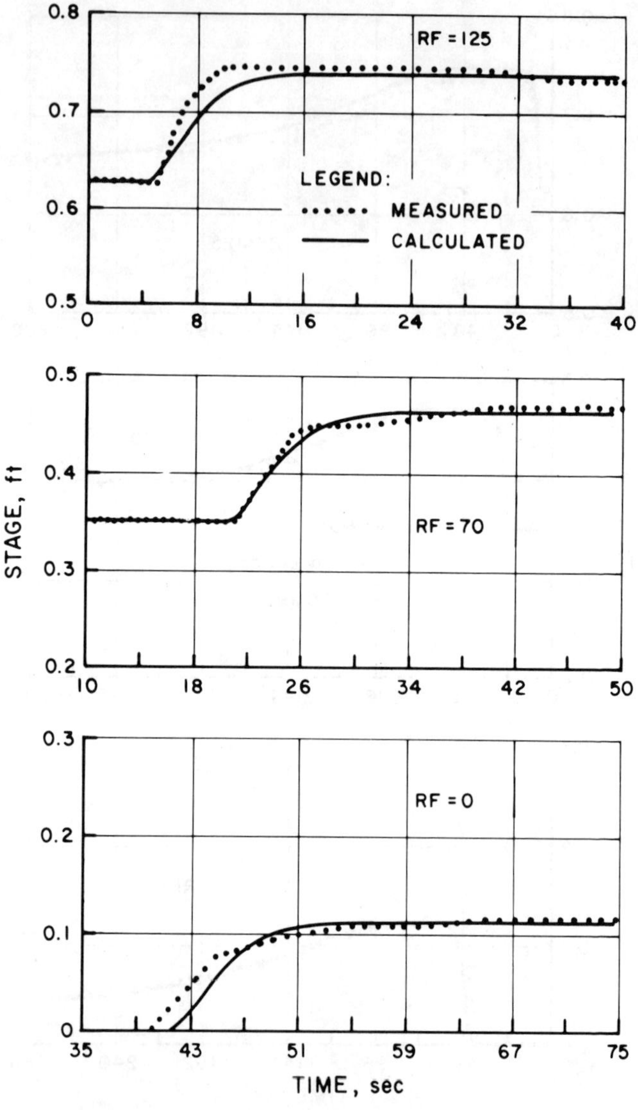

FIG. 3—Measured vs Computed Stages, Assume n = 0.008 for Computation, Exaggerated Time Scale

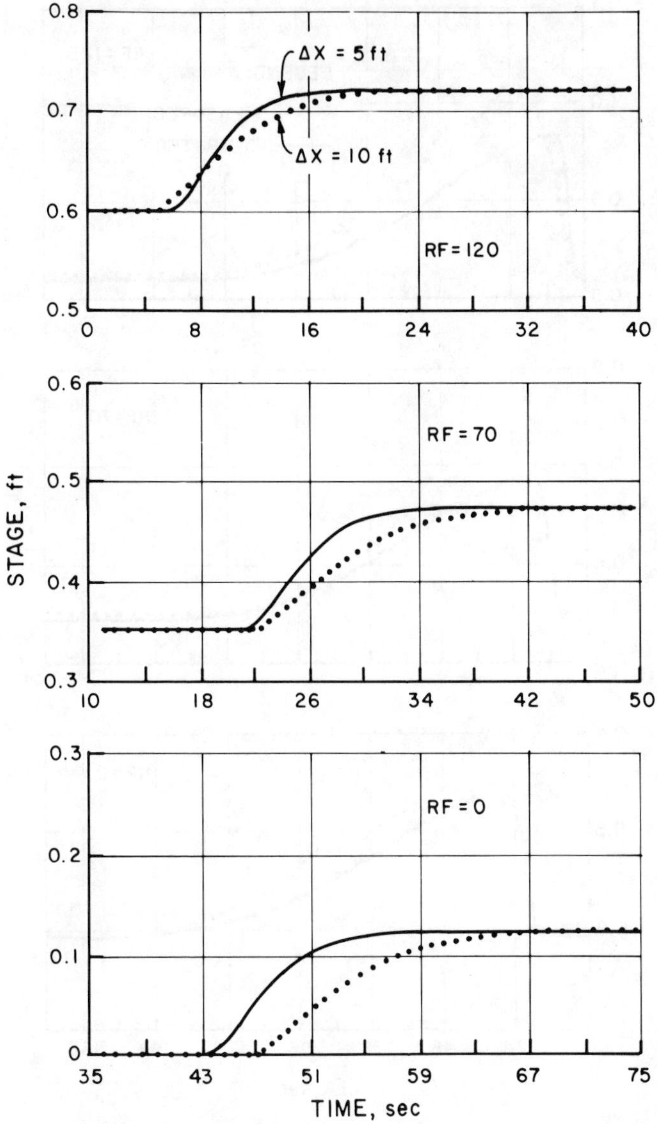

FIG. 4.—Effect of Grid Refinement on Model Predictions

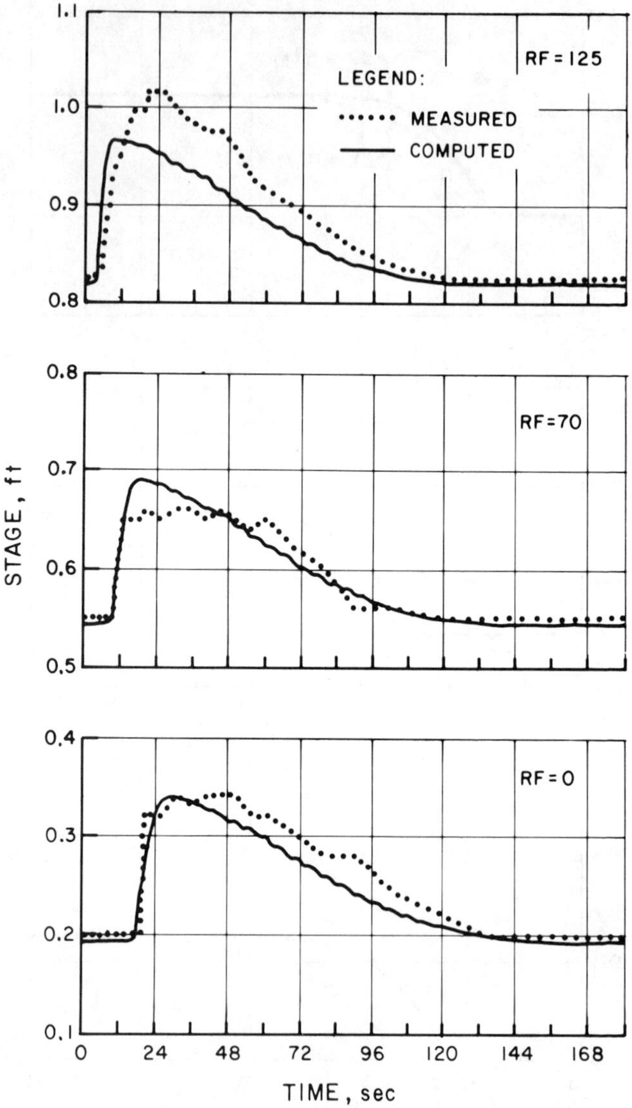

FIG. 5.—Measured vs Computed Stages COE Case 1.1 (20), Dam Break on Base Flow

Several questions regarding the integrity of the measured data make further interaction of model and data non-productive.

Model response to mesh refinement in this case is similar to that for the dry channel. The difference between the simulations using the 5- and 10-foot mesh spacing are very small for the wet channel case. Considering the approximation that "a" in Equation 9 is constant in a time step, it is not surprising that an adequately refined computational mesh is flow dependent. The effect of dropping the water surface slope term is also similar to that for the dry channel case.

Case Study II - Hiwassee River

Though much was learned from the laboratory study, and excellent model performance was noted, testing of the model against field data remains the key factor of the test program. TVA's Apalachia Dam is situated at Hiwassee River Mile (HRM) 66 near the southern end of the border of Tennessee and North Carolina. At the dam, the Hiwassee River has a drainage area of 1018 square miles. In the 13-mile study reach below the dam, the river is steep, dropping over 350 feet, and has an extremely rough stream bed. The average bed slope in the reach is equal to the slope of the flume used in Case Study I. The river has a Manning's roughness, however, which is about an order of magnitude greater than that of the flume. At present, flow in this reach occurs only during floods and as a result of local inflows from a drainage area totaling 118 square miles. The bulk of the flow is diverted through a conduit to the Apalachia powerhouse located near the end of the reach.

A field study was conducted by TVA on March 22-23, 1979, to test the model and validate it for use on the tailwater. Water was released at the dam according to the schedule given in Figure 6. The time variation of stage was recorded at river miles 62.8, 59.0, 56.9 and 53.0 during the test period. The release pattern was selected to study wave travel on dry, wet, and "partially wet" streambeds.

Often, the utility of a model is limited by large input data requirements. These data are generally difficult and/or expensive to obtain. The strategy used here was to adapt the simplest formulation of the model which could meet accuracy requirements. It was evident from available map data and limited field reconnaissance that channel width, roughness and slope must be modeled as having longitudinal variation. The study reach was discretized into one-mile reaches. Values of channel width and slope were obtained from USGS quadrangles. The channel was approximated as rectangular throughout, lateral inflow was neglected, and Manning's roughness was estimated on the basis of field observations.

A comparison of measured and computed stage variations over the two-day study are presented in Figure 7. A number of attempts were made to improve the overall agreement by varying the estimated parameters (roughness, width). Sensitivity studies were conducted to serve as a guide in this process. Increased channel roughness caused, as expected, a slowing of wave movement, an increase in steady state stages for a given flow, and an extended duration of increased stage. Also following intuition, increased channel width affected a decrease in the magnitude of stage changes, a slowing of wave movement, and a decreased period of increased flow.

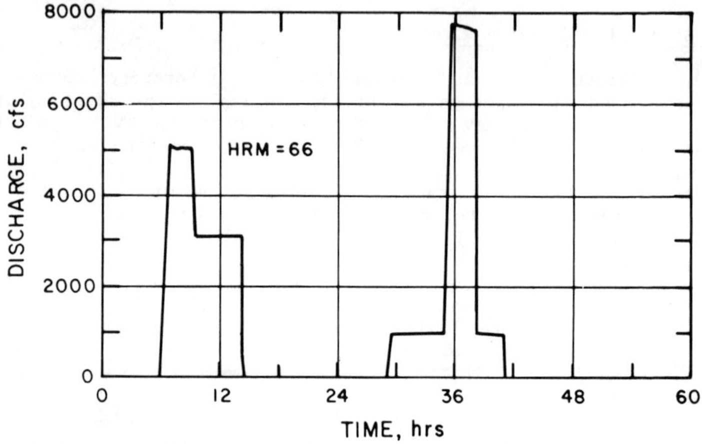

FIG. 6.—Apalachia Dam Flow Releases, March 22-23, 1979

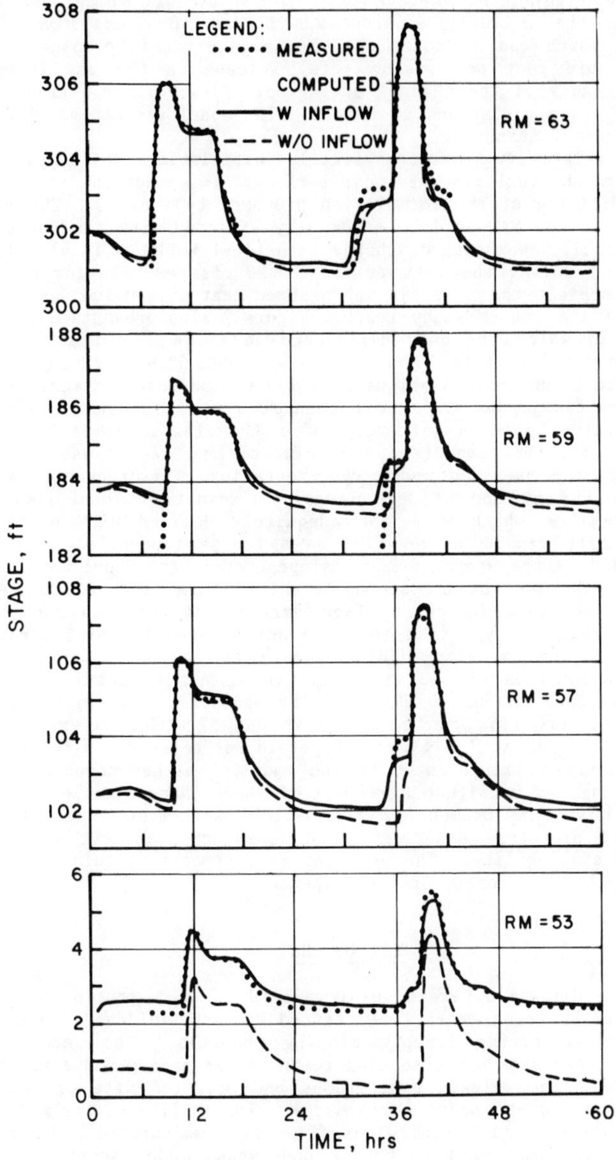

FIG. 7.—Measured vs Computed Stages, Hiwassee River Below Apalachia Dam

No significant improvements in the overall comparison were achieved in the calibration process. Obvious problems with model predictions which can be noted are the large error in propagation of the 1000 cubic foot per second (cfs) release on the dry channel, improper drainage of the channel during zero flow release periods, and exaggerated stage variations at HRM 53. The model was not meeting the required accuracy standards.

A relatively simple, physically justifiable model improvement was made with the inclusion of local inflows. This addition required a minor modification of the formulation proposed by Koussis. The model predictions at HRM 53 would be especially affected with the inclusion of the Apalachia powerhouse discharge as a local inflow. Local inflows were measured during the test period at 20 cfs per mile for the 3.5 mile reach nearest the dam. It was assumed that this value was representative of the entire study reach. Figure 7 also presents a revised comparison in which the only difference in the model input was the inclusion of local inflows. Measured data at HRM 62.8 were compared to computed data at HRM 63. A slight time lag in the measured results was expected and can be noted. Local channel size and bottom slope are well represented by the model except when the flow rate was 1000 cfs. In this instance the channel at the point of measurement was slightly smaller than the averaged model representation. Discrepancies which occurred at HRM 59 and 57 were largely a result of local pool and riffle structures which were not adequately defined with prototype data, and which cannot be resolved on a coarse (one-mile) numerical grid. At HRM 59 the averaged local slope used in the model was not as large as the actual channel slope in this riffle river reach. As a result, the decrease in river stage during low flow was much more extreme in the prototype than in the model. A small timing error in propagation of the modeled 1000 cfs wave can be attributed to the coarse representation of the river and to a lack of detailed inflow data in the reach HRM 63 to HRM 59. The opposite trend in the stage comparison at low flow can be noted at HRM 57. The river at this location was pooled at low flow. Stage did not decrease in the prototype as readily as represented in the model. Further downstream, at HRM 53, timing and magnitude agreement of stage changes was excellent.

Further improvement in the overall agreement of model predictions and prototype data was constrained by goemetric and local inflow data availability. The accuracy requirements of this study did not justify the additional work and expense.

CONCLUSIONS

The diffusion wave model developed for flow routing in tailwater streams has been extensively tested against field and laboratory case studies involving steeply sloping channels. The interaction between model and data was essential to an understanding and evaluation of the model. The release hydrographs and measured data provided a discriminating test of model performance. The ability of the model to accurately simulate field tailwater flow was demonstrated. Numerical tests were also conducted to better understand model sensitivity to input parameters and sources of error inherent to the numerical solution technique.

APPENDIX I - REFERENCES

1. Chen, C. L., "Laboratory Verification of a Dam-Break Flood Model," ASCE Journal of the Hydraulics Division, Vol. 106, April 1980.

2. Ferrick, M. G., and W. R. Waldrop, "Two Techniques for Flow Routing with Application to Wheeler Reservoir," TVA Division of Water Management, Water Systems Development Branch, Report No. 3-519, July 1977.

3. "Floods Resulting from Suddenly Breached Dams - Conditions of Minimum Resistance," Miscellaneous Paper No. 2-374, Report 1, U.S. Army Corps of Engineers, Waterways Experiment Station, Vicksburg, MS, February 1960.

4. Garrison, J. M., J. P. Granju, and J. T. Price, "Unsteady Flow Simulation in Rivers and Reservoirs," ASCE Journal of the Hydraulics Division, Vol. 95, September 1969.

5. Henderson, F. M., "Flood Waves in Prismatic Channels," ASCE Journal of the Hydraulics Division, July 1963.

6. Henderson, F. M., Open Channel Flow, The Macmillan Company, New York, 1966.

7. Koussis, A., "An Approximative Dynamic Flood Routing Method," International Symposium on Unsteady Flow in Open Channels, University of Newcastle-upon-Tyne, England, April 12-15, 1976.

8. Ponce, V. M., and D. B. Simons, "Shallow Wave Propagation in Open Channel Flow," ASCE Journal of the Hydraulics Division, Vol. 103, December 1977.

9. Ponce, V. M., Ruh-Ming Li, and D. B. Simons, "Applicability of Kinematic and Diffusion Models," ASCE Journal of the Hydraulics Division, Vol. 104, March 1978.

10. Weinmann, P. Erwin, and E. M. Laurenson, "Approximate Flood Routing Methods: A Review," ASCE Journal of the Hydraulics Division, Vol. 105, December 1979.

APPENDIX II - NOTATION

A = cross sectional area of stream

B = width of stream

$\bar{B}$ = average width in a reach

c = wave celerity

Fr = Froude number

n = Manning's roughness

O() = the order of

q = lateral inflow per unit length

$\tilde{q}$ = discharge per unit width

Q = discharge

$\dot{Q}_j$ = time derivative of Q at section j

R = hydraulic radius

S_f = slope of the energy grade line

S_o = slope of the channel bottom

Δt = time step

t = time

V = velocity

Δx = length of a reach

x = longitudinal distance

y = depth

<> = average over a time step

Θ = weighting function of time derivatives

MODELING OF SEDIMENT TRANSPORT IN COTTONWOOD CREEK

By Dr. Alan L. Prasuhn[1], F. ASCE
And Edward F. Sing[2], A. M. ASCE

ABSTRACT

Two multi-purpose dams have been authorized by Congress on the South Fork and main stem of Cottonwood Creek, a tributary of the Sacramento River in northern California. The Corps of Engineers used the computer model, HEC-6, to evaluate the impact of the project dams on many of the channel characteristics such as sediment load, armoring of the bed, and degradation of the channel. Particular emphasis was placed on evaluating the impact of the dams on the gravel beds which serve as anadromous fish spawning grounds. A gravel quarrying operation some distance below the confluence of the two forks was also incorporated in the model. Finally, the relative magnitudes of the project impact and quarrying operation are compared.

INTRODUCTION

The Cottonwood Creek project was authorized for construction by Congress in the Flood Control Act of 1970. The authorized plan provides for construction of two multiple-purpose storage lakes, Dutch Gulch and Tehama, for flood control, local irrigation, municipal and industrial water supply, recreation, and anadromous fishery enhancement. Engineering studies were conducted [1] to evaluate a number of sediment related problems including:

a. Project effects on streambed and banks along Cottonwood Creek downstream from the damsites.

b. Project effects on fish spawning size gravels (0.5 to 6 inches in diameter) in Cottonwood Creek and the transport of these gravels to the Sacramento River.

The Cottonwood Creek basin is located in Tehama and Shasta Counties, California (Fig. 1). The creek flows into the Sacramento River about midway between the cities of Red Bluff and Redding. The watershed area consists of portions of the eastern face of the Coast Range and the southern slopes of the Trinity Mountains. The lower elevations consist of foothill and valley floor area. Cottonwood Creek drains an area of 927 square miles with a length of about 46 miles and a maximum width of about 30 miles [3]. The study reach of the downstream erosion and

---

[1]Professor of Civil Engineering, South Dakota State University, Brookings, South Dakota

[2]Hydraulic Engineer, U. S. Army Engineer District, Sacramento, California

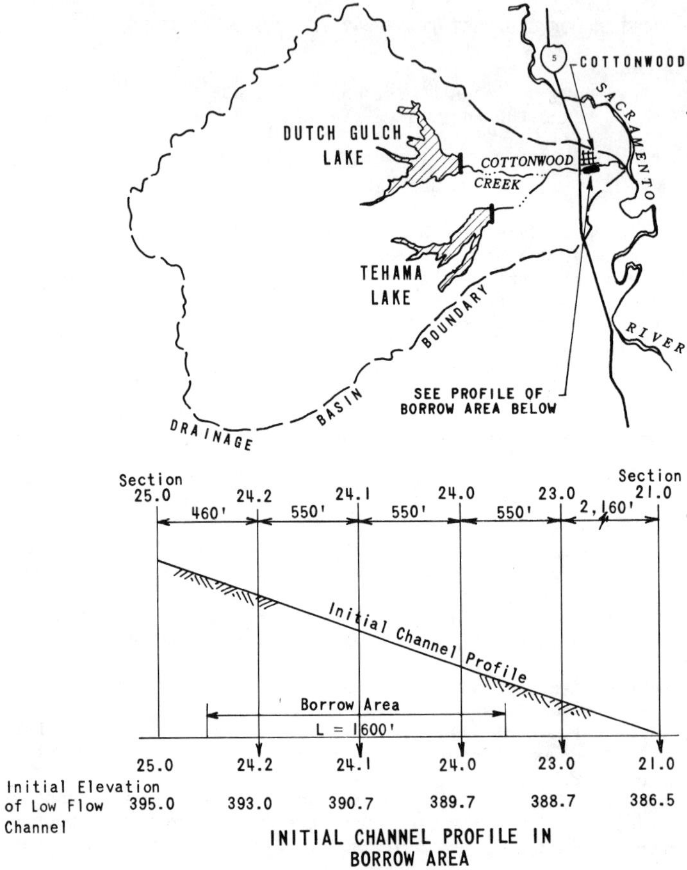

Figure 1. Location Map (Not to scale)

sedimentation study extends from the Sacramento River up the mainstem approximately 18 miles to the authorized Dutch Gulch Reservoir site and up the South Fork from the confluence of the mainstem approximately 6 miles to the authorized Tehama Reservoir site. In the past, several commercial firms have removed material from the Cottonwood Creek and its tributaries for fill and concrete processing.

COTTONWOOD CREEK MATHEMATICAL MODEL

The downstream effects of the project on the streambed, and on gravel recruitment to the Sacramento River were evaluated utilizing the computer program HEC-6, "Scour and Deposition in Rivers and Reservoirs, LGR (Lower Granite Reservoir) Version" developed at the U. S. Army Corps of Engineers' Hydrologic Engineering Center, Davis, California, [2,4,6]. The LGR Version was selected for use in this portion of the study as it has the capability to evaluate sediment transport in a branching stream network as well as simulate local water and sediment inflow into the network. The South Fork of Cottonwood Creek was incorporated into the computer model as a branch of the mainstem and all tributary water and sediment inflow into Cottonwood Creek between damsites and the mouth of Cottonwood Creek (at the Sacramento River) was incorporated as a local inflow point (hereafter referred to as tributary streams) approximately 5,000 feet downstream of the confluence of the South Fork and mainstem of Cottonwood Creek.

HEC-6 is divided into three submodels. The first, the geometric submodel, is used to calculate hydraulic properties (at each cross section for each water discharge analyzed) by backwater computations similar to those in HEC-2. Data requirements for this portion of the HEC-6 model include initial conditions of the creek's cross sections, reach lengths and Manning's "n" values for water surface profile calculations. Various data sources were used in calibration of "n" values along the study reach. Approximately 100 cross sections were used.

The sediment submodel calculates the sediment transport associated with the hydraulic properties at each cross section for each water discharge analyzed, as well as the bed gradation changes and scour or deposition. Data requirements for this portion of the HEC-6 model include grain size distribution of the material in the creek bed at each cross section, the gradation and amount of total inflowing sediment load as a function of water discharge, and the fluid and sediment properties. Bed and bank material samples were taken along the study reach. Based on these samples, "averaged" bed gradations were assigned to reaches of the creek (See Table 1 for typical data). Inflowing suspended sediment load curves were developed from analysis of previously collected data in the basin.

The model, when using the normally recommended Toffaleti transport function, was found to greatly underestimate gravel movement. After evaluating the use of several other transport functions to calculate gravel movement, the Schoklitsch bedload function [5] was found to provide a satisfactory estimate of gravel movement in the study reach. This function was incorporated into the HEC-6 model to calculate gravel transport. In addition, transport rates of small cobble size material presented herein are based upon manual calculations using the Schoklitsch bedload equation and stream hydraulic parameters from the HEC-6 runs.

Table 1
Typical Initial Bed Material Size Distribution
(gravel and small cobbles)

Size Range	Size Range (Inches)	Percent in bed — Cottonwood Creek at Sacramento River	Dutch Gulch	Tehama	Tributary Streams
VFG--Very Fine Gravel	.079- .16	.095	.125	.0483	.165
FG--Fine Gravel	.16 - .31	.125	.125	.0567	.194
MG--Medium Gravel	.31 - .63	.14	.120	.0683	.180
CG--Coarse Gravel	.63 - 1.26	.21	.110	.1267	.09
VCG--Very Coarse Gravel	1.26 - 2.52	.125	.040	.160	.09
SC--Small Cobble	2.5 - 5.0	.125	.015	.22	.14

The hydrologic submodel is used to describe the flow hydrograph on which sediment calculations are to be performed. The flow hydrograph is input as a flow histogram, a series of discrete flow events of a magnitude and duration such that the total sediment and water volume passed by the histogram is equal to that passed by the hydrograph it is simulating. For the long term simulation runs, 34 year preproject and project flow hydrographs were constructed using monthly, daily or hourly flow data, made available from previous reservoir operations studies. The peak pre-project discharge during the study was 70,000 cfs.

Model Verification. Passing of the 34-year project hydrograph through the model provided a reasonable test of the performance of the model. First, no instability is present in the output. That is, there is no oscillation of the bed elevation at any point with time. With the exception of lead-in sections which respond to too much or too little sediment inflow, there are generally only modest changes of bed elevation with time. Typical values are shown on Fig. 2 for cross section 3 (2000 ft. upstream of the mouth of Cottonwood Creek and interpreted as the mouth to avoid end section effects), cross section 95 (Dutch Gulch damsite), and cross section 127 (Tehama damsite). These results are labeled as the preproject conditions in each case. All other sections appear to be equally acceptable, with many showing almost no change after 34 years.

RESULTS OF LONG TERM SIMULATION

The project model differs from the preproject model in that the hydrograph is based on reservoir operating rules for the corresponding simulated 34 year hydrograph used previously. It was assumed that the reservoirs would trap all sediment. This would certainly be true for the sand and gravel sizes. However, inflow of both water and sediment from the tributary streams is left identical to that under preproject conditions.

Bed Elevation and Gradation Changes. The HEC-6 results of the project run showing changes in the bed as a function of time are included in Fig. 2 and are labeled as project results. The changes with time generally show a tendency for less scour than the preproject conditions. Those sections where some deposition previously occurred now have a little less deposition than before or even a small amount of scour. Exceptions occurred at sections just below the Dutch Gulch damsite and in the lower portions of the creek. Results from the HEC-6 model show that construction of the dams would result in some degradation

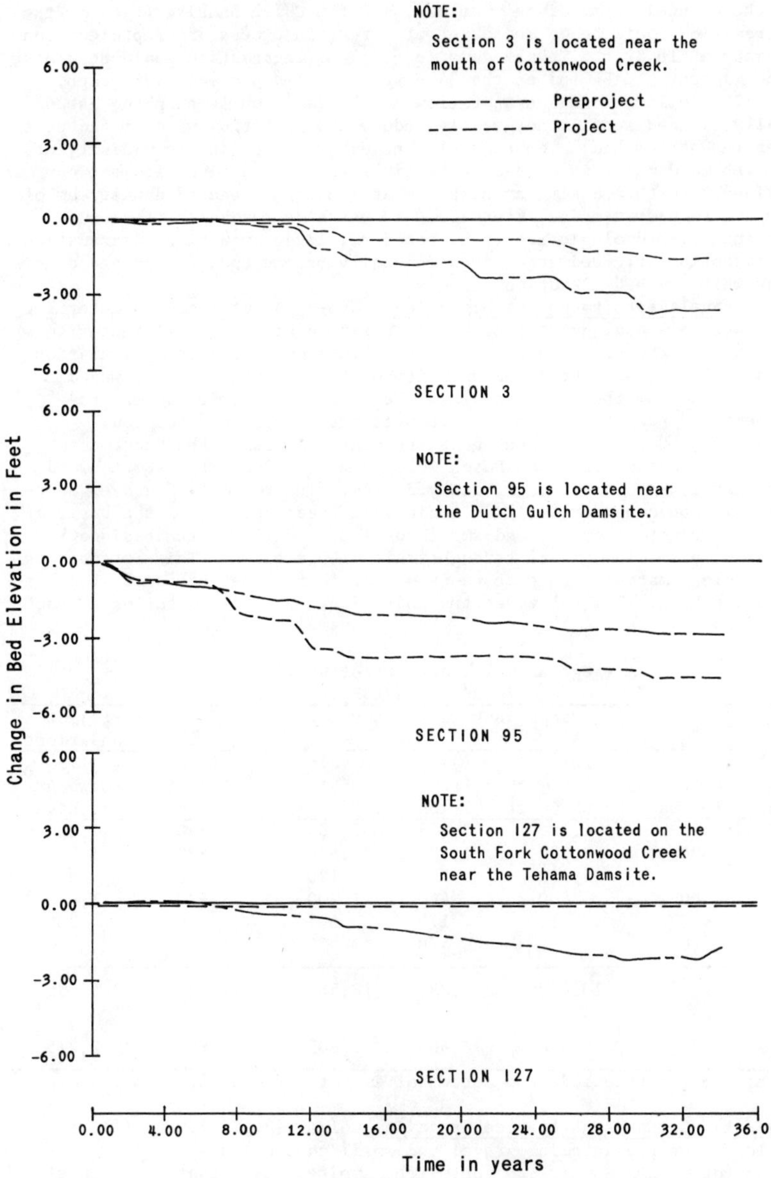

Figure 2. Bed Changes Versus Time (Long Term Simulation Runs)

of the channel. Immediately below the Dutch Gulch Damsite (Fig. 3) the degradation would be on the order of 2 feet in excess of preproject conditions while at the Tehama damsite (Fig. 4) degradation would be almost non-existent at the end of the 34 year simulated project hydrograph.

The implications are as follows: the dams, while stopping essentially all sediment, also greatly reduce the peak flow which would pass over the stream bed. Consequently the existing bed in the vicinity of the Tehama damsite is already effectively armored against project conditions. Further a similar stable state would be reached downstream of the Dutch Gulch damsite after about 12 years depending on the actual sequence of hydrologic events. Therefore, large amounts of degradation would not occur. Bed gradations generally become courser and are consistent with the bed elevation changes.

Gravel Recruitment and Migration. HEC-6 was utilized to determine the average annual quantities by grain size class of gravels entering and exiting the Cottonwood Creek basin based upon long term simulation runs of 34 years. The total quantity of the tributary inflow was also determined from the HEC-6 results. Transport quantities under preproject and project conditions are presented in Table 2 for four locations: Cottonwood Creek near the confluence with the Sacramento River, Cottonwood Creek at Dutch Gulch damsite, South Fork Cottonwood Creek at Tehama damsite, tributary inflow. The results of gravel movement will be considered (1) by grain size fraction, (2) on the basis of total movement of gravel and small cobbles, and (3) on the basis of movement of medium gravel through small cobble sizes. This latter size range close matches the size range of 1/2 to 6 inches necessary for fish spawning beds. Table 1 gives the initial bed size distribution at each location.

Table 2
Average Annual Transport of Gravel in Tons
(Simulated 34-Year Hydrograph)

	Proproject Conditions					Project Conditions
(1)	(2)	(3)	(4)	(5)	(6)	(7)
Size	At Tehama	At Dutch Gulch	From Tribs	Sum (2)(3)(4)	Ctwd At Sac River	Ctwd At Sac River
VFG	410	4,300	2,675	7,385	4,340	2,320
FG	420	3,200	1,750	5,370	3,420	1,500
MG	420	2,000	700	3,120	2,950	960
CG	580	860	460	1,900	2,970	1,050
VCG	400	120	175	695	900	265
SC	20	0	70	90	60	0
Total	2,250	10,480	5,830	18,560	14,640	6,095
Total MG/SC	1,420	2,980	1,405	5,805	6,800	2,275

Note: Does not include the potential impacts of gravel mining.

(1) Comparisons by size fraction. Columns (2), (3), and (4), of Table 2 give the incoming gravel and small cobble loads by size fractions for Tehama damsite, Dutch Gulch damsite, and tributary streams respectively. The total load for each size fraction is given in Col (5). A comparison of Col (5) and Col (6) shows that the incoming load for each size fraction is in relatively good balance with the movement out of

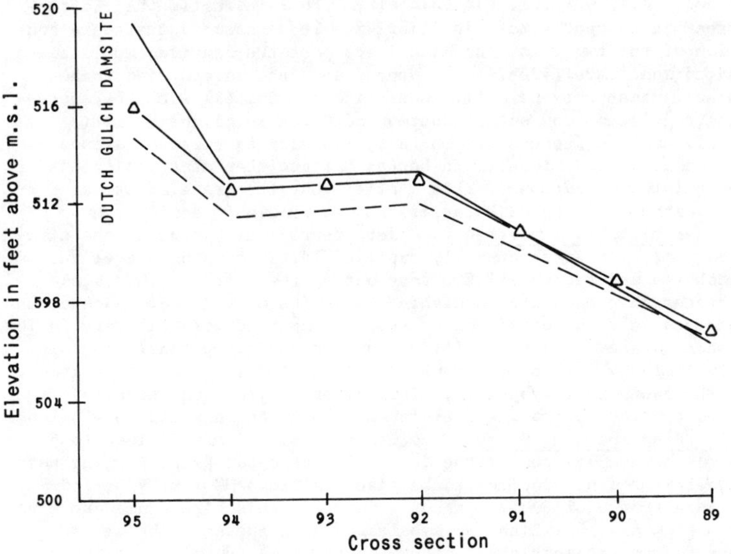

Figure 3. Channel Profile Downstream of Dutch Gulch Dam

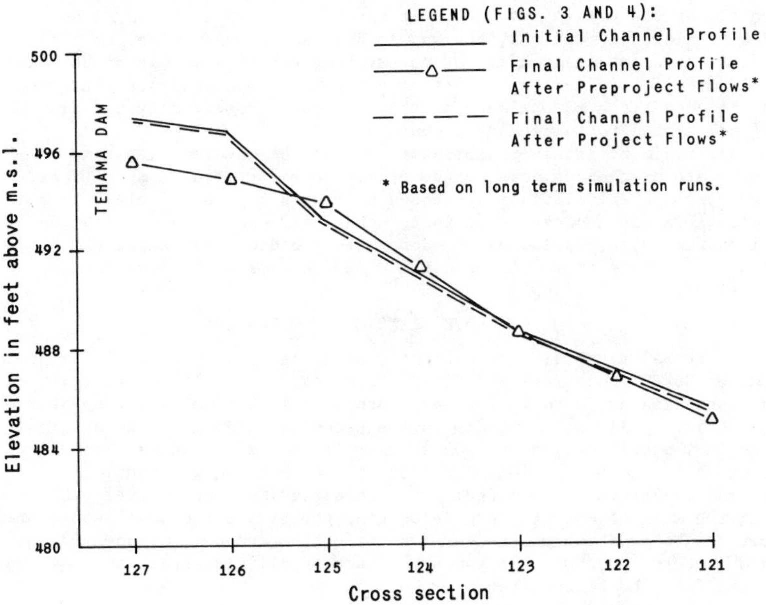

Figure 4. Channel Profile Downstream of Tehama Dam

Cottonwood Creek, Col (6), for that size. This indicates that Cottonwood Creek is in approximate equilibrium. It is assumed that upon construction of the two dams, incoming loads past the damsites would be eliminated and gravel supply will depend upon the unregulated tributaries as indicated by Col (4). Comparison of Col (4) with the corresponding regulated flow out of Cottonwood Creek as given in Col (7) shows only small differences. In fact, the size by size comparison under regulated conditions is in better balance than that indicated for the unregulated conditions. This implies that the gravel beds, at least downstream of the tributaries, should remain in equilibrium.

(2) Total gravel movement. Under preproject conditions the simulated 34-year hydrograph annually supplied 18,600 tons of gravel and small cobbles and carries 14,600 tons out of the system. This again demonstrates that equilibrium within the limits of sediment calculation accuracy prevails in Cottonwood Creek. A comparison of tributary inflow with the regulated conditions (5,800 tons/yr vs. 6,100 tons/yr) shows that the tendency will be to transport almost the same amount of gravel out of the basin as is supplied. Thus under project conditions, it is estimated that the overall gravel movement will be in reasonable balance.

(3) Fish Spawning Sizes. Bed material size from 1/2 inch to 6 inches has been suggested as the desirable range for fish spawning beds. This range is roughly duplicated by size fractions MG, CG, VCG, and SC, which range from 0.31 to 5.0 inches. Sums for this size range are given on the bottom of Table 2 and so labeled. Again the equilibrium is apparent for pre-project conditions with an annual supply of 5,800 tons and an annual movement out of Cottonwood Creek to the Sacramento River of 6,800 tons. Note that this does not represent all of the bedload nor even all of the gravel movement, but rather only the selected sizes. Under regulated conditions, the supply of fish spawning size gravel drops to 1,400 tons per year while the movement of these sizes out of Cottonwood Creek is calculated to be 2,300 tons per year. This indicates that under project conditions the balance of these sizes is essentially the same as that calculated for unregulated conditions.

All sediment transport computations must be treated with some caution. State-of-the-art does not generally permit precise computations. However, these calculations are based on reasonably good field data and should be fairly accurate. In fact, manual calculations using various bedload functions, including the Schoklitsch bedload function, the measured bed load from USGS, and the HEC-6 results, all appear quite consistent.

MODEL OF GRAVEL EXTRACTION OPERATION

Creek bed materials are currently being removed by several commercial firms on Cottonwood Creek and its tributaries. Since the gravel extraction operation on Cottonwood Creek approximately 5 miles upstream of the mouth (Fig. 1) directly impacts on the stream's sediment transport and those on the tributaries are somewhat smaller in size and have an indirect impact, only the extraction operation on the mainstem was considered. Sand and gravel is removed from a region extending approximately 1600 feet along the channel with the provision that the excavation shall not be made lower than the bottom elevation of the low flow channel. Since 1974, approximately 100,000 cubic yards per year of all size fractions have been removed from the borrow area.

Mathematical Model. The geometric model was modified in an attempt to simulate the present gravel extraction operation on Cottonwood Creek. The program, through its Dredge Option, is capable of simulating removal of all material in the bed within specified lateral limits and down to a specific excavation elevation depth when deposition causes the bed elevation to rise. The bottom elevation of the excavation was set at the bottom of the initial low flow channel at each cross section within the gravel extraction site. It was assumed that excavation would take place across the full width of the movable bed portion of the cross sections within the borrow area although in actuality the mining operations are required by their State and Federal permits to provide a low flow channel separate from the borrow area. Finally, although mining occurs on a periodic, irregular basis, the computer coding necessitated the assumption that all excavation occurred at the beginning of each water year analyzed (just before the high flow season).

The hydrologic model for preproject and project flow conditions with gravel mining consists of monthly and daily flows for the 9-year period 1970-1978, representing both high flow (1970-1974) and low flow (1975-1978) years.

RESULTS OF GRAVEL EXTRACTION MODELING

Physical observation and modeling studies have indicated that extractions may now equal or exceed annual natural replenishment volumes along Cottonwood Creek near the operation. It is surmised that current extractors are mining past deposits in addition to sediments currently being deposited at the site. However, since the trap efficiency of the operation is not 100 percent, some bedload is passing the site, especially during high flow events.

Effect on Sand and Gravel Transport. For the two periods (1970-74 and 1975-78) analyzed, the average transport (in 1000's of tons/year) past Sections 24.2 (upstream end of borrow area), 24.1, 24.0 (downstream end of borrow area), 23.0, 21.0 (immediately downstream of borrow area) and 3.0 (mouth of Cottonwood) are tabulated in Table 3 below, for preproject and project conditions. For comparison the average transport for the same time periods from the long term simulation runs are presented in Table 3. Included also is the average annual excavation in 1000's of tons/year. However, the Dredge Option was not executed in every year, as some years had flows that did not deposit sufficient material in the bed to activate the Dredge Option of the program.

The excavation quantities (in 1000's of tons/yr) are in the right hand column in Table 3. They indicate that the excavation quantities under project conditions are 71% of preproject quantities during 1970-74 and 67% during 1975-78. This range of percentages can be taken as representative of the effect of the project on the gravel mining operation. The remainder of Table 3 gives an indication of the separate and combined effects of the project and the gravel mining on sand and gravel transport in Cottonwood Creek.

Table 3 shows transport quantities that are roughly in balance with the excavation quantities. Concentrating on the period 1970-74 which is believed to contain the more reliable data, the long term simulation without Dredge Option values show a nearly constant transport of sand and gravel section by section through the model. By way of contrast, the with Dredge Option values indicate a general and significant reduction in transport quantities at the downstream portion of the borrow area and beyond. Taking Section 23 as representative of downstream conditions,

Table 3. Average annual transport of sands, gravels and small cobbles in 1000's of tons.

	Without Gravel Extraction[1/]						With Gravel Extraction[2/]						Dredge[2/] Quantities (1000's T/Yr)
Section	24.2	24.1	24.0	23.0	21.0	3.0	24.2	24.1	24.0	23.0	21.0	3.0	
PREPROJECT													
1970-74	230	243	228	225	221	221	261	219	169	172	171	130	96
1975-78	140	139	138	138	139	54	159	132	94	95	94	108	67
PROJECT													
1970-74	73	71	70	70	66	59	83	56	29	29	29	50	68
1975-78	26	26	25	25	24	23	32	22	8.6	8.3	8.4	9.8	45

Table 4. Average annual transport of MG-SC in 1000's of tons.

	Without Gravel Extraction[1/]						With Gravel Extraction[2/]					
Section	24.2	24.1	24.0	23.0	21.0	3.0	24.2	24.1	24.0	23.0	21.0	3.0
PREPROJECT												
1970-74	6.8	5.5	5.5	4.0	2.9	16.5	10.3	2.2	0.9	1.7	1.6	10.4
1975-78	2.3	2.0	2.0	2.2	2.2	6.7	4.5	0.6	0.2	0.5	0.6	7.4
PROJECT												
1970-74	4.9	3.0	3.0	2.8	2.4	7.0	6.7	0.8	0.1	0.2	0.2	1.6
1975-78	1.0	1.0	1.0	0.9	0.7	2.0	2.2	0.1	0	0	0	0.7

1/ From long term simulation runs

2/ From 9 yr gravel extraction runs

note that the long term simulation transport quantities past this section are almost identical with the upstream quantities. On the other hand, the preproject with Dredge Option values of transport of sands and gravels past Section 23 is reduced to 75% of the upstream (without gravel extraction) value during 70-74 and 68% during 75-78. Similarly, for project with Dredge Option, the transport is reduced to 40% and 32% of upstream (without gravel extraction) transport during the respective time periods. The project would therefore have certain impacts on the quantities excavated and on the replenishment of these quantities by the stream.

Effect of Gravel Mining on Movement of Fish Spawning Size Gravel. An evaluation of the effect the gravel mining has on movement of fish spawning size gravel was performed. Results are tabulated in Table 4. Values shown are 1000's of tons per year at selected stations for the two time periods analyzed (1970-74 and 1975-78).

These results indicate that the project would have less impact on rate of transport of the MG-SC sizes downstream of the gravel mining area than does the mining operation. The combined average reduction due to gravel extraction and project on transport of these sizes past the borrow area is 95%. The merging of the two effects is very complex. However, it is estimated that gravel extraction is responsible for about 60% of the combined reduction and the project for about 40%. Gravel extraction alone causes an average reduction of transport of 75% and the project alone about 45%.

In the long term, these effects should prevail downstream. However, the transport rates at Section 3, shown in Table 4, do not show this trend as it is believed the stream has not had time to adjust to the mining operation. Although the mining area effectively stops most transport of the larger sizes to the downstream channel reaches, the high water discharge for the preproject with gravel extraction conditions moves the gravel currently in the bed and hence the transport rates are as high as upstream of the mining area. On the other hand, the project either with or without the gravel mining operation would immediately impact on the transport rate as the low, project water discharges, would move much less gravel. Long term simulation of the gravel extraction would probably show that Section 3 would gradually adjust to the same trend exhibited immediately below the borrow area. In addition, the combined gravel extraction and project effects would effectively reduce transport and supply of the larger sizes and hence the bed between the mining area and Section 3 may slowly become less coarse.

CONCLUSIONS

Construction of the two dams could cause some adjustment of the stream throughout. This would generally consist of an increase in armoring with only small amounts of degradation, and lead to a reduction in the amount of sand found in the bed. Most likely this would result in an increase in the amounts of the largest sizes, particularly small cobbles and larger above the gravel mining area. However the bed below the gravel mining area may become somewhat less coarse.

Although there would be a reduction in the amount of gravel supply, both overall as well as in those sizes needed for fish spawning, the unregulated tributary inflow should be nearly sufficient to maintain the balance both overall and in the fish spawning sizes upstream of the gravel extraction operation. Present flow and sediment transport conditions on Cottonwood Creek can be classified as "preproject - with gravel extraction," but the average annual transport of fish spawning size gravels to the Sacramento River will continue at a rate close to the "preproject - without gravel extraction" rate of 6,800 Tons/Yr until the effects of the gravel mining travel downstream to the mouth of Cottonwood Creek. The gravel extraction will eventually reduce the average annual transport of these gravel sizes to a minimum of 25% of its present level. No estimate of the length of time necessary for the stream transport rate to adjust to the gravel extraction could be made with the available data and computer runs.

Continued gravel extraction and flow regulation by the proposed Tehama and Dutch Gulch dams would significantly reduce gravel movement to the mouth of Cottonwood Creek. It is estimated that the movement of gravel in the fish spawning sizes would be reduced by flood flow regulation to about 33% of its present level of 6,800 Tons/Yr and as the effects of the gravel mining eventually migrate downstream, to a minimum level of about 5% of its current rate.

ACKNOWLEDGMENT

This study was conducted by the Department of the Army, Corps of Engineers, Sacramento District. Dr. Prasuhn was a consultant to the study. The authors wish to express their appreciation to W. A. (Tony) Thomas, U. S. Army Waterways Experiment Station (Vicksburg, MS) and Dr. Michael Gee, U. S. Army Hydrologic Engineering Center (Davis, CA).

REFERENCES

1. Downstream Erosion and Reservoir Sedimentation Study, Office Report, Sacramento District, United States Army Corps of Engineers, Sacramento, CA, Jan., 1980.

2. Emmett, W. W., and Thomas, W. A., "Scour and Deposition in Lower Granite Reservoir, Snake and Clearwater Rivers near Lewiston, Idaho, U.S.A.", Journal of Hydraulic Research, IAHR, Vol. 16, No. 4, 1978, pp. 327-345.

3. Hydrology Design Memorandum No. 1, Cottonwood Creek Ca, Sacramento District, United States Army Corps of Engineers, Sacramento, CA, July 1977.

4. "Scour and Deposition in Rivers and Reservoirs", Computer Program HEC-6, Hydrologic Engineering Center, United States Army Corps of Engineers, Davis, CA, Mar., 1976.

5. Sedimentation Engineering, ASCE Manual No. 54, ASCE, 1975.

6. Thomas, W. A., and Prasuhn, A. L., "Mathematical Modeling of Scour and Deposition", Journal of Hydraulics Division, ASCE, Vol. 103, No. HY8, Proc. Paper 13132, Aug. 1977, pp. 851-863.

Soap Films, Steiner Trees and Water Systems

Chan F. Lam[1] and Pramod R. Bhave[2]

ABSTRACT

In designing water distribution systems, one minimizes the cost which is a nonlinear function of the pipe length as well as satisfying the system of nonlinear equations describing the water system. Due to the nonlinear nature of the problem, iterative optimization algorithms are used. All of these algorithms require some initial estimates of the parameters (pipe length in this case). This paper presents a method for obtaining the optimal geometric layout that give the minimal total pipe length based on the location of source and demand nodes. These estimates would then be refined so as to satisfy the draw-offs.

Introduction

In designing water distribution systems, the main objective is to minimize the cost of the network by using the smallest possible pipe size that still satisfy the system demands in terms of draw-offs and head pressures. In the case of a new system or in a rural setting, the geometrical layout of water distribution network is not fixed a priori. Thus, not only the pipe diameters but the geometrical layout, hence the pipe length, can also be manipulated so as to achieve the minimal cost.

The objective function or cost function is a nonlinear function of the pipe length and pipe diameter (3,5,8):

$$C = K_1 \; L \; D^y \qquad 1$$

where L and D are respectively the length and diameter of a pipe, and K_1, and Y are constants. The constraints are a set of nonlinear algebraic equations relating the heads, flows, and the nonlinear resistance (such as the Hazen-Williams formula) of the network (12). Since both the objective function and the constraints are nonlinear functions, nonlinear optimization techniques must be used. Most of

[1]Department of Biometry, Medical University of South Carolina, Charleston, South Carolina.

[2]Department of Civil Engineering, Viservaraya Regional College of Engineer, Naypur, India.

these optimization techniques require initial estimates of the parameters being adjusted. Thus, the secondary but important problem is to find the initial estimates of parameters.

Steiner Tree Approach

In the case of water distribution systems, the adjustable parameters are the length and diameter of the pipes. However, the diameter parameters can be eliminated from the objective function using the headloss relationship:

$$H_L = \frac{K_2 \ L \ Q^P}{D^r} \quad (2)$$

in which H_L = head-loss through a link of length L and diameter D while carrying a discharge Q; K_2 = a coefficient that depends upon the pipe materials, diameter and pipe conditions, type of flow and the units of the other terms; p = an exponent lying between 1.7 and 2.0; and r = and exponent lying between 4.7 and 5.0. Combining Eqs. 1 and 2 and eliminating D, we get

$$C = K_1 \ K_2^{y/r} \ L^{1+y/r} \ Q^{py/r} \ H_L^{-y/r} \quad (3)$$

or

$$C = K \ L^{1+y/r} \ Q^{py/r} \ H_L^{-y/r} \quad (4)$$

in which $K = K_1 \ K_2^{y/r}$

However, depending on the layout of the network, the length of the pipes vary widely. For example there are four ways of connecting three nodes, with one being source and the other two demand nodes, as shown in Figure 1. In a multi-nodes system, the total cost would be the sum of the cost of each pipe given by Eq. 4. As a first estimate in the optimization process, the connection between the demand nodes and the source nodes can be obtained from the network that produces the minimal total pipe length.

To obtain the optimal layout for a multiple-node system it will be necessary to first decide the general nature of the optimal layout giving the number of nodes. This can be achieve if the multiple-node distribution system is treated as a Steiner problem. The Steiner problem (9,10,11,13) is to find the network N connecting all nodes X so that the total length of N is minimal. There exist procedures for finding the minimum spanning tree whose network N is formed by placing straight lines between appropriate pairs of nodes. However, if one is allowed to add additional points to X (these points are called Steiner points, e.g. point J in Figure 1 d) as in the case of distribution network whose layout is not fixed a priori, the resulting minimum spanning tree is called a minimum Steiner tree for X (6). Computerized techniques for finding the minimum Steiner tree of a generalized network are available (4,6,7). By using the Steiner tree approach, one can obtain the minimum length layout of the system in

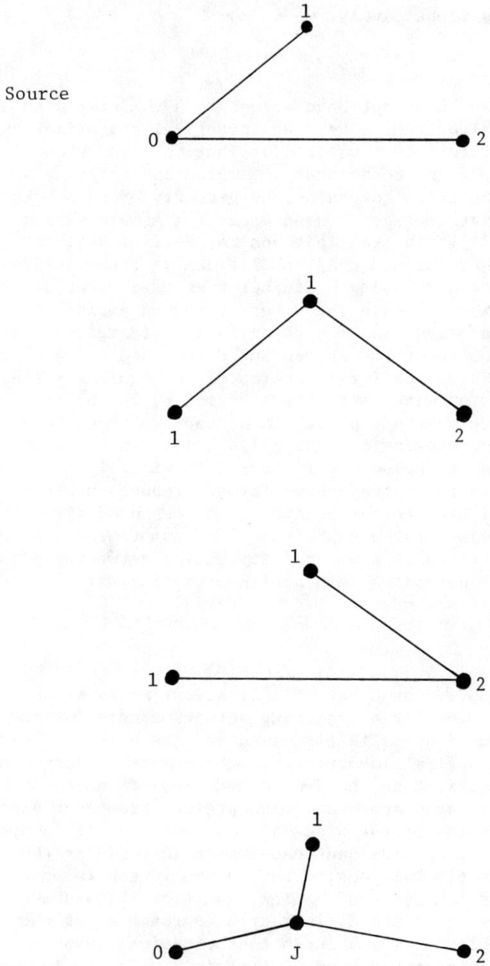

Figure 1. Four ways of connecting one source to two demands nodes.

Figure 2, for example, as that of Figure 3. Note that the total length of the layout in Figure 2 and the minimum Steiner tree in Figure 3 are 32.25, and 27.97 units respectively.

Soap Films Approach:

For investigators who do not have access to a computer with a minimum Steiner tree algorithm, a good and practical approximation of the minimum Steiner tree of a water distribution network can be obtained by using soap films techniques. Almgrem and Taylor have recently shown that the rules governing the geometry of soap bubbles and soap films are a mathematical consequence of a simple area minimizing principles (1,14). An excellent non-mathmetical description of the geometry of soap films and soap bubbles can be found in (2). The soap film procedure of finding a Steiner tree of a water distribution system is to draw to scale the water system on a piece of wood (preferably transparent plexiglass) with nails (or plexiglass rods) perpendicular to the board at the source and demand nodes. A piece of plexiglass is then attached firmly on top of the equidistant nails (rods) to form a sandwich with empty space formed by the nails (rods) in between. By submerging the sandwich in a soapy solution, one then observes a soap film connecting all the nails (rods) in the sandwich space. Since the area of the soap film formed is minimal, its cross-section which resembles the water system layout produces a Steiner tree. The location of the Steiner points can be measured directly from the plexi glass side of the sandwich. The geometric parameters from a minimum Steiner tree of a water distribution network can then be used as the initial estimates in a nonlinear optimization procedure to satisfy the draw-off demands.

SUMMARY

A procedure for obtaining a set of initial estimates of an optimal geometrical layout for a branching network consisting of several sources and demand nodes is discussed in this paper. The general nature of the optimal layout, i.e., the number of junction points and their connections to the demand and draw-off nodes is obtained by the Steiner tree approach. The Steiner tree approach considers only the lengths of the pipes but does not take into considerations the draw-off demands and head-losses in deciding the general nature of the optimal layout. But as the length is more predominant in the cost function of a link, as it can be assumed, for all practical purposes, that the Steiner tree approach gives the general nature of the layout which leads to the optimal layout. Further, the Steiner tree approach gives the minimum-length layout which also serves as a fairly good initial layout for further optimiztion to satisfy the various optimality criteria. Soap films techniques can also be used when computerized minimum Steiner tree algorithm is not available.

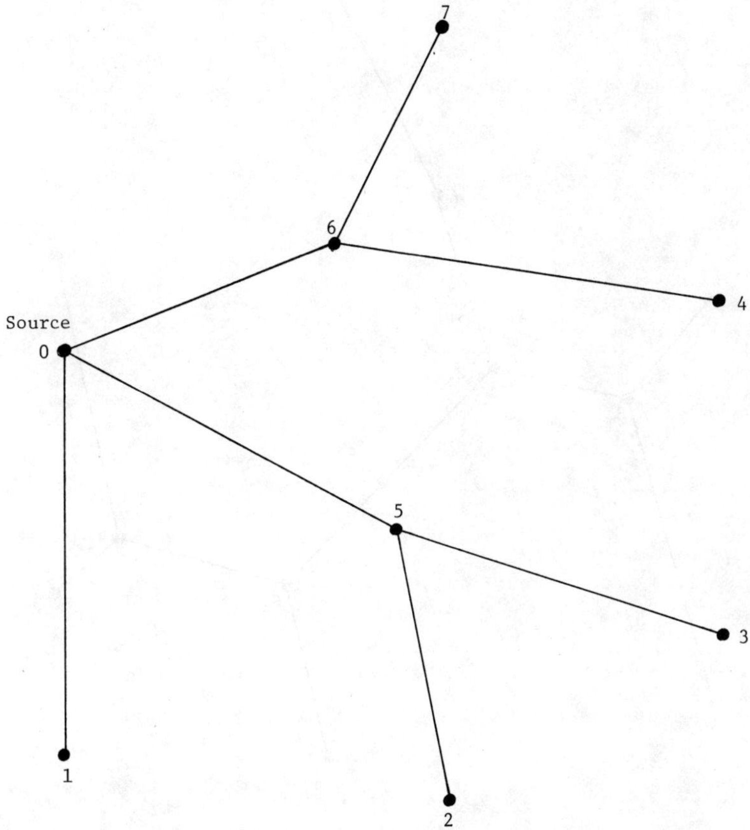

Figure 2. A water distribution system with one source and seven demand nodes.

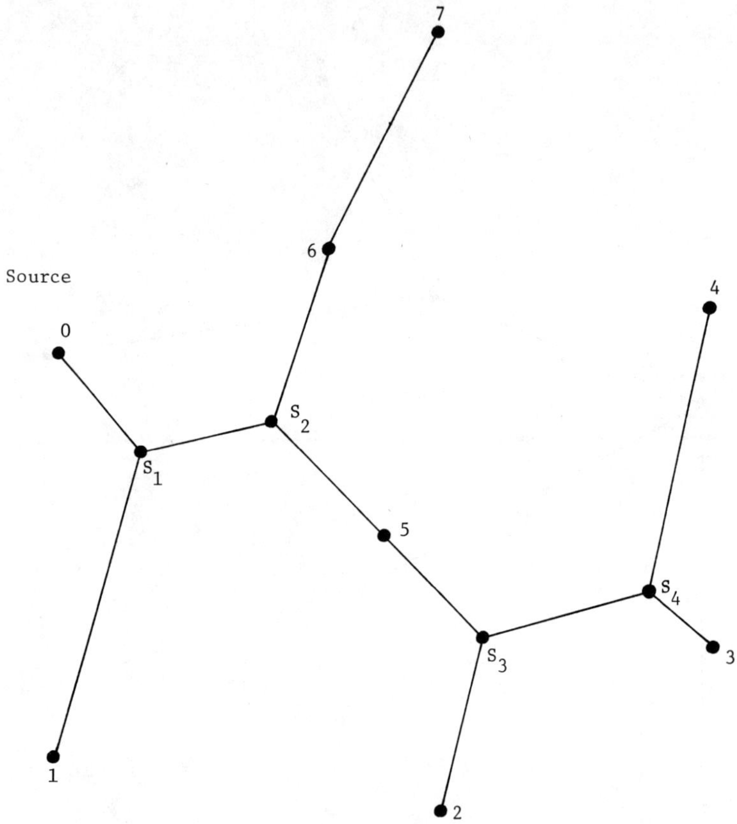

Figure 3. The minimum Steiner tree of the system shown in Figure 2.

REFERENCES

1. Almgrem, F.J. Jr., "Existence and Regularity Almost Everywhere of Solution of Elliptic Variation Problem with Constraints," Memior of American Mathematical Society, Vol. 4, No. 165, 1976.

2. Almgrem, F.J., and Taylor, J.E., "The Geometry of Soap Films and Soap Bubbles," Scientific American, July, 1976, pp. 82-93.

3. Bhave, P.R., "Noncomputer Optimization of Single-Source Networks," Journal of the Environmental Engineering Division, ASCE, Vol. 104, No. EEr, Proc. Paper 13956, Aug., 1978, pp. 799-814.

4. Boyce, W.M., and Seery, J.B., "Steiner 72, An Improved Version of Cockayne and Schiller's Program STEINER for the Minimal Network Problem," Bell Laboratories Technical Memorandum, Murray Hill, J.J., 1973.

5. Cembrowicz, R.G., and Harrington, J.J., "Capital Cost Minimization of Hydraulic Networks," Journal of the Hydraulics Division, ASCE, Vol. 99, No. HYe, Proc. Paper 9606, Mar., 1973, pp. 431-440.

6. Chung, F.R.K., and Graham, R.L., "Algorithm Aspects of Combinations - Steiner Trees for Ladders," Ann. of Discrete Mathematics, 1978, pp.173-200.

7. Cockayne, E.J., and Schiller, D.G., "Computation of Steiner Minimal Trees," Combinatories, D.J.A. Welsh and D.R. Woodall, eds., Inst. Math. Appl., 1972, pp. 53-71.

8. Deb, A.K., "Least Cost Design of Branched Pipe Network System," Journal of the Environmental Engineering Division, ASCE, Vol. 100, No. EE4, Proc. Paper 10711, Aug., 1974, pp. 821-835.

9. Garey, M.R., Graham, R.L., and Johnson, D.D., "The Complexity of Computing Steiner Minimal Trees," SIAM J. Appl. Math., 32 , 1977, pp. 835-859.

10. Gilbert, E.N., and Pollak, H.O., "Steiner Minimal Trees," SIAM J. Appl. Math., 16. 1968, pp. 1-29.

11. Graham, R.L., "Some Results on Steiner Minimal Trees," Bell Laboratories Technical Memorandum, Murray Hill, N.J., 1967.

12. Lam, C.F. and Wolla, M.L., "Computer Analysis of Water Distribution Systems: Part I. Formulation of Equations, J. of the Hydraulics Division, ASCE, Vol. 98, No. 412, Feb. 1972, pp. 335-344.

13. Shamos, M.I., and Hoey, D., "Closet-Point Problems," 16th Annual Symposium on Foundations of Computer Science, IEEE, 1975, pp. 151-152.

14. Taylor, J.E., "The Structure of Singularity in Soap-Bubbles-Like and Soap-Film-Like Minimal Surfaces," Ann. of Math., Vol. 103, 1976, pp. 489-539.

UNSTEADY FLOWS IN WATER DISTRIBUTION SYSTEMS

SÜHA SEVÜK

ABSTRACT

A general, flexible computer program in FORTRAN language is developed for the simulation of gradually and rapidly varying unsteady flows in municipal water distribution systems. The systems may include; network of pipes, valves, reservoirs, air chambers, surge tanks and pumps, as the elements. There exists no limitation on the number of elements, and no restriction on the network topology. In the mathematical model pipes are represented by the one dimensional unsteady flow equations of compressible fluids in slightly deformable conduits. Other elements are represented by their respective mass and energy conservation equations. These element equations are coupled at the junction points by the nodal continuity and compatibility equations. The element equations,node equations, and the appropriate boundary conditions are transformed into an unconditionally stable implicit finite difference form. These difference equations are then simultaneously solved using a modified Newton-Raphson procedure, and an efficient sparsity oriented programming technique. The utility of the proposed model is illustrated by solving a gradually varied unsteady flow, mass oscillations, and a water hammer problem in three hypothetical systems.

INTRODUCTION

In pipe networks, steady, gradually varied unsteady, rapidly varying transient flows, and periodic mass oscillations can be observed. For the analysis of steady pressurized flow in pipe networks various computer oriented numerical solution techniques are proposed (1,5,7) in the literature. Gradually varied unsteady flows in water distribution networks are simulated by quasy-steady solution techniques (4). Rapidly varying transient flows or the so called water hammer problems are solved numerically by using the Method of Characteristics (2,6) generally, and mass oscillation problems are solved by special finite difference techniques (3).

In this study an attemptis made for the simulation of all of the above flow occurences by a single mathematical model, and their solution by a single computer program.

MATHEMATICAL MODEL

From topological view point pressurized piping systems are either tree-type systems as in the case of water transmission lines and petroleum pipelines or network type systems with closed loops as in the case of

Assoc. Prof., Dept. of Civ. Engrg., Middle East Tech. Univ., Ankara, Turkey.

municipal water distribution systems. A node-link representation is used for the definition of system topology. In this representation system elements such as pipes, valves, pumps, reservoirs, air chambers, and surge tanks are considered as the links, and junction points of these elements are considered as the nodes, by assigning a distinct number to each link and node, the system topology is defined by two connectivity matrices. The first matrix shows the element numbers connected to each node while the second matrix gives the upstream and downstream node numbers of each link.

Mathematical model of flow at any particular time t, consists of:

1. Two link equations expressing the principles of conservation of mass and energy for each element,
2. A node equation expressing the principle of conservation of mass at the junction of elements, and,
3. An external boundary condition expressing either the inflow to the system, or the piezometric head at each node.

The primary variables in this mathematical model are;

1. Inflow to and outflow from each link, i.e., link flows,
2. Inflow to or outflow from each node, i.e., node flows, and
3. Piezometric head at each node.

If the values of these primary variables are known at time t (either as the initial conditions or as the results of previous computations), their unknown values at time $t+\Delta t$ can be computed by simultaneous numerical solution of link equations, node equations and boundary conditions.

As an example, the system shown in Figure 1a, consists of P=6 links and N=9 nodes. Mathematical model of this system consists of 2P=16 link equations, N=9 node equations, and N=9 boundary conditions, altogether 2P+2N=34 equations. These equations contain 2P=16 link flows, N=9 node flows, and N=9 node heads, altogether 34 primary variables. Starting with the known values of the variables at time t, the unknown values of 2P+2N variables at time $t+\Delta t$(Figure 1c), can be obtained by the simultaneous numerical solution of 2P+2N equations. The equations and their numerical solution technique are described in the following sections.

LINK EQUATIONS

Pipes.-In the mathematical model pressurised flow of slightly compressible fluids through slightly deformable pipes is represented by the one dimensional mass and momentum conservation equations (2,6):

$$\frac{\partial Q}{\partial x} + \frac{gA}{a^2}\frac{\partial h}{\partial t} = 0 \quad \text{.......} (1)$$

$$\frac{\partial Q}{\partial t} + gA\frac{\partial h}{\partial x} + f\,\frac{Q|Q|}{2DA} = 0 \quad \text{.......} (2)$$

In these quasilinear hyperbolic partial differential equations dependent variables Q, and h, are the flow rate and the piezometric head respectively, independent variables x and t are the distance along the pipe axis and the time, respectively. The coefficients; g=gravitational acceleration, D=diameter, f=friction factor, and

$$a = (K/\rho)^{\frac{1}{2}}/(1\ KD/Ee)^{\frac{1}{2}} \quad \text{.......} (3)$$

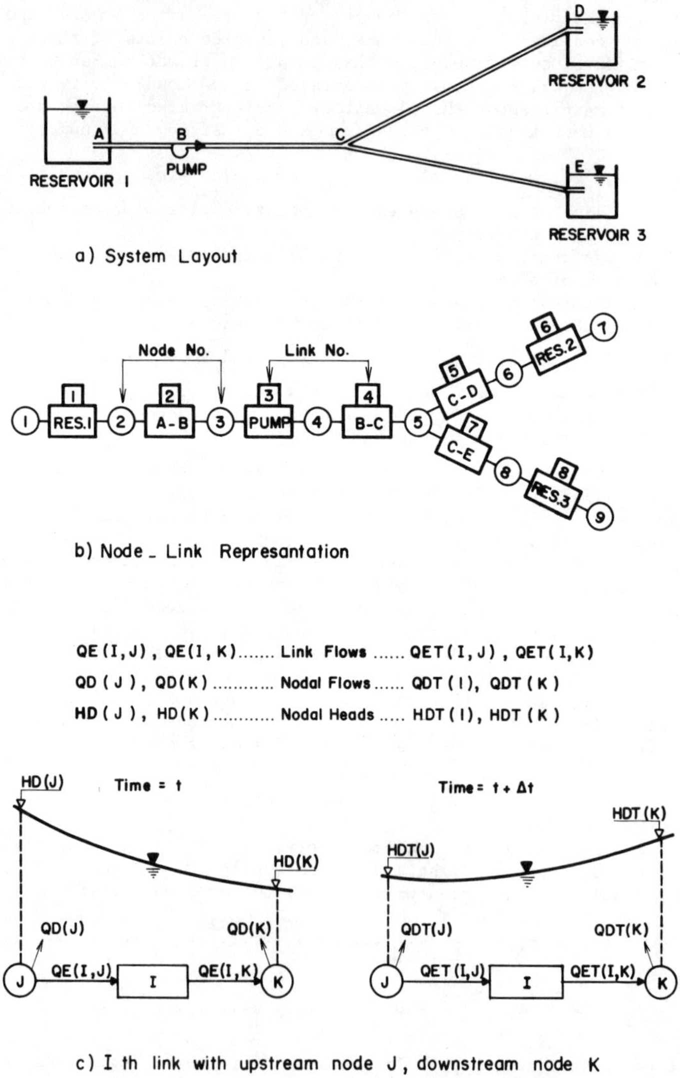

Figure 1- NODE - LINK REPRESENTATION AND NOTATION

is the velocity of propagation of elastic waves through the fluid. In Equation 3, K and E are the elastic modulus of the fluid and pipe walls, respectively, ρ is the fluid density, and e is the wall thickness of the pipe.

Equations 1 and 2 are put into the following implicit finite difference form (Figure 1c);

$$S(I)=\theta\left[QET(I,K)-QET(I,J)\right]/\Delta x + (1-\theta)\left[QE(I,K)-QE(I,J)\right]/\Delta x$$

$$+(gA/a^2)\left[HDT(J)+HDT(K)-HD(J)-HD(K)\right]/2\Delta t = 0 \quad \ldots\ldots\ldots\ldots\ldots\ldots (4)$$

$$M(I)=\left[QET(I,J)+QET(I,K)-QE(I,J)-QE(I,K)\right]/2\Delta t$$

$$+(gA/\Delta x)\left[\theta(HDT(K)-HDT(J)) + (1-\theta)(HT(K)-HT(J))\right]$$

$$+(f/4DA)\ \theta\left[QET(I,J)|QET(I,J) + QET(I,K)|QET(I,K)|\right] \quad \ldots\ldots\ldots (5)$$

$$+(f/4DA)\ (1-\theta)\ \left[QE(I,J)\ QE(I,J)| + QE(I,K)|QE(I,K)|\right] = 0$$

where; S(I), and M(I), denote respectively, the mass and momentum conservation equation of the pipe with link number I, Δx and Δt are the distance and time increments, respectively, and θ is a weighing factor. For stability and convergence of the solutions the optimum value of θ is found to be between 0.5 and 0.6.

Assuming that at time t, inflow to the pipe, QE(I,J), outflow from the pipe, QE(I,K), piezometric heads HD(J) and HD(K) are known, the corresponding values at time $t+\Delta t$, i.e.,QET(I,J) QET(I,K), HDT(J), and HDT(K) are the four unknowns in Equations 4 and 5.

Valves.-Valves are the links with no storage capability. Therefore conservation of mass principle;

$$S(I) = QET(I,J) - QET(I,K) = 0 \quad \ldots\ldots\ldots\ldots\ldots\ldots\ldots\ldots\ldots (6)$$

expresses the equality of inflow, QET(I,J) and outflow, QET(I,K). Conservation of energy principle, can be expressed by the orifice equation;

$$QET(I,J)=C_vA_v\sqrt{2g\ HDT(J)-HDT(K)} \qquad HDT(J)>HDT(K) \quad \ldots\ldots\ldots\ldots (7a)$$

$$QET(I,J)=C_vA_v\sqrt{2g\ HDT(K)-HDT(J)} \qquad HDT(K)>HDT(J) \quad \ldots\ldots\ldots\ldots (7b)$$

in which, A_v is the open cross sectional flow area, and C_v is the valve discharge coefficient which is a function of the valve type and valve opening, A_v.

Valve operation if any can be described (6) by a function, $\tau(t)=C_v\ A_v$, and thus the energy conservation equation takes the following form:

$$M(I)=QET(I,J)-\tau(t+\Delta t)\sqrt{2g\left[HDT(J)-HDT(K)\right]} \qquad HDT(J)>HDT(K) \quad \ldots.. (8a)$$

$$M(I)=QET(I,J)-\tau(t+\Delta t)\sqrt{2g\left[HDT(K)-HDT(J)\right]} \qquad HDT(K)>HDT(J) \quad \ldots.. (8b)$$

Reservoir.-Conservation of mass principle in reservoirs can be expressed as;

$$QDT(J)-QET(I,K)=A_d\ \frac{dHd}{dt} \simeq A_d\left[H_d(t+\Delta t)-H_d(t)\right]/\Delta t \quad \ldots\ldots\ldots\ldots\ldots (9)$$

where; QDT(J) inflow to the reservoir, QET(I,K) outflow from the reservoir, A_d horizontal area of the reservoir, and H_d reservoir fluid level.

Neglecting the kinetic energy and the local energy losses, conservation of energy equation takes the form of a kinematic compatibility condition, i.e.

$$H_d(t+\Delta t) = HDT(J) = HDT(K) \quad \text{(10a)}$$

$$H_d(t) = HD(J) = HD(K) \quad \text{(10b)}$$

eliminating H_d from equations 9 and 10;

$$S(I) = QDT(J)-QET(I,K)-[HDT(J)-HD(J)]\, A_d/\Delta t \quad \text{(11)}$$

$$M(I) = HDT(J)-HDT(K) = 0 \quad \text{(12)}$$

the link equations for a reservoir can be obtained in terms of the primary unknown variables.

Air Chambers.-Conservation of mass principle in air chambers can be expressed as:

$$QET(I,J)-QET(I,K) \simeq Q_k \simeq -[V(t+\Delta t)-V\ (t)]/\Delta t \quad \text{(13)}$$

where; Q_k is the fluid inflow into the chamber and V is the air volume in the chamber, which is assumed to follow reversible polytropic law, given by,

$$PV^{1.2} = K_o = \text{constant} \quad \text{(14)}$$

The absolute air pressure P in Equation 14 can be expressed in terms of height of fluid column as follows,

$$P = H_{at} - [H_t-(V(t+\Delta t)/A_k)]\, H_k \quad \text{(15)}$$

in which, H_{at}=atmospheric pressure head in meters of fluid, H_t=topographic elevation of the chamber ceiling, A_k=cross sectional area of the chamber and H_k is the piezometric head at the air-fluid interface, expressed by the orifice equation,

$$H_k = HDT(J) - C_o Q_k |Q_k| \quad \text{(16)}$$

in which C_o is the coefficient of differential orifice at the chamber entrance.

Eliminating the variables P, Q_k, and $V(t+\Delta t)$ from Equations 13 through 16, the conservation of mass principle can be expressed in terms of the primary unknowns as follows,

$$\begin{aligned} S(I) &= HDT(J)-K_o/[V(t)-QET(I,J)-QET(I,K)\Delta t] \\ &+ [V(t)-(QET(I,J)-QET(I,K))\Delta t]/A_k - H_t + H_{at} \\ &- C_o[QET(I,J)-QET(I,K)]\ |QET(I,J)-QET(I,K)| = 0 \quad \text{(17)} \end{aligned}$$

Conservation of energy principle states the equality of piezometric heads at the upstream and downstream nodes of the chamber, i.e.

$$M(I) = HDT(J) - HDT(K) = 0 \qquad (18)$$

and together with Equation 17, constitutes the link equations for an air chamber.

Surge Tanks.-Conservation of mass principle in surge tanks can be expressed as,

$$QET(I,J)-QET(I,K)=A_b\, dH_b/dt \simeq A_b\left[H_b(t+\Delta t)-H_b(t)\right]/\Delta t \qquad (19)$$

in which, A_b is the cross sectional area of the tank, and H_b is the fluid level in the tank.

Conservation of energy principle, is given by;

$$HDT(J)=HDT(K)=H_b(t+\Delta t)+\Delta h(t+\Delta t) \qquad (20a)$$

$$HD(J)=HD(K)+\Delta h(t) \qquad (20b)$$

in which Δh is the local energy loss, which can be neglected in the case of simple surge tanks, and which can be expressed by the orifice equation,

$$\Delta h(t+\Delta t)=C_b\left[QET(I,J)-QET(I,K)\right]\left|QET(I,J)-QET(I,K)\right| \qquad (21a)$$

$$\Delta h(t)=C_b\left[QE(I,J)-QE(I,K)\right]\left|QE(I,J)-QE(I,K)\right| \qquad (21b)$$

in case of differential surge tanks, with a differential orifice coefficient C_b.

Eliminating H_b and Δh from Equations 19, 20 and 21 the two link equations for a surge tank,

$$\begin{aligned} S(I)=&QET(I,J)-QET(I,K) - A_b\left[HDT(J)-HDT(K)\right]/\Delta t \\ &+A_bC_b\left[QET(I,J)-QET(I,K)\right]\left|QET(I,J)-QET(I,K)\right|/\Delta t \\ &-A_bC_b\left[QE(I,J)-QE(I,K)\right]\left|QE(I,J)-QE(I,K)\right| \end{aligned} \qquad (22)$$

$$M(I) = HDT(J)-HDT(K) \qquad (23)$$

can be obtained in terms of the four primary unknown variables.

Pumps.-Conservation of mass principle through pumps can be expressed as,

$$QET(I,J)=QET(I,K) = QP \qquad (24)$$

in which QP is the pump discharge. The energy balance across a pump yields,

$$HDT(J)+HP = HDT(K) \qquad (25)$$

in which HP is the head supplied by the pump. Pump discharge QP, pump head HP, pump torque TP and rotational speed NP are related to each other by the so called pump characteristic curves. The curves can be plotted in a dimensionless-homologous form, and straight line segments of the curves can be used to approximate them:

$$\phi/\alpha^2=a_o+a_1\, \eta/\alpha \qquad (26)$$

$$\beta/\alpha^2=b_o+b_1+\eta/\alpha \qquad (27)$$

in which; ϕ=HP/HR, α=NP/NR. β=TP/TR, η=QP/QR, and HR, NR, TR, and QR are respectively, the rated head, rated speed, rated torque and rated discharge of the pump. The values of the coefficients a_o, a_1, b_o and b_1 can be determined from the piecewise linearized characteristic curves.

The sudden interruption of power to a pump causes a reduction in rotational speed. The rate of change of rotational speed is given by,

$$TP=-IP\ dw/dt=-IP(2\pi/60)dNP/dt \qquad (28)$$

in which IP is the polar moment of inertia, and dW/dt is the angular acceleration of the pump. Equation 28 can be put into the dimensionless form,

$$TP/TR=-(IP/TR)\ (2\pi/60)\ NR\ d(NP/NR)/dt \qquad (29)$$

and hence a relation between the dimensionless variables β and α can be obtained as follows;

$$\beta=-KP\ d\alpha/dt \qquad (30)$$

where, KP=IP 2π NR/60 is the fly-whell coefficient of the pump.

Eliminating β from Equation 27 and 30,

$$b_o\alpha^2 + b_1\eta\alpha=-KP\ d\alpha/dt \qquad (31)$$

and integrating this equation numerically;

$$b_0\alpha^2(t)+b_1\eta(t)\alpha(t)\simeq-KP\left[\alpha(t+\Delta t)-\alpha(t)\right]/\Delta t \qquad (32)$$

the unknown rotational speed at time t+Δt i.e, $\alpha(t+\Delta t)$ can be expressed in terms of the known variables $\alpha(t)$ and $\eta(t)$. Having thus computed the value of α, substituting into Equation 26,

$$\phi = a_o\alpha^2+a_1\eta\alpha \qquad (33)$$

and hence;

$$HP = HR\left[a_o\alpha^2+a_1\alpha QET(I,J)/QR\right] \qquad (34)$$

can be obtained, in which HP is expressed in terms of the primary unknown variable QET(I,J). Therefore the conservation of mass and energy equations (Equations 24 and 25) can be reduced into the following form

$$S(I) = QET(I,J)-QET(I,K) = 0 \qquad (35)$$

$$M(I) = HDT(J)+HR\left[a_o\alpha^2+a_1\alpha QET(I,J)/QR\right]-HDT(K) = 0 \qquad (36)$$

which include the primary variables, QET(I,J), QET(I,K), HDT(J), and HDT(J) as the only unknowns.

NODE EQUATIONS

These equations express the conservation of mass principle, or in other words, the equality of the inflows and outflows at each node.

$$D_j = QDT(J)+\Sigma QET(I,J) = 0, \quad j=1,.....N \quad .... \quad (37)$$

D_j denotes the node equation at the jth node, QDT(J) is the nodal flow, and the last term is the algebraic sum of the link flows joining to the jth node. According to the selected sign convention inflows to the node are assumed to be positive and outflows from the node are assumed to be negative.

BOUNDARY CONDITIONS

For a unique solution of the link and node equations an external boundary condition must be specified at each node. These boundary conditions express either the nodal flow;

$$QD(J) = G_j(t), \; QDT(J) = G_j(t+\Delta t) \quad .... \quad (38)$$

or the piezometric head;

$$HD(J) = F_j(t), \; HDT(J) = F_j(t+\Delta t) \quad .... \quad (39)$$

at each node. In these expressions G_j and F_j are known functions of time.

NUMERICAL SOLUTION TECHNIQUE

Mathematical model of a system with P links and N nodes consists of, 2P link equations;

$$S_I\left[QET(I,J),QET(I,K),HDT(J),HDT(K)\right] = 0,$$

$$M_I\left[QET(I,J),QET(I,K),HDT(J),HDT(K)\right] = 0, \quad I=1,2,.....,P$$

N node equations;

$$D_j\left[QDT(J),QET(I,J)\right] = 0 \quad J=1,2,.....,N$$

and N boundary conditions;

$$F_j(t+\Delta t) = HDT(J), \; j=1,2,.....N_1$$

$$G_j(t+\Delta t) = QDT(J), \; j=n_1+1,....N$$

These 2P+2N equations contain 2P link flows; QET(I,J), QET(I,K), I=1,2,..P, and N nodal flows; QDT(J), J=1,2,...N, as unknowns

An investigation of the structure of the equations indicates that; the link equations are in general nonlinear, the node equations are linear, and the equations expressing boundary conditions each contain a single unknown. Therefore at the first step of the numerical solution, equations expressing the boundary conditions are solved one at a time for a single unknown, and hence the total number of unknowns to be determined simultaneously is reduced to 2P+N. At the second step then, these 2P+N unknowns are obtained by simultaneous numerical solution of the 2P link equations and N node equations, using Generalized-Newton-Raphson technique. Starting with the known values of the unknowns at time t, the solution is thus advanced to time $t+\Delta t$.

COMPUTER PROGRAM

Computer program of the mathematical model is written in FORTRAN and does not contain any system-specific-instructions. Therefore it can be executed on any computer system which includes a FORTRAN compiler.

Memory requirement is approximately; $70000+4|18N+2PN+3(2P+N)^2|$ bytes. In this expression; P is the number of links, and N is the number of nodes of the pipe system to be analyzed.

Execution time of the program likewise depends on the size of the pipe system. The most time consuming computation in the program is the simultaneous numerical solution of the 2P+N algebraic equations at each time step. Therefore execution time increases approximately with $(2P+N)^2$. For instance for two systems with 2P+N=14, and 62, the observed execution times on IBM 370/145 for advancing the numerical solutions one time step, Δt, are 1.33 secs, and 27.08 secs. respectively.

The computer program is general and flexible. Any piping system consisting of pipes, valves, pumps, reservoirs, surge tanks, and air chambers can be analyzed without any restriction on the system topology and any limitation on the number of system elements.

APPLICATIONS OF THE MATHEMATICAL MODEL-EXAMPLES

The proposed mathematical model can simulate unsteady as well as steady pressurised flows in complicated pipe systems.

Steady flows can be simulated by specifying time independent external boundary conditions. This dynamic relaxation technique is particularly usefull for the determination of steady initial flow conditions in comlicated pipe networks at the first step of unsteady flow computations.

Unsteady flows can be simulated by specifying time dependent external boundary conditions and/or a variation in a link behaviour (such as a pump failure or a valve operation) with time. Depending on the time rate of change of these conditions, flows can be classified as gradually varied unsteady flow, mass oscillations, or water-hammer. In the following sections a solved example is given for each of these types of fluid flow.

Gradually Varied Unsteady Flow.- A simple hypothetical municipal water distribution system is shown in Figure 2a. Water is supplied from a fixed head reservoir at node 1 and is consumed at nodes 2, 3, 4, 5, 6, 7, and 8, at equal rates (Table 1). There exists a small (250 m^2 area) equalizing reservoir at node 9.

Table 1- HOURLY VARIATION OF NODAL CONSUMPTION (lt/sec)

Hour	00	02	04	06	08	10	12	14	16	18	20	22	24
Cons	15	10	10	15	30	30	40	30	20	20	40	30	15

Simulation of the system behaviour for 24 hours is realized by numerical solution of the mathematical model using Δt=1 hr. time increments. Part of the results are shown in Figure 2b and 2c.

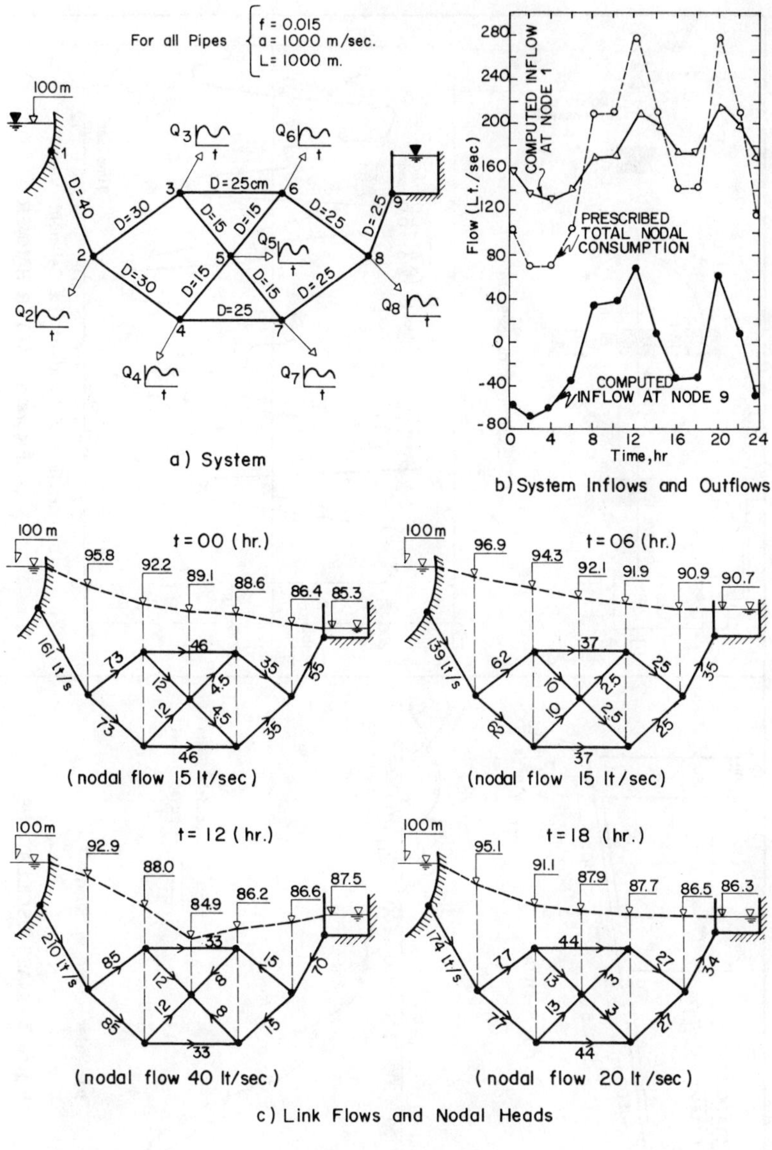

Figure 2- GRADUALLY VARIED UNSTEADY FLOW IN A PIPE NETWORK

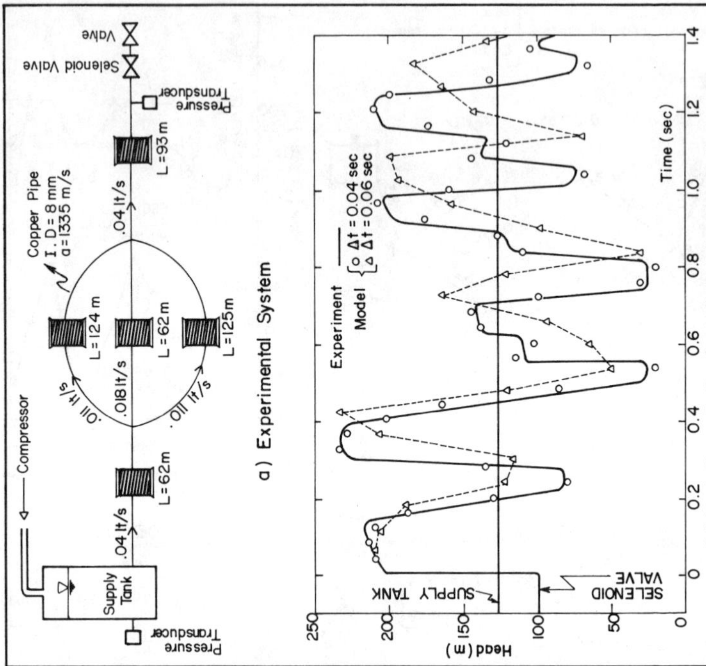

b) Variation of Head at the Selenoid Valve

Figure 4 - WATER HAMMER

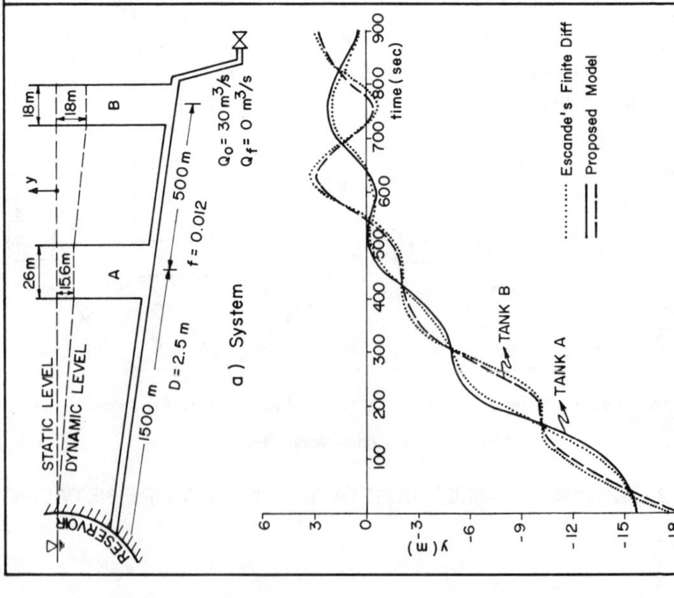

b) Water Level in Surge Tanks

Figure 3- MASS OSCILLATIONS

Mass Oscilations.- A hypothetical reservoir-pressure tunnel-surge tank-penstock combination of a hydroelectric installation is schematically shown in Figure 3a. A constant initial steady flow of 30 m^3/sec through the system is assumed to be stopped suddenly by an instantaneous closure of the turbine valves located at the outlet of the penstock. The pressure waves thus set up propagate up the penstock and create water level fluctuations (mass oscillations) in the surge tanks. Numerical solution of this oscillation problem is obtained by using Δt=5 sec. time increments, and the computed water level fluctuations in the surge tank are shown in Figure 3b. Also shown in Figure 3b for comparison are the results obtained by the Escande's finite difference technique (3).

Water Hammer.- In the simple experimental system shown in Figure 4a, initial steady flow is disturbed by an instantaneous closure of the selenoid valve located at the system outlet. The pressure waves thus set up are recorded by the help of a pressure transducer, located next to the selenoid valve. Numerical solution of this transient flow is obtained by using Δt=0.04 and Δt=0.06 sec. time increments. Recorded and computed piezometric heads at the selenoid valve are shown in Figure 4b. Comparison of these experimental and numerical results indicate that; the time increment Δt to be used in numerical computations must be smaller than the smallest pipe period (T=L/a 62/1335=.046 sec) in the system, as is the case in the Method of Characteristics (2, 6).

REFERENCES

1. Cross, H. "Analysis of Flow in Networks of Conduits or Conducters". Univ. of Illinois, Bulletin, No. 286.

2. Fox, J. A. "Hydraulic Analysis of Unsteady Flow in Pipe Networks". MacMillan Press, London 1977.

3. Pickford, J. "Analysis of Water Surge". Gordan and Breach Pub., New York 1969.

4. Rao, H. S., and, Bree, D. W., "Extended Period Simulation of Water Systems-Part A" Journal of the Hydraulics Div., ASCE, Vol. 103, No. HY2.

5. Shamir, U. and Howard, C. D. D. "Water Distribution System Analysis". Journal of the Hydraulics Div., ASCE, Vol. 94, No. HY1.

6. Streeter, V. L., and Wylie, B. E., "Hydraulic Transients". McGraw Hill Inc., New York 1967.

7. Wood, J. D., and Charles, C. A., "Hydraulic Network Analysis Using Lineer Network Theory". Journal of the Hydraulics Div., ASCE, Vol. 98, No. HY7.

USE OF "CORPS" FOR HYDRAULIC DESIGN

By Bobby J. Brown,[1] and John J. Ingram,[2] A. M. ASCE

ABSTRACT: The Hydraulic Analysis Division, Hydraulics Laboratory, U. S. Army Engineer Waterways Experiment Station (WES), has the responsibility of developing and disseminating hydraulic design guidance for the Corps of Engineers. In the last 10 years, a computer-aided design system, Conversationally Oriented Real-Time Program-Generating System (CORPS), has been developed and is operational to supplement other guidances. The computer programs in CORPS are executed in the interactive mode and can be easily used by noncomputer-oriented hydraulic designers. Input data requests are presented in an easy to understand format while output may be tabular and/or graphical on CRT type terminals. This paper presents the utility and practicality of this computer-aided hydraulic design system.

The Hydraulic Analysis Division, Hydraulics Laboratory, U. S. Army Engineer Waterways Experiment Station (WES), has the responsibility of developing and disseminating hydraulic design guidance for the Corps of Engineers. Since 1952, this program has been a continuous effort and the guidance has been published in the form of "Hydraulic Design Criteria," "Engineer Manuals," "Engineer Technical Letters," etc. In the last 10 years, a computer-aided design system, Conversationally Oriented Real-Time Program-Generating System (CORPS), has been developed and is operational to supplement these other guidances. Presently, 60 computer programs for hydraulic design (listed in Table 1) are available to Corps offices through regional computer facilities in Seattle, Washington; Macon, Georgia; and Vicksburg, Mississippi. Some current areas covered by these programs are: spillways, stilling basins, outlet works, locks, closed conduit flow, open-channel hydraulics, stable channel design, and sediment transport. Each computer program, as developed, is thoroughly checked to be as foolproof as possible. A documentation for each program is prepared according to category "A" standards established by the Office, Chief of Engineers (OCE), with emphasis on the step-by-step engineer rationale of each program. A copy of the documentation, when approved by OCE, is filed in the Engineer Computer Program Library (ECPL) at the WES.

---

[1] Hydraulic Engineer, Hydraulic Analysis Division, U. S. Army Engineer Waterways Experiment Station, Vicksburg, MS.

[2] Hydraulic Engineer, Hydraulic Analysis Division, U. S. Army Engineer Waterways Experiment Station, Vicksburg, MS.

TABLE 1

HYDRAULIC COMPUTER PROGRAMS

H0011	KINEMATIC VISCOSITY OF WATER-EFFECTS OF TEMPERATURE
H0114	BRIDGE PIER LOSSES-RECT OR TRAP SECT-ANY FLOW CLASS
H0920	TOTAL SEDIMENT TRANSPORT RATE IN SAND BED STREAMS BY COLBY'S METHOD
H0921	BED-LOAD TRANSPORT IN RIVERS BY EINSTEIN'S METHOD
H0922	TOTAL SEDIMENT LOAD BY MODIFIED EINSTEIN PROCEDURE
H0923	BED-LOAD TRANSPORT RATE BY MEYER-PETER MULLER'S METHOD
H0924	COMPUTATIONS OF SEDIMENT DISCHARGE IN RIVERS BY SHEN AND HUNG'S METHOD
H0925	TOTAL SEDIMENT DISCHARGE BY YANG'S METHOD
H0926	SAND DISCHARGE BY TOFFALETI'S METHOD
H0941	STABLE CHANNEL DESIGN
H1102	*STANDARD HIGH SPILLWAY CREST COORDINATES
H1103	*STAGE-DISCH RELATION FOR STANDARD SPILLWAY (UNGATED)
H1105	*STAGE-DISCH RELATION-SPILLWAY CREST-UNCONTROLLED FLOW
H1111	*STANDARD SPILLWAY CREST WATER SURFACE ELEV-HIGH DAM
H1116	*STANDARD SHAPE SPILLWAY CREST PRESSURE-HIGH DAM
H1121	OPTIMUM STILLING BASIN DESIGN
H1170	*HIGH SPILLWAY CREST COORDINATES- 3-1 UPSTREAM FACE
H1180	*HIGH SPILLWAY CREST COORDINATES- 3-2 UPSTREAM FACE
H1190	*HIGH SPILLWAY CREST COORDINATES- 3-3 UPSTREAM FACE
H2030	*DISCH-PRESSURE CONDUITS-DARCY-WEISBACH FORMULA
H2033	*DISCH-PRESSURE CONDUITS-MANNING FORMULA
H2040	*GEOMETRIC ELEMENTS OF A HORSESHOE CONDUIT
H2041	*GEOMETRIC ELEMENTS OF A RECTANGULAR CONDUIT
H2042	*GEOMETRIC ELEMENTS OF AN OBLONG CONDUIT
H2043	*DISCHARGE IN HORSESHOE COND (MAN OR DAR FORM)
H2044	*DISCHARGE FOR RECT CONDUIT (DAR OR MAN)
H2045	*DISCH IN AN OBLOG OR CIRC CONDUIT (MAN OR DAR FORM)
H2250	RESERVOIR STORAGE DRAWDOWN TIME
H2251	OUTLET WORKS LOSS COEFF FROM PROTOTYPE MEASUREMENTS OF DRAWDOWN
H2261	STILLING BASIN DESIGN FOR CONDUIT OUTLET WORKS
H3106	*STANDARD DISCH RELATION-TAINTER GATE ON CURVE CREST
H3201	*DISCH FOR PARTLY OPEN VERTICAL LIFT CONDUIT GATES
H5310	*SURGES IN NAVIGATION CHANNELS
H5320	LOCK FILL AND EMPTY--SYMMETRICAL SYSTEMS
H6001	*GEOMETRICAL ELEMENTS-TRAP, RECT, TRIA - CHANNEL
H6002	*GEOMETRIC ELEMENTS OF CIRCULAR CONDUIT
H6005	GEOMETRIC ELEMENTS OF A GENERAL CHANNEL CROS-SECT
H6110	*NORMAL DEPTH-TRAP, RECT, TRIA SECT - MANNING FORMULA
H6111	*NORMAL DEPTH AND VELOCITY - CIRC CONDUIT - MANN FORMULA
H6112	*NORMAL DISCHARGE - MANNING FORMULA
H6113	*NORMAL DEPTH - TRAP, RECT, TRIA SECT - CHEZY FORMULA
H6114	*NORMAL DISCHARGE - CHEZY FORMULA
H6115	*NORMAL DEPTH AND VELOCITY - CIRC CONDUIT - CHEZY
H6116	*NORMAL DISCHARGE - COLEBROOK WHITE FORMULA
H6117	*NORMAL DEPTH - TRAP, RECT, TRIA SECT - COLEBROOK WHITE
H6118	*NORMAL DEPTH AND VEL - CIRC CONDUIT - COLEBROOK WHITE
H6122	CONJUGATE DEPTH - CIRCULAR SECTION
H6123	CONJUGATE DEPTH - TRAP, RECT, TRIA - SECTION

continuation of Table 1

H6124	SPECIFIC FORCE AND ENERGY - TRAP, RECT, TRIA - SECTION
H6125	SPECIFIC FORCE AND ENERGY - CIRCULAR SECTION
H6140	*CRITICAL DEPTH AND VELOCITY - TRAP, RECT, TRIA SECT
H6141	*CRITICAL DEPTH AND VELOCITY FOR CIRCULAR CONDUIT
H6201	*FRICTION SLOPE - ANY FLOW SECT - MANNING, CHEZY, OR COLEBROOK WHITE FORMULA
H6208	*FLOW PROFILE - CIRC CONDUIT - MANNING, CHEZY, OR COLEBROOK WHITE FORMULA
H6209	*FLOW PROFILE - PRISMATIC CHANNEL - MANNING, CHEZY, OR COLEBROOK WHITE FORMULA
H6210	LDR-I BACKWATER PROFILES
H6602	SPIRAL BANKED CURVE-SUPERCRITICAL FLOW-RECT X SECT
H7010	RIPRAP REQUIREMENTS FOR OPEN CHANNELS
H7220	*EROSION AT CULVERT OUTLETS AND RIPRAP REQUIREMENTS
H7780	WAVE RUNUP AND WIND SETUP, COMPUTATION MODEL

* Denotes OCE approved category "A" programs.

Some advantages of a computer-aided design system such as this are as follows: (a) it provides a library of thoroughly tested and completely documented hydraulic design programs with an effective means of communication for the noncomputer-oriented designer; (b) the programs are written for Engineers by Engineers, thus providing a means of computer communication in the user's own disciplinary language; (c) the programs are stored in machine language, thereby eliminating expensive compilation and debugging time, plus, the programs are executed in real time providing immediate results; (d) a rerun option allows the user to evaluate alternative solutions by changing any or all of the input variables after each run; (e) a plot capability is available on many of the programs, whereby the output may be displayed graphically for users with cathode ray tube (CRT) equipment; and (f) the system is easily accessible to all Corps offices through regional facilities and the process of higher authority review is thereby enhanced.

A potential user of "CORPS" can easily run any of the hydraulic design programs as long as he knows how to access "CORPS." Once the system had been accessed, any of the available programs can be immediately run or the following information may be retrieved:

(a) a description of "CORPS,"
(b) a listing of the available programs,
(c) a brief description of any of the programs in the system, and
(d) an execute command for any of the programs.

The practicality of this computer system may best be described by its application to some possible problems.

PROBLEM STATEMENT 1

A free overflow concrete spillway stilling basin is to be designed for a crest elevation of 463 ft, a design head of 17 ft, a face slope factor of 0.78, and a crest width of 150 ft. The fraction of D_2 that is to intersect the tailwater surface elevation is 0.9.

With these given conditions, H1121 - Optimum Stilling Basin Design is selected to design the spillway stilling basin. A brief description of this program, as found in the "CORPS," is as follows:

H1121 - OPTIMUM STILLING BASIN DESIGN

THIS PROGRAM IS USED TO DETERMINE AN OPTIMUM STILLING BASIN DESIGN FOR A FREE OVERFLOW SPILLWAY WITH A VERTICAL UPSTREAM FACE.

INPUTS REQUIRED ARE SPILLWAY CREST ELEVATION, DESIGN POOL SURFACE ELEVATION, SPILLWAY FACE SLOPE FACTOR, SPILLWAY SURFACE ROUGHNESS, PERCENT OF D2 TO INTERSECT THE TAILWATER SURFACE, BASIN WIDTH, EFFECTIVE WIDTH OF THE SPILLWAY CREST, AND TAILWATER SURFACE ELEVATION FOR A COMPUTER DISCHARGE BASED UPON THE PREVIOUS INPUT VALUES.

OUTPUT PRESENTS A LISTING OF THE OPTIMUM DESIGN CHARACTERISTICS FOR THE INPUT VARIABLES.

To run H1121, the following command is entered at the user's terminal and the computer will then respond. (Note: All input provided by the user is underlined.)

```
*RUN WESLIB/CORPS/H1121,R

INPUT-H1121

AA-ENTER THE CREST ELEVATION OF THE FREE OVERFLOW SPILLWAY
IN FT.
=463
AB-ENTER THE DESIGN POOL SURFACE ELEVATION IN FT.
=480
AC-ENTER THE FACE SLOPE FACTOR. THIS FACTOR IS EQUIVALENT
TO THE LATERAL CHANGE OF THE FACE SLOPE PER UNIT CHANGE
IN THE VERTICAL.
=.78
AD-ENTER THE SURFACE ROUGHNESS (K). A VALUE OF 0.002 FT
IS RECOMMENDED FOR CONCRETE.
=.002
AE-ENTER THE PERCENT OF D2 (.85 THROUGH 1.00) THAT IS TO
INTERSECT THE TAILWATER SURFACE ELEVATION. NOTE: FOR
DESIGNS WITH THE TAILWATER AT 0.85 D2, THE VELOCITY OF FLOW
AT THE FIRST ROW OF BAFFLES WILL BE VERY LITTLE LESS THAN
THE VELOCITY OF THE FLOW ENTERING THE BASIN. THUS, IN HIGH
DAMS SOME CAVITATION DAMAGE TO THE BAFFLE PIERS SHOULD BE
EXPECTED AT FLOWS WHICH RESULT IN TAILWATER DEPTHS LESS THAN
D2. THEREFORE, THE SELECTION OF A PERCENT OF D2 TO INTERSECT
THE TAILWATER SHOULD BE DEPENDENT UPON THE FREQUENCY OF THESE
HIGHER RATES OF FLOW.
=.9
AF-ENTER THE WIDTH OF THE SPILLWAY AND BASIN IN FT.
=150
```

Continuation of INPUT-H1121.

```
AG-ENTER THE EFFECTIVE WIDTH OF THE SPILLWAY CREST IN FT.
(SEE HDC 111-3/1)
=150
ENTER THE TAILWATER SURFACE ELEVATION FOR A COMPUTED DISCHARGE
OF 42371.1 CFS.
=425

OUTPUT-H1121

DISCHARGE PER UNIT WIDTH                    =    282.5 CFS
SPILLWAY CREST ELEVATION                    =    463.0 FEET
POOL SURFACE ELEVATION                      =    480.0 FEET
TAILWATER SURFACE ELEVATION                 =    425.0 FEET
THE PERCENT OF D2 TO MATCH
 THE TAILWATER SURFACE                      =      0.90
WIDTH OF SPILLWAY AND BASIN                 =    150.0 FEET
FACE SLOPE FACTOR                           =      0.78 FT/FT
SPILLWAY SURFACE ROUGHNESS, K               =    0.20E-02

BASIN CHARACTERISTICS

APRON ELEVATION                             =    395.1 FEET
VERTICAL DIFFERENCE OF APRON TO CREST       =     67.9 FEET
HEIGHT OF BAFFLE PIERS                      =      4.0 FEET
HEIGHT OF END SILL                          =      2.8 FEET
WIDTH OF BAFFLE PIERS                       =      3.9 FEET
NUMBER OF BAFFLES IN FIRST ROW              =     19
DISTANCE FROM TOE TO FIRST BAFFLE           =     54.7 FEET
DISTANCE FROM THE FACE OF THE FIRST
 BAFFLE TO THE SECOND                       =     10.0 FEET
MINIMUM APRON LENGTH FOR LESS THAN
 SPILLWAY DESIGN FLOW                       =     88.0 FEET
CONNECTING RADIUS FROM SPILLWAY
 TO STILLING BASIN APRON, R                 =     50.0 FEET

HYDRAULIC CHARACTERISTICS

CRITICAL DEPTH OF FLOW                      =     13.5 FEET
FLOW DEPTH AT SPILLWAY TOE                  =      4.0 FEET
SEQUENT FLOW DEPTH, D2                      =     33.2 FEET
FLOW DEPTH OVER THE END SILL                =     27.2 FEET
CRITICAL FLOW VELOCITY                      =     20.9 FPS
FLOW VELOCITY AT SPILLWAY TOE               =     72.9 FPS
FLOW VELOCITY AT END SILL                   =     10.4 FPS
SPILLWAY HEAD LOSS                          =      3.4 FEET
FROUDE NUMBERS AT SPILLWAY TOE              =      6.2
ENERGY HEAD AT SPILLWAY TOE                 =    476.6 FEET
ENERGY HEAD AT END SILL                     =    426.7 FEET
ENTER  RERUN OR STOP
=STOP
```

PROBLEM STATEMENT 2

A computation is required to determine the water surface elevation which must be held at the downstream end of a one-mile long trapeqoidal channel to produce a 30-ft flow depth at the upstream end of the channel for a discharge of 40,000 cfs. The channel has a slope of .001 ft/ft, a bottom width of 200 ft, a side slope factor of 1.5 and the Manning n value is 0.25. The channel invert elevation at the downstream end of the reach is 300 ft msl.

With these conditions, H6209-Flow Profile for Prismatic Channels will be used to make the necessary computations. A brief description of the program, as found in "CORPS," is as follows:

H6029 - FLOW PROFILE FOR PRISMATIC CHANNELS

THIS PROGRAM COMPUTES THE FLOW PROFILE IN A STRAIGHT PRISMATIC CHANNEL (TRAP., TRIA., OR RECT.) USING EITHER THE MANNING, CHEZY, OR COLEBROOK-WHITE RESISTANCE FORMULA. HORIZONTAL AND ADVERSE SLOPES ARE ACCEPTABLE.

INPUTS REQUIRED ARE: DISCHARGE IN CFS, VALUE OF RESISTANCE COEF., CHANNEL INVERT SLOPE IN FT/FT, CHANNEL BOTTOM WIDTH IN FT, CHANNEL SIDE SLOPE IN HORIZONTAL RUN PER UNIT VERTICAL RISE, CHANNEL LENGTH IN FT, INTERVAL ALONG CHANNEL AT WHICH DEPTHS ARE TO BE COMPUTED IN FT, INITIAL DEPTH IN FT, INITIAL DEPTH STATION NUMBER IN FT FROM POINT OF BEGINNING, CHANNEL INVERT ELEVATION AT INITIAL DEPTH STATION IN FT ABOVE MSL, ENERGY COEFF. (ALPHA), AND WATER TEMP. IN DEG. F., IF COLEBROOK-WHITE FORMULA IS USED.

OUTPUT INCLUDES GIVEN DATA, FLOW PROFILE CLASSIFICATION, NORMAL AND CRITICAL DEPTHS, CRITICAL SLOPE, AND FLOW PROFILE. FLOW PROFILE INCLUDES STATION NUMBER, INVERT ELEV., WATER DEPTH, WATER SURFACE ELEV., FLOW VELOCITY, VELOCITY HEAD, AND SPECIFIC ENERGY FOR EACH STATION.

NOTE: STATION NUMBER MAY RUN EITHER UPSTREAM OR DOWNSTREAM.

The computation is initiated by assuming a downstream flow depth of 30 ft and entering the following commands.

```
*RUN WESLIB/CORPS/H6209,R

INPUT H6209-FLOW PROFILES IN PRISMATIC CHANNEL

AA-ENTER DESIRED RESISTANCE LAW AS: MANN=MANNING, CHEZ=CHEZY,
OR COLE=COLEBROOK-WHITE.
=MANN
AC-ENTER DISCHARGE IN CFS.
=40000
AD-ENTER VALUE OF RESISTANCE COEFFICIENT.
=.025
AE-ENTER CHANNEL INVERT SLOPE IN FT/FT.
=.001
AF-ENTER CHANNEL BOTTOM WIDTH IN FT. ENTER 0.0 IF TRIA. CROSS-
SECTION
=200
```

Continuation of INPUT H6209

```
AG-ENTER CHANNEL SIDE SLOPE AS COTANGENT OF ACUTE ANGLE WITH
HORIZONTAL.  ENTER 0.0 IF RECTANGULAR SECTION.
=1.5
AH-ENTER CHANNEL LENGTH IN FT.
=5280
AI-ENTER INTERVAL FOR WHICH DEPTHS ARE TO BE COMPUTED IN FT.
NOTE:  CHANNEL LENGTH/INTERVAL CANNOT EXCEED 500.  FURTHERMORE,
THE INTERVAL IS AUTOMATICALLY ADJUSTED SMALLER TO OBTAIN
ACCURACY OF 0.01 FT. IN COMPUTED DEPTH.
=300
AJ-ENTER THE INITIAL DEPTH IN FT.
=30
AK-ENTER INITIAL DEPTH STATION NUMBER IN FT FROM POINT OF BEGINNING.
=5280
AL-ENTER DIRECTION STATION NUMBERS INCREASING AS DS = DOWNSTREAM,
US = UPSTREAM.
=DS
AM-ENTER ELEVATION OF CHANNEL INVERT AT INITIAL DEPTH STATION
IN FT ABOVE MSL.
=300
AN-ENTER ENERGY COEFFICIENT (ALPHA).
=1
AO-ENTER STORE OUTPUT FOR GRAPHICS AND/OR OTHER USE?
=N
CHANGE ANY INPUT BEFORE RUN,Y OR N?
=N

OUTPUT H6209

FLOW PROFILE IN TRAP CHANNEL USING MANN RESISTANCE LAW WITH
THE FOLLOWING GIVEN DESIGN DATA.
DISCHARGE:              40000.00 CFS
CHANNEL WIDTH:            200.00 FT
INVERT SLOPE:               0.00100 FT/FT
SIDE SLOPE:                 1.50H:1V
RESISTANCE COEF:            0.02500
CHANNEL LENGTH            5280.00 FT
ALPHA:  1.00

FLOW PROFILE WITH FOLLOWING CHARACTERISTICS:

CHANNEL SLOPE CLASSIFICATION:  MILD
ZONE DESIGNATION:  M1
CURVE TYPE:  BWATER
FLOW TYPE:  SUBCRT

NORMAL DEPTH:  16.24 FT.
CRITICAL DEPTH:  10.46 FT.
CRITICAL SLOPE:  0.004404 FT/FT.
```

STATION NO.	INVERT ELEV (MSL)	WATER DEPTH (FT)	WATER SURF ELEVATION (MSL)	FLOW VELOCITY (FPS)	VELOCITY HEAD (FT)	SPECIFIC ENERGY (FT)
52+80.00	300.000	30.000	330.000	5.442	0.460	30.460
49+80.00	300.300	29.727	330.027	5.501	0.470	30.197
49+00.	300.380	29.654	330.034	5.517	0.473	30.127
46+00.	300.680	29.382	330.062	5.578	0.483	29.866
43+00.	300.980	29.112	330.092	5.639	0.494	29.605
40+00.	301.280	28.842	330.122	5.701	0.505	29.346
37+00.	301.580	28.573	330.153	5.764	0.516	29.089
34+00.	301.880	28.305	330.185	5.828	0.528	28.833
31+00.	302.180	28.039	330.219	5.894	0.539	28.578
28+00.	302.480	27.773	330.253	5.960	0.552	28.325
25+00.	302.780	27.509	330.289	6.027	0.564	28.073
22+00.	303.080	27.247	330.327	6.095	0.577	27.823
19+00.	303.380	26.985	330.365	6.164	0.590	27.575
16+00.	303.680	26.725	330.405	6.234	0.603	27.329
13+00.	303.980	26.467	330.447	6.305	0.617	27.084
10+00.	304.280	26.210	330.490	6.377	0.632	26.841
7+00.	304.580	25.954	330.534	6.450	0.646	26.600
4+00.	304.880	25.701	330.581	6.524	0.661	26.362
1+00.	305.180	25.449	330.629	6.599	0.676	26.125
0+00	305.280	25.365	330.645	6.625	0.681	26.046

Note that the water depth at the upstream station has been computed as 25.365 ft and a depth of 30 ft is desired. The provided rerun option is used to recompute this depth for another initial depth at station 52+80 as follows:

```
ENTER RERUN OR STOP
=RERUN
INPUT H6209-FLOW PROFILES IN PRISMATIC CHANNEL.

RERUN OPTION PERMITS YOU TO CHANGE ANY OR ALL INPUT VARIABLES.
AT >>> = QUE TYPE IN THE TWO LETTERS (AA,AB,ETC.) CORRESPONDING
TO THE VARIABLES YOU WISH TO CHANGE.  THEN AT NEXT = QUE, ENTER
THE NUMERICAL VALUE.  TO TERMINATE DATA ENTRY, TYPE A CARRIAGE
RETURN AT >>> = QUE.
>>>
=AJ
=34.92
>>>
=

OUTPUT-H6209

FLOW PROFILE IN TRAP CHANNEL USING MANN RESISTANCE LAW WITH THE
FOLLOWING GIVEN DESIGN DATA.
DISCHARGE:             40000.00 CFS
CHANNEL WIDTH:           200.00 FT.
INVERT SLOPE               0.00100 FT/FT
SIDE SLOPE:                1.50H:1V
RESISTANCE COEF:           0.02500
CHANNEL LENGTH:         5280.00
ALPHA:  1.00
```

Continuation of H6209

FLOW PROFILE WITH FOLLOWING CHARACTERISTICS:

CHANNEL SLOPE CLASSIFICATION: MILD
ZONE DESIGNATION: M1
CURVE TYPE: BWATER
FLOW TYPE: SUBCRT

NORMAL DEPTH: 16.24 FT.
CRITICAL DEPTH: 10.46 FT.
CRITICAL SLOPE: 0.004404 FT/FT.

STATION NO.	INVERT ELEV (MSL)	WATER DEPTH (FT)	WATER SURF ELEVATION (MSL)	FLOW VELOCITY (FPS)	VELOCITY HEAD (FT)	SPECIFIC ENERGY (FT)
52+80.00	300.000	34.920	334.920	4.539	0.320	35.240
49+80.00	300.300	34.635	334.935	4.584	0.326	34.962
49+00.	300.380	34.560	334.940	4.596	0.328	34.888
46+00.	300.680	34.276	334.956	4.642	0.335	34.610
43+00.	300.980	33.992	334.972	4.688	0.341	34.333
40+00.	301.280	33.709	334.989	4.736	0.348	34.057
37+00.	301.580	33.426	335.006	4.784	0.355	33.782
34+00.	301.880	33.145	335.025	4.833	0.363	33.507
31+00.	302.180	32.863	335.043	4.882	0.370	33.233
28+00.	302.480	32.582	335.062	4.933	0.378	32.960
25+00.	302.780	32.302	335.082	4.984	0.386	32.688
22+00.	303.080	32.023	335.103	5.036	0.394	32.417
19+00.	303.380	31.744	335.124	5.089	0.402	32.146
16+00.	303.680	31.466	335.146	5.143	0.411	31.876
13+00.	303.980	31.188	335.168	5.197	0.419	31.608
10+00.	304.280	30.912	335.192	5.252	0.428	31.340
7+00.	304.580	30.636	335.216	5.309	0.438	31.073
4+00	304.880	30.361	335.241	5.366	0.447	30.808
1+00.	305.180	30.087	335.267	5.424	0.457	30.543
0+00.	305.280	29.995	335.275	5.443	0.460	30.455

ENTER RERUN OR STOP
=STOP

As observed, the upstream depth has now been computed close to the desired 30 ft depth.

CONCLUSION

The ease of using the "CORPS" for hydraulic design has been presented. The use of this system, being interactive, provides an effective means of computer communication for the user. This communication is enhanced with the provision of a rerun option for each program giving the user the ability to evaluate alternate solutions by changing any or all of the input variables. Hydraulic design using an interactive system such as this also provides a reduction in computational errors and improved response times.

The tests described and the resulting data presented herein, unless otherwise noted, were obtained from research conducted under the Civil Works Program of the U. S. Army Corps of Engineers by the Waterways Experiment Station. Permission was granted by the Chief of Engineers to publish this information.

NUMERICAL SIMULATION OF TURBULENT STRATIFIED FLOWS IN SURFACE WATER IMPOUNDMENTS

by

Angelos N. Findikakis* and Robert L. Street
Department of Civil Engineering
Stanford University
Stanford, California 94305

ABSTRACT

This paper discusses the potential and some of the problems involved in using the Finite Element (FE) method to simulate turbulent stratified flows. The conceptual basis of the Large Eddy Simulation (LES) method is briefly reviewed. A simple model for Sub-Grid-Scale (SGS) turbulence is described. Two examples of flow in a stratified basin are presented. In the first example the driving shear force is constant in time, while in the second it contains a randomly varying component. The spectral characteristics of the generated velocity field in the second case are examined.

1. INTRODUCTION

Turbulent flows contain motions of a wide range of scales. The largest scales are determined by the geometry of the flow region and the smallest by the viscosity of the fluid. Great computer resource requirements make it practically impossible to perform numerical simulations of the development in space and time of all scales, even in the simplest of turbulent flows. Therefore, one is forced to describe these flows in terms of averages of the flow variables.

The conventional approach for the description of turbulent flows is to use time or ensemble averages of the flow variables to describe a mean flow field. The fluctuations from these averages play an important role in shaping the mean flow field. This is expressed by the turbulent momentum and heat flux terms in the flow equations. In many flows the scales of the mean flow are determined by the dimensions and the geometry of the flow regime. Thus, all motions of smaller scales are treated as fluctuations from the averaged field and their effect must be accounted for by the turbulence model. The prospects for success of any single such model on a variety of flows are limited by the fact that the structure of turbulent motions of all scales is not the same in every flow. It has been observed that in many flows the large turbulent structures, or eddies, are not totally random but "are repeating but not precisely so" and that they are anisotropic and flow dependent [Ferziger, 1977]. In contrast, motions of smaller scales tend to be isotropic with very similar structure in many classes of flows. This is

*Now with Hydraulics/Hydrology Group, Bechtel Inc., San Francisco.

one of the basic reasons for the success of an alternative approach to the numerical simulation of turbulent flows, the Large Eddy Simulation (LES), which originated in meteorological applications and has been developed over the last fifteen years.

The LES approach consists of working with space-averaged variables and simulating their time history. The explicitly simulated motions range from those whose scale is determined by the geometry of the flow regime to those whose scale is determined by the size of the averaging operator. It is limited by the spatial discretization used for the numerical solution of the averaged-variable equations. One basic difference between the two approaches is that in the LES method we need a turbulence model for the description of the effect of turbulent motions of scales smaller than the scale of the averaging operator, while in the conventional approach we need turbulence models able to account for the effect of all turbulent motions smaller than the scale of the mean flow. This enhances the chances for success of the LES, especially in cases where it is possible to use an averaging operator such that the motions of scales smaller than the size of the operator are nearly isotropic.

The promise of success in simulating different flows using a single model makes the LES very attractive for the simulation and prediction of environmental flows, where in many instances little is known in advance about the nature of the flow. Many environmental flows of interest take place in irregularly shaped domains. In such cases the most efficient way to solve the flow equations numerically is the Finite Element (FE) method. Therefore, a combination of the LES approach with the FE method has the potential of providing a powerful tool for the numerical simulation of environmental flows. In this paper we discuss some aspects of this combined approach and present an example of its application in the case of a turbulent, temperature-stratified flow.

2. FORMULATION OF THE LARGE-SCALE FIELD EQUATIONS AND CLOSURE APPROXIMATIONS

We review briefly the formulation of the basic equations used in a LES. A more extensive treatment of this subject can be found in Findikakis and Street [1979] and Findikakis [1980]. The first step is to define the explicitly simulated flow variables. Each flow variable is decomposed into a large-scale (LS) and a sub-grid-scale (SGS) component. Leonard [1974] has proposed a formal definition for the LS component of the flow variables as the convolution of the total value of the variables with a filter function. In the present work we use the box filter which is equal to a constant within a box centered at the point of application of the filter and zero everywhere else. Different choices of the filter function are possible.

Equations for the LS field are obtained by applying the filtering operator to the Navier-Stokes, continuity and heat-balance equations. Filtering the nonlinear terms gives rise to SGS Reynolds stresses and turbulent heat flux terms. In addition, since the averages of the variables are not constant in space, the filtered value of the product of any two LS components is only approximately equal to the product of the filtered components. This observation was first made by Leonard [1974], who proposed additional terms for a better approximation of the filtered products of LS variables. For the same reason filtering the product of a LS component and a SGS component does not give zero, as in the conventional approach. Approximations for these terms have been proposed by Clark et al. [1977]. Thus, the equations describing the LS field of stratified flow are:

$$\frac{\partial \overline{U}_i}{\partial x_i} = 0 \tag{1}$$

$$\frac{\partial \overline{U}_i}{\partial t} + \overline{U}_j \frac{\partial \overline{U}_i}{\partial x_j} + \frac{1}{\rho_o} \frac{\partial \overline{P}}{\partial x_i} - \delta_{i3} g \beta \overline{\Theta} + \frac{\partial}{\partial x_j}\left(\overline{u_i u_j} + L_{ij} - \nu \frac{\partial \overline{U}_i}{\partial x_j}\right) = 0 \tag{2}$$

$$\frac{\partial \overline{\Theta}}{\partial t} + \overline{U}_i \frac{\partial \overline{\Theta}}{\partial x_i} + \frac{\partial}{\partial x_i}\left(\overline{u_i \theta} + L_{i\theta} - \kappa \frac{\partial \overline{\Theta}}{\partial x_i}\right) = 0 \tag{3}$$

where the overbars indicate the LS components of a variable, i.e., $\overline{U}_i$, $\overline{P}$ and $\overline{\Theta}$ are the LS components of the ith component of the velocity, the pressure and the temperature, respectively, while u_i, p, and θ are the SGS components of the same variables, correspondingly. Also, ρ_o is a reference density; β is the coefficient of thermal expansion; ν is the kinematic viscosity; κ is the thermal diffusivity of the fluid; and g is the acceleration of gravity; t is the time and x_i is the coordinate in the ith direction (i = 1,2 is used for the two horizontal directions and i = 3 for the vertical direction). The terms L_{ij} and $L_{i\theta}$ in Eqs. (2) and (3) represent the combined approximations proposed by Leonard and Clark for the extra terms which arise when we filter the nonlinear terms and they do not have any counterparts in the conventional approach. These approximations are:

$$L_{ij} = \frac{\Delta_k^2}{12} \frac{\partial \overline{U}_i}{\partial x_k} \frac{\partial \overline{U}_j}{\partial x_k}, \qquad L_{i\theta} = \frac{\Delta_k^2}{12} \frac{\partial \overline{U}_i}{\partial x_k} \frac{\partial \overline{\Theta}}{\partial x_k} \tag{4}$$

where Δ_k is the dimension of the filter in the kth direction.

To close the system of Eqs. (1-3), we follow an approach which lies midway between the eddy-viscosity approximation and the use of the full differential equations for the SGS Reynolds stresses and turbulent heat fluxes. The first step in this approach is to consider the full differential equations for the SGS second-order moments. We then make proper closure approximations for this system and certain simplifying assumptions to reduce the differential to algebraic equations. Solving this system of algebraic equations we obtain explicit expressions of the SGS Reynolds stresses and turbulent heat fluxes. These expressions have the form of eddy viscosity - eddy diffusivity approximations, with the eddy coefficients expressed as functions of the local structure of the velocity and temperature fields. This approach was first presented by Deardorff [1972], described in more detail by Sommeria [1976] and extended by Findikakis and Street [1979]. The derivation of the expressions and a discussion on the assumptions involved can be found in the last two references. Here we present only the final expressions for the Reynolds stresses and the turbulent heat fluxes:

$$\overline{u_i u_j} - \frac{1}{3} \delta_{ij} \overline{u_i u_j} = -\varepsilon S_{ij}$$

$$\overline{u_i\theta} = -\varepsilon_\theta \frac{\partial\overline{\Theta}}{\partial x_i} \qquad i = 1,2$$

$$\overline{u_3\theta} = -\varepsilon_{\theta k} \frac{\partial\overline{\Theta}}{\partial x_k} - \varepsilon_{\theta 3} \frac{\partial\overline{\Theta}}{\partial x_3}, \qquad k = 1,2$$

where the eddy coefficients are given by:

$$\varepsilon = f(c_1\Lambda)^2 \left(\frac{1}{2} S_{ij}S_{ij}\right)^{1/2} \tag{5}$$

$$\varepsilon_\theta = \frac{\varepsilon}{a_T}\;; \quad \varepsilon_{\theta k} = \frac{\varepsilon}{a_T}\phi_k, \quad k = 1,2\;; \quad \varepsilon_{\theta 3} = \frac{\varepsilon}{a_T}\phi_3 \tag{6}$$

and

$$f^2 = \frac{1}{2}\left[1 - (c_2 + c_3)Ri + \left[(1 - (c_2 + c_3)Ri)^2 + 4c_3Ri + 4c_2c_3Ri'^2\right]\right]^{1/2} \tag{7}$$

$$\phi_3 = \frac{1}{1 + c_3 \frac{Ri}{f^2}}\;; \quad \phi_k = -\phi_3 c_3 \frac{\beta g\, \partial\overline{\Theta}/\partial x_k}{f^2\left(\frac{1}{2} S_{ij}S_{ij}\right)}, \quad k = 1,2 \tag{8}$$

$$S_{ij} = \frac{\partial\overline{U}_i}{\partial x_j} + \frac{\partial\overline{U}_j}{\partial x_i} \tag{9}$$

$$Ri = \beta g \frac{\partial\overline{\Theta}/\partial x_3}{\left(\frac{1}{2} S_{ij}S_{ij}\right)} \tag{10}$$

$$Ri' = \beta g \frac{\left[(\partial\overline{\Theta}/\partial x_1)^2 + (\partial\overline{\Theta}/\partial x_2)^2\right]^{1/2}}{\left(\frac{1}{2} S_{ij}S_{ij}\right)} \tag{11}$$

Here Λ is a characteristic length scale of SGS turbulence and the constants c_1, c_2, c_3, a_T represent combinations of the constants used in the closure approximations and the simplifying assumptions made to reduce the differential equations for the SGS quantities to a system of algebraic equations [Findikakis, 1980]. The values of the constants used in the simulations presented in this paper are $c_1 = 0.20$, $c_2 = 2.50$, $c_3 = 2.94$ and $a_T = 0.40$. These values were selected in accordance with the experience of other investigators, primarily in the atmospheric sciences. Further work to determine the value of these constants with more confidence, especially in connection with the type of flows discussed in Section 5, is definitely needed. Experimental work in this direction is under way at Stanford University.

The length scale Λ is related to the size of the filter and is often simply chosen as the cube root, or the square root in two-dimensional applications, of the product of the dimensions of the filter. Caution is needed in defining Λ near solid boundaries. There, it is more appropriate to select Λ as the minimum of a filter-related scale and a scale which is characteristic of near-the-wall turbulence, such as Prandtl's mixing length.

3. FORMULATION AND SOLUTION OF THE FE EQUATIONS

We employ the FE method to solve the system (1)-(3) for two-dimensional flow in a vertical plane. A quadratic approximation function with eight-node isoparametric elements is used for the LS velocity and temperature fields and a linear approximation with four-node elements is used for the LS pressure field. The time derivatives are approximated by finite differences. All the other terms in the equations are evaluated at a point in the time interval between time steps, the location of which characterizes the time-stepping scheme. We use the Galerkin method to formulate the FE equations weighting the continuity equation with the linear approximation function and the momentum and heat balance equations with the quadratic approximation function.

We solve the system of the resulting FE equations iteratively. To reduce the computational effort in solving this system we break it into two smaller systems. We solve first the continuity and the momentum equations, simultaneously. Then we use the obtained velocity field in the solution of the heat-balance equations. The nonlinear convective terms, the eddy viscosity, the eddy diffusivity, and the Leonard terms are evaluated from the velocity and the temperature fields of the just previous iteration. To accelerate iteration convergence at the end of each iteration cycle we use an underrelaxation scheme to obtain the values of the unknowns which are going to be used in the evaluation of the nonlinear terms at the next iteration.

4. A PROBLEM AND SOURCE OF ERROR IN FE SIMULATIONS OF TURBULENT FLOWS

Filtering of the flow equations precedes the numerical solution and is based solely on the decision to simulate explicitly only scales larger than a certain cutoff value. However, before selecting the filter size and introducing the terms proposed by Leonard and Clark, we must consider that any numerical method gives only an approximation to the exact solution of equations and, therefore, has an inherent filtering effect which shows up as an effective error in the numerical solution when compared to the exact solution. This error depends on the density of the grid points, the order of the piecewise approximation of the unknowns between grid points, and the principle used for the formulation of the equations of the numerical solution. For LES based on filtering to be meaningful, the numerical error must be negligible (at least in the explicitly simulated large-eddy scales) when compared to the LES filter effect.

The filter operator may be anisotropic and/or inhomogeneous, reflecting the local characteristics of the FE discretization of the flow regime. However, caution is needed in selecting the size of the filter. The shortest waves which can be represented on certain grid have wave lengths equal to twice the spacing of the grid points. Therefore, it is meaningless to use a filter smaller than twice the distance between nodes in the vicinity of the application of the filter. There is no upper limit on the size of the filter. However, since in most cases it is desired to extend as much as possible the range of scales included in the explicitly simulated LS field, the size of the filter is usually selected equal to twice the node spacing.

Findikakis [1980] has compared, on a norm basis, the magnitude of the effect of filtering the analytic solution of several one-dimensional equations to the numerical error in their FE solution. For linear elements and the minimum meaningful filter size, i.e., twice the node spacing, the effect of filtering is of the same order of magnitude as the error in the numerical solution. For quadratic elements, the error of the numerical solution is much smaller than the effect of filtering, especially for waves of substantially greater length than the FE node spacing. In both cases, the error in the numerical solution becomes relatively more important compared to the effect of filtering, as the ratio of filter width to the FE node spacing decreases.

5. APPLICATIONS

As an example we simulate the process of thermal mixing and the development of the flow in a stratified basin under the action of a wind stress at the surface. It is assumed that the fluid is initially linearly stratified and that the temperature at the water surface and bottom is maintained constant. The driving force is imposed suddenly and remains constant in time. The flow pattern developed in response to the action of the surface shear consists of one or more circulation cells, depending on the intensity of the driving force, the temperature difference between top and bottom, and the rate of turbulent momentum and heat transport at the shear layers between cells. The number, size and shape of simulated large eddies for any given conditions depends on the value of the coefficients involved in the turbulence model and it may be limited by the size of the filter used to define the LS field. As discussed in Section 2 the minimum allowable filter size is dictated by the discretization of the solution domain for the numerical solution of the hydrodynamic equations. In the present example there are eight elements in the horizontal direction and twelve in the vertical direction. Backward time differencing and a time step of 200 sec were used.

We consider the case of flow in a cavity of square cross section L = 10 m by H = 10 m, driven by a shear stress τ_o = 0.1 dynes/cm^2. A Reynolds number for this flow, based on the shear velocity at the surface and the dimension of the cavity and the kinematic viscosity is 31,600. We start the simulation with the fluid at rest and an initial stratification which yields a Brunt-Väisälä frequency

$$N = [\beta g(\partial\Theta/\partial z)]^{1/2} = 0.002 \text{ rad/sec}.$$

The reason for selecting a regime of such idealized geometry instead of a more realistic impoundment shape is that experiments on a similar flow in a cavity are under way at Stanford University. Nevertheless, the present model by its very nature as a FE model can handle flows in regimes of any geometrical shape.

The transient period for the development of the flow pattern may be rather long compared to the corresponding transient period in the thermally uniform case. Figure 1 shows the flow pattern at different points

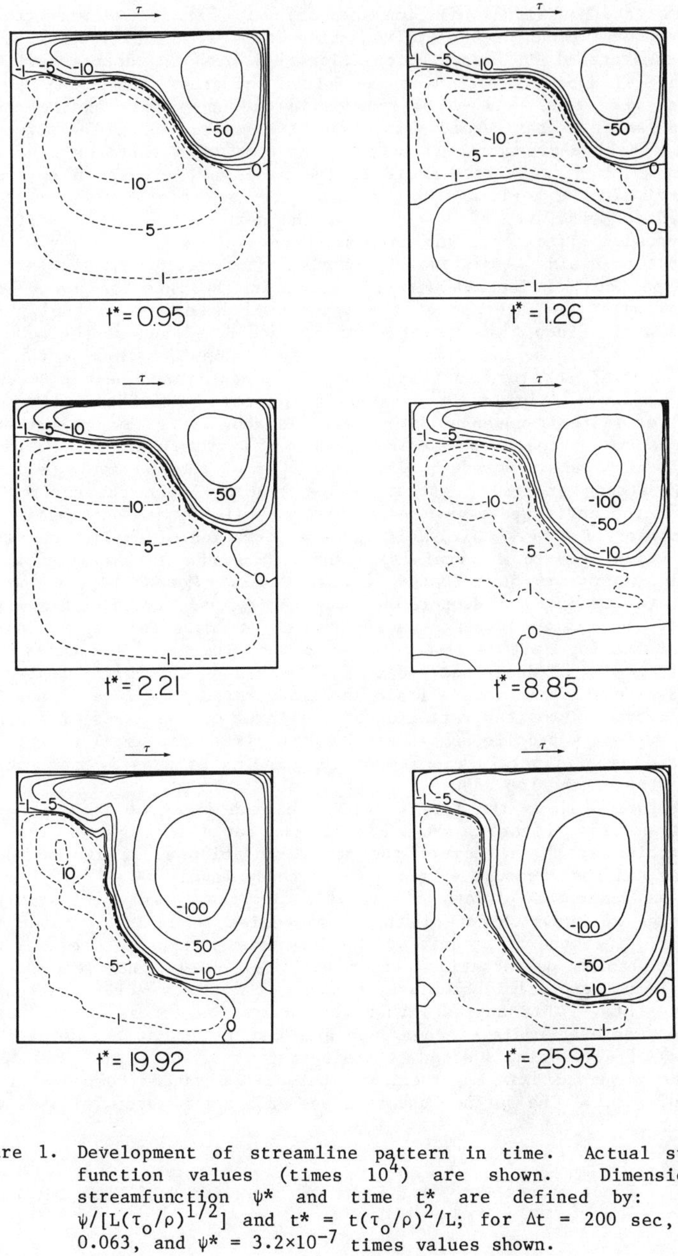

Figure 1. Development of streamline pattern in time. Actual streamfunction values (times 10^4) are shown. Dimensionless streamfunction ψ^* and time t^* are defined by: $\psi^* = \psi/[L(\tau_o/\rho)^{1/2}]$ and $t^* = t(\tau_o/\rho)^2/L$; for Δt = 200 sec, Δt^* = 0.063, and $\psi^* = 3.2\times10^{-7}$ times values shown.

in time (the nondimensional time step $\Delta t^* = 0.063$). We observe the successive development of three circulation cells. Initially the flow pattern consists of one large circulation cell over the entire cavity. A second cell appears first near the bottom and grows, expanding upwards, forcing the first cell to be reduced in the upper part of the cavity. As the center of the second cell rises, the velocities in the lower part of this cell decrease allowing the formation of a third cell near the bottom which grows subsequently in the same way. These cells are not symmetric in the horizontal direction. The center of the upper cell is shifted downwind, while the center of the next cell below is shifted in the opposite direction. The cell structure interacts strongly with the temperature field. A distinct thermocline is developed at the bottom of the upper cell. Thermal stratification in the interior of the cells weakens with the interior of the upper cell becoming eventually fully mixed, while steep temperature gradients are developed at the zones between cells. These temperature gradients in combination with the local shear control the turbulent transport of momentum and heat between the large eddies. This factor determines the strength of the lower cells which derive their momentum from the cells above. Following further the development of the flow pattern we see that the third cell which is quite weak breaks. However, since the lower of the two major cells left is not strong enough to establish its presence over the entire lower part of the cavity permanently, a third cell is formed once again. The development of a weak third cell and the formation of smaller corner eddies, which grow to a certain size and then gradually shrink until they finally disappear, is repeated several times. Meanwhile the top cell penetrates deeper and deeper into the cavity and occupies the largest part of the cavity leaving space for just one more cell. The penetration of the top cell is associated with the sinking of the thermocline. The effect of the described sequence of events on the vertical structure of the flow and temperature field is demonstrated in Figures 2 and 3.

Figure 2 shows the variation of the level of the zeros of the horizontal velocity profile at $x/L = 0.25$ with time. The level of the upper zero of the horizontal velocity profile indicates grossly the level of the center of the top eddy.

Figure 3 shows the time variation of the level of the isothermals at $x/L = 0.25$. Figure 3 shows clearly the rapid development of a steep thermocline at the bottom of the top circulation cell, its subsequent sinking and the eventual approach to a steady level.

The described pattern of evolution of the large-scale structures resembles the development of the large eddies in a LES. In our case, however, since the force driving the flow is kept constant in time without any random perturbations which might stimulate turbulent fluctuations in the explicitly simulated motions, the flow reaches eventually a steady state. This suggests that the generation, breaking and merging of the simulated eddies in our case are just transient phenomena in the process of developing a steady-state pattern.

As a second example, then, we introduce a random component in the driving force. The surface shear stress at the time step (n+1) is given by:

$$\tau^{n+1} = \tau_o + r(\tau^n - \tau_o) + s(1 - r^2)^{1/2} (-2 \log R_1)^{1/2} \cos(2\pi R_2) \quad (12)$$

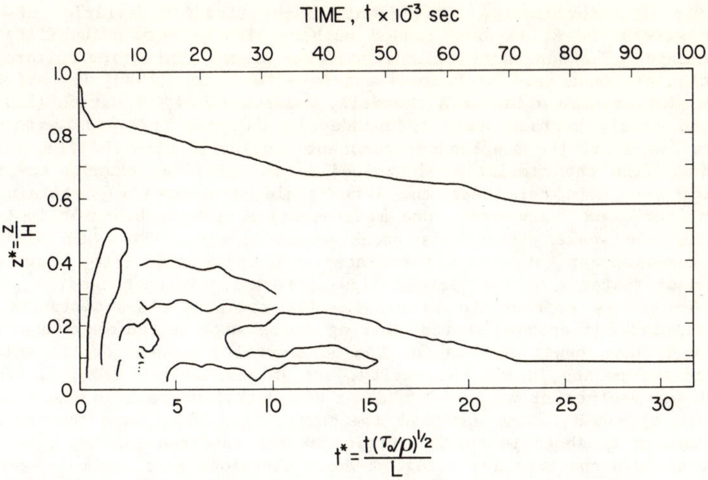

Figure 2. Variation with time of the level of zeros of the horizontal velocity profile at x/L = 0.25.

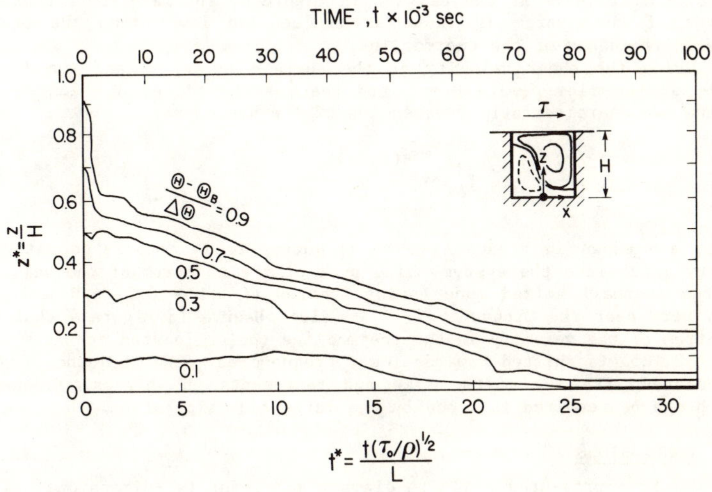

Figure 3. Variation with time of the level of isothermals at x/L = 0.25; Θ_B is the bottom temperature and $\Delta\Theta = \Theta_T - \Theta_B$ = constant, where Θ_T is the surface temperature.

where r is a correlation coefficient, s the standard deviation of the shear stress and R_1, R_2 are random numbers with uniform probability of occurrence. The superscripts indicate time steps. This form of forcing function has been used by Bedford and Babajimopoulos [1980] in performing a similar simulation in a thermally uniform fluid. Equation (12) is applied on all surface nodes independently. We use $r = 0.5$ and $s = 0.045$. We start the new random-component simulation with the flow field obtained from the simulation described in the previous example at $t = 100{,}000$ sec. In this case the large-scale structures may contain an actual turbulent component. One manifestation of this behavior is that some of the weaker structures break or merge with others and similar structures appear later in their place to have the same fate. However, the major features of the overall flow pattern remain unchanged.

Now it is appropriate to examine the spectral characteristics of the simulated response to the driving force with a random component. Figure 4 shows power spectra for the horizontal component of the velocity at two points inside the cavity, one located at $x/L = 0.500$, $z/H = 0.583$ and another at $x/L = 0.500$, $z/H = 0.125$, where L is the length ($L = 10$ m) and H is the depth of the cavity ($H = 10$ m), and the coordinate system is shown in the definition sketch inserted in Figure 4. We see that both spectra have a region where the slope is close to -3 which is the theoretical slope of the two-dimensional turbulence spectrum [Lilly, 1969]. The slope of the spectra becomes steeper near the Brunt-Väisälä frequency corresponding to the temperature gradient of the thermocline, indicating a sharp drop in the spectral density function at this frequency. This frequency is equal to 0.003 to 0.004 rad/sec (the range is given because the temperature field is not uniform in the horizontal direction). As can be seen in Figure 4, the largest part of the energy of the system is found at frequencies lower than the Brunt-Väisälä frequency of the thermocline.

Using the shear velocity at the surface $(\tau_o/\rho)^{1/2}$ and the length scale of the flow domain $H = L$ and that of the filter $\Lambda = 2\Delta x$ we can define two characteristic frequencies of the system as:

$$f_L = \frac{\pi(\tau_o/\rho)^{1/2}}{L} , \quad f_{\Delta x} = \frac{\pi(\tau_o/\rho)^{1/2}}{\Delta x}$$

which are shown in Figure 4. The frequency f_L is associated with the energy input into the system. The peak of the spectrum of the velocity in the thermally mixed zone (point located at $x/L = 0.5$, $z/H = 0.583$) lies very near the frequency f_L. We also observe in Figure 4 that the spectrum of the velocity in the thermocline (point located at $x/L = 0.5$, $z/H = 0.125$) is shifted towards lower frequencies. The frequency $f_{\Delta x}$ is associated with the filter size and represents the highest frequency which can be resolved in space by the large eddy simulation.

6. CONCLUSIONS

We have presented a finite-element model for two-dimensional simulations of turbulent stratified flows. The model predicts successfully the general spectral characteristics of such flows. Some numerical problems encountered in such simulations have been discussed. Special attention has been given to the proper formulation of the SGS model to ensure numerical stability and convergence of the solution.

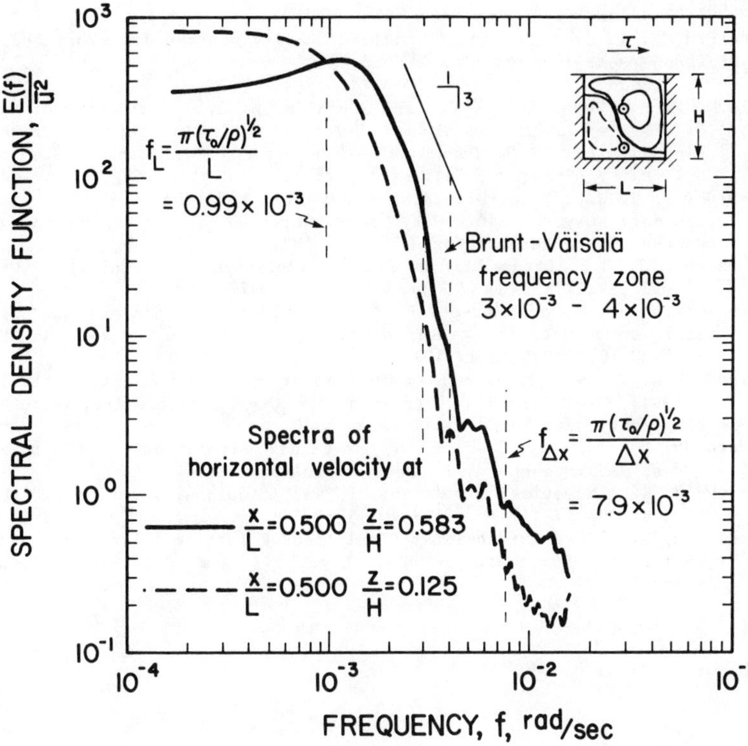

Figure 4. Spectra of the horizontal velocity at x/L = 0.500, z/L = 0.583, and x/L = 0.500, z/L = 0.125 in a flow driven by a surface shear with a randomly varying component.

ACKNOWLEDGEMENTS

This work was supported in part by the Water Resources, Urban and Environmental Engineering Program, Engineering Division, National Science Foundation under Grant ENG-77-13880. The computations were performed in part at the National Center for Atmospheric Research, which is sponsored by the National Science Foundation.

REFERENCES

Bedford, K. W., and C. Babajimopoulos, "Verifying Lake Transport Models with Spectral Statistics," Proc. ASCE, Vol. 106, No. HY1, pp. 21-38, 1980.

Clark, R. A., Ferziger, J. H., and Reynolds, W. C., "Evaluation of Sub-Grid-Scale Turbulence Models Using a Fully Simulated Turbulent Flow," Report TF-9, Department of Mechanical Engineering, Stanford University, Stanford, California, 1977.

Deardorff, J. W., "Three-Dimensional Numerical Modeling of the Planetary Boundary Layer," Am. Met. Soc. Workshop in Micrometeorology, Boston, Massachusetts, pp. 217-311, 1972.

Ferziger, J. H., "Large Eddy Numerical Simulation of Turbulent Flows," AAIA J., Vol. 15, No. 9, pp. 1261-1267, 1977.

Findikakis, A. N., and Street, R. L., "An Algebraic Model for Sub-Grid-Scale Turbulence in Stratified Flows," J. Atm. Sci., Vol. 36, No. 10, pp. 1934-1949, Oct. 1979.

Findikakis, A. N., "Finite Element Simulation of Turbulent Stratified Flows," Ph.D. Thesis, Department of Civil Engineering, Stanford University, Stanford, California, 1980.

Leonard, A., "Energy Cascade in Large-Eddy Simulations of Turbulent Flows," Adv. Geoph., Vol. 18A, pp. 237-248, 1974.

Lilly, D. K., "Numerical Simulation of Two-Dimensional Turbulence," Phy. Fluids, Suppl. II, pp. 241-249, 1969.

Sommeria, G., "Three-Dimensional Simulation of Turbulent Processes in an Undisturbed Trade Wind Boundary Layer," J. Atm. Sci., Vol. 33, pp. 216-241, 1976.

IMPACT OF VERTICAL SCALE DISTORTION ON TURBULENT MIXING IN PHYSICAL MODELS

by

Benjamin R. Roberts* and Robert L. Street
Department of Civil Engineering
Stanford University
Stanford, California 94305

ABSTRACT

Distorted, physical (hydraulic) models are widely used, but the accuracy of their predictions for mixing phenomena is open to question. A study has been made in a variable distortion facility which models a typical cooling lake with a surface discharge jet and submerged power plant intake. The reported model results of studies at 10:1 and 2:1 distortions of the prototype system show highly complex interactions between the jet and main water body which are significantly influenced by the vertical scale distortion. The results raise further doubts about the ability of distorted hydraulic models to properly represent large scale flow patterns when jet-induced mixing is an important factor.

1. INTRODUCTION

Hydraulic models are often used to predict the behavior of thermal and other contaminant discharges in water bodies, particularly in situations where stratification and complex geometries make the application of numerical models difficult. Such cases include deep lakes and cooling ponds as well as estuaries, e.g., Chesapeake Bay (Neale, 1971). Many hydraulic models are distorted; for otherwise they would be too shallow to maintain the turbulent flow found in natural environments. The effects of this distortion are not fully understood, especially with regard to dispersion and turbulent mixing.

Fischer (1976) notes that "usually in distorted models turbulent-mixing coefficients do not scale properly" and under certain conditions, "it can be shown that distorted models underdisperse vertically and overdisperse transversely." Moretti and McLaughlin (1977) describe a set of experiments on the spreading of a mid-density jet in a stratified environment. They compared jet behavior at two different model distortions under several sets of similarity criteria. Their tentative conclusion was that transverse and vertical mixing could not be simultaneously scaled properly in a distorted model. However, the similarity criteria used to create the model could be selected to properly model either one. There is a clear need for increased understanding of turbulent transport and mixing in hydraulic models. Scaling laws for

---

*Now at Anderson-Nichols & Co., Palo Alto, California 94303.

turbulence modeling in distorted models need to be developed; ideally, these laws would allow accurate prediction of both mixing and mean flow processes.

The ongoing research program at Stanford is directed to providing answers to the above issues through several courses of action: detailed turbulence measurements in a small, recirculating and stratified flow; fractional analysis studies to develop a complete set of dimensionless governing equations for turbulent, variable density flows; and experiments in a large variable-distortion model of a deep lake or cooling pond. A previous paper (Roberts, et al., 1979) describes this research program, the experimental facilities, and the research goals. The focus of the present paper is on preliminary results of experiments in the variable distortion facility (VDF), some tentative conclusions, and a prognosis for future work.

2. ANALYTICAL WORK: A BASIS FOR EXPERIMENT

One of the principal goals of the research effort is to develop and experimentally verify a set of basic hypotheses regarding the influence of vertical scale distortion on mixing processes in physical models. Fractional analysis techniques often yield new insights into the physical implications of complex equations and have been used to derive most of the presently accepted physical modeling principles. Nondimensional equations for three-dimensional turbulent flows have been derived. The set of equations includes spatially-averaged equations for momentum transport. When the averaged equations are scaled by the water body dimensions and inflow velocity and the turbulent transport equations are scaled by a turbulence integral scale and the maximum shear velocity, for example, the resulting fractional analysis is quite revealing. One sees immediately from examining the averaged equations that similarity of the averaged (or mean-flow) terms can be maintained in distorted models but that the transport terms due to turbulence contain scales which do not allow for distortion; thus, there does not appear to be any consistent set of similarity requirements which will correct for both the averaged equations and the distortion-related errors introduced by the turbulent terms. Similarly, if one looks at the transport equations, one finds that the turbulence terms are consistent and not necessarily influenced by distortion. However, the averaged-flow terms which enter into the turbulent transport equations are influenced by distortion and the same difficulty arises in trying to interpret the turbulent transport equations for distorted models.

Thus, our analysis indicates in a rigorous fashion that distortion of model geometry causes similarity to be violated in several physical processes. From our analysis we can hypothesize that the differences described above account for radically different mixing processes in distorted models. Given such difficulties, similitude requirements for model design can only minimize the effects of distortion on model behavior. Our study indicates that the following traditional conditions must be met to provide proper similitude in the <u>mean flow processes</u>:

Reynolds numbers (Re) must be in the turbulent range and as nearly equal in model and prototype as possible,
Froude numbers (Fr) must be equal in model and prototype, and
Richardson numbers (Ri) must be equal in model and prototype.

The requirement that both Ri and Fr be matched forces the density differences in a variable density flow to be the same in model and prototype. The degrees of freedom for model design are, as result, quite limited. As implied by the results of Moretti and McLaughlin (1977), one or more of the above criteria can be violated to produce mixing similarity in one dimension, but this causes similitude errors in either the mean flow processes or in mixing in other dimensions. There is no obvious way of correcting for this problem. The VDF experiments are designed to test these conclusions and provide qualitative and quantitative measures of the mixing processes under different amounts of distortion.

3. EXPERIMENT DESCRIPTION

The variable distortion facility (VDF) was designed to model a "typical" lake or cooling pond of roughly kidney-shaped plan form and with a sloping bottom. The prototype lake, if it existed, would have a length of 2000 m, a maximum width of 920 m, and a maximum depth of 30 m. A horizontal scale factor of 200:1 was used, dictating a model length of 10 m and a maximum model width of 4.6 m. Since distortions between 1:1 and 10:1 can be created, model maximum depth can be set in the range from 0.15 m to 1.5 m. Figure 1 shows a plan view of the VDF, including the position of the temperature probes and the power plant intake (where water is withdrawn from the lake) and discharge works. Figure 2 shows a cross section of the VDF along a nominal flow line from plant discharge channel to intake at the two distortions (10:1 and 2:1) discussed in this paper.

The model circulation system and intake and discharge works were conceived to represent a moderate-size power plant using the lake for once-through cooling. A temperature rise of 8°C and a cooling water flow rate of 25 m^3/s are assumed to be prototype conditions (corresponding to a power plant size of approximately 500 MW). A prototype discharge channel of 5 m x 5 m with a flow velocity of 1 m/s and a "top hat" intake are modeled, as shown in Figure 2.

The present set of experiments encompass vertical scale distortions of 10:1 and 2:1. Based on the analytical results, Ri and Fr are maintained while Re is kept in the turbulent range. The following parameter values obtain.

Parameter	"Prototype"	Distorted Models 2:1	Distorted Models 10:1
Length L_L [m]	2000	10.0	10.0
Width W_L [m]	920	4.6	4.6
Depth D_L [m]	30	0.3	1.5
Discharge Velocity U_j [m/s]	1	0.10	0.224
Discharge Temperature Rise $(\theta_j - \theta_L)$ [°C]	8	8	8
Discharge Channel Geometry $(W_j \times D_j)$ [m × m]	5 × 5	0.025×0.05	0.025×0.25
Jet Reynolds No. (Re_j)	5×10^6	5×10^3	5.6×10^4
Lake Froude No. (Fr_L)	0.058	0.058	0.058
Jet Richardson No. (Ri_j)	0.11	0.11	0.11
Characteristic Time $\tau (= L_L/U_j)$ [s]	2000	100	44.6

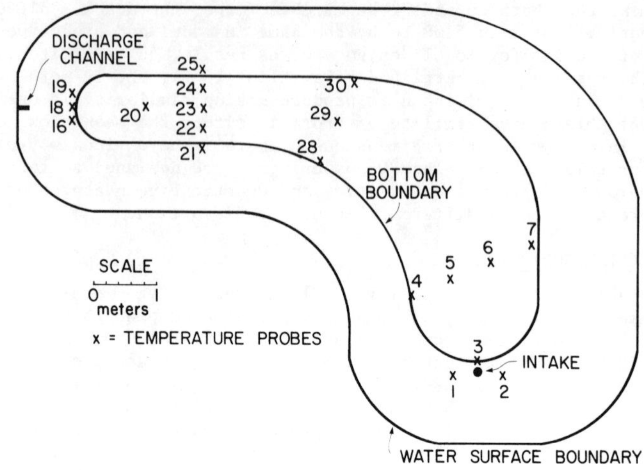

Figure 1. Plan View of Facility and Instrumentation

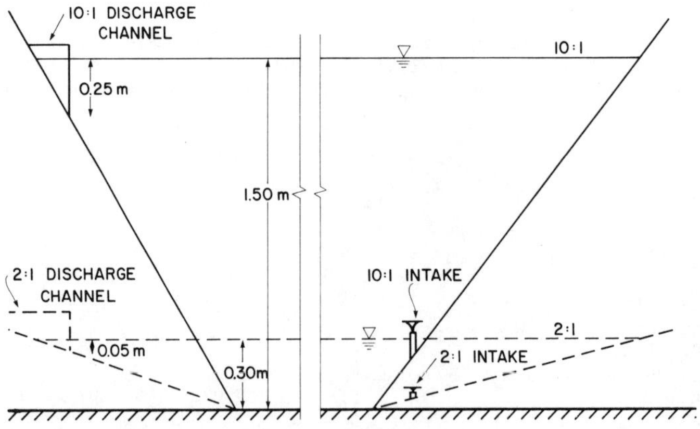

Figure 2. Cross Section of Facility at 10:1 and 2:1 Distortions

Here, it is assumed that a cartesian coordinate system lies on the lake with its origin at the discharge channel and x-axis along the discharge jet centerline. Then L_L = the maximum x-value or the lake length, W_L = the lake width, D_L = the maximum lake depth, U_j = the discharge jet's average velocity, D_j = the discharge channel depth, W_j = the discharge channel width. Accordingly,

$$Re_j = \frac{U_j D_j}{\nu}$$

$$Ri_j = \frac{g \frac{\Delta\rho}{\rho} D_j}{U_j^2}$$

$$Fr_L = \frac{U_j}{(gD_L)^{1/2}}$$

In these equations g is the acceleration due to gravity and ρ is the fluid density. We take $\Delta\rho/\rho = (\rho_L - \rho_j)/\rho_L$ where ρ_j = the density of the inflow jet (at temperature θ_j) and ρ_L = the initial lake density (at temperature θ_L) for the heated inflow/isothermal-initial-lake cases reported on here.

The jet is always in the turbulent range, but Re_j varies between the models by an order of magnitude. In addition, flows in the lake which are not in the near-field jet region will have Reynolds numbers less than 1000 for the 10:1 distortion and less than 100 for the 2:1 case. Hence, the regime will not be turbulent everywhere, as it essentially is in the prototype. Based on Reynolds number effects alone, it would be unreasonable to expect complete mixing similarity.

A full set of experiments has been run at each distortion, including steady state, isothermal cases; heated discharge with an initially isothermal lake; and mid-density discharge into a linearly stratified lake. Only the heated discharge cases will be discussed here. For each case the needed discharge velocity and temperature were set and maintained during the experiment. An equal amount of water was withdrawn through the intake structure to simulate a closed system. Measurements were then made at suitable intervals over an equivalent prototype time of more than 60 hours beginning with a still, isothermal lake. Photographic studies of plume mixing and transport were made in separate runs using Rotamine dye for flow visualization.

Temperature measurements were made at 19 stations over the lake plan, as shown in Figure 1, using thermister probes. Two-dimensional turbulent velocity measurements were made at the jet centerline and in the jet mixing region (at thermisters 18 and 19) using crossed hot-film probes. All 21 probes were mounted on a horizontal lattice which was moved vertically to take simultaneous measurements at all stations for each depth. The resulting profiles can, as a result, be considered to reflect a picture of lake conditions at a single point in time. An HP 2100 computer system was used to sample, digitize, store, and reduce the large amounts of data coming from each experiment. Our discussion here focuses on the temperature measurements as being representative of both near-field jet mixing and overall flow patterns.

4. EXPERIMENT RESULTS

Two representations of the temperature regimes in the models are used. Vertical profiles at selected points near the discharge channel (Figures 3-5) are used to document vertical mixing rates in the jet and the shape and position of the jet in the near-field region. Plots of isotherms at a single depth (Figures 6-9) provide a picture of the overall circulation pattern and temperature distribution over the entire lake plan.

Since the two distortions have different velocity, depth, and time scales and were not made under identical thermal conditions (due to variations in ambient laboratory temperatures), dimensionless quantities are used to report the results. These quantities are defined as follows:

Temperature -- $\theta^* = (\theta - \theta_L)/(\theta_j - \theta_L)$;

Depth -- $D^* = D/D_L$; and

Time -- $T^* = T/\tau$

where θ, D and T are the measured values at a given point.

The vertical temperature profiles at the jet centerline for $T^* \approx 12$, 40, and 88 (Figure 3) show jet expansion and mixing over time for both distortions. While the total heat content of the profiles (integral of θ^* over depth) for a given T^* are nearly the same for the two distortions, the vertical profiles is quite different. The dimensionless mixing rate in the 2:1 case is about twice that of the 10:1 case. The profile 20 cm to the right of the jet centerline (Figure 5) shows a similar behavior, with even greater vertical mixing in the 2:1 case. The greater penetration of the jet at this station may also be due to recirculation induced by jet entrainment of ambient temperature fluid.

Figure 4 shows the vertical temperature profile 20 cm to the left of the jet centerline. Here the two cases show markedly different distributions, with the 2:1 distortion showing much larger θ^* throughout the profile. This is most likely due to the observed movement of the jet to the left of the centerline due to large scale flow patterns. Note that the disparity in θ^* increases substantially with time.

The isotherm plots in Figures 6-9 confirm and amplify the dissimilar behavior shown in the vertical profiles. Figures 6 and 7 show the dimensionless temperature field at $T^* \approx 12$ and $D^* = 0.067$ for the 10:1 and 2:1 distortions, respectively, while Figures 8 and 9 show the fields at the same depth with $T^* \approx 88$. It is obvious from these figures that the horizontal distribution of temperature is radically different in the two cases. The jet disperses more rapidly and uniformly in the transverse direction for the 10:1 distortion than for the 2:1 case (see small arrows which schematically show dye-traced patterns). Even as early as $T^* \approx 12$, a very pronounced circulation pattern is apparent for the 2:1 distortion, as evidenced by the lobes of elevated θ^* extending down either side of the lake.

Dye studies done for both distortions confirm this behavior. In the 10:1 case, the jet expands symmetrically and transverse mixing is rapid. Transport towards the intake appears to be symmetric and no obvious horizontal circulation cells develop. Dye studies for the 2:1 case show very plainly a strong circulation pattern which causes the jet

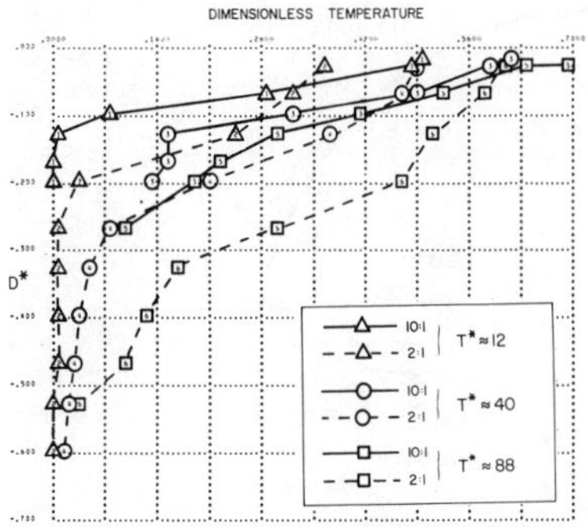

Figure 3. Probe 18 (on Jet Centerline)

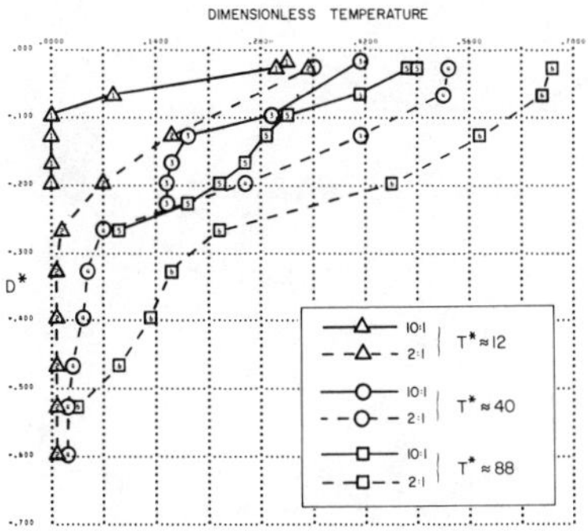

Figure 4. Probe 19 (Left of Jet Centerline)

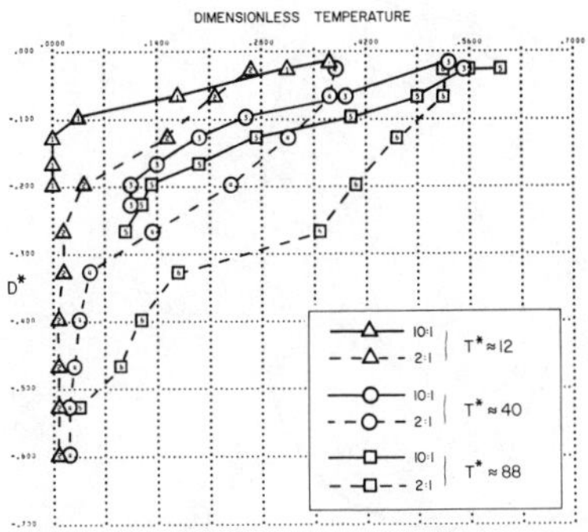

Figure 5. Probe 16 (Right of Jet Centerline)

Figures 3, 4, 5. Dimensionless Vertical Temperature Profiles at $T^* \approx 12, 40, 88$

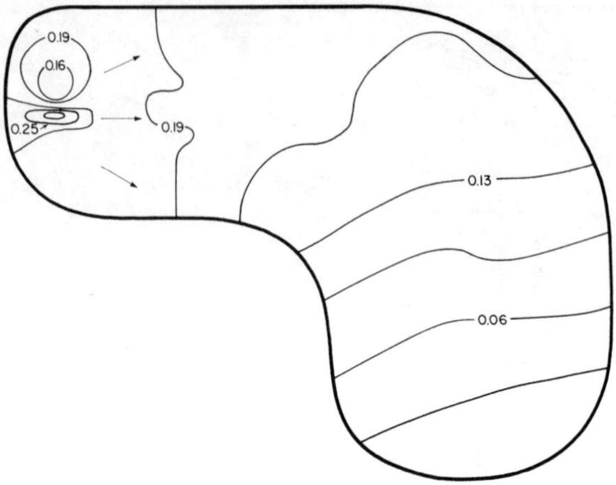

Figure 6. Plot of Dimensionless Isotherms for 10:1 Distortion at $T^* \approx 12$ and $D^* = 0.067$

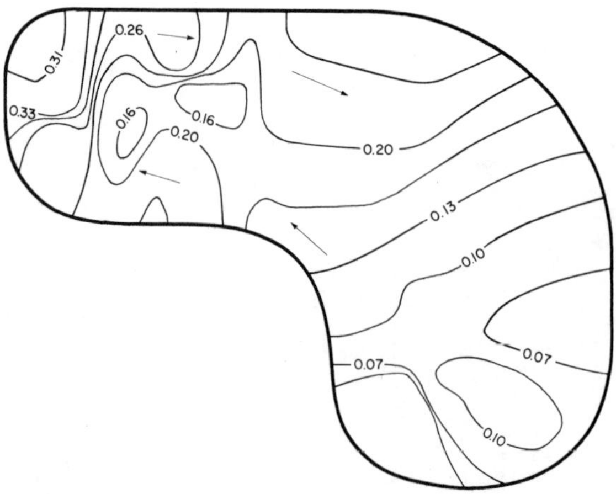

Figure 7. Plot of Dimensionless Isotherms for 2:1 Distortion at $T^* \approx 12$ and $D^* = 0.067$

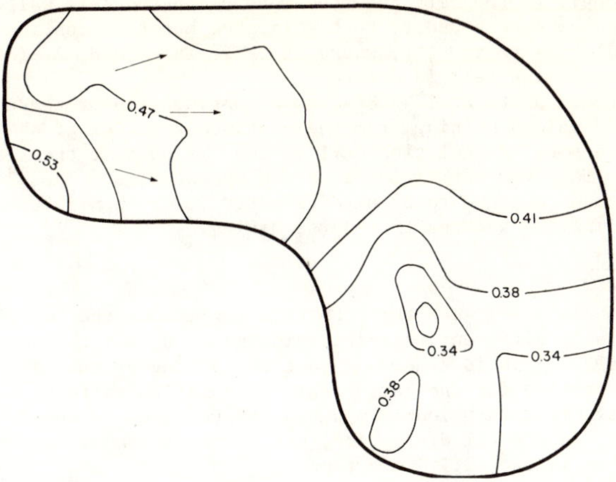

Figure 8. Plot of Dimensionless Isotherms for 10:1 Distortion at $T^* \approx 88$ and $D^* = 0.067$

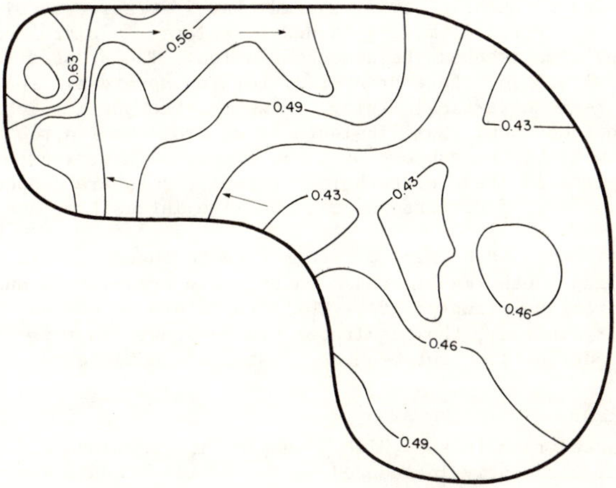

Figure 9. Plot of Dimensionless Isotherms for 2:1 Distortion at $T^* \approx 88$ and $D^* = 0.067$

to skew strongly to the left and then move along the left wall. The plume follows the lake boundary to the intake, but then continues on along the wall back to the jet, where it is re-entrained. A single, clockwise circulation cell is created.

The isotherm plots at $T^* \approx 88$ show a continuation of these patterns, with higher θ^* values arising from the continuous influx of heat. There is a hint of a small circulation cell in the 2:1 case at the intake end of the lake. This was not explored in the dye studies. The θ^* values in the 10:1 case continue to be significantly lower than in the 2:1 distortion, implying greater transverse mixing.

5. DISCUSSION

There is clear evidence that the flow regimes at the two distortions are substantially different, despite maintenance of all the accepted similarity laws. This is true, not just for mixing of the jet in the near field, but also for the large scale circulation patterns in the lake. Dimensionless turbulent mixing in the vertical is approximately twice as great in the 2:1 distortion, while transverse mixing appears to be smaller than in the 10:1 distortion.

There is likely to be substantial interaction between the jet mixing rates and the circulation patterns. In the 2:1 distortion case, one possibility is that higher dimensionless vertical mixing rates cut off return flow to the jet along the bottom and, in that way, force development of a horizontal return flow along the side. This return flow eventually leads to a fully established circulation which is strongly influenced by the size and shape of the model. The deeper 10:1 model with a lower dimensionless vertical mixing rate allows return flow to the jet origin along the bottom, circumventing the horizontal circulation.

Moretti and McLaughlin (1977) showed that if transverse mixing is properly modeled, vertical mixing is underpredicted in distorted models. Fischer (1976) also reached the same conclusion. The present results are in agreement with this, since greater distortion apparently leads to (relatively) reduced vertical mixing. However, the physics behind this behavior is not obvious. Some insights arise from the fractional analysis and will be the subject of later papers. Further experiments and more complete analyses of turbulent velocity data are needed to document the effect of distortion on turbulence intensities and structures.

A caveat should be presented with regard to these results. The profiles and isotherms are based on only two experiments and a limited number of sampling points. While there are no obvious reasons to doubt their accuracy, the results should be viewed as a preview of, rather than a definitive statement on, distorted model behavior.

6. CONCLUSIONS

The reported results show highly complex interactions which are, apparently, significantly influenced by the vertical scale distortion in the hydraulic models. While the results are clearly related to turbulence and mixing processes, the details of the mixing processes and how they vary with distortion are unknown. The results raise serious doubts regarding the ability of distorted hydraulic models to properly represent large scale flow patterns when jet-induced mixing is an important factor (e.g., see Neale, 1971). Further weight is also lent to existing doubts regarding the accuracy of dispersion modeling in hydraulic models. What remains now is to further describe and document these effects and develop means for correcting physical model results to account for such errors.

ACKNOWLEDGMENT

This material is based upon work supported by the National Science Foundation under Grant Nos. ENG-7713880 and CME-7921324.

REFERENCES

1. H. B. Fischer (1976). "Mixing and Dispersion in Estuaries," Annual Review of Fluid Mechanics, 8, pp. 107-134.

2. P. M. Moretti and D. K. McLaughlin (1977). "Hydraulic Modeling in Stratified Lakes, J. Hydr. Div., Proc. ASCE, 103, HY4, pp. 368-380.

3. L. C. Neale (1971). "Chesapeake Bay Model Study for Calvert Cliffs," J. Pwr. Div., Proc. ASCE, 97, PO4, pp. 827-839.

4. B. R. Roberts, A. N. Findikakis and R. L. Street (1979). "A Program of Research on Turbulent Mixing in Distorted Hydraulic Models, Proc. Spec. Conf. on Conserva. and Uililiz. Water and Ener. Resources, ASCE, pp. 57-64.

APPLICATION OF A CONSISTENT VORTEX SCALING LAW

Eugene Chang*

ABSTRACT

This paper describes work carried out under a general research program to determine satisfactory scaling laws for modeling free surface vortices at hydraulic intakes. A new, consistent mathematical form of scaling law has been derived, based on a Froude and Reynolds number contribution, which assumes that the model to prototype velocity ratio is a function of scale ratio only. Analysis of results from four geometrically similar rigs, based on a rectangular channel, shows that other parameters are also involved. However, a series of curves has been obtained giving the appropriate scale velocity for a limited range of model sizes.

1. INTRODUCTION

Hydraulic scale models are undoubtedly of great value to the design engineer. However, when complex flow phenomena are being modeled, the operation of the model and the analysis of the test data have to be carried out with care. It is usually not possible to achieve complete geometric, kinematic, and dynamic similarity between model and prototype if the same fluid, i.e., water, is used in both, since the relevant dimensionless parameters require conflicting operating conditions.

The problem then becomes one of choosing the appropriate model velocity to reproduce or predict prototype flow patterns. The choice of the velocity scale is an important one. In scaling free surface vortices, for example, if the model velocity is too low, then severe vortices may not be observed whereas too high a velocity may exaggerate the prototype phenomena and hence lead to over-conservative design modifications.

In practice, a compromise has to be made, and for all but the simplest of situations, this is still a matter of argument, and personal judgment. Refs. 1 and 2 describe previous work in the field of modeling vortices, and show that various scaling laws have been proposed, ranging from once Froude scale velocity ($V_m/V_p = s^{1/2}$ where s is the linear scale ratio) up to once Reynolds scale velocity ($V_m/V_p = s^{-1}$). If gravity and inertial forces predominate, then the effect of other forces is usually allowed for by increasing the once Froude scale velocity by a multiplying factor, k say, so that

*Senior Engineer, BHRA Fluid Engineering, Cranfield, Bedford, England, presently at the Alden Research Laboratory, Worcester Polytechnic Institute, Holden, Massachusetts, 01520, USA, on an exchange program.

$$\frac{V_m}{V_p} = k\ s^{1/2}$$

The choice of the exact value for k is a subjective decision, based on past experience of different scales and geometrical configurations.

The following section provides a basis for determining the required value of the multiplying factor, k, when both the Froude and Reynolds numbers are taken into account. Experimental results are then discussed in subsequent sections in an attempt to determine the appropriate value of k for modeling free surface air-entraining vortices. Further details can be found in ref. 3.

2. DERIVATION OF A CONSISTENT SCALING LAW

In order to derive a general scaling law which concurs with accepted practice, it is of most practical use to consider the ratio of model to prototype velocity, and to express it as a function of scale ratio, s, only, i.e.,

$$\frac{V_m}{V_p} = f(s)$$

This agrees with accepted practice whereby the same modeling procedures are adopted for a wide range of intakes, regardless of their geometry.

Hence, various forms of f(s) are possible, and some examples are:

$f(s) = s^{1/2}$ for once-Froude scaling

$f(s) = k_o s^{1/2}$ for multi-Froude scaling, where k_o is a constant multiplying factor

$f(s) = s^{-1}$ for once-Reynolds scaling

Another form which may apply for more complex flow phenomena is:

$f(s) = s^{1/2} + a_o s^{-1}$ which represents a combination of Froude and Reynolds scaling, a_o being a constant to represent the Reynolds number contribution. This combination would be a more realistic form for phenomena where viscous as well as gravitational forces should be considered.

However, one important requirement which has not been considered before, as far as the author is aware, is that there must be "internal consistency" or "invertibility" in the form of the function f(s), such that the scaling law can be used both for scaling up from a model to a prototype (scale ratio s) and for scaling down from a prototype to a model (scale ratio 1/s). Mathematically, this means that if

$$\frac{V_m}{V_p} = f(s)$$

then the inverse scaling requires that

$$\frac{V_p}{V_m} = f(\frac{1}{s})$$

From the above two equations, it can be seen that f(s) must satisfy the relationship

$$f(s) = \frac{1}{f(\frac{1}{s})}$$

Referring to the examples of f(s) given above, $f(s) = s^{1/2}$ and $f(s) = s^{-1}$ both satisfy the consistency relationship. In fact, s raised to any power, including zero (which gives the "equal model-prototype velocity" law) would be acceptable. However, the multi-Froude law is not consistent since

$$f(s) = k_o\ s^{1/2} \neq \frac{1}{k_o\ s^{-1/2}} \quad \text{(unless } k_o = 1\text{)}$$

Similarly, the combination Froude and Reynolds law is not internally consistent.

One general class of functions which is internally consistent is

$$f(s) = \frac{g(s)}{g(\frac{1}{s})}$$

For once-Froude scaling, $f(s) = s^{1/2}$, so that $g(s) = s^{1/4}$. Similarly, for once-Reynolds scaling, $g(s) = s^{-1/2}$. Thus, a consistent combination law can now be written in the form

$$f(s) = \frac{a\ s^{1/4} + b_o\ s^{-1/2}}{a\ s^{-1/4} + b_o\ s^{1/2}} = (\frac{a + b_o\ s^{-3/4}}{a + b_o\ s^{3/4}})\ s^{1/2}$$

$$f(s) = \frac{V_m}{V_p} = (\frac{1 + b\ s^{-3/4}}{1 + b\ s^{3/4}})\ s^{1/2}$$

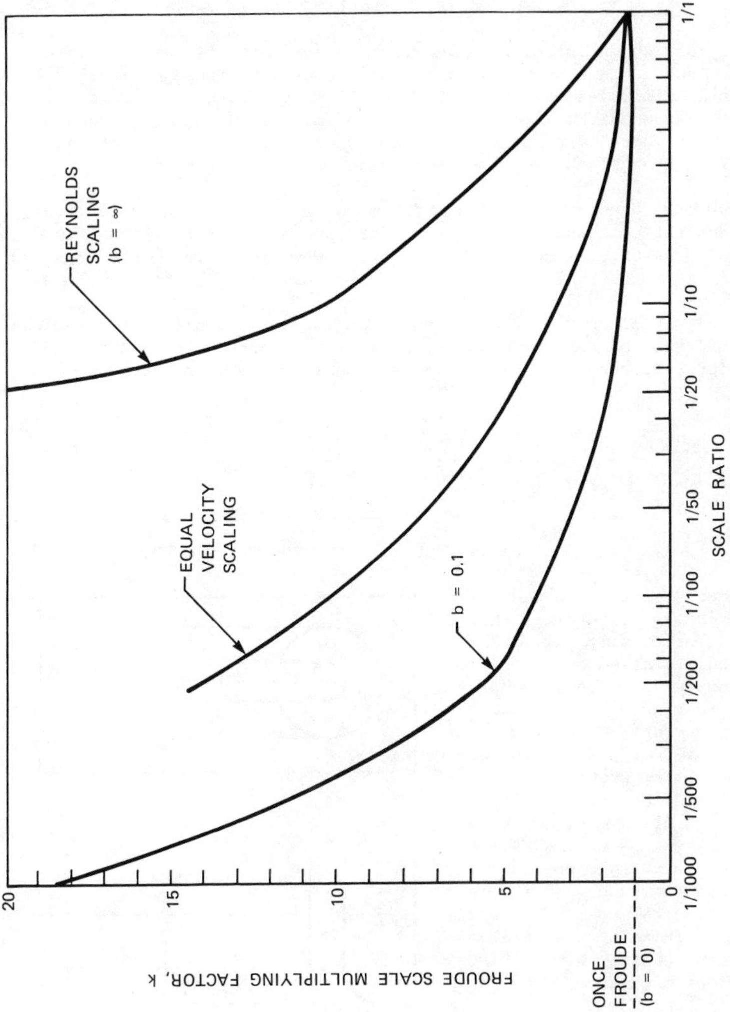

Fig. 1 Variation of Froude Scale Multiplying Factor, k, with Scale Ratio, s, for Various Values of Reynolds Number Contribution, b

3. APPLICATION OF THE CONSISTENT SCALING LAW TO VORTEX FORMATION

The expression in brackets in the last equation for f(s) is equivalent to the Froude scale multiplying factor, k, used to increase model velocities as discussed earlier. It depends on the value of b, and this could be found either from correlation of model and prototype data or from experimental data from models built to different scale ratios. This latter approach is discussed in the following sections.

To give an idea of the results which might be obtained, a typical curve of k against s is shown in Figure 1 for a value of b=0.1. There is a whole family of such curves for different positive values of b, ranging from once Froude scaling, when b=0 to once Reynolds scaling when b=∞.

The analysis of section 2 could be extended, if appropriate, to include the influence of surface tension, by adding a further term to the combination law, representing the Weber number contribution, i.e.,

$$f(s) = \frac{s^{1/4} + bs^{-1/2} + cs^{-1/4}}{s^{-1/4} + bs^{1/2} + cs^{1/4}}$$

4. BRIEF DESCRIPTION OF THE EXPERIMENTAL PROCEDURE

In order to determine the value of b, and hence the k factor, four geometrically similar rigs were built, based on the common rectangular pump sump geometry shown in Figure 2. The four sizes were called the 1.5", 3", 6", and 12" rigs according to their nominal suction pipe diameter, and the corresponding bellmouth diameters, D, were 61mm, 122mm, 244mm, and 488mm, respectively. In each case, the channel width, W/D = 2, end wall clearance, X/D = 0.25, and the length of the straight approach channel, L, was 8.8D. The floor clearance, C/D, was varied from 0.4 to 0.8.

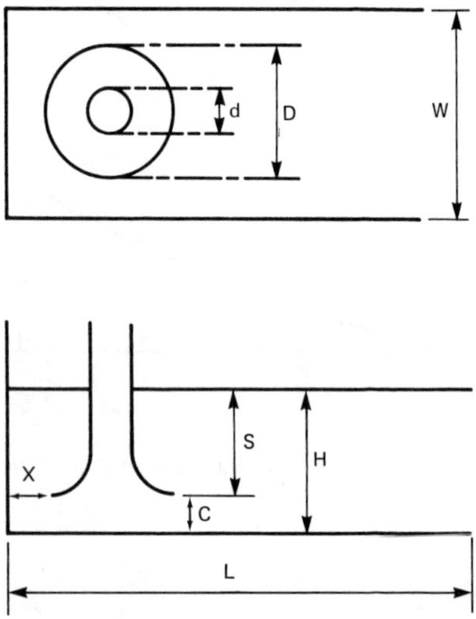

Fig. 2 Definition Diagram of Rectangular Pump Sump Geometry

The occurrence of free surface vortices was observed for a range of submergences and velocities. Graphs of critical submergence, S_c/D, against bellmouth velocity, V_D, were determined for each rig. An example of these for one particular geometry is shown in

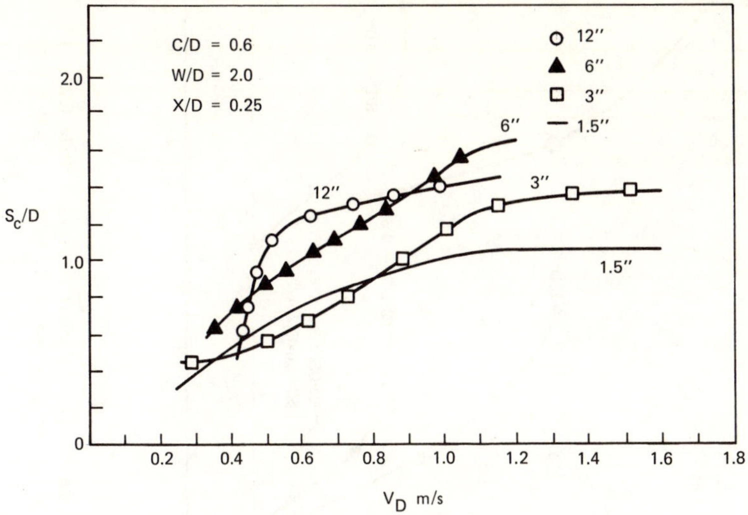

Fig. 3 Graph of Critical Submergence S_c/D Against Bellmouth Velocity V_D for the Four Rigs, C/D = 0.6

Figure 3. The critical submergence for a given bellmouth velocity was defined as the submergence at which a fully air-entraining vortex was first observed. Thus, in Figure 3, the curve for a particular size rig divides the submergence-velocity field into two regions, one below the curve in which fully air-entraining vortices occur, and one above the curve where weaker vortices occur. The value of Reynolds number, based on V_D and D, ranged from a minimum of 10^4 on the 1.5" rig up to a maximum of 6×10^5 on the 12" rig.

5. RESULTS, ANALYSIS, AND DISCUSSION

In order to correlate the results with a particular value of b, the 12" rig was taken as the prototype. Then, using plots like the one shown in Figure 3 for a given sump geometry, values of bellmouth velocity $V_{1.5}$, V_3, V_6, and V_{12} for the four rigs were determined for a particular value of S_c/D (to maintain geometric similarity). Taking V_{12} as the prototype velocity, values of $f(s) = V_m/V_p$ were obtained from $V_{1.5}/V_{12}$, V_3/V_{12}, and V_6/V_{12}, corresponding to values of scale ratio, s, of 1/8, 1/4, and 1/2. These values were then substituted into the last equation of section 2 to determine b. This analysis was repeated for several values of S_c/D, and also subsequently for the other sump geometries tested.

The analysis showed, firstly, that for a given geometry and scale ratio, s, the value of V_m/V_p and hence b was not constant, but generally increased with submergence. Various alternative forms of plotting the results were tried, but it was not possible to obtain constant values of

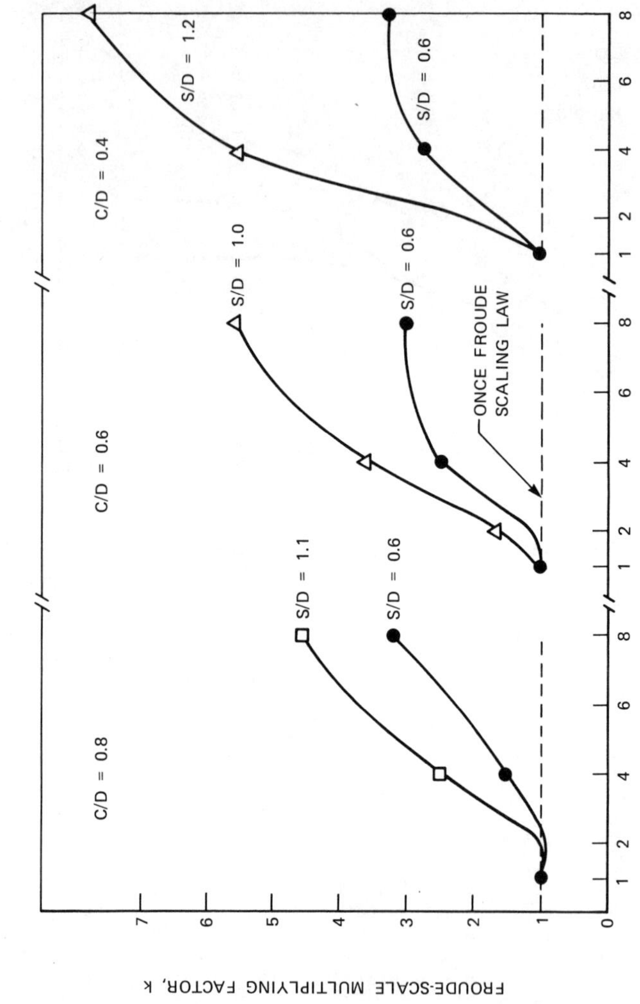

Fig. 4 Graph of the Froude-Scale Multiplying Factor, k, Against the Inverse Scale Ratio, 1/s for a Range of Submergences S/D and Floor Clearances C/D, W/D = 2, X/D = 0.25

f(s) over the range of S/D and V_D tested. Thus, it is deduced from this that V_m/V_p is not simply a function of scale ratio only, but also depends on submergence. Denny and Young (ref. 4) obtained similar results which could be interpreted as indicating a dependence of V_m/V_p on submergence. However, they attributed the form of their results to scale effects, and proposed a scaling law independent of submergence.

Secondly, the value of V_m/V_p varied with C/D, the one sump parameter that was varied. Thus, it should be noted that a scaling law used for a particular hydraulic model could also vary with sump parameters such as the floor clearance.

While it was not possible to derive a general scaling law because of the variation of b, it was possible to obtain the variation of the Froude scale multiplying factor, k, over the range of scale ratios and geometries tested. This is shown in Figure 4 for three values of C/D and selected values of S/D, with 1/s along the abscissa. The curves have been drawn to pass through the point (1, 1), and their general trend agrees with those obtained by Zajdlik (ref. 5). The trend of increasing k with increasing S/D is the opposite of that reported in ref. 6. However, an offset approach channel was used in ref. 6, showing again the importance of the intake geometry in influencing vortex formation.

The curves shown in Figure 4 have direct application over the range of sizes and Reynolds number tested. However, it is uncertain to what extent they apply outside of this range. The curves seem to flatten out to an asymptotic k value as 1/s increases. However, further work with larger sizes used as the prototype would be needed to confirm this.

6. CONCLUSIONS

It is possible to generate a mathematical form of scaling law which takes into account both Froude and Reynolds number contributions. However, the experimental results show that the original form assumed

$$\frac{V_m}{V_p} = f(s)$$

does not hold in practice. Instead, the velocity ratio depends not only on scale ratio but on other parameters, i.e.,

$$\frac{V_m}{V_p} = f(s,\ b,\ \text{geometric parameters})$$

where b is a measure of the influence of Reynolds number. While values of the Froude scale multiplying factor have been obtained, which are appropriate for scale ratios up to 1:8, their application to larger prototype sizes needs to be verified by further work.

ACKNOWLEDGEMENTS

The author is grateful to Dr. A.J. Cochran for valuable discussions on several ideas presented in the paper. Thanks are due to Alden Research Laboratory, Worcester Polytechnic Institute, for their assistance in the preparation of the paper. The author acknowledges the permission given by the Director and Council of BHRA to publish this paper.

REFERENCES

1. Chang, E., "Review of Literature on Drain Vortices in Cylindrical Tanks," BHRA Report TN 1342, 32pp., March 1976.

2. Chang, E., "Review of Literature on the Formation and Modeling of Vortices in Rectangular Pump Sumps," BHRA Report TN 1414, 30pp., June 1977.

3. Chang, E., "Scaling Laws for Air Entraining Vortices," BHRA Report RR 1519, 30pp., January 1979.

4. Denny, D.F., and Young, G.A.J., "The Prevention of Vortices and Swirl at Intakes," Proc. 7th IAHR Meeting, Lisbon, pp. C_1-1 to C_1-14, 1957.

5. Zajdlik, M., "New Checking Mode of Model Parameters for Vortex Formation in Pump Tanks," Proc. 17th IAHR Meeting, Baden-Baden, Vol. 5, pp. 379-386, 1977.

6. Lertsachanant, C., "Scale Effects on Free Surface Vortices and Vortex Suppression," M.S. Thesis, Worcester Polytechnic Institute, December 1979.

SIND MODELING FOR UNSTEADY FLOW ASSESSMENTS

by

Joel L. Caves[a] and Robert L. Johnson[b], Members, ASCE

ABSTRACT

Mathematical modeling often requires judgements on the relative importance of coefficients and terms in the governing equations. In problems involving simultaneous, interactive, unsteady flow in more than one system there is the additional difficulty of deciding how the problem can best be modeled within given time and cost constraints.

With the exception of some groundwater models that can model both confined and unconfined aquifer flow, nearly all off-the-shelf unsteady flow models simulate hydraulic transients in a single type of flow system, e.g., piping system, open channel system, unconfined aquifer system. Thus use of traditional packaged models requires selection of the dominate system for modeling and either neglect or approximation of effects from secondary systems on the modeled primary system. A third alternative of developing a special purpose model for simulating conditions within each flow system and the interactions between the systems can be time consuming and expensive.

The system-independent (SIND) technique, a departure from the traditional system-dependent approach to unsteady flow modeling, offers another alternative for making these assessments. This paper briefly describes this new method and illustrates with two case studies how it can be advantageously used.

SYSTEM-INDEPENDENT EQUATIONS

The system-independent method is based on a single set of equations, called system-independent (SIND) equations, which provide a generic description of one- or two-dimensional liquid flow through conduits. A "conduit", as defined in the SIND concept, is a liquid conveyance with or without porous media. Pipes, tunnels, river channels, coastal waters and aquifers are examples of conduits in this sense.

The control volume for deriving the SIND equations is shown in Fig. 1. Flow takes place between the upper and lower flow surfaces

---

[a]Consulting Hydraulic Engineer, Reading, PA 19606
[b]Professor of Civil Engineering, Fritz Engineering Laboratory, Lehigh University, Bethlehem, PA 18015

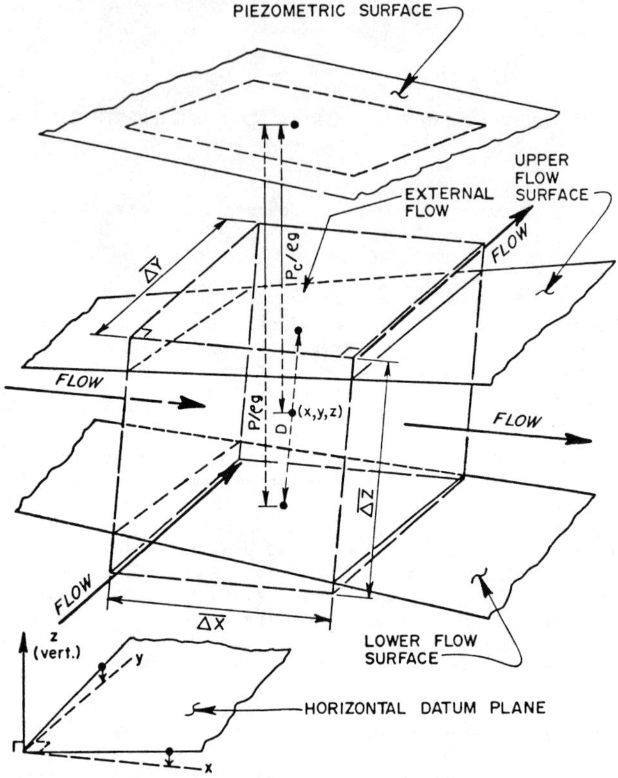

Fig. 1 Control Volume for System-Independent Development

of the conduit which are located within the control volume. The space between these flow surfaces may be occupied entirely by liquid, or by porous media and liquid. In two-dimensional flow, the sides of the control volume are normal to the respective x and y velocity components, U* and V*, whereas in one-dimensional flow the control volume sides parallel to the x-axis are located outside the lateral confines of the conduit. The velocities U* and V* are the integrated velocities through the area of liquid in the cross sections, and are related to the gross-area velocities (U and V) by U* = U/n and V* = V/n, where n is the porosity of the media.

The pressure P is measured relative to the lower flow surface, and Z is the elevation of the lower flow surface relative to the datum plane. These define the piezometric surface which may be either separate from the upper flow surface (e.g., closed conduit and confined aquifer systems) or coincident with the upper flow surface (e.g., unconfined aquifer and all free surface systems).

Assuming that pressures are hydrostatic normal to the main flow direction(s) and that steady flow relationships may be used for viscous losses, the SIND equations in terms of three unknowns (U*,V*,P) and three independent variables (x,y,t) can be derived (2). All notations in these and the other equations are defined in the text.

Continuity Equation -

$$C_{C1}\left(\frac{\partial P}{\partial t} + U^*\frac{\partial P}{\partial x} + V^*\frac{\partial P}{\partial y}\right) + \frac{\partial U^*}{\partial x} + \frac{\partial V^*}{\partial y} + C_{C3}U^* + C_{C5}V^* + C_{C2} + C_{S1} = 0 \quad (1)$$

Equation of Motion in X-Direction -

$$C_{P1}\frac{\partial P}{\partial x} - gS_{ox} + C_{U1} + (C_{L1}+C_{S1})U^* - f_c V^* + \frac{\partial U^*}{\partial t} + U^*\frac{\partial U^*}{\partial x} + \epsilon_x V^*\frac{\partial U^*}{\partial y}$$
$$+ (\epsilon_x - 1)U^*\left[\frac{\partial V^*}{\partial y} + V^*\left(C_{C1}\frac{\partial P}{\partial y} + C_{C5}\right)\right] = 0 \quad (2)$$

Equation of Motion in Y-Direction -

$$C_{P1}\frac{\partial P}{\partial y} - gS_{oy} + C_{U2} + (C_{L2}+C_{S1})V^* - f_c U^* + \frac{\partial V^*}{\partial t} + V^*\frac{\partial V^*}{\partial y} + \epsilon_y U^*\frac{\partial V^*}{\partial x}$$
$$+ (\epsilon_y - 1)V^*\left[\frac{\partial U^*}{\partial x} + U^*\left(C_{C1}\frac{\partial P}{\partial x} + C_{C3}\right)\right] = 0 \quad (3)$$

In these SIND equations, f_c is the Coriolis coefficient, S_{ox} and S_{oy} are the respective slopes of the x and y velocity components, ϵ_x and ϵ_y are factors representing the relative velocities of mass conveyed through the lateral control volume boundaries, and the "C" coefficients are called SIND coefficients.

The SIND coefficients have the following physical meanings:

C_{C1} - Relates relative change in liquid volume (**V**) to a change in liquid pressure

C_{C2}, C_{C3}, C_{C5} - Relative changes in liquid volume (**V**) caused by changes in external loads and/or atmospheric pressure

C_{L1}, C_{L2} - ratio of conduit wall-related shear force (per unit mass of liquid) to the respective velocity component, U* or V*

C_{P1} - inverse of the liquid density ($1/\rho$)

C_{S1} - rate of spatially-varied mass flow to or from the external surroundings, per unit mass of internal flow

C_{U1}, C_{U2} - wind-related shear force (per unit mass of liquid) on upper surface in x and y-directions, respectively.

The various C_C SIND coefficients represent the combined contributions of area, density, and porosity variations to the system capacitance. Mathematically, the C_C SIND coefficients evolve from using the following expression for relating changes in the liquid volume within the conduit to pressure changes:

$$\frac{1}{\mathbf{V}} \frac{\partial \mathbf{V}}{\partial \zeta} = C_{C1} \frac{\partial P}{\partial \zeta} + C_{Ci} \qquad (4)$$

where **V** = liquid volume contained in the conduit
ζ = x, y or t
C_C = capacitance-related SIND coefficients
i = i(ζ) = 2, 3 or 5 corresponding to ζ = x, y or t, respectively

The magnitude of each SIND coefficient is system-dependent. For many flow systems, analytical expressions can be derived for the SIND coefficients from basic principles of fluid and solid mechanics (2). Alternatively, values for most SIND coefficients may be determined experimentally, an aspect that can be especially useful in modeling unusual flow systems that defy development of analytical expressions for the coefficients.

SIND ANALYSIS OF UNSTEADY FLOW

Most transient flow problems involve multiple, interactive flow systems to some extent. A river flood, for instance, actually affects not only conditions in the river but also the pressures and flow in the material comprising the channel unless the river perimeter is completely impervious. Many other multisystem examples exist in nature and in the works of man.

The flow systems comprising a multisystem problem can be categorized as primary and secondary systems, their distinction being in the purpose of the hydraulic analysis. A primary system contains the unknowns of principal interest in a particular problem, whereas a secondary system is a flow domain for which the major interest is the influence upon the primary system. By these definitions, all multisystem problems consist of one or more primary systems; secondary systems may or may not exist.

The conventional approaches to analysis of multisystem unsteady flow problems can be generally classified as simplification-by-neglect, simplification-by-separation and combined-analysis. Simplification-by-neglect involves reduction of the multisystem problem to a single primary flow system after assuming or determining that the event of interest will have negligible effect on connected flow systems. Many problems can be treated very adequately by this approach.

Simplification-by-separation entails separate treatment of the various flow systems in a problem. The results of the analyses of the separated systems may then provide a basis for approximating the effects of interactions between the systems. The third approach to analyzing a multisystem problem, combined-analysis, involves developing a special purpose computer model by using coupling relationships to link the appropriate system-dependent equations for unsteady flow.

The SIND equations, Eqs. 1, 2 and 3, provide the theoretical relationships for a fourth approach. A major advantage of these equations is that they facilitate development of highly versatile computer models capable of simulating innumerable problems involving flow in single and multiple flow systems. SIND analysis represents a departure from the other methods and techniques described which use system-dependent equations. A SIND-based model that exploits the generality of the SIND equations can thoroughly simulate the dynamic behavior of each primary and secondary system together with their coupling boundary condition relationships. Thus, SIND analysis avoids the assumptions inherent in the simplification-by-neglect and simplification-by-separation methods. It also eliminates the time and cost required for developing a special purpose model for each combined-analysis.

The examples described in the following two sections illustrate use of SIND analysis. The numerical model applied in the various simulations is based on a finite difference formulation of the SIND equations using the weighted four point implicit method. Only input data changes were involved in the simulations of these widely different problems.

WETTINGEN TUNNEL PROBLEM

Initiation of flow in the tail race tunnel system (Fig. 2) at the Wettingen Power Station near Zurich, Switzerland, is a classical case of transitory free surface-pressurized flow. The problem, originally described in 1932 by Meyer-Peter and Favre (4), involves analysis of unsteady flow occurring in the system when the turbine gates are fully opened in five seconds. In addition to Meyer-Peter and Favre's mathematical analysis and physical model study, the Wettingen conditions have been investigated by Calame (1), Cunge and Wegner (3) and, most recently, by Wiggert (6).

Initially the water in the tunnel is at rest, partially filling the cross section. As inflow to the surge tank begins, unsteady free surface flow is initiated. When the surge chamber water level reaches the top of the tunnel, a transitional front is created that moves downstream as the chamber water level continues to rise. During the time

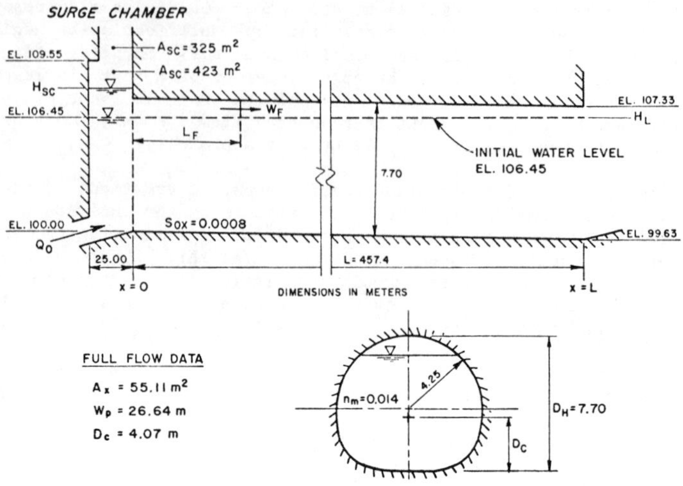

Fig. 2 Wettingen Tunnel Problem

that the front is travelling through the tunnel, pressurized flow occurs in the lengthening region behind the front while free surface conditions continue ahead. Once the front reaches the downstream end, unsteady closed conduit flow exists throughout the tunnel.

In the mathematical modeling by Wiggert (6), the most thorough of the previous analyses, it was assumed that the water and the tunnel behave inelastically in the pressurized region and thus can be described by the surge equation. To assess the validity of this assumption, the importance of waterhammer phenomena in the pressurized region was investigated in the present study by using different values for the bulk modulus of liquid elasticity in calculating SIND coefficients. In the simulations, the SIND equations, which describe the unsteady flow in both the free surface and pressurized flow regions, were solved with an internal boundary condition for the transitional front and a boundary condition for the surge chamber.

The results for the elastic and inelastic cases (Figs. 3 and 4) were found to be virtually identical. Although probably apparent in the physical experiments of Meyer-Peter and Favre and of Wiggert, the generality of the SIND equations and the versatility of the SIND-based model facilitated expeditious confirmation that the pressurized region behind the moving front can be accurately represented as a surge.

FLOOD MODIFICATION PROBLEM

Special purpose models have been developed by Pinder and Sauer (5) and Zitta and Wiggert (7) to investigate the effects of an aquifer in modifying a river flood wave. Since both models are based on traditional separate system-dependent equations for the river and for the

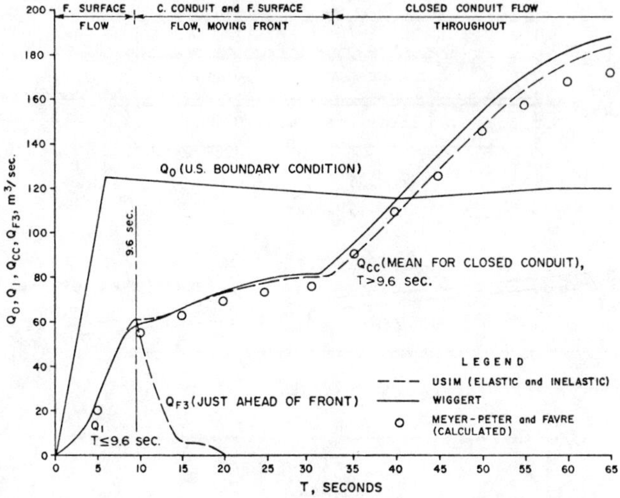

Fig. 3 Wettingen Tunnel Problem - Flows

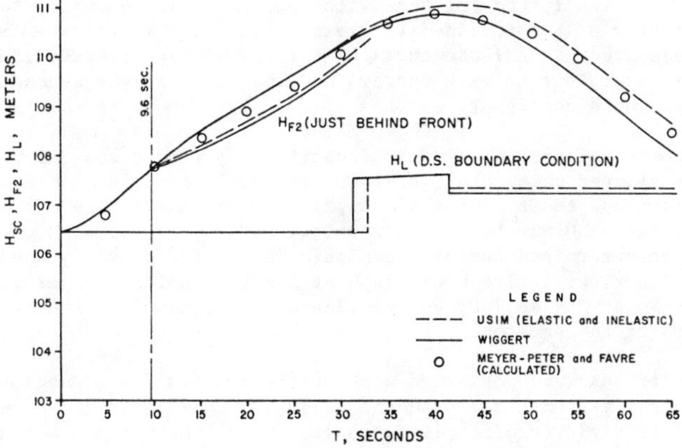

Fig. 4 Wettingen Tunnel Problem - Piezometric Heads

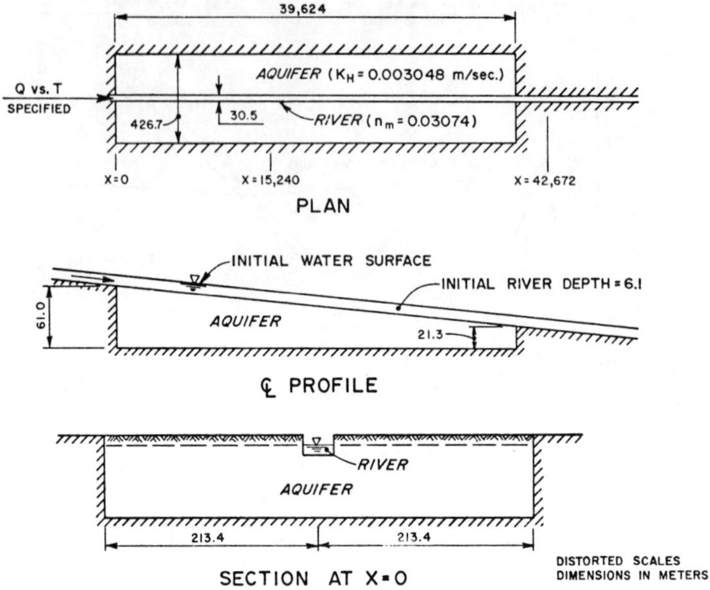

Fig. 5 Flood Modification Problem

aquifer which are linked together with coupling relationships, they are representative of the combined-analysis approach to multisystem modeling. These studies indicate that the effects of flow interaction between the river (the primary system) and the aquifer (the secondary system) can be significant.

The physical problem (Fig. 5) described by Pinder and Sauer has been investigated using the SIND-based model. The one-dimensional flow in the river and the two-dimensional flow in the aquifer were represented by the SIND equations; appropriate values for the SIND coefficients were determined during the simulations. The boundary conditions included the river inflow hydrograph at X = 0, a uniform river flow relationship at X = 48,768, and no-flow specifications at the exterior boundaries of the aquifer.

Simulations were performed without the aquifer (i.e., neglecting the secondary system), and with the aquifer. Two different values for the specific yield or effective porosity (n) of the homogeneous aquifer, were considered.

The two-way interchange of flow between the primary and secondary system is shown in Fig. 7. As is evident in Fig. 6, this interaction can significantly modify the attenuation, dispersion and speed of flood wave.

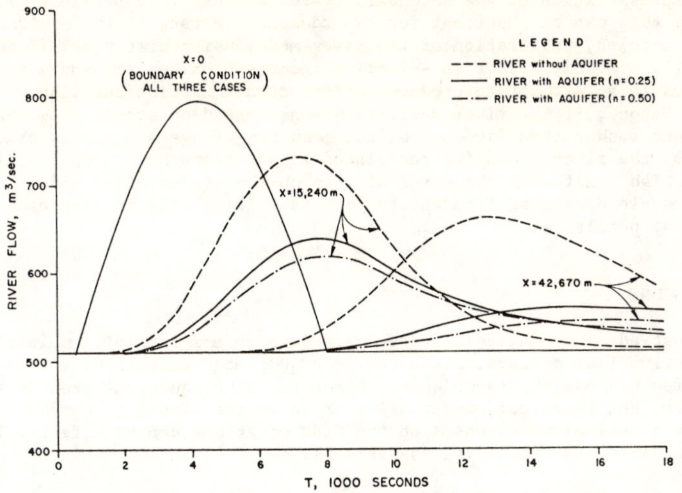

Fig. 6 Flood Modification Problem - Aquifer Effects on River Flows

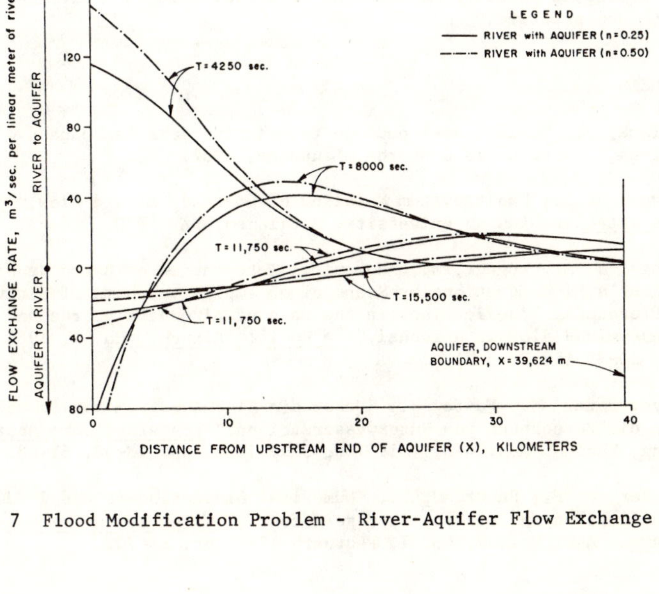

Fig. 7 Flood Modification Problem - River-Aquifer Flow Exchange

Representation of the secondary system effects in a pervious river such as this can be important for two reasons. First, if the aquifer is not modeled, calibration of the river model using historical flood data will probably result in selecting incorrect values for model coefficients in an attempt to produce sufficient attenuation and dispersion. Second, if the river coefficients are based on steady flow measurements rather than flood data, the resulting flows and depths calculated by the river model for postulated flood inflow hydrographs will be too high. Although the error will be on the conservative side, their use in designing flood protection facilities will incur unnecessary cost penalties.

CONCLUSIONS

Unsteady flow analysis, especially of problems involving multiple, interactive flow systems, often requires judgement concerning the application of modeling techniques. Since the SIND equations provide a complete, but practical, description of an entire class of "conduit" flow, a numerical model based on the SIND equations can be effectively used in making many modeling assessments.

As illustrated by the Wettingen Tunnel application, the completeness and accuracy of a SIND-based model facilitate investigations of the importance of terms in governing system-dependent equations. The versatility of a sind-based model, demonstrated in the flood modification problem, can be exploited in readily assessing the effects of secondary systems upon hydraulic conditions in the system(s) of primary interest. The advantages of SIND analysis compared to traditional system-dependent modeling methods can be of value in both engineering research and practice.

REFERENCES

1. Calame, J., "Calcul de l'onde de translation dans les canaux d' usines," Editions la Concorde, Lausanne, 1932.

2. Caves, J. L., "Multisystem Modeling of Unsteady Liquid Flow," Ph.D. Dissertation, Lehigh University, Bethlehem, PA, 1980.

3. Cunge, J. A., Wegner, M., "Numerical Integration of Barre de Saint-Venant's Flow Equations by Means of an Implicit Scheme of Finite Differences. Applications in the Case of Alternately Free and Pressurized Flow in a Tunnel," La Houille Blanche, No. 1, 1964, pp. 33-39 (in French).

4. Meyer-Peter, E., Favre, H., "Ueber die Eigenschaften von Schwällen und die Berechnung von Unterwasserstollen," Schweizerische Bauzeitung, Zürich, Nos. 4-5, Vol. 100, July 1932, pp. 43-50, 61-66.

5. Pinder, G. F., Sauer, S. P., "Numerical Simulation of Flood Wave Modification due to Bank Storage Effects," Water Resources Research, AGU, Vol. 7, No. 1, February 1971, pp. 63-70.

6. Wiggert, D. C., "Transient Flow in Free Surface, Pressurized Systems," Journal of the Hydraulics Division, ASCE, No. HY1, Paper 8628, January 1972, pp. 11-27.

7. Zitta, V. L., Wiggert, J. M., "Flood Routing in Channels with Bank Seepage," Water Resources Research, AGU, Vol. 7, No. 5, October 1971, pp. 1341-1345.

A ONE-DIMENSIONAL FLOW MODEL OF THE POTOMAC RIVER

by Raymond W. Schaffranek [1/] and Robert A. Baltzer [1/], M. ASCE

ABSTRACT

A generally applicable one-dimensional model for simulation of flow in a system of interconnected open channels is applied to a portion of the tidal Potomac River in the Washington, D.C., metropolitan area extending upstream from the Woodrow Wilson Bridge to the head of tide near Chain Bridge. The flow model itself is broadly applicable to a wide range of hydrologic conditions and varying field situations. It accommodates tributary inflows and diversions, includes the effects of wind shear on the water surface as a forcing function, and incorporates a momentum coefficient which can be used to compensate for any nonuniform velocity distribution present in the channel cross section. Water-surface elevation and discharge are computed at any desired location throughout the network using an implicit, finite-difference technique. Weighting factors can be specified to vary the implicit-solution technique from box-centered to fully forward. A unique equation transformation technique is employed to reduce machine memory and execution time requirements.

The modeled portion of the Potomac River and its tributaries is schematized as a network of 14 interconnected, one-dimensional channels. Preliminary flow simulations are stable and reasonably approximate prototype conditions. Further evaluation of model performance awaits water-surface elevation and flow velocity data presently being measured throughout the network.

INTRODUCTION

Like estuarine systems in many regions of the world, the Potomac River and Estuary are being subjected to stresses caused by the pervasive activities of man: natural upland river flows are being altered, on occasion channels are dredged and tidal marshlands reclaimed, and the waters are used for industrial cooling, for effluent transport and dispersal, and for recreational purposes. Such activities--and their resultant stresses--will continue and are likely to intensify in the future. Until recently, little thought was given to the ecological consequences or to the ultimate, longer-range economic and social impacts of these activities. However, a growing consciousness of the need to protect the riverine and estuarine environment of the Potomac is now beginning to emerge. One can reasonably anticipate that, as more knowledge of the Potomac is made available, the concept of protection will gradually evolve into one of management. Effective and enlightened

[1/] Research Hydrologist, National Center, U.S. Geological Survey, Reston, Virginia management, however, requires a basic level of understanding of the overall riverine and estuarine processes of the Potomac that does not now exist. It also requires the availability of data and operational methods for appraisal of the impact of management decisions (3).

Specific questions concerning the Potomac estuarine system are being asked by both private-citizen groups and governmental agencies. What impact would increased withdrawal of fresh water have on the system? If such withdrawals were to be made, where best should intakes be located? How much sediment is being transported throughout the Estuary? What is the excursion range of the turbidity maximum? How do nutrient and particulate-matter budgets vary throughout the Estuary? Any attempt to address these questions underscores the need for hydrodynamic information about the Estuary. Because circulation and mixing patterns dominate almost all other aspects of estuarine behavior, knowledge of estuarine hydrodynamics is required not only to answer specific questions regarding the Potomac Estuary but is fundamental to all estuarine science in general. Yet circulation and mixing patterns in estuaries are difficult both to measure and to analyze for they do not evolve to steady-state phenomena, but rather undergo ceaseless change in response to the ebb and flood of the tides, variations in fresh-water inflow, erratic effects of winds, movements of weather fronts, irregularities in both riverine and estuarine geometry, and to density differences between fresh and salt water.

As part of its River Quality Assessment program, the U.S. Geological Survey is currently undertaking a comprehensive investigation of the Potomac Estuary. The purpose of the study is to attain a better understanding of the interrelationship of the predominant estuarine processes. The study is designed to provide both information and technology with which others may appraise management alternatives. One important aspect of this investigation is determination of the flow dynamics occurring throughout the extent of the Estuary. Various types of flow models, calibrated and verified with prototype data, are being developed to achieve this objective. This paper briefly describes one of these models, specifically a one-dimensional, branch-network flow model (5). It discusses the implementation and application of this model to a riverine portion of the Potomac Estuary extending from the head of tide to a location 19.7 km (12.3 mi) downstream.

METHODOLOGY

Numerous approaches exist for solution of the equations of unsteady open-channel flow. In the branch-network flow model a weighted, four-point, finite-difference approximation of the nonlinear, partial-differential, flow equations is employed, thus providing the modeler the flexibility to vary the solution technique from a box-centered scheme on the one hand to a fully forward scheme on the other. After the flow-equations are linearized and appropriate boundary conditions are defined, coefficient matrices are generated. In order to reduce the size of the coefficient matrices and thereby speed up the computational process, a unique equation transformation process is employed. The model, being so formulated, is broadly applicable to a variety of prototype flow conditions and is readily adaptable to various network configurations.

Equation Formulation

The branch-network flow model is based upon a comprehensive set of the nonlinear, partial-differential equations governing one-dimensional, unsteady flow in channels for which the dependent variables are the water-surface elevation, Z, and the flow rate, Q. The distance, x, in the longitudinal direction along the channel thalweg and the elapsed time, t, are the independent variables. For a channel with negligible lateral inflow the equation of continuity can be written as

$$B\frac{\partial Z}{\partial t} + \frac{\partial Q}{\partial x} = 0$$

in which B is the channel top width. When wind effect is taken into consideration the equation of motion for one-dimensional open-channel flow can be expressed as

$$\frac{\partial Q}{\partial t} + \beta \frac{Q}{A}\frac{\partial Q}{\partial x} + Q\frac{\partial(\beta Q/A)}{\partial x} + gA\frac{\partial Z}{\partial x} + \frac{gk}{AR^{4/3}} Q|Q| - \xi\, BU_a^{\,2} \cos \alpha = 0,$$

in which g is the acceleration of gravity, A is the cross-sectional area, R is the hydraulic radius and U_a is the wind velocity vector making an angle α with the positive x-axis. The coefficient k is a function of the flow-resistance coefficient, η (similar to Manning's n), which can be expressed in the inch-pound system of units as

$$k = \left(\frac{\eta}{1.49}\right)^2 \qquad \text{(or in the metric system as } k = \eta^2\text{).}$$

The coefficient ξ is a dimensionless wind-resistance coefficient which can be expressed as a function of the water-surface drag coefficient, C_d, the water density, ρ, and the air density, ρ_a, as

$$\xi = C_d \frac{\rho_a}{\rho} .$$

The coefficient β, known as the momentum or Boussinesq coefficient, which can be expressed as

$$\beta = \frac{\int u^2\, dA}{U^2 A} ,$$

is present to adjust for any nonuniform velocity distribution over the channel cross section. In this expression u represents the velocity of water passing through some finite elemental area, dA, U is the mean velocity in the cross section, and A is the cross-sectional area as previously defined.

Approximate solutions of the unsteady, open-channel, flow equations can be obtained by replacing the partial-differential equations by appropriate finite-difference expressions. In the branch-network flow model a weighted, four-point, finite-difference scheme is employed. The finite-difference formulation treats time derivatives of the dependent variables, stage (water-surface elevation) and discharge, as centered in both space and time, and spatial derivatives of the dependent variables as centered in space and positioned in time according to a variable weighting factor. As can be seen from the space-time grid system of figure 1, the four points used in the approximations are identified by the intersections of the vertical lines at distances x_i and x_{i+1} with the horizontal lines representing the time levels at t^j and t^{j+1}. The time and spatial derivatives of the functional value $f(I)$, which represents the dependent variables, stage and discharge, are approximated, respectively, as follows:

$$\frac{\partial f(I)}{\partial t} \approx \frac{f_{i+1}^{j+1} + f_{i}^{j+1} - f_{i+1}^{j} - f_{i}^{j}}{2\Delta t} ,$$

and

$$\frac{\partial f(I)}{\partial x} \approx \theta \frac{f_{i+1}^{j+1} - f_{i}^{j+1}}{\Delta x_i} + (1 - \theta) \frac{f_{i+1}^{j} - f_{i}^{j}}{\Delta x_i} .$$

The spatial derivative of the cross-sectional area in the equation of motion is approximated by a forward-difference technique as

$$\frac{\partial A}{\partial x} \approx \frac{A_{i+1}^{j+1} - A_{i}^{j+1}}{\Delta x_i} .$$

In the finite-difference expressions, Δt represents the time step and Δx_i represents the length of the ith segment (reach) of any given branch (channel). In the expression for the spatial derivative, θ is a weighting factor used to determine the time between the t^j and t^{j+1} time levels at which the spatial derivatives are evaluated. In a manner similar to the treatment of the spatial derivatives, the cross-sectional area, top width, hydraulic radius, and the discharge in nonderivative form in the equation of motion, represented by $f(I)$, are approximated as

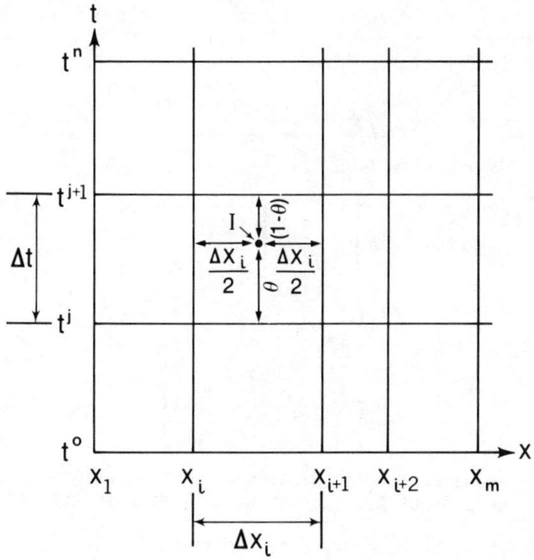

Fig. 1.--Space-Time Grid System for Finite-Difference Approximations.

$$f(I) \approx \chi \frac{f_{i+1}^{j+1} - f_i^{j+1}}{2} + (1 - \chi) \frac{f_{i+1}^{j} - f_i^{j}}{2}.$$

In this expression χ is a weighting factor, similar to θ, specifying the time at which these functional quantities are evaluated between the t^j and t^{j+1} adjacent time levels at the midpoint of the ith segment. Updating of these functional values and the spatial derivative of the cross-sectional area lags the computation of the dependent variables by one iteration.

In the four-point, finite-difference scheme, θ is a real constant generally throught of as lying in the range $0.5\leq\theta\leq1$. A θ value of 0.5 yields the traditional box scheme used by Preissman (4) and by Amein and Fang (1), whereas a value of 1.0 for θ represents the fully forward scheme presented by Baltzer and Lai (2). In channels where flow resistance is small, oscillations, known as psuedo-instability or computational mode, may appear for θ values near 0.5. Such oscillations disappear as θ is increased, consequently experience has shown that it is desirable to treat θ as lying in the interval $0.6\leq\theta\leq1$.

The weighting factor χ may be assigned in the range $0\leq\chi\leq1$. If χ is zero, function values are determined exclusively from previously computed quantities, whereas a χ weighting factor of one produces a fully forward approximation of the applicable functions.

By utilizing these finite-difference expressions and the notation $\tilde{f}(I)$ to signify function values derived from the previous iteration, it is possible to write the partial-differential flow equations in the following finite-difference forms; for the continuity equation,

$$\tilde{B}\left[\frac{Z_{i+1}^{j+1} + Z_i^{j+1}}{2\Delta t} - \frac{Z_{i+1}^{j} + Z_i^{j}}{2\Delta t}\right] + \theta \frac{Q_{i+1}^{j+1} - Q_i^{j+1}}{\Delta x_i} + (1 - \theta) \frac{Q_{i+1}^{j} - Q_i^{j}}{\Delta x_i} = 0,$$

and for the equation of motion,

$$\frac{1}{g\tilde{A}}\left[\frac{Q_{i+1}^{j+1} + Q_i^{j+1}}{2\Delta t} - \frac{Q_{i+1}^{j} + Q_i^{j}}{2\Delta t}\right] + \frac{2\beta\tilde{Q}}{g\tilde{A}^2}\left[\theta \frac{Q_{i+1}^{j+1} - Q_i^{j+1}}{\Delta x_i} + (1 - \theta) \frac{Q_{i+1}^{j} - Q_i^{j}}{\Delta x_i}\right]$$

$$- \frac{\beta\tilde{Q}^2}{g\tilde{A}^3} \frac{\tilde{A}_{i+1}^{j+1} - \tilde{A}_i^{j+1}}{\Delta x_i} + \theta \frac{Z_{i+1}^{j+1} - Z_i^{j+1}}{\Delta x_i} + (1 - \theta) \frac{Z_{i+1}^{j} - Z_i^{j}}{\Delta x_i}$$

$$+ \frac{k|\tilde{Q}|}{\tilde{A}^2\tilde{R}^{4/3}}\left[\chi \frac{Q_{i+1}^{j+1} + Q_i^{j+1}}{2} + (1 - \chi) \frac{Q_{i+1}^{j} + Q_i^{j}}{2}\right] - \frac{\xi\tilde{B}}{g\tilde{A}} U_a^{\,2} \cos\alpha = 0.$$

Solution of these equations results in determination of the stage and discharge at the ends of the segment whose length is defined as Δx_i.

Using appropriate boundary conditions and initial values, one may effect a solution for the set of flow equations for all segments within a network. In the branch-network flow model formulation, however, branch-transformation equations are developed from the segment-flow equations which relate the unknowns at the ends of the branches (5). Significant savings in computer memory and execution time accrue through use of these branch-transformation equations in place of segment-flow equations. Initial solution of the branch-transformation equations results in determination of the stage and discharge at the ends of the branches of the network. Stages and discharges at intermediate cross sections delineating segments within each branch are subsequently derived by back substitution of the branch-transformation equations.

Boundary Conditions

In order to solve the flow equations it is necessary to define boundary conditions at the ends of all branches, i.e., at all junction locations.

By neglecting velocity differences and energy losses due to turbulence, boundary conditions can be expressed for internal junctions. At a junction of "n" branches, discharge continuity requires that

$$\sum_{m=1}^{n} Q_m = W_k ,$$

where W_k is zero or some specified nodal flow value at junction "k". Stage compatibility requires that

$$Z_m = Z_{m+1} \quad m = 1, 2, \ldots (n-1) .$$

Equations describing these internal boundary conditions at branch junctions are automatically generated by the model.

Boundary condition equations for the extremities of the network are derived from user-supplied time histories of stage or discharge data or formulated from user-specified functions.

Solution Technique

In order to effect a solution of the system of flow equations and specified boundary conditions, initial values of the dependent variables are required. These may be measured, computed from some other source, or computed from previous simulations. By use of these initial values, the computation is initiated, and then, proceeds step-by-step to the end of the simulation through successive solution of the coefficient matrices. The system of equations is solved by Gauss elimination using maximum pivot strategy. Although iteration within the time step is not required this feature is available as an option in the branch-network flow model and can be employed to produce results of a higher order accuracy. Such iterative improvement of computed results is optional. The effect of the iteration process is to align the function values with the computed unknowns.

THE TIDAL POTOMAC RIVER

The Potomac Estuary extends from its confluence with Chesapeake Bay 188 km (117 mi) to the head-of-tide near Chain Bridge in the northwest quadrant of the District of Columbia. Based upon its hydrodynamic regime, its geomorphic basin characteristics, and on the biogeochemical properties of its waters, the Estuary may be conveniently separated into three distinct zones or segments. These segments are the coastal estuary, the transition zone, and the tidal river. The tidal-river segment of the Estuary--sometimes referred to as its riverine segment--extends from river kilometer 137.4 (statute mile 85.4) to the head of tide, a distance of nearly 50 km (31 mi). It is this riverine segment of the Potomac Estuary--particularly the portion shown in figure 2 and extending from Wilson Bridge located immediately south of the District of Columbia to the head-of-tide--that is of primary concern in this report.

Description

As it wends its way to the ocean, the Potomac River transverses the Fall Line separating the Piedmont from the Atlantic Coastal Plain. A series of falls and rapids indicate where the River intersects the Fall Line. The last of these falls and rapids is located immediately upstream of Chain Bridge. It is the abrupt, almost discontinuous change in channel gradient that identifies this site as the head-of-tide.

Flow downstream of Chain Bridge is confined for a short distance (approximately 5 km (3 mi)), to a narrow, deep, but gradually expanding channel bounded by steep, rocky banks and high bluffs. From this point downstream the channel changes to a broad, shallow, and rapidly expanding section confined between low banks having low-to-moderate relief. Several small inlets exist where small streams discharge into the tidal river, thus creating irregularities and indentations in the channel.

Two distinct flow patterns are exhibited within the tidal river segment of the Potomac Estuary; namely, unidirectional pulsating flow and bidirectional flow. The former typically occurs in the narrow channel of the upstream portion of the segment, whereas the latter occurs in the broader downstream channel. However, the location of the transition from one flow pattern to the other varies in response to changing inflow at the head of tide, to changing tidal and meteorologically induced flows, and to changing channel characteristics. For typical inflow and tidal conditions, the transition is usually located in the vicinity of Key Bridge. Although channel geometry is irregular and channel cross-sectional area increases rapidly with downstream distance, the flow is vertically well mixed and nominally tangential to the thalweg. Amplification of the tide wave as it propagates upstream is generally observed throughout the segment. Brackish water seldom penetrates into the tidal-river segment.

Data For Modeling

Three basic types of prototype data are required to implement and use a comprehensive model such as the branch-network flow model; namely, channel geometry data, boundary condition data, and initial condition data. Careful planning and close supervision of efforts to collect and process each of these data types are required in support of an effective modeling effort.

As an initial task in the overall comprehensive investigation of the Potomac Estuary by the U.S. Geological Survey, available hydrologic, geodetic, and hydrographic information were identified, assembled, and

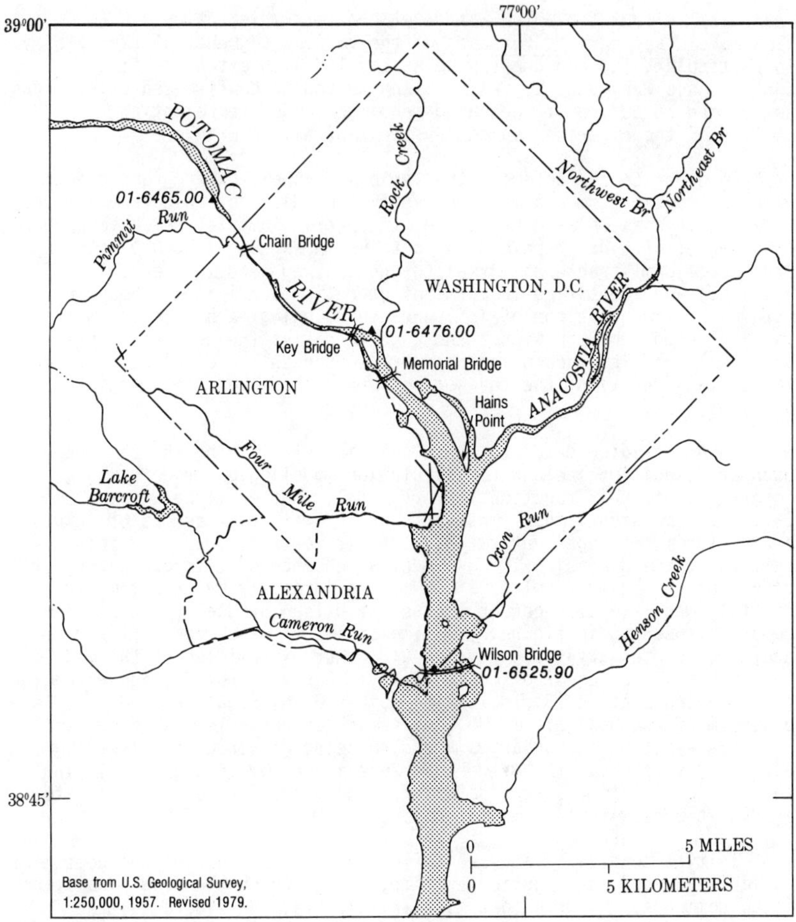

Fig. 2.--Map of Potomac River Near Washington, D.C.

reviewed. A tentative plan was developed for implementing one (or more) flow models appropriate for each of the various segments of the estuary. In particular, map and chart studies coupled with extensive field reconnaissance were used to refine planning and to confirm tentative organization and structure of the one-dimensional model representing the portion of the riverine segment upstream of Wilson Bridge.

In order to characterize the various channels, hydrographic data were abstracted from a comprehensive file of data obtained from National Ocean Survey, National Oceanic and Atmospheric Administration (NOS/NOAA). This subset of hydrographic data contains in excess of 30,000 soundings. Cross-sectional transects normal to the thalweg were selected, as required, and identified in terms of geodetic coordinates. Soundings within a narrow envelope of influence surrounding each transect were then selected from the subset and used to derive the desired cross-sectional profile. Reach lengths were subsequently determined from distances measured along the thalweg between successive cross-sectional profiles.

Boundary-value data and their role in establishing the time-dependent boundary conditions needed for simulation modeling underscore the importance of their function. Every model, whether it be a one-, two-, or three-dimensional model must be supplied with time series of values of one (or more) dependent variables at its extremities. For this particular one-dimensional flow model a sequence of discrete water-surface elevations is being supplied at 15-minute intervals by an automatic, digital, water-level recorder located at Wilson Bridge (field station number 01-6525.90 in figure 2). Time series of discharges (flux rates) are used as boundary-value data at all other extremities of the model. One such time series of discharge data is derived from a stage-discharge-rated gaging station (field station number 01-6465.00) located immediately upstream of the head-of-tide. Time series of water levels, flow vectors, and cross-sectional discharges are also being gathered at selected locations throughout the network for use in calibrating and verifying the model. All time-series data are being processed, edited, and stored in direct-access-type, computer files.

Initial condition data are defined as those values of the dependent variables needed at a particular time, t_o, at which the flow simulation is to commence. As with most modeling efforts, these data must be approximated from available prototype data (or otherwise estimated) and provided at branch and segment locations within the model in order to commence the simulation.

POTOMAC RIVER FLOW MODEL

The modeled portion of the Potomac River is shown in figure 2. One-third of the tidal-river portion of the Potomac Estuary is treated in the one-dimensional flow model. This reach extends upstream from Wilson Bridge to the head of tide near Chain Bridge, a distance of 19.7 km (12.3 mi) as shown in figure 2. The tide-affected extent of the Anacostia River, as well as Washington Channel, the Washington Tidal Basin, and a channel to the west of Theodore Roosevelt Island are also included in the network schematization.

Network Schematization

A schematic of the modeled river system is shown in figure 3. The network is composed of 14 branches identified by the Roman numerals. These branches join and(or) terminate at 14 junction locations identified by the numbered boxes.

Bathymetric data depicting the bottom geometry of the various channels of the network system were obtained from the National Ocean Survey and utilized to generate cross-sectional profiles of the river channels for the one-dimensional model. A total of 33 cross-sectional profiles, located at unequal intervals along the various channels, were used to describe the network. Cross-sectional profile locations were chosen in order to accurately characterize the network channel geometry for model implementation. Thirteen of these cross sections, located at intervals ranging from 0.91 km (0.56 mi) to 2.51 km (1.57 mi), were used to describe the Potomac River geometry from Chain Bridge to Wilson Bridge. Of the remaining 20 cross-sectional profiles, 10 were needed to depict the channel geometry of the Anacostia River; the remaining 10 were utilized to describe the geometry of the other channels of the network.

The highly irregular and drastically changing cross-sectional properties of the reach of the Potomac River from Chain Bridge to Wilson Bridge are illustrated by plots of selected cross sections shown in figure 4. As these profiles illustrate, the channel cross section expands rapidly in the downstream direction. From a minimum width of 44.2 m (145 ft) at Chain Bridge the channel expands to a width of 1,160 m (3800 ft) at Wilson Bridge. The corresponding cross-sectional area increases from approximately 232 m^2 (2500 ft^2) to 3800 m^2 (41,000 ft^2). As indicated in the plot of figure 4, the deepest point in any of the five cross sections is 8.78 m (28.8 ft). However, the channel bottom in the reach from Chain Bridge to Key Bridge is particularly irregular with occasional scour-hole depths in excess of 24 m (80 ft). It was, nevertheless, deemed unnecessary to account for these local bottom variations in the flow model.

As the network schematization of figure 3 illustrates, boundary conditions must be specified for the Potomac River at Chain Bridge (junction 1), for the Potomac River at Wilson Bridge (junction 7), for the upstream extent of the Anacostia River (junction 8), for the end of Washington Channel (junction 13), and for the Washington Tidal Basin (junction 14). For this purpose the model uses discharges derived from a rating curve for a gaging station located 1.9 km (1.2 mi) upstream of Chain Bridge (station number 01-6465.00 in figure 2) as its upstream Potomac River boundary condition. Water-surface elevations continuously recorded at a gaging station located at Wilson Bridge (station number 01-6525.90 in figure 2) are utilized as the downstream Potomac River boundary condition. Presently the Anacostia River is treated as a deadend channel with a zero discharge boundary condition at its upstream tidal extent. It is also possible to use computed upriver discharges for the Anacostia River as boundary values for conducting the model simulations. The Washington Tidal Basin and the Washington Channel are also treated as deadend channels with a zero discharge boundary condition. Those boundary conditions required at the internal junctions of the network system, consisting of discharge continuity and stage compatibility conditions, are automatically generated by the model.

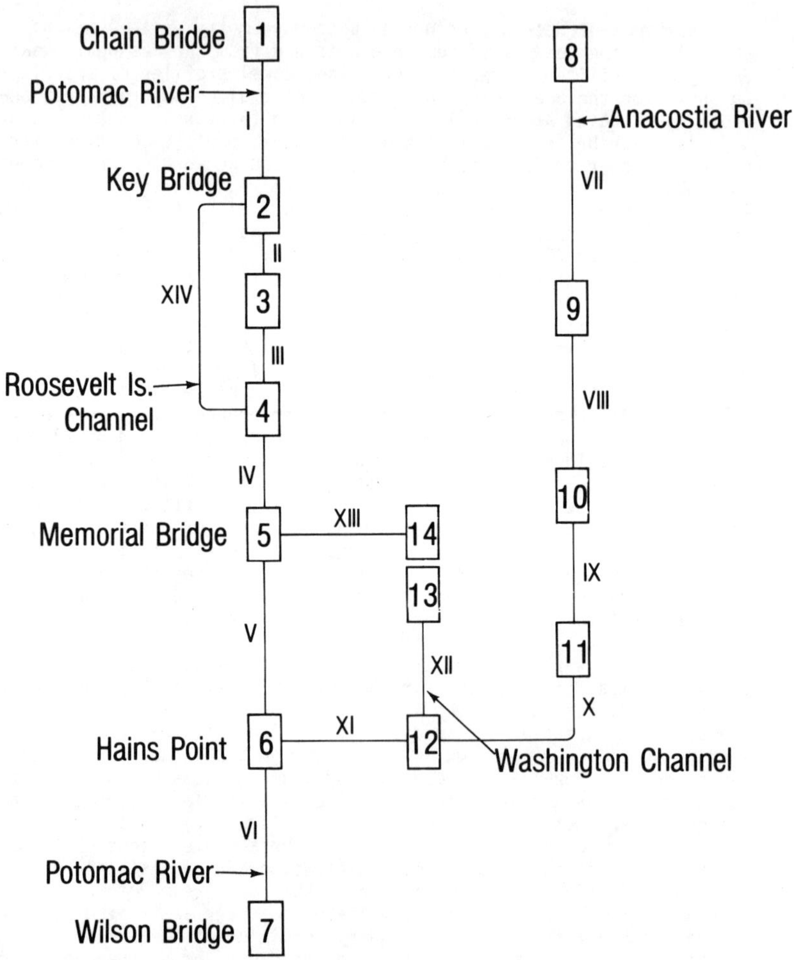

Fig. 3.--Schematization of the Potomac River System for the One-Dimensional Model.

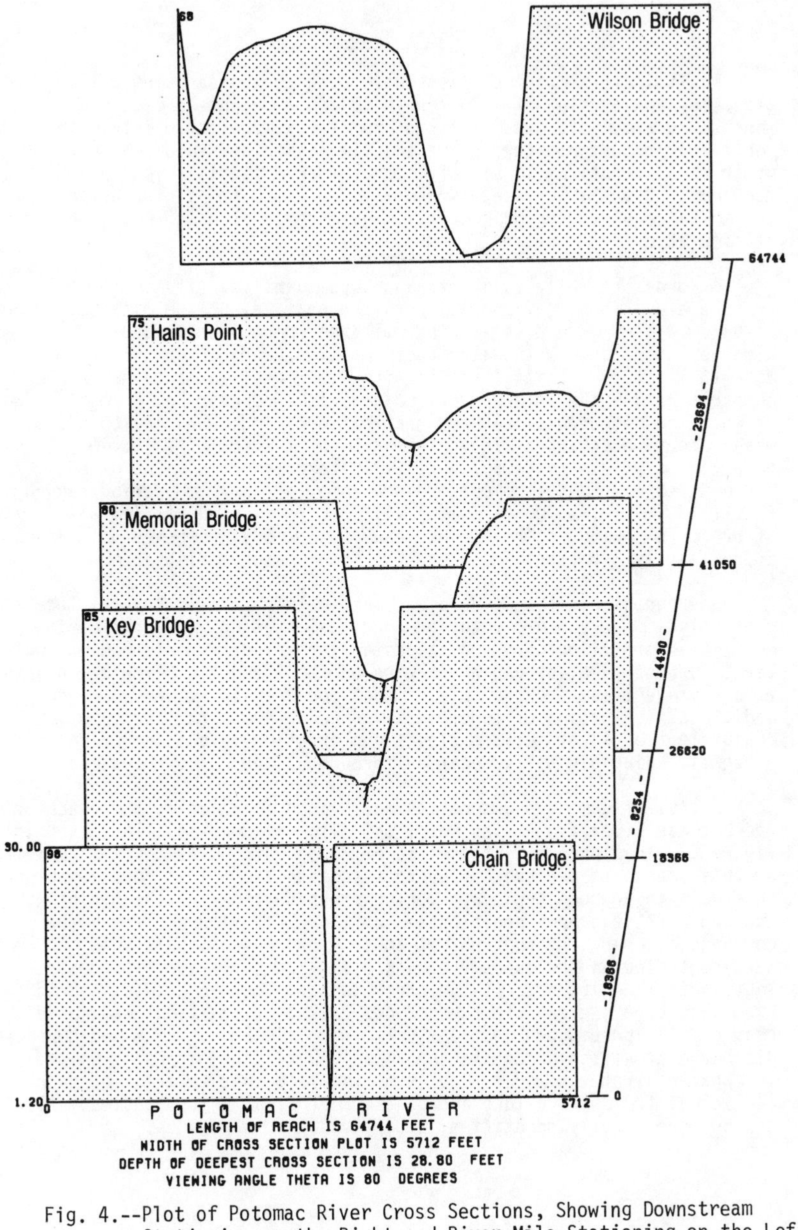

Fig. 4.--Plot of Potomac River Cross Sections, Showing Downstream Stationing on the Right and River Mile Stationing on the Left.

Model Results

Preliminary flow simulations have been conducted using a 15-minute time step and a value of 0.75 for the χ and θ weighting factors. A month-long simulation was performed to assess model stability. The model remained stable throughout the simulation and required on the order of 3 iterations to satisfy convergence criteria which were set at 0.46 cm (0.015 ft) for stage and 0.71 m^3/s (25.0 ft^3/s) for discharge. The model required approximately 1 minute of IBM 370/155 CPU time per day of simulation.

In order to fully calibrate and subsequently verify the model, water-surface elevations continuously recorded near Key Bridge (station number 01-6476.00 in figure 2) are being utilized. Model computed discharges are also being compared with discharges measured near Wilson Bridge on June 13-14, 1979. Initial results from the model indicate favorable comparison with the intermediate stages recorded near Key Bridge, as well as with the discharges measured near Wilson Bridge. During the June measurement, discharges ranged from a high of 1,460 m^3/s (51,400 ft^3/s) to a low of -840 m^3/s (-29,700 ft^3/s) with an average value of 538 m^3/s (19,000 ft^3/s). Model computed discharges for the same period ranged from 1,290 m^3/s (45,700 ft^3/s) to -1,000 m^3/s (-35,200 ft^3/s) with a mean of 524 m^3/s (18,500 ft^3/s). The mean of the model computed discharges is approximately 3 percent less than the mean measured discharge.

Sparseness or lack of required bathymetric data in several locations presently impede further detailed model calibration and verification. Additional profiling work and the establishment of an intermediate water-level, boundary-value, gaging station on the Anacostia River are planned to alleviate these shortcomings so that model calibration can be completed. Additional evaluation of model performance also awaits the synoptic measurement of water-surface elevations and flow velocities which are presently being digitally recorded throughout the network system.

A plot of computed discharges for branch 6, which is the reach of the Potomac River from Hains Point to Wilson Bridge, is illustrated in figure 5. This branch is comprised of 3 segments. The channel geometry of this branch is depicted by 4 cross-sectional profiles located at uneven intervals throughout the branch. The cross sections at the beginning and end of this branch are plotted in figure 4. Figure 5 shows the computed discharge hydrographs at these 4 cross-sectional locations which are identified as curves 6-1 through 6-4. The curve identified as 6-4 represents the discharge at the cross section located at Wilson Bridge. From this plot one can see the reduced effect of the tidal wave on discharge as it propagates upstream. The significant upstream and downstream discharge at Wilson Bridge is appreciably diminished at Hains Point. It is apparent from the model results of figure 5 that by means of such analytical tools the nature and behavior of tidal-driven unsteady flows can be more readily identified, understood, and explained.

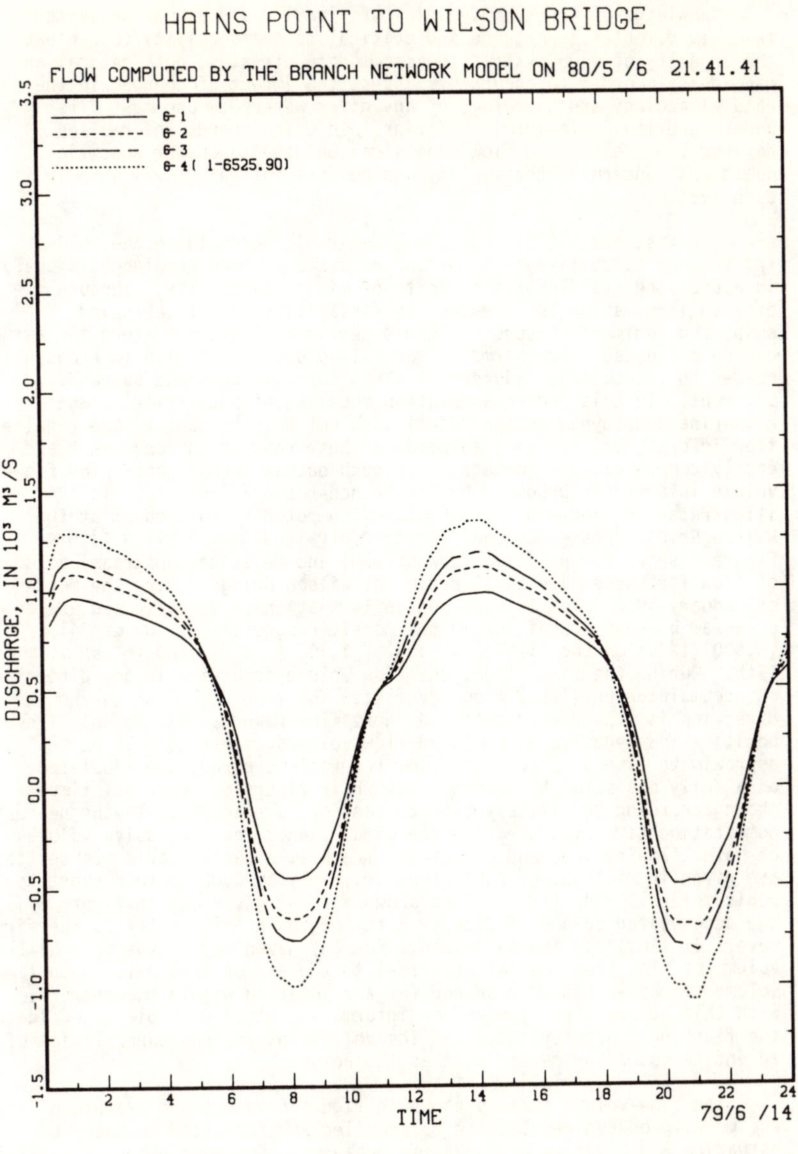

Fig. 5.--Model-Generated Plot of Computed Discharges for the Potomac River.

MODEL UTILIZATION

Knowledge of the hydrodynamics of riverine and estuarine systems is of fundamental importance and critical to one's ability to evaluate the capacity of such systems to accommodate stresses, both natural and manmade. In order to accurately assess the impact of changes in the natural ecology and hydrology of any given waterbody one must first understand its hydrodynamic behavior. In this regard, information derived from calibrated flow models can be used to answer specific questions concerning the quality and quantity of water available in such systems.

For instance, it is frequently desirable to evaluate the volume interchange capability of an estuarine system. Such knowledge is useful in appraising its flushing capacity or mixing capability. Through use of such information the tidal-cycle variability of nutrients and suspended sediment at specific cross-sectional locations along the estuary can be evaluated. Furthermore, control-volume information is frequently needed to locate major sources or sinks for nutrients and suspended sediment. In this regard simulation modeling of both riverine and estuarine hydrodynamics can be utilized not only to compute the required flow information, but also to produce these results in convenient and easily comprehensible formats. One such output format containing flow-volume information produced by the branch-network flow model is illustrated in figure 6. Flow volumes computed by this model at the Wilson Bridge cross-sectional location are tabulated in this figure. Figure 6 shows the positive (downstream) and negative (upstream) volumes of flow for successive tidal cycles at Wilson Bridge during the month of January 1980. For this period inflows at Chain Bridge, used as the upstream boundary condition in the model, ranged from a low of 241 m^3/s (8,500 ft^3/s) on the 11th to a high of 1,190 m^3/s (42,100 ft^3/s) on the 17th. During the simulation, the flow volume occurring in one direction is accumulated until the flow reverses. The flow volume in the new direction is accumulated until the next flow reversal and so on. These positive and negative accumulated flow volumes, separated by the approximate time at which each flow reversal occurred, are tabulated on a daily basis for the month as shown in figure 6. Reversal times, shown according to military time convention, are accurate to the nearest computation time step used for the simulation. The successive volumes of water flowing into and out of the reach at a particular cross section are given in millions of cubic feet (multiply by 0.02832 to convert to cubic meters). The first volume shown for any given day in figure 6 is the accumulated volume of flow from the beginning of the day to the first reversal; similarly the last volume for any given day is the accumulated volume of flow from the last reversal to the end of that day. Such flow-volume summaries can be produced for any location within the network. With this accumulated flow-volume information it is possible to evaluate the flushing characteristics and the volume interchange capabilities of an entire estuarine system such as the Potomac.

The flow-volume summary shown in figure 6 is but one example of the way model-produced results can be compiled and formatted to describe estuarine and riverine hydrodynamic behavior. Numerous other illustrative techniques ranging from discharge hydrographs, such as shown in figure 5, to more sophisticated mass curve, particle movement, and time-of-travel plots, to name but a few, are available as reported elsewhere (3). These examples emphasize the role of model utilization as a viable and economical means of producing the basic flow information needed to conduct the analysis, aid the decision-making process, and thereby assess the impact of water-management alternatives.

January 1980 Flow in Millions of Cubic Feet at Station 01-6525.90

Day No.	Flow Vol.	(Rev.) (Time)	Flow Vol.	(Rev.) (Time)	Flow Vol.	(Rev.) (Time)	Flow Vol.	(Rev.) (Time)	Flow Vol.
1	254	(0300)	-327	(0630)	915	(1445)	-345	(1845)	642
2	379	(0345)	-237	(0715)	967	(1600)	-371	(2000)	411
3	486	(0445)	-274	(0815)	801	(1615)	-382	(2030)	369
4	553	(0500)	-329	(0900)	800	(1700)	-344	(2115)	248
5	626	(0545)	-285	(0915)	809	(1745)	-380	(2145)	166
6	765	(0700)	-330	(1115)	537	(1745)	-356	(2215)	89
7	674	(0630)	-379	(1045)	750	(1900)	-227	(2245)	48
8	787	(0830)	-105	(1115)	720	(2030)	-216		
9	-2	(0015)	654	(0815)	-333	(1245)	566	(2015)	-231
10	-23	(0030)	673	(0845)	-309	(1315)	640	(2145)	-153
11	-67	(0130)	623	(0930)	-327	(1400)	517	(2130)	-87
12	-103	(0230)	752	(1100)	-227	(1415)	884		
13	22	(0045)	-122	(0345)	558	(1130)	-286	(1545)	685
14	41	(0115)	-214	(0445)	579	(1200)	-314	(1615)	680
15	58	(0115)	-204	(0445)	946	(1400)	-182	(1730)	768
16	268	(0245)	-127	(0515)	1235	(1445)	-160	(1715)	1105
17	583	(0400)	-62	(0545)	1846	(1600)	-84	(1800)	1129
18	803	(0500)	-109	(0700)	1554	(1630)	-181	(1900)	828
19	932	(0530)	-80	(0730)	1949	(1845)	-58	(2030)	463
20	1126	(0645)	-54	(0815)	1692	(1845)	-90	(2045)	424
21	1388	(0745)	-60	(0915)	1783	(2015)	-58	(2145)	202
22	1411	(0815)	-86	(1015)	1491	(2030)	-86	(2245)	81
23	1333	(0845)	-149	(1115)	1371	(2130)	-103	(2400)	3
24	1174	(0945)	-122	(1215)	1082	(2200)	-129	(2400)	2
25	-44	(0130)	1097	(1045)	-175	(1330)	1080	(2315)	-23
26	-127	(0215)	991	(1130)	-249	(1430)	999		
27	4	(0015)	-170	(0330)	901	(1215)	-202	(1545)	879
28	80	(0130)	-197	(0445)	788	(1245)	-256	(1630)	793
29	141	(0200)	-233	(0530)	818	(1345)	-270	(1730)	747
30	253	(0315)	-163	(0600)	802	(1500)	-207	(1900)	428
31	274	(0315)	-253	(0715)	796	(1545)	-205	(1930)	465

Fig. 6.--Model-Produced Table of Accumulated Flow Volumes at Wilson Bridge for January 1980.

FURTHER MODEL DEVELOPMENT

In order to provide flow information for the furthermost extent of the riverine segment of the Potomac Estuary, a second one-dimensional model, or an extention of the present model, is planned for later implementation. This model (or extension) will represent the tide-affected reach of the Potomac River from Wilson Bridge to Indian Head, Maryland (a distance of 30 km (19 mi)). Water-surface elevations continuously recorded at Indian Head will be used to supply the boundary conditions for conducting the model simulations. Because wind is known to have a significant influence on the flow in this segment, a major objective of this extension of the one-dimensional modeling effort will be to define the relative importance of the wind stress term as a forcing function on the flow. Data recorded by a self-contained weather station located at Indian Head, Maryland, augmented by weather data obtained from National Airport, will be used to define the wind-resistance coefficient and to evaluate the wind stress effects.

Additional Calibration and Verification

Considerable work remains with respect to model calibration and verification. In order to properly calibrate a flow model, discharges or velocity vectors depicting the flow field in the network throughout several extended periods of time must be obtained. It is also desirable that these data reflect a range of flow conditions. As mentioned previously, initial model calibration has been dependent upon a tidal-cycle discharge measurement made near Wilson Bridge and upon water-surface elevations collected near the midpoint of the Potomac reach in the vicinity of Key Bridge. More such tidal-cycle discharge measurements are planned to supplement stage data being continuously recorded throughout the network. These data are to be augmented and extended in duration by concurrently operated, bottom-moored, current meters which provide speed and direction-of-flow data at selected locations. Further detailed calibration of the model awaits the synoptic measurement of the water-surface elevations and flow velocities presently being digitally recorded throughout the network. A portion of these data, not used for calibration, will also be used to evaluate and verify model performance.

SUMMARY

A comprehensive, one-dimensional, branch-network flow model has been implemented to depict the flow regimen occurring throughout a portion of the riverine segment of the Potomac Estuary. The modeled network of channels extends upstream from Wilson Bridge to the head of tide immediately upstream of Chain Bridge in the District of Columbia and includes Washington Channel, the Tidal Basin, and the Anacostia River. A time series of discrete, digitally recorded water levels occurring at Wilson Bridge is used as the downstream boundary condition. Time series of discharges (or discharge equations) are used to satisfy necessary boundary condition requirements at all other model extremities. Although calibration of the model has begun, a data base comprised of measured time series of water levels, flow vectors, and cross-sectional discharges is presently being compiled which will permit thorough calibration and utlimately provide data for model verification. Preliminary flow simulations indicate that the model is stable and that it is able to satisfy convergence criteria with minimal iterative computation.

The primary objective of the implementation and utilization of this model is to provide basic, overall flow information for the network system which can be used to better understand the interdependencies of the geochemical, biological, and other processes. The capability of the model to offer flow-volume summaries, as well as, discharge hydrographs, particle movement, time-of-travel, and other plotted results at any of its 33 cross sections underscores its ability to provide operational information with which to appraise the impact of management decisions.

REFERENCES

1. Amein, M., and Fang, C. S., 1970, Implicit flood routing in natural channels: American Society of Civil Engineers Proceedings, Journal of the Hydraulics Division, v. 96, no. HY12, p. 2481-2500.

2. Baltzer, Robert A., and Lai, Chintu, 1968, Computer simulation of unsteady flows in waterways: American Society of Civil Engineers Proceedings, Journal of the Hydraulics Division, v. 94, no. HY4, p. 1083-1117.

3. Lai, C., Schaffranek, R. W., and Baltzer, R. A., An operational system for implementing simulation models, A case study: Seminar on Computational Hydraulics, Proceedings, ASCE Hydraulics Division, 26th Annual Hydraulics Specialty Conference, 1978, p. 415-454.

4. Preissmann, Alexander, 1960, Propagation des intumescences dans les canaux et riviers (Propagation of translatory waves in channels and rivers): First Congress of the French Association for Computation (1er Congrès) d'association Française de calcul, Grenoble, France, p. 433-442.

5. Schaffranek, R. W., Baltzer, R. A., and Goldberg, D. E., A model for simulation of flow in singular and interconnected channels: Chapter 3, Book 7, Automated Data Processing and Computations, TWRI, U.S. Geological Survey, 1980 (in press).

DIVERGENT FLOW IN THE VICINITY OF A BREACHED DAM

by Nikolaos D. Katopodes[1], A.M. ASCE

ABSTRACT

The wave emerging from a partially breached dam is studied using both computational and experimental data. The location of the wave front is determined by means of high speed photography in the laboratory, and the results are compared to the solution of a mathematical model in the form of two-dimensional axonometric projections of the water free surface. Following verification and calibration of the model, the extent of the divergent flow downstream of the dam is determined, and the two-dimensional computation is allowed to degenerate to a one-dimensional unsteady flow model, which is used for the computation of the remaining part of the wave. This results in a relatively inexpensive means of simulation of partial dam break problems having two desirable features: satisfactory accuracy of stage hydrographs a long distance downstream of the dam and a detailed picture of the flow in the vicinity of the breach, which is of importance to designers of power plants and related structures.

INTRODUCTION

The assumption of total dam collapse in predicting the extent and rapidity of flooding has been found to result in gross exaggeration of the catastrophic potential of the dam break flood wave. A more realistic picture of the floodwave is obtained if the dam is assumed to fail partially and perhaps that the breach is gradually enlarged as a result of the flow through it. Of the many possible combinations of breach extent and location and channel characteristics, some present little additional difficulty in modeling, as compared to the classical, complete failure approach, while others require higher dimension models, which can only be achieved with substantial difficulty and computational expense.

Satisfactory results have been obtained for partial failure in a long valley with well defined boundaries, by use of internal boundary conditions and one-dimensional unsteady flow models (c.f. Ref. 6). The success of this approach on a macroscopic scale is due to the fact that the internal conditions at the breach ensure conservation of mass and energy (for the free flowing breach) and mass and momentum (for the submerged breach). The length of the breach and the divergent flow region in its vicinity is considered very short compared to the length of the channel, and the entire section is represented by a discontinuity in depth.

---

[1]Asst. Professor of Water Resources and Civil Engineering, University of Louisville, Louisville, KY.

Observation of laboratory experiments indicates that the loss of information in the vicinity of the dam on the microscopic scale can be of enormous importance to the determination of potential flooding of this area. In fact, this one-dimensional approach underestimates seriously the depth of flow downstream of the breach, since it entirely neglects the standing hydraulic jumps that are formed at the sides of the channel and which can bring the water surface two to four times higher than its value at the center of the channel.

In order to simulate in detail the divergent flow downstream of the dam a two-dimensional model was developed (c.f. Ref. 3), with attention focused on the floodwave spreading on an unbounded plain of arbitrary topography. While use of such a model in the above situation is necessitated by the two-dimensional character of the entire flow domain, the expense of its application to long valleys is not warranted in the light of evidence that a short distance downstream of the breach the lateral component of flow is diminished and the wave is essentially one-dimensional. It is obvious that an intelligent model should take advantage of both approaches, and perhaps even utilize directly the results of physical simulation, where justified. In fact, this is a perfect example of mathematical and physical modeling complementing rather than antagonizing each other.

THEORETICAL BACKGROUND

From the computational hydraulics point of view, the most important difference between complete and partial dam failure is the presence of substantial vertical acceleration in the vicinity of the breach, long after collapse of the structure. This raises a serious question with respect to the applicability of the shallow water theory, which has provided the basis for one-dimensional mathematical models. The answer can be only heuristic: preliminary results obtained by assuming hydrostatic pressure distribution in the vertical gave satisfactory results. On the other hand, the complexity and expense of the two-dimensional shallow-water model hardly encourages addition of another dimension to the simulation process. In the following, vertical accelerations are neglected, so that conservation of mass and momentum leads to

$$\frac{\partial h}{\partial t} + \frac{\partial (uh)}{\partial x} + \frac{\partial (vh)}{\partial y} = 0 \tag{1}$$

$$\frac{\partial u}{\partial t} + u\frac{\partial u}{\partial x} + v\frac{\partial u}{\partial y} + g\frac{\partial h}{\partial x} = g(S_0{}^x - S_f{}^x) \tag{2}$$

$$\frac{\partial v}{\partial t} + u\frac{\partial v}{\partial x} + v\frac{\partial v}{\partial y} + g\frac{\partial h}{\partial y} = g(S_0{}^y - S_f{}^y) \tag{3}$$

in which $h(t,x,y)$ = depth of flow; $u(t,x,y)$ = velocity component in the x direction; $v(t,x,y)$ = velocity component in the y direction; t = time; g = gravitational force per unit mass; $S_0{}^x$ and $S_0{}^y$ = slope of the channel bed in x and y directions, respectively; and $S_f{}^x$, $S_f{}^y$ = hydraulic resistance slope in the x and y directions, respectively, as follows

$$S_f{}^x = \frac{n^2 u \sqrt{u^2+v^2}}{C_u{}^2 \; h^{4/3}} \tag{4}$$

$$S_f{}^y = \frac{n^2 v \sqrt{u^2+v^2}}{C_u{}^2 \; h^{4/3}} \tag{5}$$

in which n = Manning coefficient of roughness; and C_u = the cube root of the ratio of g in any given system of units to its value in the SI system.

In a flow system described by Eqs. 1-3, information propagates exclusively along directions determined by the local depth and velocity, known as characteristic directions. It can be shown (3,4) that these directions coincide with the generators of a quadric cone, with vertex at the source of information. Furthermore, a short distance away from the origin, depth and velocity are no longer the same and the generators of the cone are smoothly curved and twisted in space, forming the so-called characteristic conoid. Defining the celerity with which elementary gravity waves propagate by $c = \sqrt{gh}$, the equation describing this characteristic conoid reads

$$(dx - udt)^2 + (dy - vdt)^2 = c^2 \, dt^2 \tag{6}$$

With the conoid defined precisely by Eq. 6, the governing equations of flow can be transformed into directional differential operators valid along the generators of the conoid, as follows

$$\frac{dh}{dt} + \frac{c}{g}\left(\cos\theta \frac{du}{dt} + \sin\theta \frac{dv}{dt}\right)$$
$$= c \cos\theta (S_0{}^x - S_f{}^x) + c \sin\theta (S_0{}^y - S_f{}^y) - \frac{c^2}{g} \, S \tag{7}$$

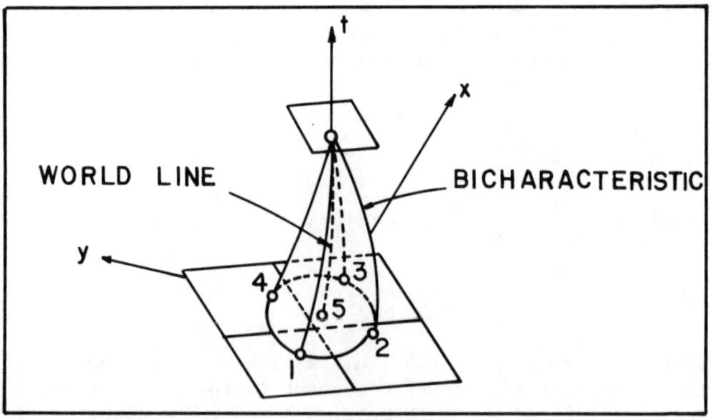

FIG. 1 - Bicharacteristic network of computation.

in which θ is the polar angle of the conoid and

$$S = \sin^2\theta \frac{\partial u}{\partial x} - \sin\theta \cos\theta \left(\frac{\partial u}{\partial y} + \frac{\partial v}{\partial x}\right) + \cos^2\theta \frac{\partial v}{\partial y} \qquad (8)$$

Although there is an infinity of equations 7, each corresponding to a different value of θ, only three of them are linearly independent, if a unique solution for h, u, and v is to be obtained. Reference 4 presents several methods of numerical solution of Eq. 7 of varying accuracy and complexity. It is shown that the most efficient is the one suggested by Butler (1), which is shown schematically in Fig. 1. Equation 7 is applied at values of θ separated by an interval of π/4 as indicated by the curves labeled bicharacteristics. In addition, the continuity equation is recast in directional form along the direction of fluid travel in space and time, labeled world line, as follows

$$\frac{dh}{dt} + \frac{c^2}{g}\left(\frac{\partial u}{\partial x} + \frac{\partial v}{\partial y}\right) = 0 \qquad (9)$$

This provides five equations for the determination of h, u, v, $\frac{\partial u}{\partial x}$ and $\frac{\partial v}{\partial y}$ at the vertex of the conoid, provided that flow conditions are known at points 1 through 5. These are obtained by interpolation from previous computation nodes, not necessarily forming a rectangular grid. The solution can be easily made correct to second order, which improves significantly the efficiency of computation.

In the absence of flow in the y direction, the flow is uniform across the channel and Eq. 7 assumes only two independent forms,

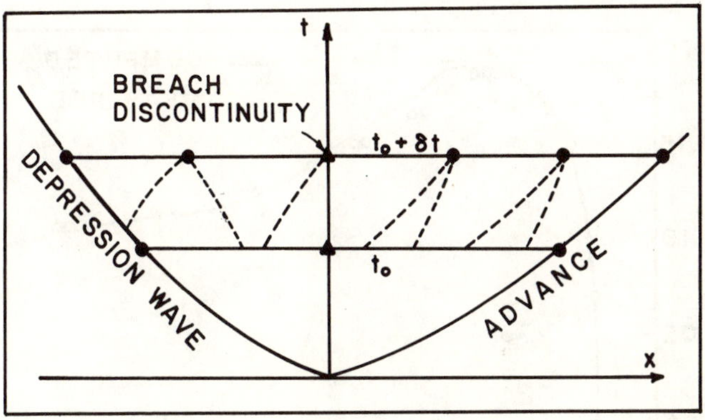

FIG. 2 - Computation pattern on x-t plane.

i.e., for $\theta = 0$, and $\theta = \pi$. Then

$$\frac{du}{dt} \pm \frac{g}{c}\frac{dh}{dt} = g(S_0{}^x - S_f{}^x) \tag{10}$$

valid along the intersection of the conoid and a plane identified by y = const. This of course is the well-known method of characteristics in one space dimension. Figure 2 shows this y = const. plane and the characteristic curves between two levels of computation. At x = 0 there exists a discontinuity in depth and velocity and the characteristic curves are not allowed to transmit information through it. Instead, the two sides of the discontinuity carry the same value of flow rate and energy, for a free flowing breach. Forward characteristics emanating from the reservoir can reach the t axis and provide one equation for the solution at x = 0. If, in addition, the flow at breach is assumed critical, i.e.,

$$u_b{}^2 = gA_b/B_b \tag{11}$$

where A_b and B_b are the cross-sectional area and top width of the breach, respectively, the depth and velocity at the upstream side of the breach can be computed independently of the downstream flow. In fact, the solution there depends only on channel and breach geometry and reservoir depth. With the high side of the discontinuity determined the constancy of discharge and energy suffices to solve for the flow conditions at the low side and the computation is completed.

There are several crude assumptions in this one-dimensional approach which, however, requires only about 20% of machine processing time per nodal point compared to the two-dimensional one. In addition, the latter requires at least as many as the square of the number

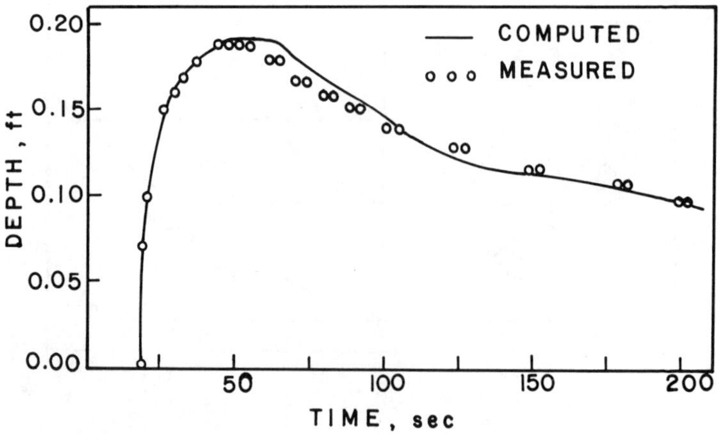

FIG. 3 - Stage hydrograph at x = 80 ft.

of one-dimensional nodes to cover the same area. Furthermore, Fig. 3 shows, as a pleasing surprise, the stage hydrograph obtained using this one-dimensional model as compared to an experimental study performed by the Corps of Engineers (7).

The channel was of rectangular cross section, 4 ft wide, laid on a slope $S_0^x = 0.005$ and characterized by a Manning n = 0.009. The breach was also rectangular 1.2 ft wide and the initial depth behind the dam was 1 ft. The hydrograph at a distance of 80 ft downstream compares excellently with the measured values. At x = 0 the WES data provide values for the section upstream of the breach, which also compares satisfactorily with computed stages and discharges. Lest the reader dismiss the two-dimensional approach as unrealistic and impractical, he is urged to consider Fig. 4, which shows the depth profiles downstream of the breach corresponding to the centerline and the side wall of the channel, as computed by the two-dimensional model and compared to the theory just presented. The profiles are plotted at time t = 1 sec after instantaneous rupture of the dam at t = 0. The one-dimensional model completely ignores the standing waves at the side walls. In practice one should expect these waves to be somewhat lower, since the method of characteristics cannot account for the energy dissipated by the turbulent action of these waves. Nevertheless, the maximum depth of flow at the side walls of the channel is of importance if nearby structures are to be protected against a potential disaster. In fact, the peak of the hydrograph at the side wall is almost as high as the peak of the hydrograph at that location corresponding to complete failure of the dam.

A second important observation is that a short distance downstream of the strongly two-dimensional flow section the standing waves decay and the flow becomes relatively uniform in the y direction.

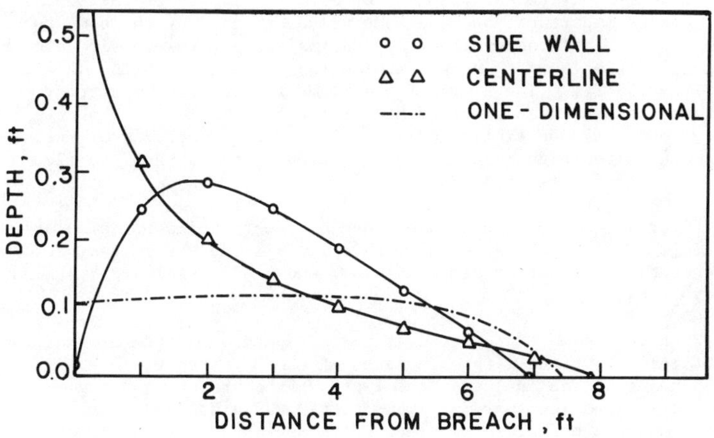

FIG. 4 - Free surface profiles for partial breach.

EXPERIMENTAL INVESTIGATION

Although the Waterways Experiment Station reports (7) present one of the most comprehensive experimental studies of the dam break problem, very little attention was paid to the flow region of interest to this work. For the two-dimensional computer model there remain the questions regarding the validity of hydrostatic pressure distribution at the breach, the accounting of energy dissipation by the standing hydraulic jumps and, of course, the reliability of Whitham's assumption (3, 8) on which the speed of propagation of the wave front is determined. The extent of this divergent flow is also of interest, since it designates the boundaries of the region where two-dimensional modeling is necessary.

The parameters controlling the solution of the dambreak problem are dam height h_0, breach width b, bottom slope S_0, hydraulic resistance (given by Manning's n), and channel cross-section geometry. It can be shown (5) that for complete failure of the dam, the flow is governed by a single dimensionless parameter, namely the normal Froude number based on the initial depth behind the dam. For partial breach an additional parameter, i.e., the dimensionless width of the breach should be employed.

A laboratory flume was constructed, following the exact specifications of the WES study for reasons of simplicity, economy, and with the intention to utilize as much as possible the already available data. The material was plastic-coated plywood, the width 4 ft, but the length only 24 ft compared to the 400 ft of the WES flume. The short length permitted easy and accurate changes of the bed slope, while it did not affect the purpose of the study. Only conditions of minimum resistance were studied, since the standing waves appear only when the breach is flowing free. The flow downstream is supercritical and the length of the channel immaterial. The length of the reservoir is of course important, but the same value was used by the computer model in the verification process. Partial breaches were simulated by a gate of variable width hinged at the top and able to swing quickly away from the dam. The floor of the flume was covered by a grid, so that when the wave was photographed from above the location of the surge front could be easily and accurately determined. An electronically controlled high speed camera was used to photograph the flow at very short time intervals.

Figures 5 and 6 show the sequence of events following simulation of a 1-ft wide breach extending vertically all the way to the bed of the channel. The bottom slope is 0.002 and the initial depth behind the dam 0.8 ft.

When the film is projected at normal speed accurate information is obtained about the rate of advance of the wave front, the height of the standing hydraulic jumps, the depth at the breach, and angle of expansion of the jet emerging from it, and the distance at which the two-dimensional effects disappear. Different breach sizes b, dam heights h_0 and bed slopes S_0 were studied, but channel width and roughness remained constant for all of the experiments. However, in view

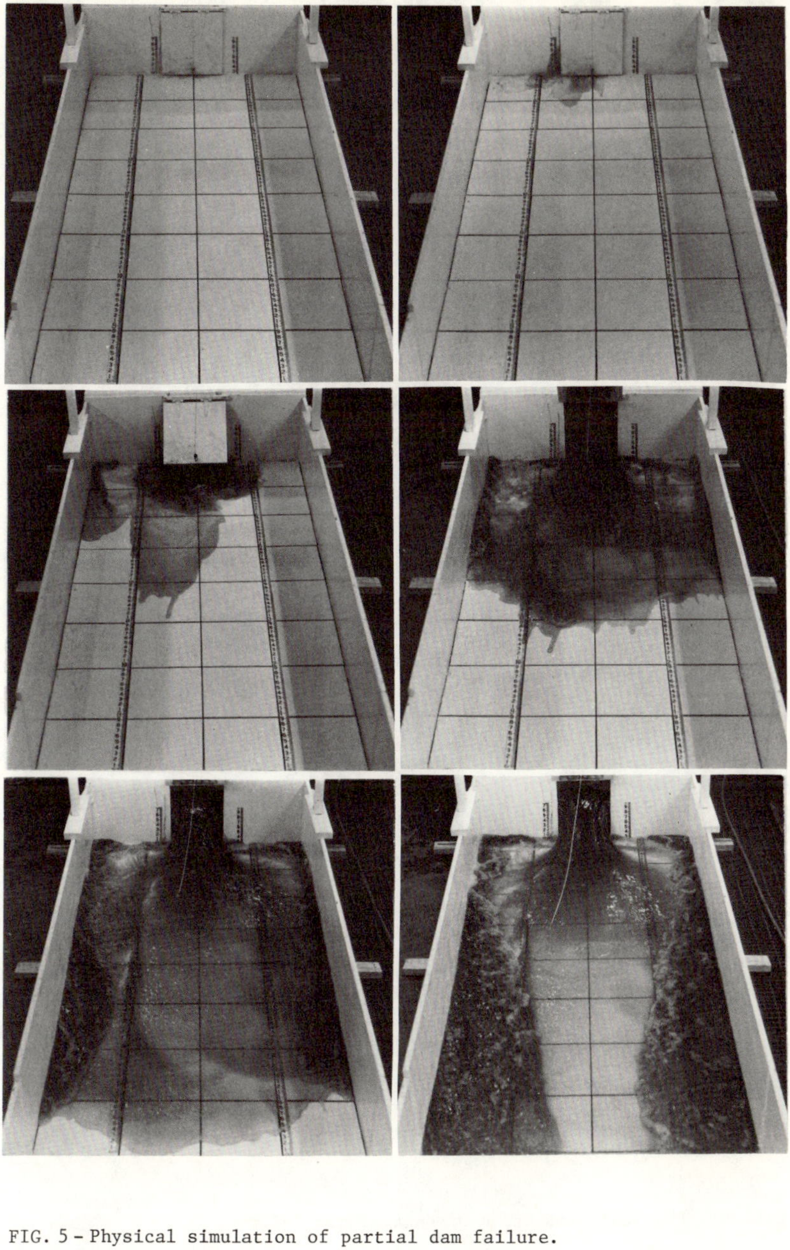

FIG. 5 - Physical simulation of partial dam failure.

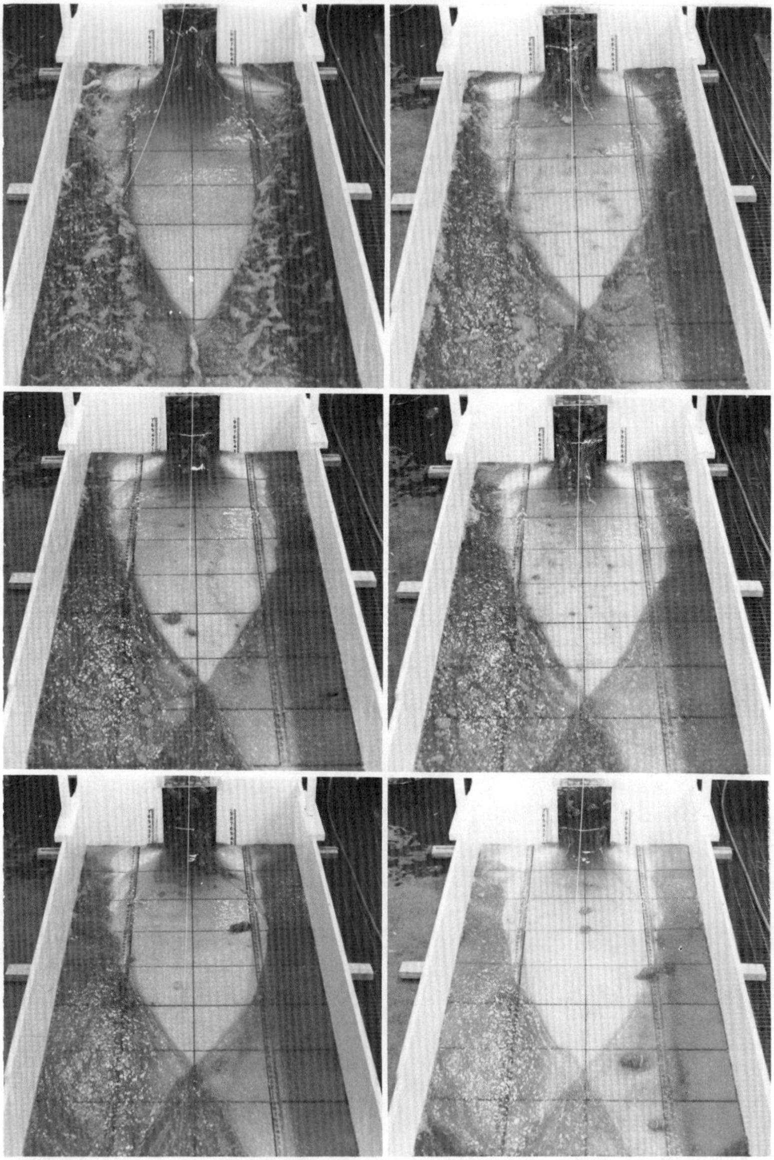

FIG. 6 - Physical simulation of partial dam failure.

of the fact that these two parameters are combined with h_0 and S_0 in the normal Froude number, which governs the flow together with dimensionless breach size, a sufficient range of values for F_n can be obtained even with n and B constant. Most of this information is presented elsewhere, since the emphasis of this work is on calibration of the mathematical model. Preliminary comparisons of the results obtained by the mathematical and physical models indicate the following. The depth of flow at the breach and the central area of the channel agree quite well. The same is true for the angle of expansion. The height of the side waves was in error, but the discrepancy there was not in agreement with the theoretical prediction. For some runs the computed values were higher and for others lower than the observed. The highly turbulent nature of the wave and the fluctuation of its peak may have caused some lack of precision in the laboratory measurements. On the other hand, the second order least-squares interpolation used by the mathematical model apparently introduces a dissipative mechanism by smoothing small disturbances and jumps in the flow. Since the maximum difference in wave height was less than 5%, it was decided to accept the model without calibration. Where calibration was found necessary, however, was the rate of advance of the wave front. Use of Whitham's method (8), that is, uniform velocity near the wave front, resulted in 5-8% error in estimating the location of the front. The computations were carried to a depth behind the front equal to 2% of the initial dam height, which increased significantly the cost of computation without any obvious improvement. Instead, very good results were obtained when the following adjustments were made to the model. Computations were stopped at 5% of h_0 and the velocity was considered uniform thereafter. When interpolation was performed in the tip area, however, a first order least-squares polynomial was used in place of Whitham's 3/7 power law. In addition, the wave front location was never corrected to second order after the new integral surface was calculated (c.f. Ref. 2). This resulted in a slightly greater speed of propagation and some improvement in the cost of computation too.

A HYBRID DIMENSION MODEL

The theory and experiments described earlier make it obvious that if a model is to be developed with predictive ability of potential flood damage for the entire length of the channel, neither the two- nor the one-dimensional model can lead to both efficient and accurate results independently. In the vicinity of the breach two-dimensional modeling is absolutely essential if potential damage in that area is of importance. On the other hand, using this model for the entire length of the WES flume, for example, requires massive computations, which are entirely unjustifiable, since the flow is clearly one-dimensional over 95% of the length of the channel. The computation can become very efficient, however, if the two models are interfaced by the following simple technique.

Following breach of the dam computations begin using the bicharacteristic model in the fashion described previously. As soon as the wave reaches the side walls and occupies the entire width of the channel, the Laplacian grid generation routine, which determines the location of the computation nodes (c.f. Ref. 2), is forced to generate one row of nodes having the same x coordinates and as close as possible to the wave front. The nodes on this x(t) = cont. line are specially labeled, so that they can be followed as the grid moves as a function of the propagation of the wave front. At the end of each time step a least-squared straight line is computed from the values of depth at the nodal points of this line. The squared error is monitored and if it is smaller than a specified tolerance, this special line becomes a model interface.

Points up to and including the interface are computed by the bi-characteristic model. Downstream of it, however, the two-dimensional grid is eliminated and replaced by a one-dimensional string of nodes, in the fashion indicated by figure 2. The characteristic model is used for the computation of u and h, as dictated by Eq. 10, and no special boundary conditions are required since the flow is supercritical and the flow at the interface is dependent only on upstream information. If a characteristic from the one-dimensional region happens to intersect the interface, the bivariate interpolation routine is called for the required value of x coupled with the y coordinate of the centerline.

Figure 7 shows an isometric projection of the water surface resulting from this hybrid model. The two-dimensional water surface is artifically extended over the width of the channel for demonstration purposes, but the actual two-dimensional computation ends where indicated. Using the above method the hydrograph of Fig. 4 is reproduced without difficulty, which shows that the one-dimensional model preserves mass and energy on the macroscopic scale, although completly ignores the flow conditions in the vicinity of the dam.

Obviously, the same hydrograph can be reproduced should one decide to utilize a laboratory measured hydrograph just a few feet downstream of the breach, that is, using a relatively small and inexpensive physical model to simulate the flow near the dam, and then simply route the outflow hydrograph to the desired distance using the one-dimensional model.

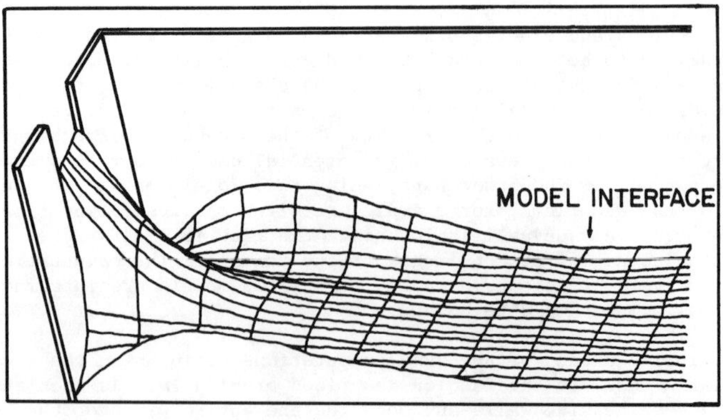

FIG. 7 - Computer simulation of partial dam failure.

SUMMARY AND CONCLUSIONS

There is recently a trend to estimate the potential flooding due to dam failure by assuming partial rather than total collapse of the structure, with the justification that this results in more realistic predictions of the time and extent of the flooding. This was shown to be correct, except for the case where there exists in the vicinity of the dam power plants or related structures, whose damage may lead to additional disasters. Under conditions of low hydraulic resistance the lateral component of flow becomes extremely important and may create waves comparable to those corresponding to complete failure of the dam.

A model consisting of a two-dimensional part used near the breach and a one-dimensional part used for the remaining of the wave was found to produce satisfactory results at reasonable cost and complexity.

ACKNOWLEDGEMENTS

This work was supported in part by a grant from the Graduate School of the University of Louisville. The writer also expresses his gratitude to the University of Louisville Computer Center for making its facilities available.

REFERENCES

1. Butler, D. S., "The Numerical Solution of Hyperbolic Systems of Partial Differential Equations in Three Independent Variables," Proceedings of the Royal Society, London, England, Vol. A225, 1960, pp. 232-252.
2. Herrmann, L. R., "Laplacian-Isoparametric Grid Generation Scheme," Journal of the Engineering Mechanics Division, ASCE, Vol. 101, No. EM5, Proc. Paper 12456, Oct. 1976, pp. 749-756.
3. Katopodes, N. D., and Strelkoff, T., "Computing Two-Dimensional Dam Break Flood Waves," Journal of the Hydraulics Division, ASCE, Vol. 104, No. HY9, Proc. Paper 14036, September 1978, pp. 1269-1288.
4. Katopodes, N. D., and Strelkoff, T., "Two-Dimensional Shallow Water-Wave Models," Journal of the Engineering Mechanics Division, ASCE, Vol. 105, No. EM2, Proc. Paper 14532, April, 1979, pp. 317-334.
5. Sakkas, J. G., and Strelkoff, T., "Dimensionless Solution of Dam-Break Flood Waves," Journal of the Hydraulics Division, ASCE, Vol. 102, No. HY2, Proc. Paper 11910, Feb. 1976, pp. 171-184.
6. Strelkoff, T., Schamber, D. R., and Katopodes, N. D., "Comparative Analysis of Routing Techniques for the Floodwave from a Ruptured Dam," Proceedings of Dam-Break Flood Routing Model Workshop held in Bethesda, Maryland on October 18-20, 1977. U. S. Department of Commerce, National Technical Information Service, PB-274537.
7. WES (Waterways Experimental Station), "Floods Resulting from Suddenly Breached Dams," Miscellaneous Paper No. 2-374, U. S. Army Corps of Engineers, Vicksburg, Mississippi, Report 1, Conditions of Minimum Resistance, 1960.
8. Whitham, G. B., "The Effects of Hydraulic Resistance in the Dam-Break Problem," Proceedings of the Royal Society, London, England, Vol. A227, 1955, pp. 399-407.

SENSITIVITY OF THE PRYCH-DAVIS-SHIRAZI MODEL

By

Barry A. Benedict,* Member ASCE, and

John W. Preston, Jr.**

The Prych-Davis-Shirazi (PDS) model is one of the most widely used models for analysis of near-field temperature distributions resulting from heated water discharges. The model is an integral model, developed originally by Prych (based on earlier work at M.I.T.) and modified by Davis and Shirazi through an extensive data-fitting program. Because of its wide use, work was undertaken to review the influence of various input parameters, especially those in which uncertainty exists in the specification. This paper reports the findings.

Parameters reviewed included the velocity ratio, entrainment coefficient, initial densimetric Froude numer, channel aspect (width-to-depth) ratio, and two parameters introduced by Prych, EH and EV. EH represents the importance of D_y, the transverse mixing coefficient, while EV is the ratio of vertical to horizontal mixing coefficients.

The findings conclusively show that EH, and, to a lesser degree, EV, are the most important parameters. While the Froude number and aspect ratio exhibit an impact, they are reasonably well defined numbers. Similar findings are true about the velocity ratio, although bank interference at higher ambient velocities may distort this. The entrainment coefficient has relatively little impact.

The transverse mixing term, EH, not only has the most impact on the results, but it is the most difficult to specify because of uncertainties in knowledge of diffusive mixing at individual sites. When a reasonable range of uncertainty for D_y is assumed, model output gives widely different results. In addition, typical values of EH are quite different (an order of magnitude or more) from the value of 0.02 suggested by Shirazi and Davis. Data taken by Weil was used to find this 0.02 value, and a review of his data shows that this value is to be expected, but that it is much lower than for natural streams.

The way in which the diffusion term is cast in the model is also reviewed. The diffusion is cast as added entrainment, rather than mass exchange, and this has been criticized.

*Associate Professor of Civil Engineering, University of South Carolina, Columbia, South Carolina, 29208.

**Department of Health and Environmental Control, State of South Carolina, Columbia, South Carolina, 29201.

Guidance is provided for potential model users or regulators who must review model results, with guides for estimating errors in isotherm areas due to errors in the diffusion coefficient.

INTRODUCTION

Substantial concern continues to exist over the possible impact of heated water discharges from electric power generating plants. The assessment of potential biological or other damage due to these heated effluents requires estimation of the distribution of temperatures in the receiving water body. A common approach to temperature estimation in the near-field region is use of an integral model treating the discharge as a jet. This paper will focus only on surface discharges. For such cases, possibly the most commonly used model is the Prych-Davis-Shirazi model (the PDS model). Due to its widespread use, a systematic review of its behavior was undertaken, and this paper reports on some of the findings.

PDS MODEL FEATURES

The PDS model had as its primary beginning the three-dimensional integral model developed by Stolzenbach and Harleman (1). Prych (2) sought to increase computational efficiency of their model and to improve predictive capabilities. Then, Shirazi and Davis (3) made numerous changes, all intended to further bring model results into line with available data. Changes and model calibrations were supported by extensive fitting, although the authors themselves acknowledge the sparsity of data for many situations. In addition, the process of fitting each of the several empirical parameters separately has been questioned.

Space precludes a complete description of the PDS model, but a few pertinent features and equations will be presented to facilitate later discussions. The model is an integral model, in which specific mathematical forms are chosen for lateral and vertical profiles of temperature and velocity. Selection of these forms enables one to integrate the basic conservation equations over the cross section of the jet and reduce the problem to a one-dimensional one in which all variables are a function only of the distance along the jet axis. Several generic difficulties exist with any integral model, chief among which is the failure to account for physical boundaries, leading to difficulties when the river bottom or bank are neared. Discussions of specific difficulties with integral models can be found in Dunn, et al (4), Jirka, et al (5), and Benedict, et al (6). Specific model features include the following:

1. The zone of flow establishment (ZOFE) is treated in a very simplified way, with the length defined by an empirical equation as function of jet aspect ratio, A, and initial densimetric Froude number, F_i; but not a function of the velocity ratio, R. Here, A = discharge channel width/discharge channel depth, and R = U_a/U_o = ambient velocity/initial jet velocity.
2. Buoyant and non-buoyant jet spreading are assumed directly additive.
3. Diffusion due to ambient turbulence can be treated as a form of

added entrainment, as proposed by Prych (7). It is further assumed that diffusion and jet-induced entrainment are added directly.

The similar profiles for temperature and velocity are Gaussian, as given by

$$\frac{\Delta T}{\Delta T_o} = \frac{\Delta T_c}{\Delta T_o} \exp(-\eta^2/B^2) \exp(-Z^2/H^2) \tag{1}$$

$$U = U_c \exp(-\eta^2/B^2) \exp(-Z^2/H^2) + V\cos\theta \tag{2}$$

in which ΔT_c = temperature rise at jet centerline

ΔT = temperature rise at distance η and Z from jet axis

ΔT_o = initial temperature rise

η = horizontal distance from jet axis

Z = vertical distance from jet axis

B = plume half-width

H = plume half-depth

U = jet velocity at distance η and Z from axis, ÷ by U_o

U_o = initial jet velocity

U_c = jet centerline velocity divided by U_o

V = U_a/U_o

U_a = ambient velocity

θ = angle between jet axis and ambient velocity

In establishing the basic equations, the change of jet discharge with distance, dQ/ds, must be evaluated and is assumed due to jet entrainment and ambient diffusion as

$$\frac{dQ}{ds} = \left.\frac{dQ}{ds}\right|_j + \left.\frac{dQ}{ds}\right|_a \tag{3}$$

To simplify matters, only the horizontal components of the terms in Equation 3 will be presented here. The vertical terms are similar, but the magnitude is limited in the existence of a density gradient.

$$\left.\frac{dQ}{ds}\right|_{j,h} = 2E_o \int_{-\sqrt{2}H}^{o} \Delta U dZ \tag{4}$$

in which $\Delta U = (U^2+V^2 \sin^2\theta)^{\frac{1}{2}} \exp(-Z^2/H^2)$ (5)

and E_o is an entrainment coefficient. By inserting Equation 5 into Equation 4, one obtains

$$\left.\frac{dQ}{ds}\right|_{j,h} = \sqrt{\pi}\, HE_o(U^2+V^2 \sin^2 \theta)^{\frac{1}{2}} \tag{6}$$

The effective entrainment due to ambient turbulent mixing, due to an analysis by Prych (7), is given as

$$\left.\frac{dQ}{ds}\right|_{a,h} = 11.0 \ \frac{H}{B} \ \frac{D_y}{U_o H_o} = 11.0 \ \frac{H}{B} \ EH \tag{7}$$

in which D_y = horizontal turbulent diffusion coefficient

H_o = initial discharge canal depth

The PDS writes equations for conservation of momentum and heat flux, which also include the expressions as in Equation 3, 6, and 7. The development yields seven simultaneous equations which are integrated stepwise along the plume axis to give values of jet coordinates, width, velocity, temperature rise, angle, and volumetric flux. Details are given by Shirazi and Davis (3).

SCOPE OF SENSITIVITY STUDIES AND RESULTS

Results of a comprehensive sensitivity analysis are presented by Preston (8). Some pertinent findings are given here. Table 1 shows variables included in the review. The variable EV is the ratio of D_Z, the vertical turbulent diffusion coefficient, to D_y. The variable F_o is the initial jet densimetric froude number. Values called Standard values were used as base conditions. For example, if EH is being varied, all other parameters take on the standard values unless otherwise noted. Note that not all possible combinations shown were evaluated.

TABLE 1 - PARAMETERS IN SENSITIVITY ANALYSIS

Parameter	Value Suggested [Ref.(3)]	Standard Value	Range Studied
E_o	0.05	0.05	0.0 - 0.075
EH	0.02	0.02	0.0 - 0.96
EV	0.2	0.2	0.0 - 0.6
R	*	0.375	0.0 - 0.8
A	*	8.0	6.0 -12.0
F_o	*	8.7	2.0 -25.0

*Calculated based on specific discharge condition

Influences of parameter changes were noted on isotherm areas, lengths, widths, and depths, as well as plume trajectory. It was found that the effect on the axial distance to a given ΔT was representative of all other changes. Figures 1 and 2 show typical results, with effects due to F_o and EH shown. The findings conclusively show that EH, and, to a lesser degree, EV, are the most important parameters. The Froude number and aspect ratio exhibit a significant impact, but they are usually well defined numbers. Similar findings are true about the velocity ratio, although bank interference at higher ambient velocities may distort this. The entrainment coefficient has relatively little influence.

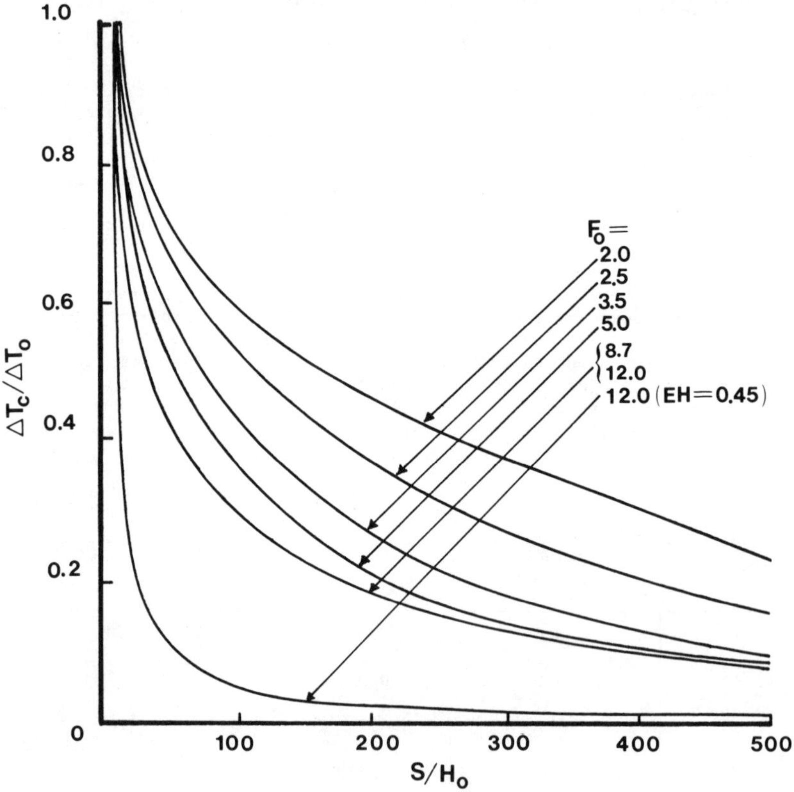

Figure 1. - Influence of Froude Number on Model Predictions

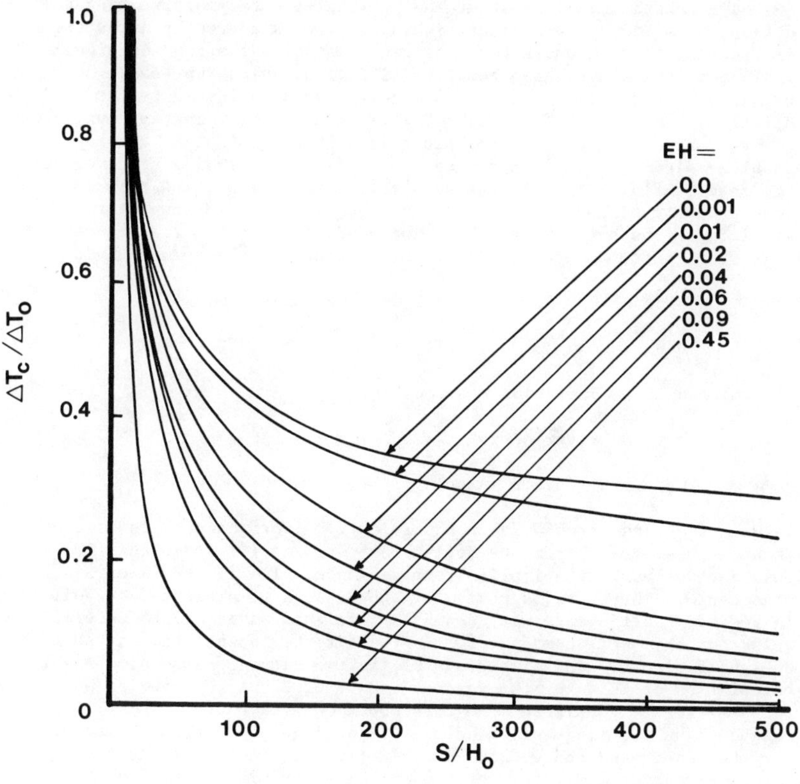

Figure 2. - Influence of EH on Model Predictions

DISCUSSION OF DIFFUSION TERM, EH

At the outset, it is disconcerting to see a jet model possibly dominated by ambient diffusion terms. Dunn, et al (4) and Jirka, et al (5) have raised questions about the physical accuracy of treating the diffusion as added entrainment. The latter work correctly notes the distinction between entrainment as a process of gathering additional fluid into the plume, as opposed to diffusion, which represents an exchange of particles. This is probably less troublesome in the dilution terms proper (Equations 3 - 7) than in the momentum flux terms. However, Dunn, et al (4) note that this probably is not as telling a problem as some of the other model assumptions, including treatment of the zone of flow establishment and buoyant spreading. The term is more properly placed in the heat flux equation, but so many other simplifications have already been included in the model that it may well still prove an acceptable way of reproducing observed behavior.

The value of D_y for river is frequently expressed as

$$D_y = \alpha h u_* \tag{8}$$

in which α = dimensionless constant

h = river depth

u_* = friction velocity

Values of α range from 0.15 in a straight, uniform channel to 4 or more around bends, as reported by Yotsukura and Sayre (9) and Chang (10). This sort of variation implies that a single value for EH is not reasonable. Therefore, a review of the basic data presented by Weil (11) is needed. Weil's data was used by Shirazi and Davis (3) to determine a value for EH. Weil used a jet oriented directly downstream in the form of a half circle. The dimensions of the two discharges he used enable calculation of EH using the value for $D_y = 0.027hu_*$ reported by Weil for his flume. Runs were made for smooth channel bottoms and a roughened bed. Using Weil's data gives a range of EH from .01 to 0.019 for the smooth bed and 0.0253 - 0.0294 for the rough bed. Average values are about 0.017 for the smooth bed and 0.027 for the rough bed, bracketing the 0.02 Shirazi and Davis claim provides a best fit. A review of the latter work (3) indicates some question as to the actual best fit value of EH, although it appears clearly within the range of Weil's data. Of greater significance is the extremely low value of D_y reported by Weil (11). This value of $0.027hu_*$ is much lower than this author has seen elsewhere, with most other reported lab flume data being no lower than $0.10hu_*$, with most closer to $0.15hu_*$, or about 5 times Weil's value. Whatever flow peculiarities may have yielded this low value, it seems that EH is not only consistent with the expression D_y/U_oH_o, it is very, very low for practical systems.

To further investigate the significance of EH, a review of the relative magnitudes of jet and diffusion horizontal dilution (as given by Equations, 6 and 7) is given. To simplify the comparison, the V sin θ term in Equation 6 was neglected as it was small for the case investigated. The system is the same as the standard case shown in Table 1 except that the ambient velocity is decreased to yield R = 0.159. This, should represent a stronger jet. The simplified ratio is given by

$$\text{jet/ambient} = \frac{\sqrt{\pi}}{11} \frac{E_o}{EH} \frac{U_c}{U_o} \frac{B}{H_o} \tag{9}$$

This ratio is presented for several different EH values in Figure 3. The values of EH can be translated to $D_y = 0.19\ hu_*$ (EH = 0.02) up to $D_y = 1.36\ hu_*$ (EH = 0.144) for h = 50 ft, H_o = 5.0 ft, ambient velocity U_a = 1.1 ft/sec and a reasonable bottom friction. As expected, the dilution contribution due to jet entrainment decreases along the jet axis as the excess velocity decreases. As EH (i.e., D_y) increases, the region of jet domination is shortened. In fact, for EH = 0.072 and higher, diffusion dilution exceeds jet contributions even at the end of the zone of flow establishment. For higher EH values, such as those in a bond, jet influence is never substantial. This raises further questions about direct addition of the jet and diffusion contributions. In fact, it raises questions as to whether or not a jet model should be used, despite the high ratio of initial jet velocity to ambient velocity. It still further raises the question of the meaning of D_y, and hence EH, for the PDS model. Perhaps all the model assumptions have caused D_y to take the role of an empirical "tuning" coefficient not necessarily having the same meaning as the turbulent diffusion coefficient defined by Equation 8. Several questions arise in conjunction with D_y. The value of D_y for the PDS model should not include the additional lateral spreading due to buoyancy as outlined by Prych (12), as the PDS model explicitly includes buoyant spreading. If, however, this is improperly cast, as some argue, then the value of D_y required for model use could be affected. The inclusion of ambient diffusion terms, on the surface, appears to yield a model applicable beyond the end of the jet region. However, questions about the diffusion term make this less certain. While numerical results show PDS predictions approaching the expected X^{-1} dependence at larger distances, failure to include boundary influences by images or the like further limits this extended use of the PDS model.

SIGNIFICANCE IN PREDICTIONS

While it is apparent that further study and data fitting is needed to clarify the diffusion term, the PDS model continues to be used and performs as well or better than other models when compared with data. The question might well be raised as to the impact of errors in D_y (and EH) on model predictions insofar as evaluating regulatory compliance. Figure 2 shows impact on length to various isotherms. A commonly used measure is the area contained within temperature isotherms. Figure 4 shows the effect of EH on predicted areas for the standard case presented in Table 1. One problem in specifying EH is the uncertainty of D_y without field data. In addition, as shown by Chang (10), D_y varies through a bend. If it is believed that D_y can be estimated within a narrow range, then values for areas of interest do not vary greatly. On the other hand, where D_y may vary within a factor of 3-4 or more, considerable uncertainty exists in predicted areas. One can always use EH = 0.02 as a conservative estimate. If plume sizes are acceptable at

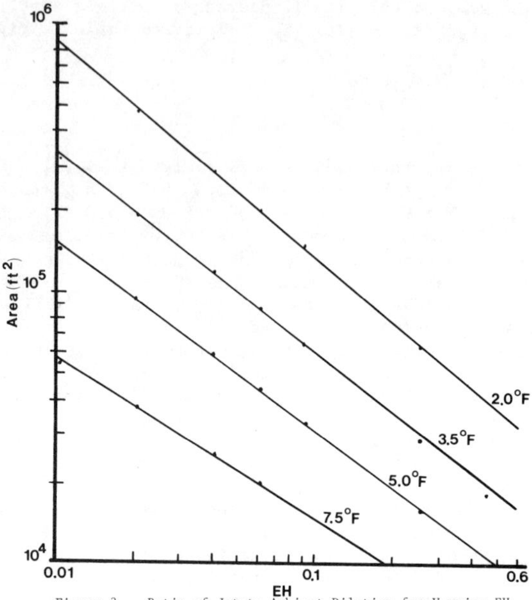

Figure 3. - Ratio of Jet to Ambient Dilution for Varying EH

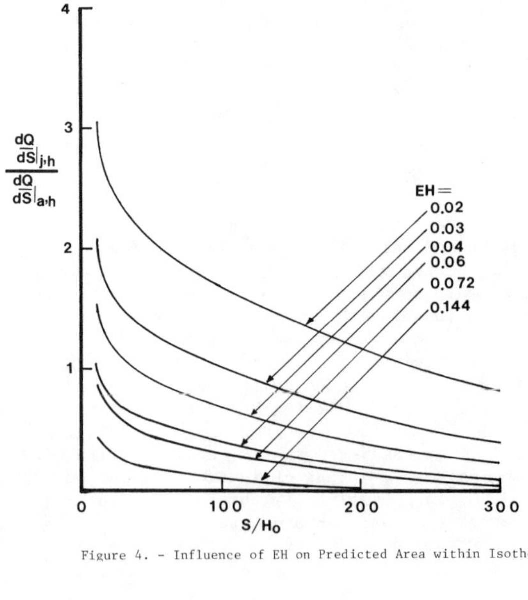

Figure 4. - Influence of EH on Predicted Area within Isotherms

that level, then no further work is needed. If, however, this causes an apparently unacceptable mixing zone, then further refinement and possible field studies are called for. Figure 4 is strictly valid only for the standard case mentioned. It does give one some ability to estimate relative magnitudes of effects for other cases. Area changes are not as rapid as EH (D_y) changes. For example, a fourfold increase in EH from 0.05 to 0.20 yields an increase in area within the 5°F isotherm from 19,000 ft^2 to 48,000 ft^2. Figure 4 gives some ability to assess the likelihood of a reasonable D_y value bringing a mixing zone within regulatory guidelines.

CONCLUSIONS

The PDS model has been shown to be strongly influenced by the ambient diffusion coefficient embodied in the model parameter EH. In fact, specifying reasonable values of the turbulent diffusion coefficient, D_y, can lead to cases dominated by diffusion even where the initial jet-to-ambient velocity ratio meets the criteria for a jet. Questions exist on the manner in which the diffusion term is cast in the PDS model, and added empirical evidence is needed to adeqately define the meaning of the parameter EH under various conditions. In the meantime, work is proceeding to consider the effect of including the diffusion terms differently. In the long term, of course, it must be recognized that continued manipulation of integral models must have a limit due to other simplifying assumptions. Some results have been presented to enable estimates of likely impact of errors in D_y (EH) on model results. Typical uncertainties in D_y at a given site may result in uncertainties in isotherm areas, for example, by a factor of 2-3 or more. It is hoped these can be useful as further model developments proceed.

ACKNOWLEDGMENTS

This investigation resulted as a part of work supported by Middle South Services, Inc., in New Orleans, La. The cooperation of Joel Patterson, Robin Kuckyr, Bill Lemon, and their colleagues is appreciated. The use of the University of South Carolina Computer Center is also gratefully acknowledged.

REFERENCES

1. Stolzenbach, K. D., and D. R. F. Harleman, "An Analytical and Experimental Investigation of Surface Discharges of Heated Water," Massachusetts Institute of Technology, R. M Parsons Lab., Report No. 135, February, 1971.

2. Prych, E. A., "A Warm Water Effluent Analyzed as a Buoyant Surface Jet," Svergis Meterologiska Och Hydrologiska Institut, Serie Hydrologi, Nr. 21, Stockholm, Sweden, 1972.

3. Shirazi, M. A., and L. R. Davis, "Workbook of Thermal Plume Prediction, Vol. 2: Surface Discharge," U. S. Environmental Protection Agency, Report EPA-R2-72-005b, May, 1974.

4. Dunn, W., A. J. Policastro, and R. A. Paddock, "Surface Thermal Plumes: Evaluation of Mathematical Models for the Near and Complete Field," Argonne National Lab., Argonne, Ill., Report ANL/WR-75-3, Part 1-May, 1975; Part 2-August, 1975.

5. Jirka, G. H., G. Abraham, and D. R. F. Harleman, "An Assessment of Techniques for Hydrothermal Prediction, "Massachuseets Institute of Technology, R. M. Parsons Lab., Report 203, 1975.

6. Benedict, B. A., J. L. Anderson, and E. L. Yandell, Jr. "Analytical Modeling of Thermal Discharges: A Review of the State of the Art," Argonne National Lab., Argonne, Ill., Report ANL/ES-18, 1974.

7. Prych, E. A., "An Analysis of a Jet into a Turbulent Ambient Fluid," Water Research,7, 647-657, 1973.

8. Preston, J. W., Jr., "Thermal Surface Jet Discharge Effects from Varying Diffusion Coefficients and Certain Physical Characteristics," M. S. Thesis, University of South Carolina, Columbia, S. C., 1979.

9. Yotsukura,N., and W. W. Sayre," Transverse Mixing in Natural Channels, Water Resources Research, 12, 695-704, 1976.

10. Chang, Y. C., "Lateral Mixing in Meandering Channels," Ph.D. Dissertation, University of Iowa, Iowa City, 1971.

11. Weil, J. "Mixing of a Heated Surface Jet in Turbulent Channel Flow," Report WHM-1, Hydraulic Engineering Lab, University of California, Berkeley, 1972.

12. Prych, E. A., "Effects of Density Differences on Lateral Mixing in Open-Channel Flows," W. M. Keck Lab. of Hydraulics and Water Resour Resources, California Institute of Technology, Pasadena, Report No. KH-R-21, 1970.

A PREDICTIVE MODEL OF HURRICANE WIND-WAVES

by

Rodney J. Sobey[1] M.ASCE
Ian R. Young[2]

ABSTRACT

A numerical model describing the generation, propagation, interaction and decay of wind-waves forced by a hurricane is presented. It is based on air-sea interaction theories and a field solution of the Radiative Transfer Equation. The nodal variable is the directional energy spectrum and the model is applicable in shallow as well as deep water. Results for a deep water prediction are included.

1 INTRODUCTION

Recently, increased construction activity in both coastal and offshore regions has highlighted the potentially destructive and almost explosive nature of the marine environment. The sea response to extreme meteorological events is itself extreme, and no more so than the surface gravity wave response to hurricanes (or tropical cyclones or typhoons). The forced gravity wave spectrum is largely concentrated in two bands, corresponding to storm surge and wind waves respectively. Storm surge is more intense in nearshore waters but wind waves of potentially destructive magnitudes are generated in both offshore and coastal waters.

Offshore mining activity is frequently located in notorious hurricane regions, such as the Gulf of Mexico or Australia's North West Shelf, and associated offshore constructions and operations must contend with hurricane wind waves. Nearshore and coastal facilities as well as numerous coastal settlements in tropical waters must withstand the combined effects of hurricane storm surge and the superimposed wind waves, the storm surge providing the vehicle to intensify wave damage in coastal regions through the sustained super-elevation of the mean water level. The human and material devastation during Hurricane Camille in 1969 along the U.S. Gulf Coast and in 1970 in Bangladesh bear witness to the destructive potential.

Common practice in the prediction of hurricane wind waves is often empirical, being based on the well known forecasting curves of Sverdrup, Munk and Bretschneider. In the long term however, an approach based on

---

1. Senior Lecturer, Department of Civil and Systems Engineering, James Cook University, Townsville 4811, Australia.
2. Post-Graduate Research Student, Department of Civil and Systems Engineering, James Cook University, Townsville 4811, Australia.

wave action conservation and air-sea interaction theories is the only real alternative. A suitable model should be applicable to continental shelf and nearshore regions where engineering interest is concentrated, not just to deep water regions remote from the coast. Specific nearshore influences such as fetch limitations and shoaling and refraction must therefore be considered. A predictive numerical model along such lines is described in this paper. It is called SPECT.

2 RADIATIVE TRANSFER EQUATION

Using the Gaussian random wave model a complex sea state is described in terms of the variance spectral density $F(\underline{k})$ of the surface gravity waves in directional wave number space. At each position and time, F represents the superposition of free linear wave components of all wave numbers and from all directions. Assuming F to be a slowly-varying function of position and time, wave energy conservation may be written (Hasselmann 1968) as

$$\frac{dF}{dt}(k_1,k_2;x,y,t) = \frac{\partial F}{\partial t} \qquad \text{(temporal accumulation)}$$

$$+ \frac{\partial F}{\partial x}\frac{dx}{dt} + \frac{\partial F}{\partial y}\frac{dy}{dt} \qquad \text{(propagation)}$$

$$+ \frac{\partial F}{\partial k_1}\frac{dk_1}{dt} + \frac{\partial F}{\partial k_2}\frac{dk_2}{dt} \qquad \text{(refraction/shoaling)}$$

$$= Q(k_1,k_2;x,y,t), \qquad \text{(source/sink)} \tag{1}$$

This equation, known as the Radiative Transfer Equation, formally summarises all the various physical processes that contribute to the evolution of the directional spectrum. The associated characteristic equations defining the wave orthogonals are

$$\frac{dx}{dt} = \frac{\partial\omega}{\partial k_1} = C_g \cos\phi \tag{2a}$$

$$\frac{dy}{dt} = \frac{\partial\omega}{\partial k_2} = C_g \sin\phi \tag{2b}$$

$$\frac{dk_1}{dt} = -\frac{\partial\omega}{\partial x} \tag{2c}$$

$$\frac{dk_2}{dt} = -\frac{\partial\omega}{\partial y} \tag{2d}$$

where ω and k are related through the linear wave theory-dispersion relationship

$$\omega^2 = gk \tanh kd \tag{3}$$

Substituting Equations 2 into Equation 1 gives

$$\frac{dF}{dt} = Q \tag{4}$$

along wave rays defined by Equations 2. Should the forcing term Q be zero (i.e. no generation or dissipation), the directional wave number spectrum is conserved along appropriate wave rays. Generally however, generation and/or dissipation are non-zero and Equation 3 describes the evolution of the directional wave number spectrum relative to a wave group moving along wave orthogonals at the group velocity C_g. In deep water the right hand sides of Equations 2c and 2d are identically zero and the wave rays are straight lines, but this is not generally the case in shallow water where the bathymetry bends or refracts the rays. Current refraction has been specifically excluded from the above equations. There is little formal difficulty in its retention but a very substantial computational penalty.

Conventional practice in coastal and ocean engineering considers the directional frequency spectrum $E(f,\theta)$ rather than the directional wave number spectrum $F(k_1,k_2)$. These alternative spectral representations are related through

$$\iint F(k_1,k_2)dk_1\, dk_2 = \iint E(f,\theta)df\, d\theta = \sigma^2 \tag{5}$$

where σ^2 is the variance of surface gravity waves. From Equation 5,

$$F(k_1,k_2;x,y,t) = \frac{CC_g}{2\pi\omega} E(f,\theta;x,y,t) \tag{6}$$

where $\omega = 2\pi f$. As a consequence Equation 1 becomes

$$\frac{\partial}{\partial t}(CC_gE) + C_g \cos\phi \frac{\partial}{\partial x}(CC_gE) + C_g \sin\phi \frac{\partial}{\partial y}(CC_gE)$$

$$+ \frac{C_g}{C}\left[\sin\phi \frac{\partial C}{\partial x} - \cos\phi \frac{\partial C}{\partial y}\right]\frac{\partial}{\partial \phi}(CC_gE) = CC_gS \tag{7}$$

where $S(f,\theta;x,y,t)$ becomes the forcing term. Likewise Equation 4 becomes

$$\frac{d}{dt}(CC_gE) = CC_gS, \tag{8}$$

the produce CC_gE being the new action variable that is conserved along wave rays for $S = 0$, and Equations 2, the wave ray equations, become

$$\frac{dx}{dt} = C_g \cos\phi$$

$$\frac{dy}{dt} = C_g \sin\phi \tag{9}$$

$$\frac{d\phi}{dt} = \frac{C_g}{C}\left[\sin\phi \frac{\partial C}{\partial x} - \cos\phi \frac{\partial C}{\partial y}\right]$$

In deep water, both C and C_g are independent of depth and constant for each frequency, the wave rays become straight lines and Equation 7 reduces to the familiar form

$$\frac{\partial E}{\partial t} + C_g \cos\phi \frac{\partial E}{\partial x} + C_g \sin\phi \frac{\partial E}{\partial y} = S \tag{10}$$

used by Pierson et al (1966), Barnett (1968), Ewing (1971) and Cardone et al (1970) for deep water numerical wave forecasting models. The present SPECT model is based on Equations 7, 8 and 9, being specifically designed to accommodate shallow as well as deep water. $S(f,\theta;x,y,t)$ in Equation 7 represents all processes which transfer energy to or from the spectrum. A numerical solution of Equations 7 to 9 would describe the detailed evolution of the wave field in both space and time.

3 ENERGY SOURCE TERMS

Following Hasselmann (1968), the energy source involves the superposition of a number of individual influences,

$$S = \sum_{i=1}^{N} S_i \tag{11}$$

The dominant terms are listed in Table 1 and represent the commonly identified physical processes contributing to the growth, decay and interaction of the separate spectral components.

Source Term	Description
S_1	Constant energy transfer to wave field through turbulent atmospheric pressure fluctuations
S_2	Increasing energy transfer to wave field through instability in coupling between waves and the mean atmospheric boundary layer flow
S_5	Energy transfer within wave field through weakly nonlinear wave-wave interactions
S_6	Energy dissipation in shallow water through turbulent bottom friction
S_7	Energy dissipation through wave breaking or white capping

TABLE 1 ENERGY SOURCE TERMS

A comprehensive review and evaluation of these five source terms is given elsewhere (Young and Sobey 1980) but the general character of the terms is illustrated through Figures 1 to 4. The Phillips mechanism S_1 is a resonance effect, peaking at spectral components with a wave celerity equal to the velocity of convection of the atmospheric turbulence. As is indicated in Figure 1 the magnitude of the effect declines and the frequency of the peak transfer increases as the angle to the wind direction is increased. It is a linear growth term and is largely responsible for the initial growth of the spectrum. The Miles term S_2 is an instability mechanism leading to continued exponential growth. Figure 2 shows the form of the energy transfer evaluated for a Pierson-Moskowitz spectrum corresponding to a wind velocity of 25 m/sec. S_5 represents the nonlinear interaction among spectral components. As indicated in Figure 3, it results in a redistribution of wave energy from the peak, to both lower and higher frequencies. The term S_6 represents wave energy dissipation to turbulent bottom friction. This term is important only in shallow water and is most intense for the shorter frequency spectral components, as shown in Figure 4. The final term, S_7 introduces white capping or wave breaking through the concept of the saturation spectrum. White capping is the limiting process controlling otherwise unbounded wave growth represented by the S_1 and S_2 mechanisms.

Just what is the correct representation for each of these terms is still the subject of some controversy. A number of rather empirical representations have been proposed in the literature and our present choice of terms is based on as objective an analysis as is presently possible. No claim is made that our choice from the presently available alternatives is entirely satisfactory. There are particular difficulties in representing wave growth at high wind speeds, wave decay to opposing winds and the steep spatial gradients of hurricane forcing. These aspects are the basis of our continuing research.

The meteorological forcing is directly related to the near-surface wind field and the NHRP model describing the space and time variable wind field has been adopted. The details of this model have been documented by Sobey, Harper and Stark (1977), in the context of a parallel predictive model of hurricane storm surge.

4 NUMERICAL SOLUTION OF RADIATIVE TRANSFER EQUATION

An appropriate numerical solution algorithm for Equation 7 must accommodate a number of conflicting demands. It is essentially a convective transport problem but wave energy is convected along ray paths that are straight lines only in deep water. In addition any field solution of the radiative transfer equation that allows a reasonable resolution of the directional energy spectrum at each node imposes massive demands on computer system resources. Discrete representations of convective transport problems often lead to severe numerical dispersion and associated solution oscillations. The remedy is higher order representations but this puts even heavier demands on computer system resources, just what must be avoided. Some compromise is essential. Our solution is a combination of "bit packing" of the spectral components at each node and a fractional step algorithm combining the method of characteristics for a propagation step with an analytical solution for a forcing step.

Following Richtmyer and Morton (1967), Equation 7 may be represented as

$$\frac{\partial}{\partial t}(CC_gE) = (A_1 + A_2)(CC_gE) \tag{12a}$$

where the A operators are respectively

$$A_1(CC_gE) = -C_g \cos\phi \frac{\partial}{\partial x}(CC_gE) - C_g \sin\phi \frac{\partial}{\partial y}(CC_gE)$$

$$-\frac{C_g}{C}\left[\sin\phi \frac{\partial C}{\partial x} - \cos\phi \frac{\partial C}{\partial y}\right]\frac{\partial}{\partial \phi}(CC_gE) \tag{12b}$$

and

$$A_2(CC_gE) = CC_g(S_1 + S_2 + S_5 + S_6 + S_7) \tag{12c}$$

The commonly accepted form of all five forcing terms can be represented generally as a + bE, whence Equation 12c becomes

$$A_2(CC_gE) = CC_g(a + bE) \tag{12d}$$

In the fractional step algorithm, the two operations represented by A_1 (propagation, refraction and shoaling) and A_2 (forcing) are given separate and consecutive consideration over each time step Δt. The *propagation step* advances the solution from E^n to $E^{n+½}$ through solution of the partial differential equation.

$$\frac{\partial}{\partial t}(CC_gE) = 2A_1(CC_gE) \tag{13a}$$

The *forcing step* advances the solution form $E^{n+½}$ to E^{n+1} through solution of the partial differential equation

$$\frac{\partial}{\partial t}(CC_gE) = 2CC_g(a + bE) \tag{13b}$$

The propagation step, Equation 13a, describes propagation of wave action at the group velocity and without loss along appropriate wave rays, which are defined by the characteristic Equations 9. Thus it is necessary to construct a wave ray for each frequency, direction band at each nodal point in the computational field. In a simulation involving 10 frequency and 16 directional bands at each of 225 nodes, 36000 ray diagrams must be constructed. Since current refraction is not presently considered, the ray diagrams need be computed once only. Utilising then the method of characteristics, Equation 13a becomes,

$$\frac{d}{dt}(CC_gE) = 0 \tag{13c}$$

along the respective wave rays. In discrete form for a particular frequency and direction at node ($i\Delta x$, $j\Delta y$), Equation 13c is

$$(CC_gE)^{n+1}_{i,j} = (CC_gE)^{n}_{r,s} \tag{13d}$$

where ($r\Delta x$, $s\Delta y$) is the position along the wave ray for that specific frequency and direction at time $n\Delta t$ (see Figure 5). Linear interpolation is presently utilised for CC_gE at time $n\Delta t$ because of the extensive demands on computer system resources. Higher order interpolation is readily accommodated within SPECT but at a severe computational penalty. A detailed discussion on this specific problem can be found in Young and Sobey (1980). We are currently investigating the potential of a "flux corrected transport" algorithm in which advantage is taken of the dependence of the hurricane forcing on a single spatial scale, the radius to maximum winds.

The forcing step achieves maximum computational efficiency through a local analytical solution to Equation 13b,

$$(E)^{n+1}_{i,j} = (E)^{n+\frac{1}{2}}_{i,j}\, e^{b\Delta t} + a\left[\frac{e^{b\Delta t} - 1}{b}\right] \tag{13e}$$

The term in square brackets is evaluated through a truncated power series expansion to avoid difficulties when b is near zero.

The numerical model, SPECT, has been programmed for the James Cook University's DEC System-10 computer (KL10-E processor). The model is large and computationally expensive, requiring 60,000 thirty-six bit words of storage and approximately thirty minutes for twelve hours of real time at a 30-minute time step. The program has been written in a general fashion, allowing the representation of any coastal, offshore or undersea features required for a particular application.

As in any propagation or initial value problem, it is necessary to specify both boundary and initial conditions. In SPECT, the sea is assumed to be perfectly calm initially and the central pressure of the hurricane is then gradually decreased to initialise the sea conditions prior to active simulation. Current boundary conditions are of two types: radiation boundaries, at which energy can propagate out of but not into the computational field and active boundaries, where the energy spectrum is represented in parametric form (Ross 1976, Shemdin 1980). Land boundaries are represented as points with zero energy.

5 A DEEP WATER HINDCAST

For the purpose of illustration, aspects of a deep water wave prediction are presented. The hurricane has a central pressure of 950 mb and a forward speed of 30 km/hr.

The distribution of significant wave heights is shown in Figure 6, where the superimposed vectors indicate the mean direction of propagation of the waves. Wave conditions are most severe in the crescent shaped area in the left front quadrant of the storm, consistent with the very limited field results that are available. It is almost certain that the peak magnitudes are too high and this is related to the present lack of data at the high wind speeds experienced within a hurricane. Current growth terms predict a rapid rising of the sea to the saturation limit imposed by white capping, the rate of rise increasing with the wind speed in a manner appropriate to moderate wind speeds. These results are presently extrapolated to the more severe winds within hurricanes, in the absence of more reliable information. Some eye witness reports however suggest that extreme winds might actually flatten rather than energise the wave field.

The dominant directions are closely related to the directional characteristics of the wind field as one might expect, at least initially. However recent measurements by Shemdin (1980) have cast doubt on such a prediction. These measurements were obtained from airborne synthetic aperture radar in the Gulf of Mexico and are the first really comprehensive pictures of the directional field characteristics of hurricane wind waves. The striking feature of these results is the lack of correlation between the local wind direction and the direction of dominant wave propagation. They are contradictory to the predictions of this and presumably all other published numerical models. If these results are subsequently confirmed, the energy source terms may require extensive revision. This aspect is a topic of current research.

In addition to the significant wave height distribution over the computational grid, detailed spectral predictions at specified points are also available from SPECT. Typical results of both one and two dimensional spectra are shown in Figures 7 and 8. Figure 7 is for a point within the eye of the storm whereas Figure 8 is for the point of maximum significant wave height. Although the one-dimensional spectra are similar for these two points, the two-dimensional spectra are considerably different. For the point within the eye, the two-dimensional spectrum indicates a very confused sea condition with waves propagating in a number of directions. This relationship is perhaps more consistent with Shemdin's results. Further from the eye, however, there is more order with the direction of propagation being distributed about the wind direction.

The Shemdin field results indicate that waves ahead of the storm essentially radiate from the storm centre and do not track the local wind direction. The present forcing terms, S_7 in particular, force the wave field to track the local wind direction which is reasonably appropriate for the more gradual spatial variations in non-hurricane wind fields. Waves generated to the left (Southern Hemisphere) of the storm will propagate in the general direction of forward movement of the storm. But S_7 allows only those waves that track the local wind direction to survive. Removal of the directional spread in S_7 will allow waves that do not track the local wind direction to survive. The saturation spectrum would still limit wave growth, as it is intended, but now for each separate component, not collectively. Figure 9 is the result achieved by removing the angular spread on the S_7 term. The wave directions are now perhaps closer to those of Shemdin but considerable refinement remains necessary. The contours of significant wave height have been excluded as they predict even larger wave heights than before, rather symptomatic of the problems that remain to be resolved.

7 CONCLUSIONS

Numerical models such as SPECT represent the first real attempts at developing a hurricane prediction technique with a sound theoretical base. However, the adequacy of such models remains difficult to determine as suitable field data in hurricanes is very limited.

Almost all of the data that is available is in the form of one-dimensional spectra at isolated points (e.g. the ODGP data during Camille and other hurricanes in the Gulf of Mexico). Comprehensive data sets comprising directional characteristics as well as magnitude are necessary before the present uncertainties may be answered with any confidence. SPECT has been designed for continuing research in hurricane wind waves but nonetheless remains a comprehensive and flexible state-of-the-art predictive model.

8 REFERENCES

BARNETT, T.P. (1968). On the Generation, Dissipation and Prediction of Ocean Wind Waves. Jr. Geoph. Res., 73 (2), pp 513-529.

CARDONE, V.J. et al (1976). Forecasting Hurricane Winds and Waves. Report to: NOAA Sea-Air Interaction Laboratory, 22 pp.

EWING, J.A. (1971). A Numerical Wave Prediction Method for the North Atlantic Ocean. Dtsch. Hydrogr. Z., 24 (5), pp 241-261.

HASSELMANN, K. (1968). Weak Interaction Theory of Ocean Waves. In: M. Holt (ed.), Basic Developments in Fluid Dynamics. Academic Press, New York, 2: pp 117-182.

PIERSON, W.J., TICK, L.J. and BAER, L. (1966). Computer-Based Procedures for Preparing Global Wave Forecasts and Wind Field Analysis. Proc. 6th Naval Hydrodynamics Symp. 2(1).

RICHTMYER, R.D. and MORTON, K.W. (1967). Difference Methods for Initial-Value Problems. New York, Wiley.

SHEMDIN, O.H. (1980). Prediction of Dominant Wave Properties Ahead of Hurricanes. Proc. 17th Int. Conf. on Coastal Engineering, Sydney.

SOBEY, R.J., HARPER, B.A. and STARK, K.P. (1977). Numerical Simulation of Tropical Cyclone Storm Surge. Dept. Civil and Systems Eng., Res. Bul. No. CS14, James Cook University.

YOUNG, I.R. and SOBEY, R.J. (1980). Numerical Prediction of Tropical Cyclone Wind Waves. Dept. Civil and Systems Eng., Res. Bul. No. CS20, James Cook University, in preparation.

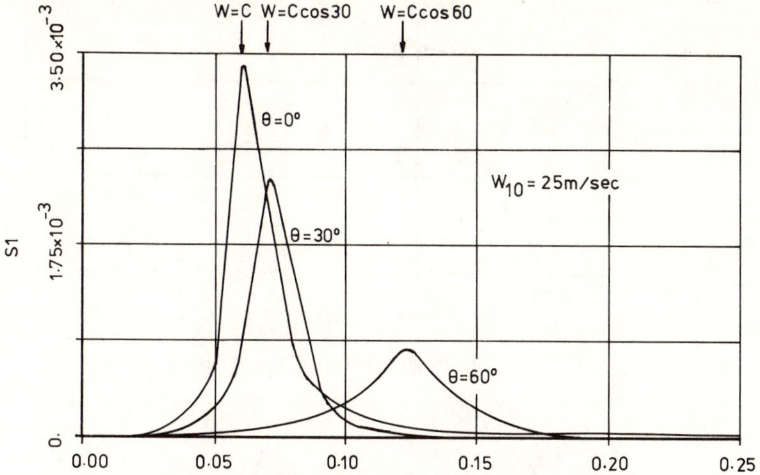

FIGURE 1 Linear Growth (Phillips) Term S_1

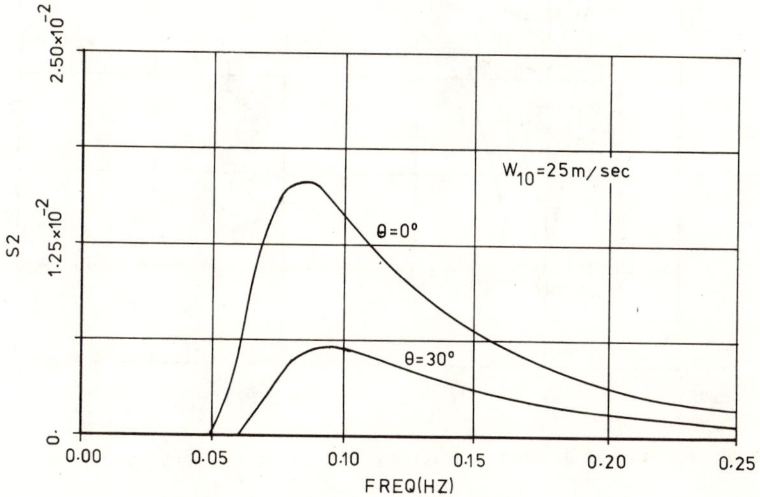

FIGURE 2 Exponential Growth (Miles) Term S_2

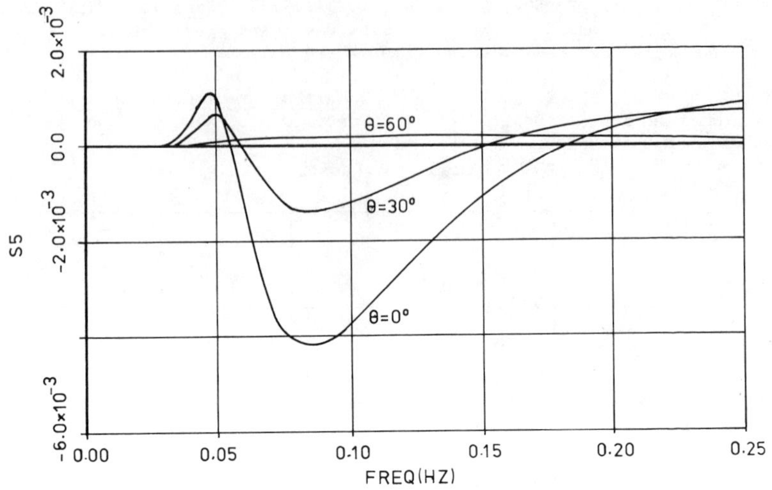

FIGURE 3 Wave-Wave Interaction Term S_5

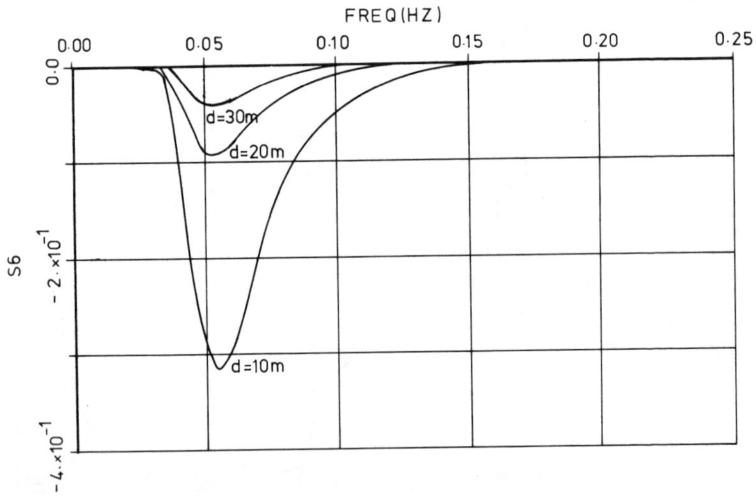

FIGURE 4 Turbulent Bottom Friction Term S_6 in Dominant Wave Direction

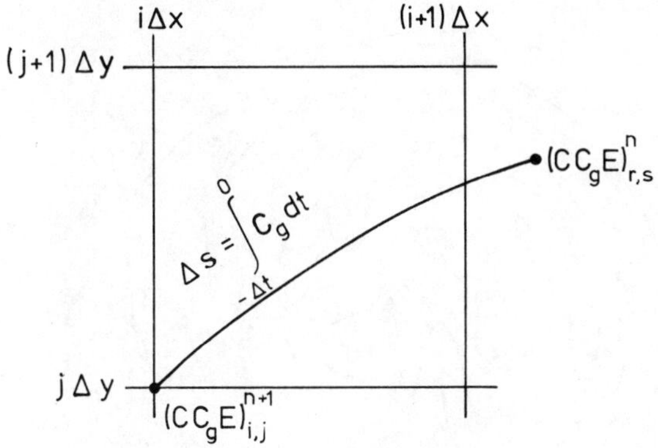

FIGURE 5 Wave-Action Propagation

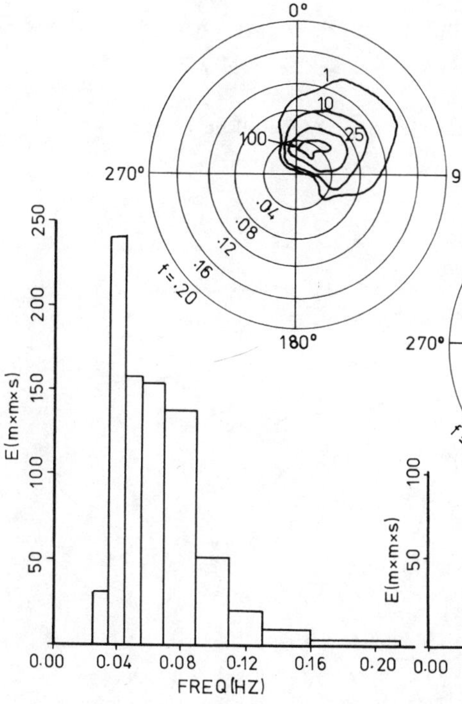

FIGURE 8 Spectra at Point P (Figure 6)

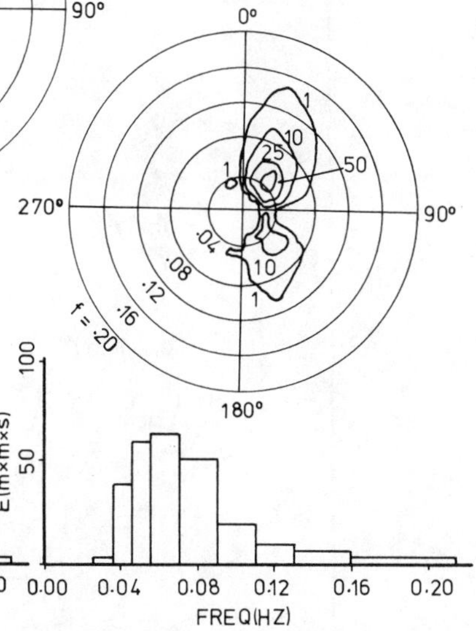

FIGURE 7 One and Two Dimensional Spectra in Eye of Hurricane

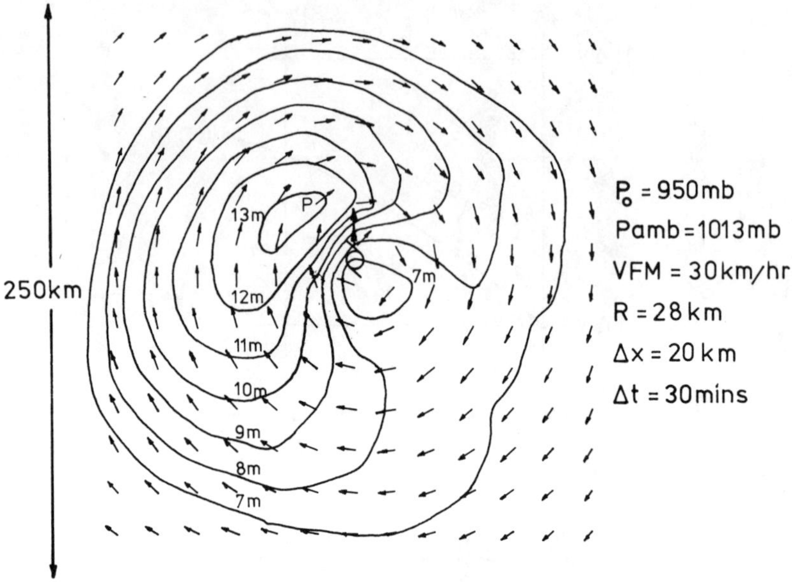

FIGURE 6 Distribution of Significant Wave Height

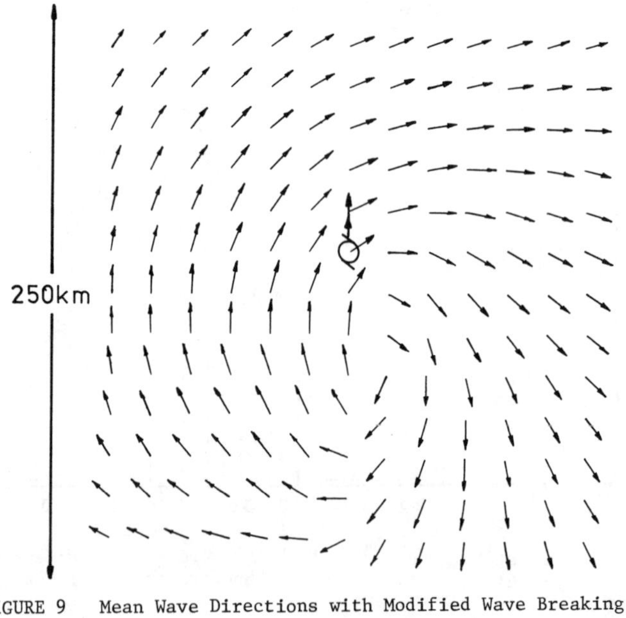

FIGURE 9 Mean Wave Directions with Modified Wave Breaking Term

CONSTANT AND VARIABLE EDDY VISCOSITY FLOW SIMULATIONS IN SETTLING TANKS

by David R. Schamber[1], A.M. ASCE and Bruce E. Larock[2], M. ASCE

ABSTRACT

Mean-flow equations (continuity and Reynolds equations) are combined with a two-equation turbulence model to compute two-dimensional velocity fields in planar sedimentation basins. This formulation allows the variation of the turbulent kinematic eddy viscosity ν_t to be computed directly, in contrast to most earlier work which assumes a constant value for ν_t, or an ad hoc algebraic variation of ν_t, or uses experimental data that are difficult to obtain. Examples compare constant ν_t simulations with the turbulence model predictions. Constant-ν_t simulations can replicate the major features of a variable-ν_t solution only if the numerical value for ν_t is carefully and appropriately chosen.

INTRODUCTION

Primary sedimentation by gravity in large basins has been used for decades in water and wastewater treatment. However, the process is still incompletely understood because detailed knowledge of basin velocity fields is very hard to obtain, either experimentally or computationally. Although mean velocities are typically only 1 to 10 ft/min, the large size of most tanks causes the flow to be turbulent. The time-averaged flow behavior may be sought via the continuity and Reynolds equations. Closure of the equation set is often obtained by an eddy viscosity turbulence model. The turbulent kinematic eddy viscosity ν_t is often selected to be a constant or allowed to vary according to some simple relation. In either situation the investigator usually does not have any foreknowledge to aid in selecting a numerical value for ν_t. In fact, ν_t varies from point to point within the flow domain and depends on the local structure of the flow field.

The present work avoids the uncertainties of this coefficient selection process by determining the variation of eddy viscosity by use of the relatively thoroughly tested k-ε turbulence model. The turbulence kinetic energy k and its rate of dissipation ε are determined from two differential transport equations which must be solved simultaneously

[1]Assistant Professor, Civil Engineering Department, University of Utah, Salt Lake City, Utah 84112

[2]Professor, Civil Engineering Department, University of California, Davis, California 95616.

with the continuity and Reynolds equations. These equations are solved by the finite element method. Constant viscosity simulations are compared with the k-ε results to show that strongly differing results can be obtained unless considerable care is exercised in choosing the correct constant value for ν_t.

MATHEMATICAL MODEL

The flow in rectangular basins is in general three-dimensional and unsteady due to corner and side effects, surface wind, variable influent temperature, solar heating, and unsteady loading conditions. For isothermal, steady flow in the absence of wind stresses the flow is approximated as two-dimensional in a vertical plane along the tank centerline from inlet to outlet. Suspended sediment in the influent is presumed not to affect the density because its magnitude is typically 2×10^{-4} by weight (11, p. 231). Although only rectangular basins are discussed herein, a similar development has also been completed for circular basins (13).

The flow is then governed by the equations of mass and momentum conservation. For two-dimensional, steady, turbulent, non-stratified flow in a planar basin these equations (13) are

$$\frac{\partial U}{\partial x} + \frac{\partial W}{\partial z} = 0 \tag{1}$$

$$U\frac{\partial U}{\partial x} + W\frac{\partial U}{\partial z} = -\frac{1}{\mathbb{F}_o^2}\frac{\partial h}{\partial x} - \frac{\partial}{\partial x}(P+\overline{uu}) - \frac{\partial}{\partial z}(\overline{uw}) \tag{2}$$

$$U\frac{\partial W}{\partial x} + W\frac{\partial W}{\partial z} = -\frac{\partial}{\partial x}(\overline{uw}) - \frac{\partial}{\partial z}(P+\overline{ww}) \tag{3}$$

The variables in Eqs. 1-3 are nondimensionalized on a characteristic length h_o and characteristic velocity U_o. The variables are defined with reference to Fig. 1 as follows (starred and characteristic variables are dimensional): $x^* = xh_o$ = horizontal coordinate; $z^* = zh_o$ = vertical coordinate; $\mathbb{F}_o^2 = U_o^2/(gh_o)$ is a representative Froude number with g = gravity; $U^* = U\,U_o$ = time-averaged mean horizontal velocity; $W^* = W\,U_o$ = time-averaged mean vertical velocity; $\rho\overline{u^*u^*} = \rho\overline{uu}\,U_o^2$, $\rho\overline{w^*w^*} = \rho\overline{ww}\,U_o^2$, and $\rho\overline{u^*w^*} = \rho\overline{uw}\,U_o^2$ are Reynolds stresses in which ρ = fluid density. The overbar denotes time averaging, and u and w are the instantaneous velocity fluctuations about their respective means. The pressure is defined with reference to hydrostatic conditions, i.e., the actual pressure is given by $P + (h-z)/\mathbb{F}_o^2$ in which $h = h^*/h_o$ is the free surface elevation and $P = P^*/(\rho U_o^2)$ is the nondimensional pressure deviation from hydrostatic conditions. Since turbulent transport dominates molecular transport, the viscous terms which normally appear in Eqs. 2 and 3 are omitted. The time-averaging process has increased the number of unknowns that appear in Eqs. 1-3 and produced the familiar closure problem associated with turbulent flows. A relation between the Reynolds stresses and other mean flow variables must be introduced to close the system of equations.

By analogy to laminar flow, Boussinesq postulated that the turbulent stresses could be represented by expressions involving products of mean velocity gradients and the turbulent kinematic viscosity ν_t. One variant of this hypothesis is the isotropic closure

$$\overline{u_i u_j} = \frac{2}{3}\delta_{ij}k - \nu_t\left[\frac{\partial U_i}{\partial x_j} + \frac{\partial U_j}{\partial x_i}\right] \tag{4}$$

in which $k = \overline{u_i u_i}/2$ is the turbulence kinetic energy. Here $\delta_{ij} = 0$ for $i \neq j$, $\delta_{ij} = 1$ for $i=j$, and the Einstein summation convention is applied to repeated indices. In Eq. 4 and throughout the paper $\nu_t = \nu_t^*/(h_o U_o)$= nondimensional turbulent viscosity. Since the effect of turbulence is to enhance momentum transport, ν_t must be several orders of magnitude larger than the molecular kinematic viscosity ν. Unlike the molecular viscosity, ν_t is not a property of the fluid but varies from point to point within the flow field, reflecting the local structure of turbulence. Computations of turbulent flows when the distribution of turbulent eddy viscosity (isotropic or non-isotropic) is known a priori have been relatively successful (5,9) and have the practical advantage that no additional differential equations are required for this turbulence closure. An isotropic closure of this type combines k appearing in Eq. 4 with P in Eqs. 2 and 3 and hence k need not be evaluated (12). Since the distribution of turbulent eddy viscosity for complicated flows is not generally known a priori, these types of closure approximations are not predictive by nature.

An alternate approach is to model via partial differential equations the variation of ν_t throughout the flow domain. Numerous closure models of varying complexity have been proposed (e.g. see 3, 6, 8). The kinetic energy-dissipation or k-ε turbulence model is used to close the present mathematical model. This model has already been tested and experimentally verified for a number of free-shear flows, boundary layer flows, internal flows, and flows with recirculation (4, 8, 10), but point velocity and turbulence measurements in full scale basins are still required to confirm the suitability of this model to the present application. (Work underway at the University of California, Davis, will soon allow this to be checked in detail.) From dimensional considerations ν_t must be related to k and its rate of dissipation ε via

$$\nu_t = c_\mu \frac{k^2}{\varepsilon} \tag{5}$$

Here $\varepsilon = \nu\left[\overline{\frac{\partial u_i}{\partial x_j}\frac{\partial u_i}{\partial x_j}}\right]$ is the isotropic dissipation rate and c_μ = constant is determined from experimental data. Equation 5 follows Kolmogorov (7, p.15) who assumed turbulence could be characterized by k, which is a scalar measure of the intensity of the velocity fluctuations, and by the frequency f of the energy-containing motions, ε being proportional to kf. The variables k and ε are computed from transport equations (13) which have the following form in steady two-dimensional flow:

$$U\frac{\partial k}{\partial x} + W\frac{\partial k}{\partial z} = P_r - \varepsilon + \mathcal{D}(k) \tag{6}$$

$$U\frac{\partial \varepsilon}{\partial x} + W\frac{\partial \varepsilon}{\partial z} = c_{\varepsilon 1}\frac{\varepsilon}{k}P_r - c_{\varepsilon 2}\frac{\varepsilon^2}{k} + \mathcal{D}(\varepsilon) \tag{7}$$

Here the production term P_r is (13)

$$P_r = -\overline{u_i u_j}\frac{\partial U_i}{\partial x_j} = \nu_t \Phi \tag{8}$$

and

$$\Phi = 2\left[\frac{\partial U}{\partial x}\right]^2 + \left[\frac{\partial U}{\partial z} + \frac{\partial W}{\partial x}\right]^2 + 2\left[\frac{\partial W}{\partial z}\right]^2 \tag{9}$$

The diffusion terms $\mathcal{D}(k)$ and $\mathcal{D}(\varepsilon)$ are given respectively by

$$\mathcal{D}(k) = \frac{\partial}{\partial x}\left[\frac{\nu_t}{\sigma_k}\frac{\partial k}{\partial x}\right] + \frac{\partial}{\partial z}\left[\frac{\nu_t}{\sigma_k}\frac{\partial k}{\partial z}\right] \tag{10}$$

and

$$\mathcal{D}(\varepsilon) = \frac{\partial}{\partial x}\left[\frac{\nu_t}{\sigma_\varepsilon}\frac{\partial \varepsilon}{\partial x}\right] + \frac{\partial}{\partial z}\left[\frac{\nu_t}{\sigma_\varepsilon}\frac{\partial \varepsilon}{\partial z}\right] \tag{11}$$

The constants in Eqs. 5-11 are (8): $c_\mu = 0.09$, $c_{\varepsilon 1} = 1.45$, $c_{\varepsilon 2} = 1.90$, $\sigma_k = 1.0$, and $\sigma_\varepsilon = 1.3$. In applying the k-ε model it is intended that these coefficients not be adjusted or "tuned" for each application since these constants possess some degree of universality.

The constants $c_{\varepsilon 2}$, c_μ, and σ_ε may be inferred from experimental data while $c_{\varepsilon 1}$ and σ_k are determined by computer optimization (i.e., many calculations were performed in which the constant was systematically varied, and the value which gave the best overall agreement with experimental results was selected). The value of $c_{\varepsilon 2}$ may be fixed by reference to the decay of grid turbulence (2). In this experiment production and diffusion are negligibly small, and Eqs. 6 and 7 can be reduced to the ordinary differential equations

$$U_1\frac{dk}{dx_1} = -\varepsilon \tag{12}$$

and

$$U_1\frac{d\varepsilon}{dx_1} = -c_{\varepsilon 2}\frac{\varepsilon^2}{k} \tag{13}$$

in which U_1 is the constant streamwise velocity. The solution of these equations shows that

$$k \propto x^{-n} \tag{14}$$

with $n = (c_{\varepsilon 2}-1)^{-1}$. Experimental measurements suggest that $n \simeq 1.0$ which fixes $c_{\varepsilon 2} \simeq 2.0$.

The constants c_μ and σ_ε may be determined from near-wall experimental data. Near a solid boundary production nearly balances dissipation (6), and Eq. 6 reduces to

$$\varepsilon = -\overline{u_1 u_2}\,\frac{dU_1}{dx_2} \simeq U_\tau^2\,\frac{dU_1}{dx_2} = c_\mu \frac{k^2}{\varepsilon}\left[\frac{dU_1}{dx_2}\right]^2 \tag{15}$$

in which x_2 = coordinate normal to the wall, U_1 = velocity component parallel to the wall, and $U_\tau = (\tau_w/\rho)^{1/2}$ is the friction velocity which is related to the wall shear stress τ_w. In this region the shear stress is nearly constant, and U_1 can be determined from the law of the wall

$$U_1 = \frac{U_\tau}{\kappa}\,\ell n\left[\frac{Ex_2}{\nu}\right] \tag{16}$$

where E equals 9 for a smooth wall and $\kappa \simeq 0.4$ is the von Karman constant. Combining Eqs. 15 and 16 yields

$$U_\tau^2 = c_\mu^{1/2}\,k \tag{17}$$

Near-wall turbulence measurements (6) indicate that the ratio $\overline{u_1 u_2}/k \simeq U_\tau^2/k$ lies between 0.24 and 0.30, indicating the c_μ should fall between 0.06 and 0.09.

The constant σ_ε may be related to the other constants by considering the near-wall simplification of the dissipation equation. Neglecting convection, setting $P_r = \varepsilon$, and noting that turbulent diffusion of k near a solid boundary is small (i.e. $\partial k/\partial x_2 \simeq 0$), Eq. 7 reduces to

$$0 = \frac{\varepsilon^2}{k}\,(c_{\varepsilon 1} - c_{\varepsilon 2}) + \frac{c_\mu}{\sigma_\varepsilon}\,k^2\left[-\frac{1}{\varepsilon^2}\left(\frac{d\varepsilon}{dx_2}\right)^2 + \frac{1}{\varepsilon}\,\frac{d^2\varepsilon}{dx_2^2}\right] \tag{18}$$

Equations 15-17 may be used with Eq. 18 to obtain

$$\sigma_\varepsilon = \frac{\kappa^2}{c_\mu^{1/2}\,(c_{\varepsilon 2} - c_{\varepsilon 1})} \tag{19}$$

With $\kappa = 0.4$, $c_\mu = 0.09$, $c_{\varepsilon 2} = 1.90$, and $c_{\varepsilon 1} = 1.45$, the computed value of σ_ε is 1.19. The constants $c_{\varepsilon 2}$, c_μ, and σ_ε, as determined by experiment, are indeed close to the actual values used by Launder and Spalding (8) in computations which explore a wide variety of flows.

It is instructive to compare the k-ε model with the simpler mixing-length turbulence model of Prandtl (7, p.23). If one neglects the convective and diffusive transport terms in Eq. 6, then the production and dissipation terms must balance. For thin shear flows where $U_2/U_1 << 1$ and $\partial U_1/\partial x_2$ is the dominant velocity gradient (U_1 being the streamwise velocity and x_2 the coordinate normal to the main stream), Eq. 6 simplifies to

$$c_\mu \frac{k^2}{\varepsilon} \left[\frac{\partial U_1}{\partial x_2} \right]^2 = \varepsilon \tag{20}$$

Since the turbulent shear stress is

$$\overline{u_1 u_2} = -c_\mu \frac{k^2}{\varepsilon} \frac{\partial U_1}{\partial x_2} \tag{21}$$

from Eq. 4, one can combine Eqs. 20 and 21 to find

$$\overline{u_1 u_2} = -c_\mu^{3/2} \left[\frac{k^{3/2}}{\varepsilon} \right]^2 \left| \frac{\partial U_1}{\partial x_2} \right| \frac{\partial U_1}{\partial x_2} \tag{22}$$

in which the term $k^{3/2}/\varepsilon$ has the units of length. If the mixing length is defined as $\ell = k^{3/2}/(\varepsilon\, c_\mu^{3/4})$, Eq. 22 is identical to the mixing length hypothesis of Prandtl. Hence, one concludes that a mixing length model will undoubtedly fail to predict the prominent characteristics of a flow dominated by convective and/or diffusive transport of turbulence.

FINITE ELEMENT SOLUTION OF THE MATHEMATICAL MODEL

For a constant turbulent viscosity simulation only Eqs. 1-4 must be solved simultaneously. In this instance there are three unknowns (U,W, P) per computational node. For a variable turbulent viscosity simulation, Eqs. 1-4, 6 and 7 must be solved simultaneously with five variables (U,W,P,k,ε) per computational node. In either case the governing equations are coupled and nonlinear.

The equations are solved numerically by the finite element method via application of Galerkin's method of weighted residuals to the appropriate equation set. This approach allows one to use a nonuniform computational mesh, to seek solutions in geometrically complex nonpolygonal domains and to avoid difficulties in applying boundary conditions on such domain boundaries.

Newton's method and variants thereof are used to solve the discretized form of the continuum equations. Variable turbulent viscosity simulations are determined by solving the k-ε equations and mean flow equations alternately and iterating between the sets of equations as required until a converged solution is obtained. A detailed account of these techniques may be found in Ref. (14).

SEDIMENTATION BASIN FLOW SIMULATION

The velocity field within a rectangular settling unit is computed for a basin of modest size: length = 40 ft (12.2 m), width = 15 ft (4.6m), inlet depth = 10 ft (3.0m), outlet depth = 9 ft (2.7m). For this example the characteristic velocity and length scale are chosen to be $U_o = 0.1$ ft/sec (0.0305 m/sec) and $h_o = 10$ ft (3.05 m). A relatively high surface loading rate of 1500 gpd/ft^2 (0.0707 cm/sec) was selected for this example(1). This rate should simulate conditions in an overloaded primary clarifier and may represent the lower limit of clarifier performance.

Figure 1 depicts the finite element grid used for this numerical simulation. The elements are graded to be dense in the inlet and exit regions (where velocity gradients are large) and relatively more sparse in the interior domain.

The inlet region for this basin is represented by the upper three elements on the left edge of the computational grid; the inflow velocity profile is approximated as half of a plane jet. A "rigid-lid" boundary condition is used along the free surface. Outflow conditions are modeled as a point sink (located in the upper right corner of the tank) whose strength matches the inflow discharge. Adjacent to the solid boundaries the shear stress is given by the law of the wall while the normal velocity component is set to zero. A detailed mathematical description of the appropriate boundary conditions may be found in Ref. (14).

The vector plot of a velocity field computed with the k-ε turbulence model is shown in Fig. 2. Approximately 2.5 tank depths downstream from the inlet the jet expands to fill the tank depth. Point S in Fig. 2 marks the location of the stagnation point on the basin floor. The fluid streamlines converge toward the overflow weir in the upper right corner. A recirculating zone is formed below the expanding jet.

Contours of the nondimensional turbulent kinematic eddy viscosity [$\nu_t = c_\mu k^2/\varepsilon = \nu_t^*/(h_o U_o)$] corresponding to Fig. 2 are shown in Fig. 3. It is apparent that the nonuniform distribution of ν_t would be very difficult to specify a priori.

For this example the numbers of elements and node points are 135 and 454 respectively. The number of mean flow equations (U,W,P) is 955, and the number of turbulence equations (k,ε) is 788. Solution of this example required 6.2 minutes of central processor time on a CDC 7600 computer.

The cost of obtaining a fully converged variable viscosity solution is substantial. One major simplification which requires less than one-tenth of the central process time of a full k-ε solution is a simulation which assumes a constant turbulent eddy viscosity. This reduces the number of unknowns per computational node by two and eliminates the need to iterate between the mean flow and turbulence equation sets (13). Using a constant value of $\nu_t = 0.01$ (an approximate mean value taken from Fig. 3), one can then recompute the flow field for the rectangular basin. The velocity-vector field for this solution is presented in Fig. 5. A comparison of Figs. 2 and 5 shows that the flow patterns are quite similar. There are, however, some differences between the two plots. The velocities predicted by the constant viscosity model in the lower portions of the basin are usually smaller (by as much as 27% along the floor). The stagnation point moves from $x \simeq 2.5$ for the k-ε model to $x \simeq 2.7$ for the constant-viscosity computation. The majority of the

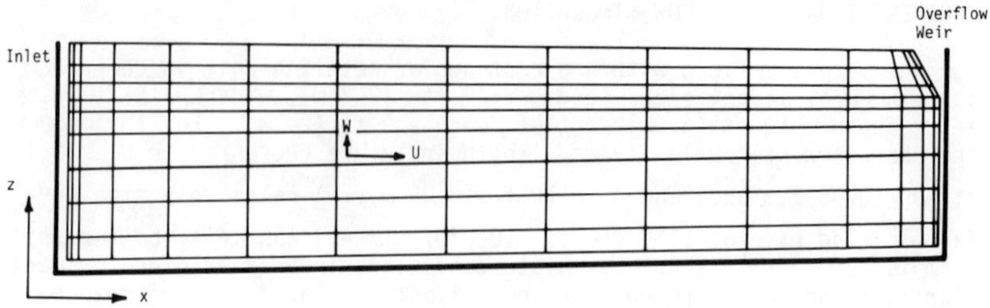

Fig. 1 Finite element discretization of tank domain.

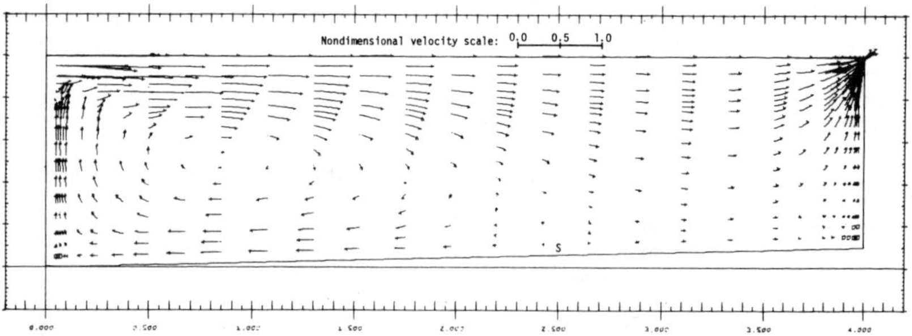

Fig. 2 Vector plot of velocity field, k-ε model.

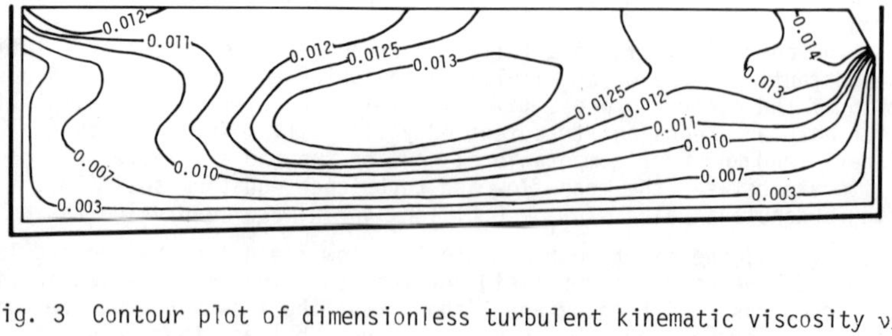

Fig. 3 Contour plot of dimensionless turbulent kinematic viscosity ν_t.

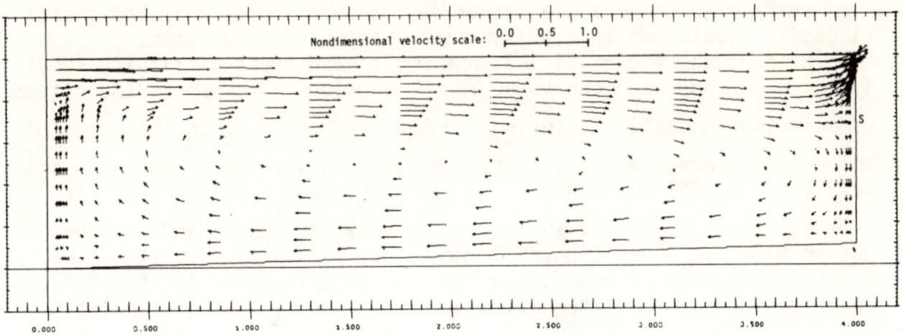

Fig. 4 Vector plot of velocity field, ν_t = 0.005.

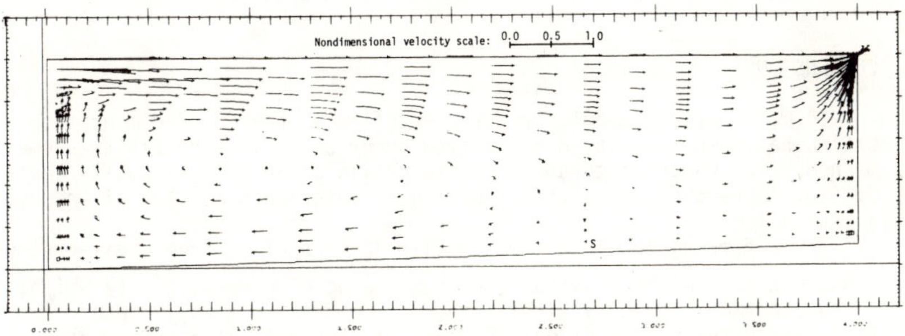

Fig. 5 Vector plot of velocity field, ν_t = 0.01.

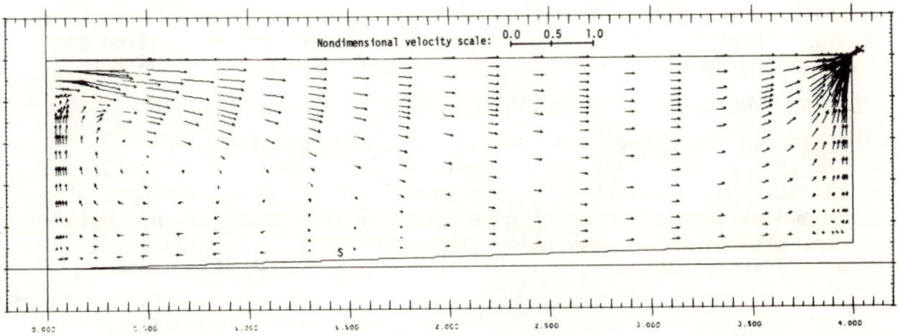

Fig. 6 Vector plot of velocity field, ν_t = 0.02.

free-surface velocities are larger (by as much as 10%) for the constant-viscosity simulation than the variable-viscosity simulation.

Without the aid of Fig. 3, one would have great difficulty specifying an appropriate mean value for ν_t a priori. To assess the effect of selecting modestly differing constant values of ν_t on the predicted velocity pattern, two additional computations were performed with one-half (ν_t = 0.005) and double (ν_t = 0.02) the mean obtained from Fig. 3. These velocity plots are shown respectively in Figs. 4 and 6. The stagnation point in these figures is marked as point S. Comparing these figures with Figs. 2 and 5, one finds a radical change in the predicted velocity fields. The stagnation point for the ν_t = 0.005 simulation (Fig. 4) is located along the right wall of the basin. The flow short-circuits from the inlet to the exit region. The stagnation point along the floor for the ν_t = 0.02 simulation (Fig. 6) is at $x \simeq 1.5$. Compared with Figs. 2 and 5, one sees that the recirculation zone below the expanding inlet jet is significantly reduced for this case.

CONCLUSIONS

A mathematical model for turbulent flow in primary sedimentation basins has been formulated and solved numerically via the finite element method. The uncertainty about the selection of an eddy viscosity coefficient is removed since this function is determined as part of the overall solution.

By referring to the predicted distribution of ν_t from the k-ε solution, one may select a suitable average value of eddy viscosity which, for a constant viscosity simulation, will predict the major mean flow features of the k-ε solution. However, a constant value of ν_t which differs <u>only</u> by a factor of two (practitioners can have difficulty selecting ν_t = constant within a factor of ten) from the proper average value is shown to yield flow patterns which differ significantly from the k-ε model flow simulations. Since the constant viscosity simulations are much less expensive to compute than the k-ε simulations, a small "library" of k-ε solutions for flow under various loading conditions would be very useful in allowing one to select a value for ν_t so that the constant -ν_t computations could be successfully used most of the time in the design and analysis of these settling units.

Finally, once the velocity field is determined it is a relatively straightforward task to solve the now-linear sediment transport equation and determine the settling efficiency of a given basin. Details and results of this computation are presented in Ref. (13) and in papers now in preparation.

ACKNOWLEDGEMENT

The work reported herein was supported by the National Science Foundation under Grant No. ENG 7618846.

REFERENCES

1. ASCE Manual No. 36, Wastewater Treatment Plant Design, 1977.

2. Batchelor, G.K., and Townsend, A.A., "Decay of Isotropic Turbulence in the Initial Period," Proc. Roy. Soc., A, Vol. 193, 1948, p. 539.

3. Hanjalic, K., and Launder, B.E., "A Reynolds Stress Model of Turbulence and its Application to Thin Shear Flows," Journal of Fluid Mechanics, Vol. 52, 1972, p. 609-638.

4. Ideriah, F.J.K., "On Turbulent Forced Convection in a Square Cavity," Numerical Methods in Laminar and Turbulent Flow, Taylor, C. et al., ed., Pentech Press, London, 1978, p. 257-269.

5. King, I.P., Norton, W.R., and Iceman, K.R., "A Finite Element Model for Two-Dimensional Flow," Finite Element Methods in Flow Problems, Oden, J.T. et al., ed., UAH press, Alabama, 1974, p. 133-137.

6. Launder, B.E., Reece, G.J., and Rodi, W., "Progress in the Development of a Reynolds-Stress Turbulence Closure," Journal of Fluid Mechanics, Vol. 68, 1975, p. 537-566.

7. Launder, B.E., and Spalding, D.B., Lectures in Mathematical Models of Turbulence, Academic Press, 1972.

8. Launder, B.E., and Spalding, D.B., "The Numerical Computation of Turbulent Flows," Computer Methods in Applied Mechanics and Engineering, Vol. 3, 1974, p. 269-289.

9. Liggett, J.A., "Cell Method for Computing Lake Circulation," Journal of the Hydraulics Division, ASCE, Vol. 96, HY3, 1970, p. 725-743.

10. McGuirk, J., and Rodi, W., "A Depth-Averaged Mathematical Model for the Near Field of Side Discharges into Open-Channel Flow," Journal of Fluid Mechanics, Vol. 86, 1978, p. 761-781.

11. Metcalf and Eddy, Inc., Wastewater Engineering: Collection Treatment, Disposal, McGraw-Hill, New York, 1972.

12. Reynolds, W.C., "Computation of Turbulent Flows," Annual Review of Fluid Mechanics, Vol. 8, 1976, p. 183-208.

13. Schamber, D.R., "Finite Element Analysis of Flow in Sedimentation Basins," thesis presented to the University of California at Davis, in 1979, in partial fulfillment of the requirements for the degree of Doctor of Philosophy.

14. Schamber, D.R., and Larock, B.E., "Numerical Analysis of Flow in Sedimentation Basins," submitted for publication to the Journal of the Hydraulics Division, ASCE, 1980.

BUOYANT JET BEHAVIOR NEAR THE MAXIMUM RISE

By Roger B. Wallace,[1] M. ASCE
and Steven J. Wright,[2] A.M. ASCE

Abstract

A two-dimensional buoyant jet discharged into a stratified fluid is considered. Simple equations are developed which relate the spreading height Z_s, minimum dilution S_m of the spreading layer, and spreading layer thickness h_s to a stratification parameter and the jet source flux of either momentum or buoyancy. Length scale analysis provides the form of the equations and conditions when each equation applies. Scale model tests verify the form of the equations and provide values of the coefficients.

Introduction

One of the more important considerations regarding waste discharges into stratified environments is the ability to predict the maximum height of rise and other features related to the stratification influence.

There has been much progress in the mathematical modeling of buoyant jet discharges including cases of stratified receiving fluids. The common approach has been through an integral analysis using similarity assumptions to describe jet velocity and concentration profiles. A major problem associated with this type of analysis near the maximum height of rise is that similarity of profiles is not a valid assumption.

Extending the analysis beyond the maximum height of rise is not possible and a different approach is required to determine additional characteristics such as the level at which the jet discharge begins to spread laterally and the corresponding thickness of the waste fluid.

There does not appear to be a straight forward approach to a formal analysis of the jet behavior between the maximum height of rise and the lateral spreading at the neutrally buoyant level. A reasonably straight forward approach is to relate the behavior in this region to information that can be readily computed by conventional methods. It is reasonable, on the basis of length scale arguments, to suggest that the distance to the neutrally buoyant level is a constant proportion of the maximum height of rise for

---

[1]Grad. Research Asst., Univ. of Mich., Ann Arbor, MI

[2]Asst. Prof. of Civil Engr., Univ. of Mich., Ann Arbor, MI

similar types of jet flows; a similar argument can be advanced for the thickness of the spreading fluid. Thus, by adequate prediction of the jet behavior at the maximum height of rise and experimental confirmation of the above assumptions, it is possible to determine the relevant conditions in the laterally spreading fluid.

Recently, experimental confirmation has been provided to demonstrate the ability to predict the maximum height of rise for buoyant jets in a stratified fluid (9). Additional experiments have been performed to extend those basic results to quantities that may be of more interest in engineering applications. These experimental results include measurements of 1) height of the horizontally spreading fluid Z_s, 2) the characteristic thickness of that fluid h_s, and 3) concentration profiles including minimum dilutions S_m; these quantities are defined in Fig. 1.

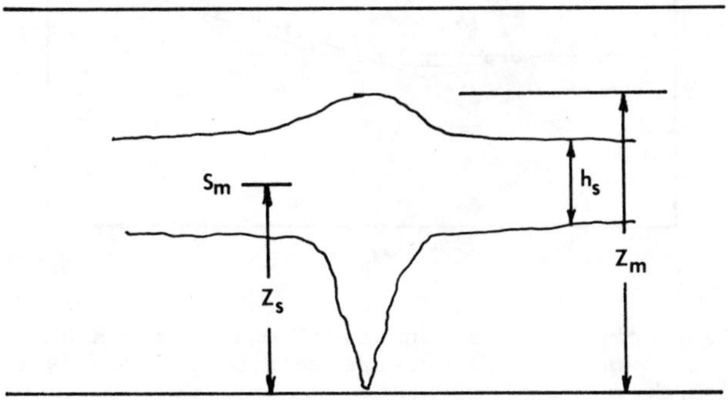

Fig. 1 - The spreading jet in a stratified ambient fluid

Theory

A two-dimensional buoyant jet discharge into a stratified ambient fluid can be characterized by any two of the following length scales if the source volume flux has no dynamic influence: l_b, $= B^{1/3}/\varepsilon^{1/2}$; l_m, $= M^{1/3}/\varepsilon^{1/3}$; $l_M = M/B^{2/3}$. Here M = the kinematic source momentum flux $(w_o Q)$; B = the source buoyancy flux $(g_o' \, Q)$, and $\varepsilon = (-g/\rho_o)(d\rho_a/dz)$, a stratification parameter characterizing the linear density gradient. w_o is the jet source velocity, $g_o' = g(\rho_o - \rho_j)/\rho_o$ with ρ_j = the jet density and ρ_o the ambient density at the source elevation, $d\rho_a/dz$ is the density gradient.

These length scales describe the functional form of the relationship for the maximum height of rise Z_m of a bouyant jet (9). They also provide the asymptotic solutions for the plume

$$Z/l_{b'} = 3.6 \tag{1}$$

and the momentum jet

$$Z_m/l_{m'} = 2.3 \tag{2}$$

which approximately describe Z_m for $l_M/l_{b'} < 3.8$ and $l_M/l_{b'} > 3.8$ respectively in Fig. 2.

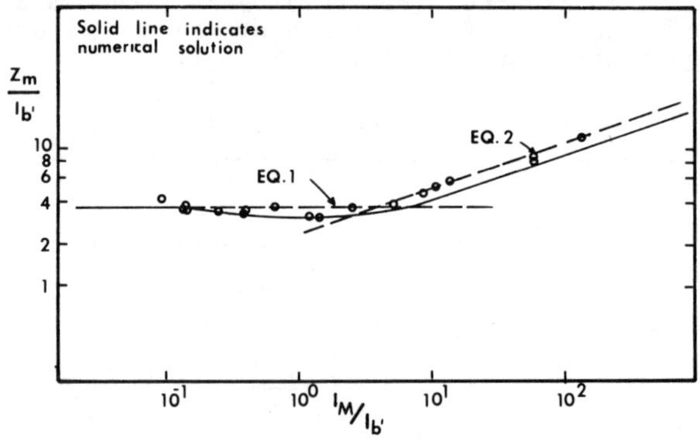

Fig. 2 - Comparison of numerical calculations and experimental measurements of Z_m from (9).

The height of the neutrally buoyant level also relates to these length scales because the significant mixing which determines the density of the spreading layer occurs before the jet reaches Z_m. The form of the solutions will be identical to those for Z_m. The thickness h_s of the spreading layer can be described similarly. The asymptotic solutions for the plume (source of buoyancy only) are

$$Z_s/l_{b'} = C_1 \tag{3}$$

$$h_s/l_{b'} = C_3 \tag{4}$$

and for the momentum jet (source of momentum only) they are

$$Z_s/l_{m'} = C_2 \tag{5}$$

$$h_s/l_{m'} = C_4 \tag{6}$$

A relationship for the dilution achieved by the time the jet spreads can be developed in a similar manner. Dilution depends upon the integrated effect of entrainment which is expressed by the volume flux at Z_m. For a plume

$$q = f(B, l_{b'}) \qquad (7)$$

since Z_m is proportional to $l_{b'}$. The appropriate relation for terminal dilution is then proportional to a ratio of the terminal volume flux and initial volume flux Q which has been assumed to have no role in the dynamics of the entrainment process

$$S_m = C_5 B^{1/3} l_{b'}/Q \qquad (8)$$

Length scale analysis suggests this relationship can be applied to a general buoyant jet if $l_M/\ell_{b'} << 1$ so that buoyancy dominates the jet behavior long before Z_m. If $l_M/l_{b'} >> 1$ the terminal volume flux can be expressed as a nondimensional function of M and $l_{m'}$

$$q = f(M, l_{m'}) \qquad (9)$$

so the minimum terminal dilution is given by

$$S_m = C_6 (M l_{m'})^{1/2}/Q \qquad (10)$$

Equations 8 and 10 are the asymptotic solutions for the buoyant jet relationship

$$\frac{S_m Q}{B^{1/3} l_{b'}} = f(l_M/l_{b'})$$

Experiments

A systematic experimental investigation was undertaken to verify the general functional form of the equations, determine the range over which the asymptotic solutions are satisfactory and to determine values for the constants in the asymptotic solutions. Experiments were performed in a tank 20 ft long, 2 ft wide, and 3 ft deep. The jets were discharged downward from the water surface from a 2-ft long source with variable slot width. The jet density was greater than ambient density. The situation is analogous to the conditions of the analysis (less dense jets discharged upwards) provided that density differences are small. This condition was always met since density differences were kept less than 2.5% of ρ_o. Density differences were created by dissolving table salt in tap water; they were measured with a conductivity probe and recorded on an oscillograph recorder. Linear density gradients were created by establishing a salt concentration gradient in the tank. Source Richardson numbers R_o were kept below 0.63 which is the limiting value for a pure plume (3).

Spreading height and minimum dilutions were determined by measuring tracer concentrations in the spreading layer. Samples were withdrawn at 5 to 8 elevations through vinyl tubes connected to a chamber where the pressure could be raised or lowered. The tubes were flushed and dried before sampling and ambient water was kept out by clamping each one closed until sampled. Sampling was to the side of the jet a distance approximately equal to Z_m. Samples were obtained after the tip of the spreading layer passed the sampler but before jet entrainment of ambient water significantly altered the stratified layer between the spreading layer and the source elevation. This allowed a short but adequate time for sampling.

Sodium fluorescein was used as a tracer in the jet fluid. Sample tracer concentration was determined with a fluorometer calibrated with dilutions prepared from jet solution and ambient water before each test. A summary of experimental results for Z_s and S_m are presented in Table 1.

TABLE 1

Summary of Experimental Results

Q $\frac{cm^2}{sec}$ (1)	b_o cm (2)	g_o' $\frac{cm}{sec^2}$ (3)	ε sec^{-2} (4)	S_m (5)	Z_s cm (6)	ℓ_M/ℓ_b' (7)	$\frac{S_m Q}{B^{1/3}\ell_b'}$ (8)	$\frac{Z_s}{\ell_b'}$ (9)
5.49	.142	.041	.0301	6.67		163	17.1	-
5.49	.142	.014	.0280	6.41		472	33.1	-
5.57	.867	3.99	.0622	5.32		.402	.96	-
5.46	.867	4.39	.0502	5.49		.322	.80	-
5.49	.867	4.54	.0613	5.5		.346	.89	-
5.51	.867	4.83	.0594	5.49		.321	.82	-
5.33	.867	3.72	.117	3.51	14.7	.57	.88	1.86
5.42	.867	6.91	.121	5.99	19.3	.315	1.01	2.01
4.59	.867	13.3	.135	7.87	-	.146	.86	-
5.46	.867	2.80	.117	3.07	13.7	.75	.91	1.85
5.46	.867	9.22	.114	6.37	21.8	.231	.86	2.00
5.52	.867	3.72	.0561	5.00	24.0	.405	.87	2.07
5.61	.867	1.75	.0705	3.08	15.5	.98	1.00	1.92
5.46	.142	.302	.0440	6.06	16.8	26.7	4.97	2.98
5.44	.142	2.28	.0591	7.41	-	4.08	1.83	-
5.46	.142	5.14	.0832	8.00	20.	2.16	1.37	1.90
5.49	.142	1.58	.123	5.29	12.8	8.58	2.41	2.18
9.11	.284	.694	.0394	5.26	-	9.17	2.79	-
9.54	.284	.591	.0472	4.18	17.0	12.3	2.74	2.08
9.79	.284	1.57	.218	3.61	11.0	10.3	2.68	2.07
9.70	.867	18.1	.156	6.94	28.0	.244	.85	1.97

S_m was determined from smoothed plots of the concentration data and Z_s was taken as the height above the source where S_m occurred. The concentration profiles provided acceptable determination of Z_s with overall accuracy of 5-10%. The measurements of S_m were more accurate. Concentration profiles defining the spreading layer thickness were obtained in some experiments. These data are presented graphically in Fig. 5 in the following section.

Analysis

The experimental results for Z_s are presented in Fig. 3. The available data verify the form of the buoyant jet equation as well as the asymptotic solutions. The experiments show Eq. 3 may be applied with $C_1 = 2.0$ if $l_M/l_{b'}$ is less than 8.5. Equation 5 may be applied with $C_2 = 1.0$ if $l_M/l_{b'}$ is greater than 8.5. The plume solutions appear correct for significantly larger values of l_M/l_b' (increased relative amounts of initial momentum) than was the case for Z_m. This results from the fact that the plume tends to spread at a height relatively closer to Z_m than is

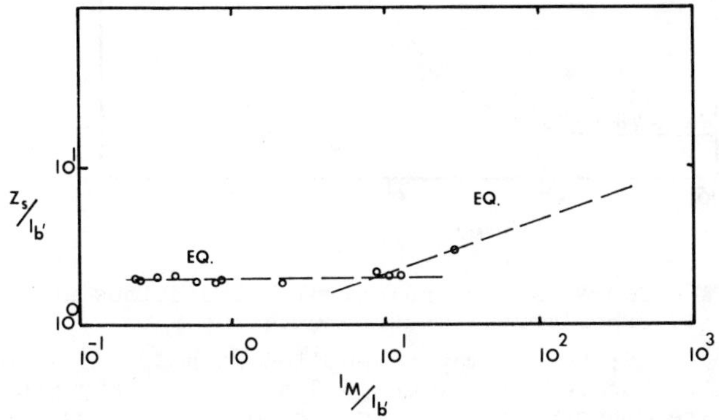

Fig. 3 - Height Z_s of the neutrally buoyant spreading layer.

the case with the momentum jet. This is reasonable because the plume must travel some distance away from the source before the stratification can reverse the buoyant forces to decelerate the initial motion; whereas with the jet this begins immediately. Thus for a plume and momentum jet with identical Z_m one expects the plume to spread closer to Z_m than the momentum jet. Although there are few data for large values of $l_M/l_{b'}$ the momentum jet asymptote appears to have been reached by $l_M/l_{b'} = 8.5$. This is more apparent if the data are plotted as Z_s/l_m' versus l_M/l_m' which allows direct determination of C_3. It is also consistent with where the asymptote is reached in the dilution data shown in Fig. 4.

The experimental results for S_m are presented in Fig. 4 along with results obtained by numerically integrating the integral equations to the maximum height of rise. The integral equations and the integration procedure were discussed in an earlier paper by Wright and Wallace (9). In the integration the initial volume flux was taken as $Q\sqrt{2}$ to account for entrainment in the zone of flow establishment as Fan and Brooks have suggested (1). Minimum dilution is $(\lambda^2/(1 + \lambda^2))^{1/2}$ of the average dilution due to the Gaussian profile shape. The experiments verify the form of the relationship for buoyant jets and the two asymptotic solutions. They show that Eq. 8 may be applied with C_5 = .88

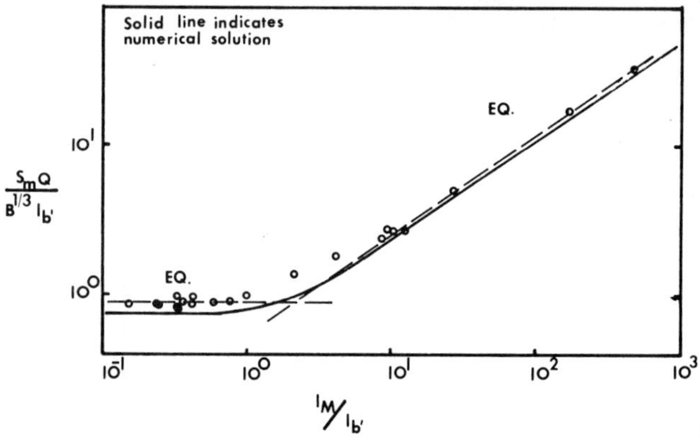

Fig. 4 - Comparison of numerical calculations and experimental measurements for S_m.

for $l_M/l_{b'} < 0.8$; Eq. 10 may be applied with C_6 = 5.7 for $l_M/l_{b'} > 10$. For $l_M/l_{b'}$ between 0.8 and 10 the asymptotic solutions are conservative predicting lower dilutions than measured. Actual dilutions may be as much as 41% greater than predicted by the asymptotic solutions; the maximum error occurs at $l_M/l_{b'}$ = 1.8. The large range of $l_M/l_{b'}$ where the asymptotic solutions are significantly in error was not seen in the data for Z_m or Z_s. It occurs in the numerical results as well as the experimental results for S_m. It can be explained by considering the dilution data in conjunction with the Z_m data. Addition of momentum to the plume produces increased entrainment and higher dilutions. This is true even in the range of $l_M/l_{b'}$ between 0.4 and 4 when Z_m changes very little. For larger values of $l_M/l_{b'}$ the total rise is also significantly increased resulting in even more dilution. The high rates of entrainment near the source, due to the source momentum, and larger Z_m both contribute to increased S_m; whereas with Z_m, high rates of entrainment near the source cause slightly smaller Z_m initially. These characteristics are seen in both the experimental and numerical results.

The numerical results for Z_m have been discussed (9). Numerical results for S_m show the same trend as the experimental values with only slightly lower S_m. This may be explained by the violation of assumptions used in the integral analysis near Z_m or the fact that experimental values for S_m were obtained after the jet had fallen back down to the equilibrium layer and started to spread.

The concentration profiles which can be used to characterize the spreading layer thickness h_s are plotted in Fig. 5. Concentration is given as a fraction of the maximum concentration which defined S_m; height Z is scaled by l_b' for experiments with $l_M/l_b' < 1$ and by l_m' for experiments with $l_M/l_b' > 8$. There are not many data but the thickness can be estimated if h_s is taken as the thickness measured between the elevations where the concentration is 10% of the maximum value. The thickness for plume-like jets is best defined because there are abrupt boundaries; in this case $C_2 = 1.8$. For the momentum dominated case $C_4 = 1.9$ and the spreading layer appears to occupy most of the depth

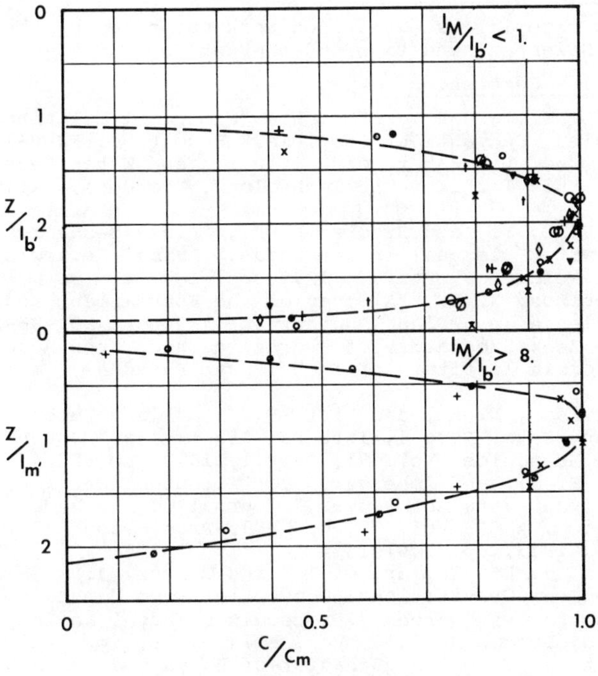

Fig. 5 - Dimensionless concentration profiles from measurements in the spreading layer.

between the spreading level and the source. This may be a result of measuring the profiles too close to the jet or because of mixing at the interface of the spreading layer and the ambient fluid; in which case, some of the fluid with tracer may be ambient water moving toward the jet. The momentum concentration profiles are clearly not symmetrical about the spreading height Z_S. Where as the plume profiles appear to be.

Conclusion

This study further demonstrates the value of length scales and the integral equations for determining approximate solutions to buoyant jet behavior. Deviations of the actual spreading height from the asymptotic predictions were less than 5%. Minimum dilution was accurately predicted by considering the influence of both momentum and buoyancy; the asymptotic solutions departed from the complete solutions by as much as 41% in some cases. Length scales were found useful in scaling the concentration profiles used to estimate spreading thickness.

Acknowledgments

This material is based upon work supported by the National Science Foundation under Grant No. ENG 78-05445.

Appendix 1 - References

1. Fan, L. N., and Brooks, N. H., "Numerical Solutions of Turbulent Buoyant Jet Problems," Report No. KH-R-18, W. M. Keck Laboratory of Hydraulics and Water Resources, California Institute of Technology, Pasadena, Calif., 1969.
2. Koh, R. C. Y., and Brooks, N. H., "Fluid Mechanics of Wastewater Disposal in the Ocean," Annual Review of Fluid Mechanics, Vol. 7, 1975, pp. 187-211.
3. Kotsovinos, N. E., "A Study of the Entrainment and Turbulence in a Plane Buoyant Jet," Report No. KH-R-32, W. M. Keck Laboratory of Hydraulics and Water Resources, California Institute of Technilogy, Pasadena, Calif., 1975.
4. Kotsovinos, N. E., and List, E. J., "Plane Turbulent Buoyant Jets, Part 1, Integral Properties," Journal of Fluid Mechanics, Vol. 81, Part 1, 1977, pp. 25-44.
5. List, E. J., and Imberger, J., "Turbulent Entrainment in Buoyant Jets and Plumes," Journal of the Hydraulics Division, ASCE, Vol. 99, No. HY9, Proc. Paper 9992, Sept., 1973, pp. 1461-1474.
6. Morton, B. R., Taylor, G. I., and Turner, J. S., "Turbulent Gravitational Convection from Maintained and Instantaneous Sources," Proceedings, Royal Society of London, London England, Vol. A234, 1956, pp. 1-23.
7. Wright, S. J., "Mean Behavior of Buoyant Jets in a Crossflow," Journal of the Hydraulics Division, ASCE, Vol. 103, No. HY5, Proc. Paper 12944, May, 1977, pp. 499-513.

8. Wright, S. J., "Effects of Ambient Crossflows and Density Stratification on the Characteristics Behavior of Round Turublent Buoyant Jets," Report No. KH-R-36, W. M. Keck Laboratory of Hydraulics and Water Resources, California Institute of Technology, Pasadena, Calif., 1977.
9. Wright, S. J., and Wallace, R. B., "Two-Dimensional Buoyant Jets in Stratified Fluid," Journal of the Hydraulics Division, ASCE, Vol. 105, No. HY11, Proc. Paper 15003, Nov., 1979, pp. 1393-1406.

Appendix II - Notation

The following symbols are used in this paper:

B = kinematic source buoyancy flux per unit length;
b_o = jet width at source;
C = constant in approximate solutions;
g' = buoyant force per unit mass $g' = g(\rho_a-\rho_j)/\rho_o$;
g_o' = source value of g';
h_s = thickness of neutrally buoyant spreading layer;
l_b' = length scale describing maximum rise for plume;
l_M = length scale describing transition from momentum to buoyant influence;
l_m' = length scale describing maximum rise for jet;
M = kinematic source momentum flux per unit length;
Q = source volume flux per unit length;
q = local volume flux per unit length;
R_o = Richardson number $R_o = BQ^3/M^3$;
S_m = minimum jet dilution;
w_o = source vertical velocity;
z_m = maximum height of rise;
z_s = height of neutrally buoyant spreading;
ρ_a = ambient density;
ρ_j = jet source density; and
ρ_o = ambient density at level of source.

FLOW COEFFICIENTS AND COMPUTER MODELS RELIABILITY

by
A'Alim A. Hannoura[1], A. M. ASCE

ABSTRACT:

Experimental studies were carried out to determine: (1) hydraulic conductivity of the common construction materials for rubble-mound breakwaters, i.e., crushed rock; (2) inertia coefficient; (3) effect of air entrainment on the hydraulic conductivity; and (4) instantaneous pressure distribution due to wave impact on a rubble-mound breakwater. Dimensionless relationships are presented to determine: hydraulic conductivity; inertia coefficient; unsaturated hydraulic conductivity; and pressure variation. A generalized Finite Difference-Finite Element Model was developed to utilize and verify the experimental results. Numerical and experimental results of wave motion in a two-layer breakwater model are in good agreement. The numerical results were compared to previous research findings to illustrate the effect of the flow coefficients.

INTRODUCTION:

Advances in digital computers have made it possible to use computer models in studying wave transmission in rubble-mound breakwaters. However, most of the available models, (1), are limited by the following approximations: (1) breakwater is homogeneous and of a rectangular cross-section; (2) inertia coefficient is negligible; and (3) air entrainment due to wave breaking and/or wave-structure interaction, i.e., internal wave breaking is not considered. Some of these assumptions are necessary because of the lack of quantitative expressions to estimate the different flow coefficients associated with wave motion in porous structures.

In this paper, the author presents the results of extensive experimental and numerical studies which were conducted in order to furnish some of the missing quantitative expressions. A brief description of the developed numerical model is also presented. The model is capable of simulating internal wave motion in trapizoidal layered breakwaters.

THEORETICAL CONSIDERATION:

---

[1]Assistant Professor, Civil Engineering Department, School of Engineering, University of New Orleans, Lakefront, New Orleans, LA 70122, USA.

i. The Flow Coefficients:

When a progressive wave collides against a rubble-mound breakwater, transmission and reflection of the wave take place on the front slope of the breakwater. The resultant internal flow is often non-Darcy and one or two phase. Two phase flow may occur due to external and/or internal wave breaking, (1). This can drastically reduce the effective conductivity of the structure, thus reducing the wave transmission and increasing reflection. At the same time the structure is subjected to unsteady wave forces. Yet, the literature on this nonlinear unsteady flow behavior in coarse granular media is very limited, (1). Moreover, it could be unsaturated, (Fig. 1).

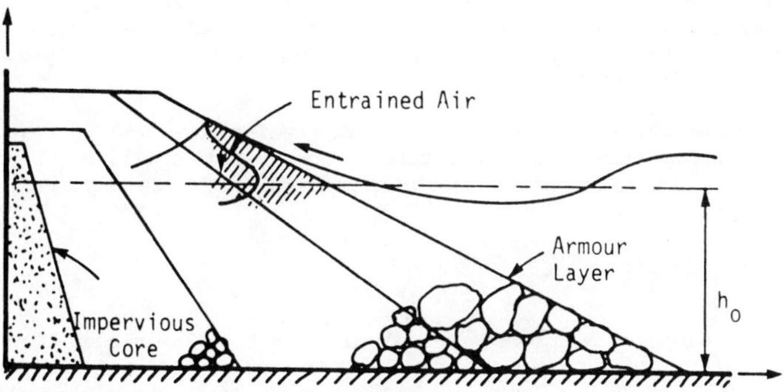

FIG. 1. Wave Interaction With a Porous Structure

The internal non-Darcy Flow can be described by a generalized resistance equation given by McCorquodale et al., (1) as:

$$Y = C_a + C_b \left(N_1 + N_2 \frac{f_\varepsilon}{f_o} X\right) \qquad (1)$$

in which Y = dimensionless resistance; C_a and C_b = constants for a particular flow regime; N_1 and N_2 = constants, f_ε = Darcy friction factor for rough surface, i.e., the construction material; f_o = Darcy friction factor for hydraulically smooth surface; and X = generalized pore Reynolds number. Equation 1 is a generalized form of Forchheimer equation, i.e.,

$$i = (a + b\,|q|)\,q \qquad (2)$$

in which i = hydraulic gradient; a and b = constants; and q = Darcy velocity.

Considering the unsteady phase of the problem, the net resistance, (applied force), on an unsteady flow past a fixed object obstruction is arbitrarily considered in three primary parts as follows, (1),

$$F_T = C_D \rho A \frac{V^2}{2} + \rho \forall_f \frac{dv}{dt} + C_M \rho \forall_s \frac{dv}{dt} \qquad (3)$$

in which C_D = drag coefficient; A = area of obstruction normal to the flow; P = fluid density; V = instantaneous representative velocity; $\forall_f$ = volume of fluid being accelerated; C_M = inertia coefficient; and $\forall_s$ = volume of the obstruction.

In a porous structure, e.g., rubble-mound breakwater, the unsteady flow problem is very complicated. Accelerating flow may affect the flow regime by causing a delay in the onset of turbulence, (1). Any change in flow regime will affect the flow resistance. The limitation on wake growth due to the limited pore volume downstream from each grain will influence the effective virtual mass volume of the particles. The possibility of soil matrix deformation (e.g., compaction) during the acceleration phase makes the exact formulation of the problem difficult, (1).

In view of these factors, the following semi-theoretical representation of one-dimensional-unsteady flow in porous media, (1), has been derived from the steady flow Forchheimer equation with the addition of the last two terms of Eq. 3, i.e.,

$$i = (a + b\,|q|)\,q + \frac{1 + C}{g}\frac{\partial q}{\partial t} \qquad (4)$$

in which i = slope of hydraulic grade line; (a + b $|q|$) = steady flow resistance at velocity q; q = Darcy velocity; a and b are empirical constants of the porous media; C = acceleration coefficient = {1 - m)/m} C_M; m = porosity; and g = acceleration due to gravity.

Although saturated flow in a porous structure can be classified as Darcy or non-Darcy, (1), which includes nonlinear laminar, turbulent transitional and fully turbulent; a similar classification does not exist for two-phase flow in porous media. Observations of air-water flow in crushed rock wave absorbers (1) indicate that the flow through the pores is either bubbly or slug flow. The relative hydraulic conductivity in the unsaturated flow region is defined by the authors, (1), as:

$$\frac{K}{K_o} = \frac{q_w\,\Delta Z\,(a + b\,|q_w|)_o}{\Delta H_f} \qquad (5)$$

in which K = two=phase conductivity; K_o = single phase, i.e., saturated, conductivity; q_w = Darcy velocity, a and b = constants; ΔH_f = two phase friction head loss through a distance of ΔZ; and (a + b $|q_w|)_o$ = single phase (saturated) resistance.

ii. <u>The Numerical Model Formulation</u>:

The problem considered in this paper is the analysis of the internal flow due to wave action on a porous structure, such as the one shown schematically in Fig. 1.

The hybrid model treats the time integration by the Method of Characteristics (MOC) and the space integration by the Finite Element Method (FEM). Thus, the major inertial effects are to be included in the MOC and the effect of nonhomogeneity, velocity and pressure variation in space are computed by the FEM and used to update the MOC.

The Characteristics Method: If the inertia and pressure coefficents are incorporated in the momentum equation, as developed by McCorquodale and Nasser (1), the following set of equations is obtained:

$$\begin{bmatrix} c_m & u/m & 0 & mgC_p \\ 0 & (h_o+\eta)/m & 1 & U/m \\ dt & dx & 0 & 0 \\ 0 & 0 & dt & dx \end{bmatrix} \begin{bmatrix} \frac{\partial u}{\partial t} \\ \frac{\partial u}{\partial x} \\ \frac{\partial \eta}{\partial t} \\ \frac{\partial \eta}{\partial x} \end{bmatrix} = \begin{bmatrix} -gFUm \\ 0 \\ du \\ dt \end{bmatrix} \tag{6}$$

in which c_m = inertia coefficient; u = horizontal intrinsic (Darcy) velocity; m = porosity; C_p = pressure distribution correction factor; t = time; g = acceleration due to gravity; F = a + b|q| is the non-Darcy friction term; q is the total intrinsic velocity; h_o = mean water depth; η = internal wave amplitude, and x is the horizontal axis.

An explicit finite difference scheme is used to discretize the equations of motion and continity in terms of u and the wave celerity, $c = \sqrt{gy}$.

The standard procedure of the characteristic method was used to develop the characteristic directions, which controls the discretization of the x-t plane (see Fig. 2).

The Finite Element Method: The momentary two dimensional internal flow is described by

$$\frac{\partial}{\partial x}\left(K(|\nabla\phi|)\frac{\partial\phi}{\partial x}\right) + \frac{\partial}{\partial y}\left(K(|\nabla\phi|)\frac{\partial\phi}{\partial y}\right) = 0 \tag{7}$$

in which $K(|\nabla\phi|) = 1/F$ = nonlinear or non-Darcy hydraulic conductivity, and ϕ = piezometric head. The corresponding variational form Eq. 7 given by McCorquodale (1) as

$$\chi = \int_A\int \{K(|\nabla\phi|)(\phi_x^2 + \phi_y^2) + G(|\nabla\phi|)\}dydx \tag{8}$$

where $G(|\nabla\phi|)$ is a continuity function. The solution of Eq. 7 is now reduced to finding the distribution of the piezometric function, ϕ, which minimizes χ, subjected to certain boundary conditions. The FEM form of Eq. 8 is

$$\sum_e \sum_j K(|\nabla\phi|)^e S^e_{ij} \phi_j = 0 \qquad (9)$$

in which $S^e_{ij} = (b_i b_j + C_i C_j)/A^e$ = the element stiffness matrix; $b_i = y_j - y_k$; $C_i = x_k - x_j$; and i, j and K are the element node numbers. The FEM nodal ordinates are ajusted to conform to the phreatic line computed by the MOC (see Fig. 2).

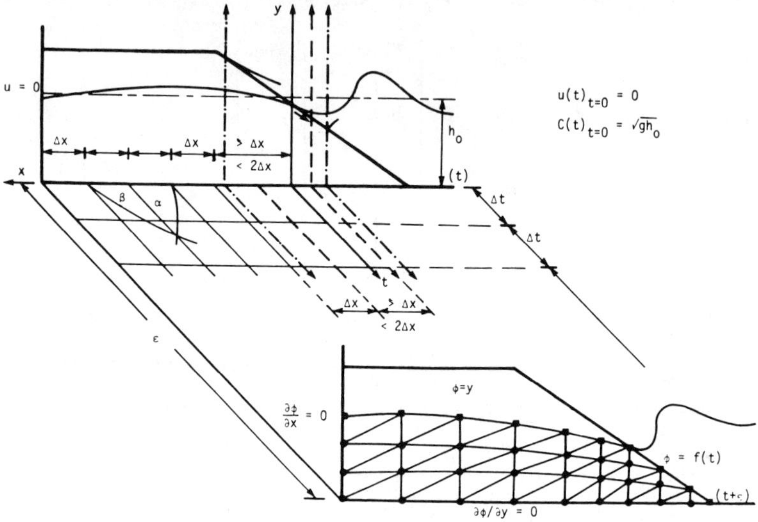

FIG. 2. Discretization of x-t and x-y Planes Showing the Interlacing of FD/MOC and FE Grids

EXPERIMENTAL PROGRAM:

i. Unsteady Flow; Estimation of Virtual Mass:

One basis for the estimation of the drag and the virtual mass coefficients is a series of steady and unsteady flow experiments. In the first set of experiments the steady flow resistance, i.e., C_D in Eq. 3 or a and b in Eq. 4, can be determined (Fig. 3). In the second set of experiments the variations with time of the velocity and pressure difference, across the porous media, can be measured and recorded so that the unsteady flow resistance can be calculated making use of the predetermined values of a, b, and C_D (Fig. 4).

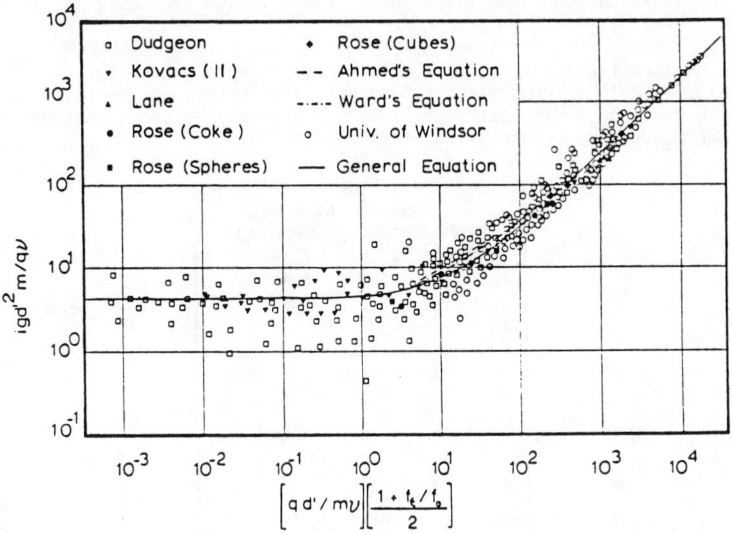

FIG. 3. Hydraulic Conductivity of Rockfill

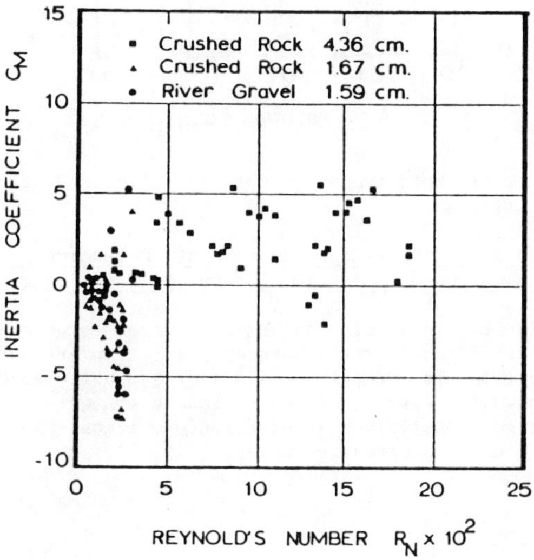

FIG. 4. Experimental Inertia Coefficient of Rockfill

ii. Unsaturated Flow; Determination of the Relative Hydraulic Conductivity:

The flow tests were made in a closed-loop and a coarse-bubble air diffuser in a 10.16 cm (4 in.) I.D. plexiglass U-tube, a removable test section and recirculating system. Both co-current and counter-current flows were investigated, (Fig. 5).

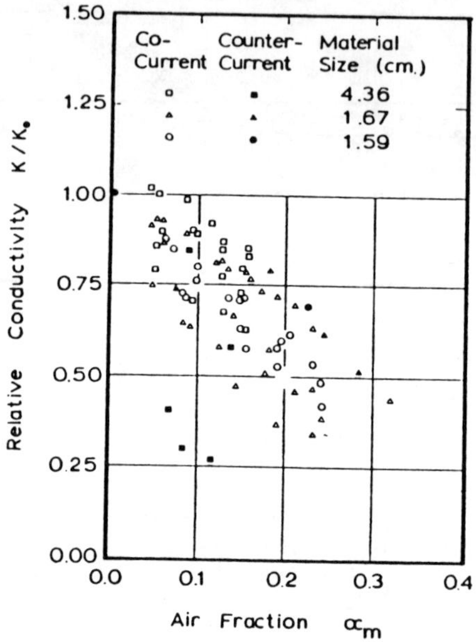

FIG. 5. Effect of Air Entrainment on the Hydraulic Conductivity of the Rockfill

iii. Wave Motion on a Rubble-Mound Embankment; Determination of Instantaneous Pressure Distribution Due to Wave Impact:

This experimental investigation was undertaken to obtain representative boundary conditions for the outcrop point movement and the pressure distribution along the front slope of a rubble-mound breakwater for shallow water waves. Pressure transducers were used to measure the instantaneous values at different locations along the interface for different progressive waves, (Fig. 6).

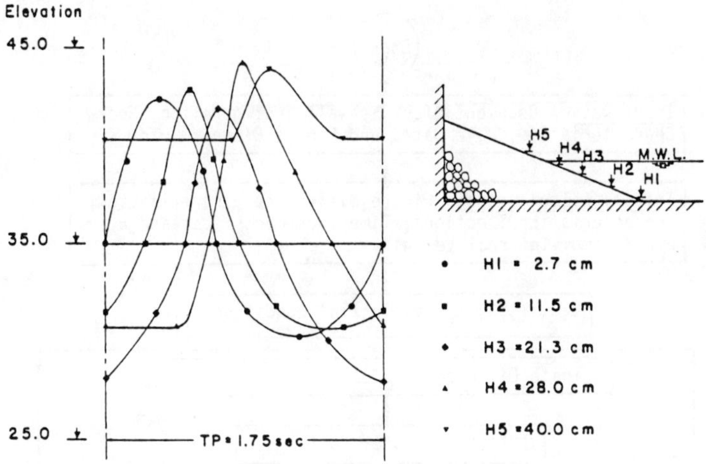

FIG. 6. Typical Piezometric-Head Distribution Test Results on a Rubble-Mound Slope

Application of multilinear regression analysis to the experimental results of the entrained air under wave action study, (1), yields the following empirical formula:

$$\frac{A_{aw}}{A_o} = 7.46\left(\frac{V_1}{V_{max}}\right)^{0.5} (1 - \sin \Theta)^{1.15} \tag{10}$$

in which A_{aw} = area of the unsaturated flow region, (mainly between the front slope and mean water level); $A_o = A_1^2/2 \tan \Theta$ = reference area; A_1 = run up height; V_1 = run up velocity; V_{max} = maximum seepage velocity through the rubble-mound; and Θ = embankment front slope.

Four different sizes, (based on geometric mean), of granular material, i.e., 1.59-cm; 7.92-cm, (river gravel); 1.67-cm; and 14.36-cm, (crushed rock) were utilized in the experimental investigations. Porosity varied from 0.34 to 0.51 and shape factor was in the range of 1.31 to 1.92.

THE NUMERICAL SOLUTION:

Typical initial and boundary conditions are shown in Fig. 2. The numerical solution is illustrated by the following simplified flow chart (Fig. 7).

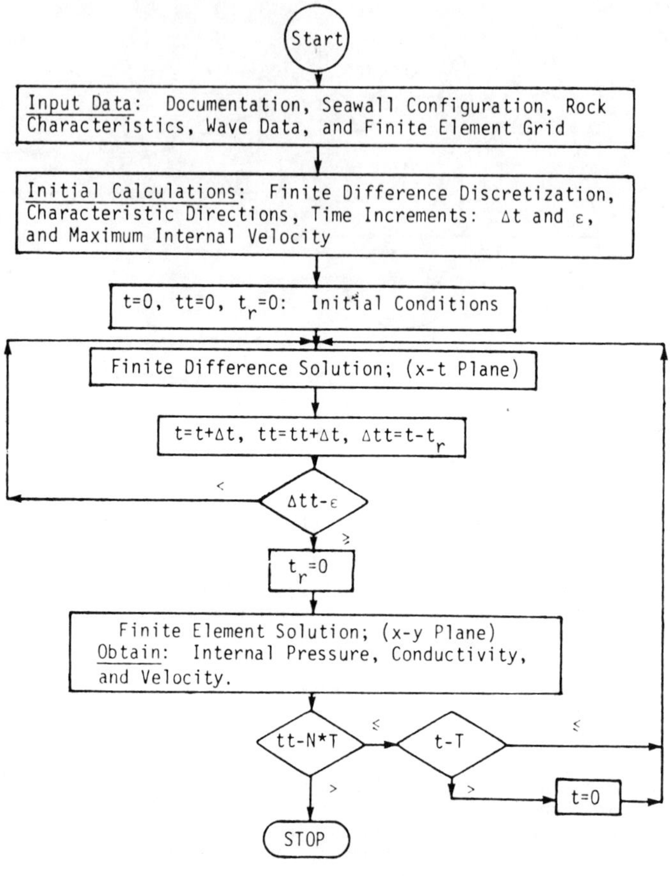

FIG. 7. Simplified Flow Chart of the FD/ MOC-FE Model.

VERIFICATION AND DISCUSSION:

Experimental and numerical results were in a good agreement (1). The following figures were chosen to show the effect of some of the flow coefficients on the numerical results.

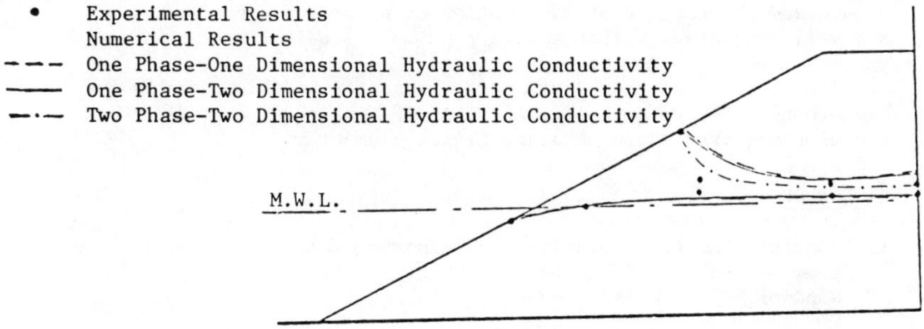

FIG. 8. Effect of Hydraulic Conductivity on the Numerical Results

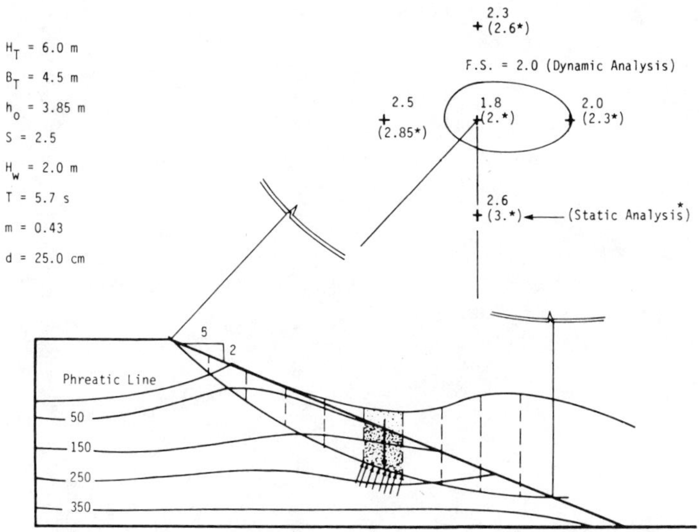

FIG. 9. Effect of Pressure Distribution on the Stability Analysis

CONCLUSIONS AND RECOMMENDATIONS:

The utilization of the experimental results in the numerical model have improved the prediction accuracy of the model. However, it is recommended to investigate the entire experimental side of this study in order to obtain statistically sound equations to estimate inertia and air entrainment effects on the flow resistance. The nature and mathematics of wave motion of a sloping embankment are not well understood either and warrant more studies.

The concept of the hybrid model presented in this paper has been found to be very efficient. The CUP time requires only 40% of the time required for a complete finite element model.

REFERENCES:

1. Hannoura, A. A., "Numerical and Experimental Modeling of Unsteady Flow in Rockfill Embankments", Ph.D. Thesis, University of Windsor, Windsor, Ont., Canada, 1978.

ON THE EFFECT OF THE VALUE OF THE EDDY COEFFICIENTS ON THE SIMULATED VERTICAL STRUCTURE OF STRATIFIED FLOWS

Angelos N. Findikakis* and Robert L. Street
Department of Civil Engineering
Stanford University, California 94305

ABSTRACT

An approximate solution of the linearized equations for thermally stratified flows is used to demonstrate the effect of the value of the eddy coefficients on the predicted vertical structure of these flows. It is shown that the number of the simulated circulation cells in a shear driven, stratified cavity flow depends on the value of the eddy coefficients, the intensity of the driving force, and the degree of stratification. The validity of some of the assumptions of the present analysis is tested using solutions of the full two-dimensional flow equations.

1. INTRODUCTION

A critical factor for the success of the numerical simulation of many flows in large water bodies is proper description of turbulent transport and mixing. Most of the numerical models for such flows solve the continuity, Navier-Stokes, and heat-balance equations and describe turbulent transport by employing the standard assumption that turbulent fluxes are linearly proportional to the local gradient of the transported quantity. The constants of proportionality are eddy coefficients which are either fixed over the entire solution domain or are allowed to vary to account for variations of the intensity of turbulence throughout the simulated flow field. In many cases the value of the eddy coefficients has a significant effect on the form of the solution, determining the shape of the boundary layers and in certain stratified flows even the overall circulation pattern. Realizing and assessing the effect of the value of the eddy coefficients on the simulated flow pattern must be one of the first steps in any attempt to establish the credibility of a numerical hydrodynamic simulation of stratified flow. This paper examines this effect for a class of flows which is closely related to the general circulation in large water bodies.

The only direct way to assess the effect of the value of the eddy coefficients on the flow pattern is to solve the hydrodynamic and heat equations, which normally must be done numerically. A multidimensional simulation, however, is always an expensive exercise. This motivated us to develop approximate solutions for the class of flows under consideration, by making certain simplifying assumptions. The central assumption is that at low Reynolds numbers the nonlinear terms in the hydrodynamic equations are small and can be neglected. The extent of the validity of

---

*Now with Hydraulics/Hydrology Group, Bechtel Inc., San Francisco.

the simplifying assumptions and some of their consequences are discussed in Section 5. We developed a numerical solution of approximate and linearized equations. To obtain further insight to the role of the eddy coefficients we also derived an analytic solution of the same equations and studied some characteristics of its behavior as the eddy coefficients vary. This analytic solution is presented in another paper by the authors (Findikakis and Street, 1980).

2. PROBLEM DEFINITION

The problem to which we apply our analysis is the response of a stratified water body of constant depth H to a shear stress applied at the surface and equal to $(\tau_A + i\,\tau_B)e^{ik_x x}$ where τ_A and τ_B are constants, $i = (-1)^{1/2}$, k_x is a wave number and 2L is the wave length of the driving force function. This representation of the driving force is made in accordance with the representation of the flow variables in terms of complex numbers, a concept which is introduced in the next section to facilitate the derivation of a solution to the flow equations. We note that the real part of this driving shear force is $\tau_R = \tau_A \cos(k_x x) - \tau_B \sin(k_x x)$ which can be used for physical interpretations. Since the problem is linear a more general surface boundary condition (BC) can be satisfied by superposition. We assume that the total velocity at the bottom is zero. Since we are primarily interested in the internal modes of the flow, we ignore the surface mode and we assume that the vertical component of the velocity at the surface is zero (a rigid lid approximation).

Under the physical situation described by the above conditions at the upper and the lower boundary one observes the development of internal waves with wave length equal to 2L in a stratified lake or the ocean. These BC combined with proper conditions in the horizontal direction also can describe the flow in a rectangular basin of depth H and length L. In this case the assumed periodic form of solution (see Eqs. 6-9 below) can satisfy the condition of zero normal velocity and zero heat flux through vertical walls. The tangential velocity at the vertical walls is, however, unrestrained. The physical problem is illustrated in the definition sketch Figure 1; the dependent variables shown there are defined in the next section. Note that the origin of the horizontal x-axis is at the center of the body, while the origin of the upward-directed vertical z-axis is at the bottom. Thus, the surface is at z = H.

We shall assume that the horizontal mean of the temperature in the water is constant and varies linearly with depth. The present numerical solution, however, can handle any form of stratification. There is also no restriction on the form of the time dependance of the driving shear force. Here, however, we shall examine only two cases, viz., the transient response of a still, stratified water body to a suddenly imposed shear which then remains constant and the response to a force which is periodic in time. These flows are classified as follows: Using the maximum shear velocity at the surface $(\tau_o/\rho)^{1/2}$, the depth H, the magnitude ε of the eddy viscosity, and the temperature difference between top and bottom $\Delta\Theta$, we can define a Reynolds number and an overall Richardson number for the flow as

$$R = \frac{H(\tau_o/\rho)^{1/2}}{\varepsilon} \quad \text{and} \quad Ri = \beta g\,\Delta\Theta\,\frac{H}{(\tau_o/\rho)}$$

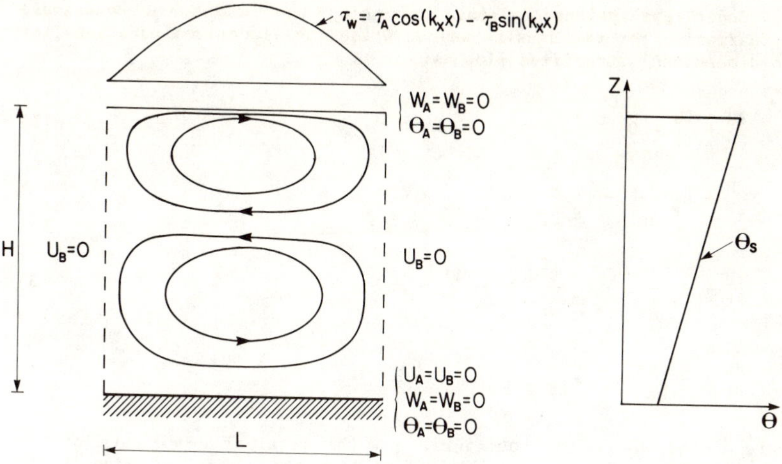

Figure 1. Schematic definition of the problem solved and boundary conditions.

where β is the coefficient of thermal expansion, g is the acceleration due to gravity, and τ_o is the maximum value of the real part of the surface shear stress.

3. BASIC ASSUMPTIONS

The basic assumptions for the development of the present solution are:

a) All nonlinear terms are small and they can be neglected. This assumption is valid only at low Reynolds numbers. The assumption of linearity has been used widely for the description of internal gravity waves in the ocean (Philips, 1977). A discussion of the applicability and some of the limitations of the linear theory in atmospheric and oceanic problems can be found in Lindzen et al. (1973). As an example, Krauss (1972) used the linearized hydrodynamic equations to study the response of a stratified ocean to wind and air pressure.

b) Turbulent momentum and heat transport can be described using constant eddy coefficients as proportionality constants in the expressions for the Reynolds stresses and the turbulent heat fluxes in terms of the local rate of (shear or normal) strain and temperature gradient, respectively.

c) The initial vertical thermal stratification is uniform in the horizontal direction and linear.

d) The solution of the linear equations for a stratified flow driven by a periodic (in the horizontal direction) force is also periodic. This assumption is connected to assumption (a) and as long as the equations are linear, this assumption is valid.

Under assumptions (a)-(c) and using the well-known Boussinesq approximation for the density we can write the linearized equations for two-dimensional stratified flow as:

$$\frac{\partial U}{\partial x} + \frac{\partial W}{\partial z} = 0 \tag{1}$$

$$\frac{\partial U}{\partial t} - \varepsilon_x \frac{\partial^2 U}{\partial x^2} - \varepsilon_z \frac{\partial^2 U}{\partial z^2} + \frac{1}{\rho_o} \frac{\partial P}{\partial x} = 0 \tag{2}$$

$$\frac{\partial W}{\partial t} - \varepsilon_x \frac{\partial^2 W}{\partial x^2} - \varepsilon_z \frac{\partial^2 W}{\partial z^2} + \frac{1}{\rho_o} \frac{\partial P}{\partial z} - \beta g \Theta = 0 \tag{3}$$

$$\frac{\partial \Theta}{\partial t} + \frac{N^2(z)}{\beta g} W - \varepsilon_{x\theta} \frac{\partial^2 \Theta}{\partial x^2} - \varepsilon_{z\theta} \frac{\partial^2 \Theta}{\partial z^2} = 0 \tag{4}$$

where U and W are the horizontal and the vertical components of the velocity, P and Θ represent deviations from the horizontal mean of the pressure P_s and temperature Θ_s; $N(z) = [\beta g(d\Theta_s/dz)]^{1/2}$ is the Brunt-Väisälä frequency corresponding to the stratification of the undisturbed temperature field; and ε_x, ε_z, $\varepsilon_{x\theta}$ and $\varepsilon_{z\theta}$ are eddy coefficients. In the above equations the kinematic viscosity and the molecular diffusivity have been absorbed in the eddy viscosity and the eddy diffusivity, respectively. The variables $P_s(z)$, $\Theta_s(z)$ must satisfy the equation of hydrostatics:

$$-\frac{dP_s}{dz} - g\rho_o + \beta g\, \rho_o(\Theta_s - \Theta_o) = 0 \tag{5}$$

where Θ_o is a reference temperature corresponding to the reference density ρ_o.

Based on assumption (d), we can express each of the flow variables as a complex number, because $\tau = (\tau_A + i\, \tau_B)e^{ik_x x}$:

$$U(x,z,t) = [U_A(z,t) + i\, U_B(z,t)]e^{ik_x x} \tag{6}$$

$$W(x,z,t) = [W_A(z,t) + i\, W_B(z,t)]e^{ik_x x} \tag{7}$$

$$P(x,z,t) = P_s(z) + [P_A(z,t) + i\, P_B(z,t)]e^{ik_x x} \tag{8}$$

$$\Theta(x,z,t) = \Theta_s(z) + [\Theta_A(z,t) + i\, \Theta_B(z,t)]e^{ik_x x} \tag{9}$$

where U_A, U_B, W_A, W_B, P_A, P_B, Θ_A, and Θ_B are real functions of z and t only. We always use the real portion of these variables for physical interpretations.

4. NUMERICAL SOLUTION

By substituting expressions (6)-(9) for the flow variables into Eqs. (1)-(4) and by setting the real and the imaginary part of each equation equal to zero, we obtain the following system:

$$\frac{\partial U_A}{\partial t} + k_x^2 \varepsilon_x U_A - \varepsilon_z \frac{\partial^2 U_A}{\partial z^2} - \frac{1}{\rho_o} k_x P_B = 0 \tag{10}$$

$$\frac{\partial W_A}{\partial t} + k_x^2 \varepsilon_x W_A - \varepsilon_z \frac{\partial^2 W_A}{\partial z^2} + \frac{1}{\rho_o} \frac{\partial P_A}{\partial z} - \beta g \Theta_A = 0 \tag{11}$$

$$-k_x U_B + \frac{\partial W_A}{\partial z} = 0 \tag{12}$$

$$\frac{\partial \Theta_A}{\partial t} + \frac{N^2}{\beta g} W_A + k_x^2 \varepsilon_{x\theta} \Theta_A - \varepsilon_{z\theta} \frac{\partial^2 \Theta_A}{\partial z^2} = 0 \tag{13}$$

$$\frac{\partial U_B}{\partial t} + k_x^2 \varepsilon_x U_B - \varepsilon_z \frac{\partial^2 U_B}{\partial z^2} + \frac{1}{\rho_o} k_x P_A = 0 \tag{14}$$

$$\frac{\partial W_B}{\partial t} + k_x^2 \varepsilon_x W_B - \varepsilon_z \frac{\partial^2 W_B}{\partial z^2} + \frac{1}{\rho_o} \frac{\partial P_B}{\partial z} - \beta g \Theta_B = 0 \tag{15}$$

$$k_x U_A + \frac{\partial W_B}{\partial z} = 0 \tag{16}$$

$$\frac{\partial \Theta_B}{\partial t} + \frac{N^2}{\beta g} W_B + k_x^2 \varepsilon_{x\theta} \Theta_B - \varepsilon_{z\theta} \frac{\partial^2 \Theta_B}{\partial z^2} = 0 \tag{17}$$

Equations (10)-(11) can be grouped into two systems with four equations each, i.e., Eqs. (10), (15), (16), and (17) form a system with U_A, W_B, P_B, and Θ_B as unknowns, and Eqs. (11), (12), (13), (14), form another system with U_B, W_A, P_A, and Θ_A as unknowns. Since one of the objectives of developing the present numerical model is to use it as an auxiliary tool in conjunction with 2-D finite element (FE) models, we set up the numerical solutions as the 1-D analog of a 2-D FE model which we have developed for simulating turbulent stratified flows (Findikakis, 1980). Thus, we use 3-node quadratic elements (12 in the vertical direction) for the velocity components and the temperature and linear elements for the pressure. The nodes of the linear elements coincide with the end (corner) nodes of the quadratic elements. The time derivatives are approximated by finite differences and all other terms are evaluated at time $t^n + \alpha\Delta t$, where t^n is the current time, Δt is the time step, and α is a parameter which characterizes the time-stepping scheme. A value of $\alpha = 1/2$ gives centered differences in time, while $\alpha = 1$ gives a backward-difference scheme, which is used herein.

The boundary conditions for the problem described in Section 2 can be expressed as:

z = 0 (Bottom)

$$U_A = U_B = 0 \; ; \qquad W_A = W_B = 0 \; ; \qquad \Theta_A = \Theta_B = 0 \qquad (18a,b,c)$$

z = H (Surface)

$$\varepsilon_z \frac{\partial U_A}{\partial z} = \tau_A \; ; \quad \varepsilon_z \frac{\partial U_B}{\partial z} = \tau_B \; ; \quad W_A = W_B = 0; \quad \Theta_A = \Theta_B = 0 \quad (18d,e,f,g)$$

and

$$x = \pm L/2 \text{ (Sides):} \quad U_A \cos\left(\frac{k_x L}{2}\right) \mp U_B \sin\left(\frac{k_x L}{2}\right) = 0 \quad \rightarrow \quad U_B = 0 \qquad (19a)$$

$$\varepsilon_{x\Theta}\left[\Theta_B \cos\left(\frac{k_x L}{2}\right) \pm \Theta_A \sin\left(\frac{k_x L}{2}\right)\right] = 0 \quad \rightarrow \quad \Theta_A = 0 \quad (19b)$$

where L is the distance between the two vertical boundaries (see Fig. 1). Here we shall solve for flows driven by shear stresses, of the form $\tau_o \cos(k_x x)$ where $k_x = m\pi/L$, $m \geqslant 1$, i.e., $\tau_A = \tau_o$, $\tau_B = 0$. In this case we find that $U_B = W_A = 0$, $P_A = 0$, and $\Theta_A = 0$. Thus BC (19a) and (19b) are satisfied automatically.

5. TEST OF THE VALIDITY OF THE ASSUMPTIONS OF THE LINEAR SOLUTION

We test the validity of the two basic assumptions made in Section 3 by comparing the results of the present model with results from the two-dimensional (2-D) numerical FE solution of the full equations for stratified flow. We consider the case that N = 0.01 rad/sec, L = H = 1000 cm and $k_x = \pi/L$ so that the surface shear stress which drives the flow is given by $\tau_o \cos \pi x/L$ with $\tau_o = 1$ dyne/cm^2. The surface stress remains constant in time. As a consequence we find $U_B = W_A = P_A = \Theta_A = 0$ everywhere. We solve Eqs. (10)-(17) to obtain the steady-state-flow solution. This solution is obtained directly by setting all the time derivatives in the system equal to zero. We assume that all eddy coefficients are equal to ε.

Figure 2 shows the steady-state solution obtained with the present numerical model, a numerical finite-element solution of the linearized flow equations with the 2-D model, and a numerical finite-element solution of the full equations with a 2-D model for ε = 100 and ε = 40 cm^2/sec, i.e., R = 10 and 25, respectively. Differences between the solution obtained with the present numerical model and the numerical solution of the linearized flow equations with the 2-D model are small in both cases. This justifies the assumption that at low Reynolds numbers the solution of the linear equations for the flow field which develops in response to a periodic driving force in the horizontal direction has the same horizontal structure as the driving force.

The effect of including the nonlinear terms in the solution depends on the Reynolds number of the flow. Figures 2e and 2f show the streamline pattern of the two flows discussed above as obtained from the FE solution of the full (i.e., including the nonlinear terms) 2-D

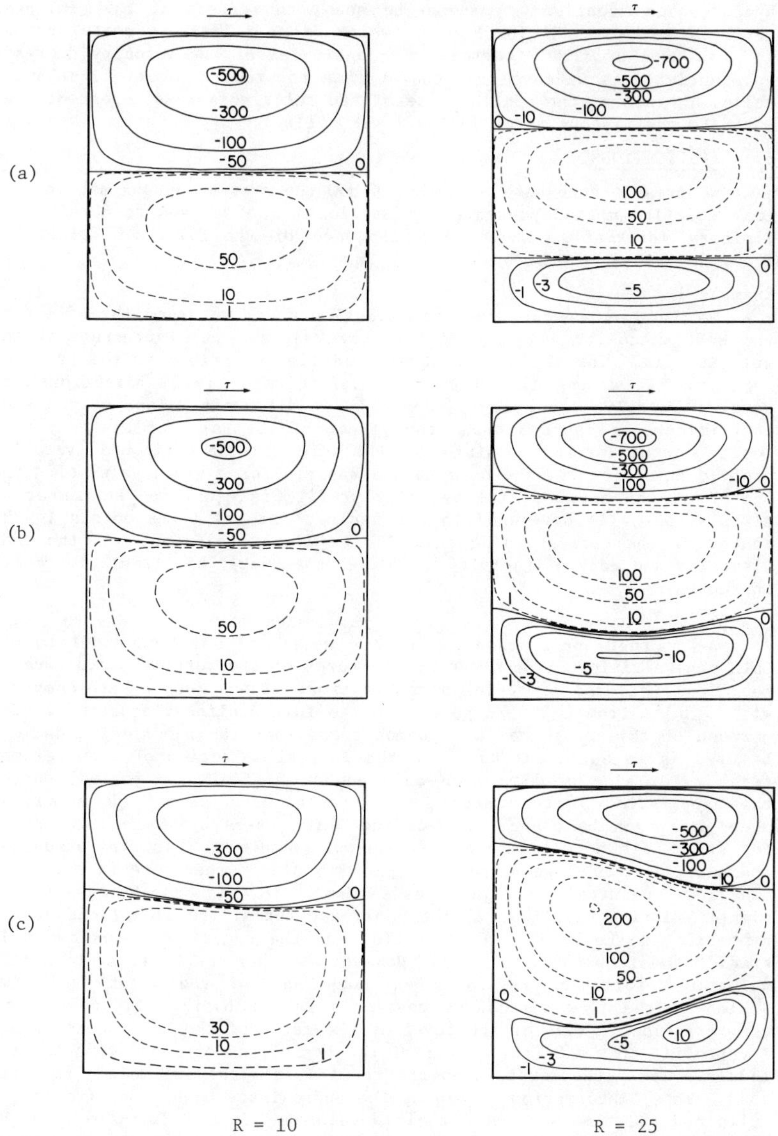

Figure 2. Circulation in a square cavity for R = 10 and R = 25 as simulated from a) the approximate linear model, b) 2-D solution of the linear equations and c) 2-D solution of the full equations.

hydrodynamic equations. As can be seen, the effect of the nonlinear terms is rather small for R = 10, while for R = 25 it is quite strong. In this type of flow, however, the magnitudes of the velocity decrease substantially as we move from the surface towards the lower circulation cells. Therefore, the nonlinear terms are relatively less important compared to the viscous terms in the lower cells.

6. APPLICATIONS

We present here some examples of how the present approximate numerical solution method presented in Section 4 can be used to obtain preliminary information about the structure of the class of stratified flows described in Section 2.

Example 1

We use the present numerical solution model to study the effect of the eddy viscosity and eddy diffusivity values on the structure of the velocity field for the stably stratified flow described in the previous section. We assume that the eddy coefficients in all directions are equal. We solve the system (10)-(17) for different values of the eddy coefficients. Figure 3 shows the effect of the value of the eddy viscosity on the vertical position of the zeros in the horizontal velocity profile at the centerline of a square basin. For a given value of ε the number of the zeros of the velocity profile is equal to the number of circulation cells developed in the basin. Each of these points is the center of the corresponding cell in the linear solution. As the eddy viscosity and eddy diffusivity decrease, the number of circulation cells increases.

Example 2

As another application we use the numerical solution model to obtain general information about key features of the vertical structure of the flow field and its development in time. These results are compared with results from a 2-D FE model for the full nonlinear equations. The purpose of this exercise is to demonstrate that the numerical model can be used as an auxiliary tool for the initial selection of some parameters (e.g., size of time step and length of simulation period) during the preparation of the input data for nonlinear runs. The numerical model can even be useful in deciding which cases are more interesting for study with a nonlinear model. As an example of this type of use of the numerical model we select a higher Reynolds number case (R = 100) of the flow presented in the previous section. In this case because of the higher relative magnitude of the nonlinear terms, the flow field is not symmetric in the horizontal direction and the numerical model can only predict the flow qualitatively. However, we can still use it to obtain valuable information regarding the magnitude of the length and time scales of the flow structures developed in the basin. Figure 4 shows the variation in time of the level of the zeros of the horizontal velocity profile at x = 0 (the center of the basin), as obtained with the two models. The agreement between the results from the two models is quite good. Thus, information regarding the approximate number of circulation cells and the time scales of their development in two dimensions can be obtained from the approximate model. This flow offers a good example of a case where although the assumption of linearity is not valid, the linear numerical solution provides useful information regarding the order of magnitude of the time and length scales of the circulation cells developed in a stratified water body.

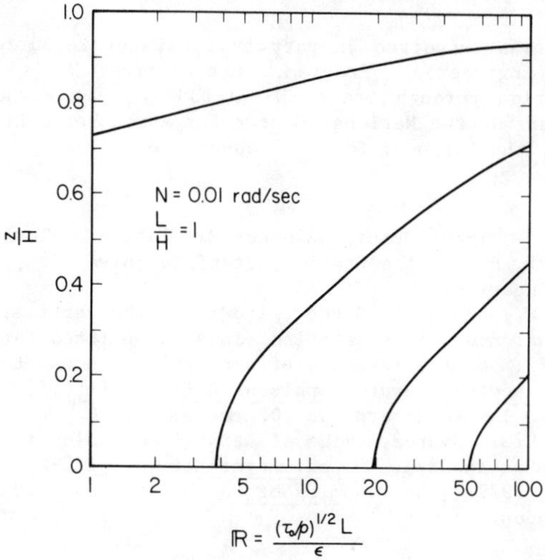

Figure 3. Effect of the eddy viscosity value on the vertical location of the zeros in the centerline profile of the horizontal velocity in a square cavity.

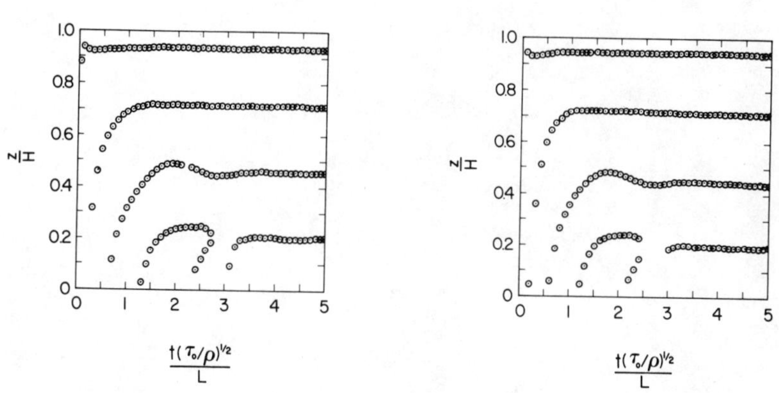

Figure 4. Variation with time of the level of zeros of the centerline velocity profile for R = 100 as obtained from a) the approximate model, and b) a full 2-D model.

ACKNOWLEDGMENT

This work was supported in part by the Water Resources, Urban and Environmental Engineering Program, Engineering Division, National Science Foundation through Grant ENG-77-13880. The computations were performed in part at the National Center for Atmospheric Research, which is sponsored by the National Science Foundation.

REFERENCES

Findikakis, A. N., 1980: Finite Element Simulation of Turbulent Stratified Flows. Ph.D. Dissertation, Stanford University, Department of Civil Engineering.

Findikakis, A. N., and R. L. Street, 1980: On the vertical structure of the internal modes in stratified flows. Submitted for publication.

Krauss, W., 1972: On the response of a stratified ocean to wind and air pressure. Deutsche Hydrographische Zeitschrift, 25, 2, 49-61.

Lindzen, R. S., L. N. Howard, A. D. Mcewan, D. P. DeLisi, and E. E. Gossard, 1973: Hydrodynamics of stratified fluids (the applicability of linear theory). Bound. Layer Met., 5, 227-231.

Philips, O. M., 1977: The Dynamics of the Upper Ocean. Cambridge Univ. Press, London.

CALIBRATING THE CHESAPEAKE BAY CIRCULATION MODEL?

By R. Walton[1] A.M. ASCE, M.A. Granat[2], and R.P. Shubinski[3] M. ASCE

INTRODUCTION

Chesapeake Bay and its tributaries (Figure 1) form the largest, and most productive, system in North America. A number of numerical and physical models (1, 2, 3, 5, 6, 7, 8, 11, 13, 14) exist, or are being developed, as predictive tools for management decision making in the Bay area. The discussion in this paper centers around the future calibration of a three-dimensional, numerical, circulation model being developed for EPA (15) to model the Bay and its tributaries. We will attempt to identify its important calibration parameters, and review similar coefficients used in previous models and measured in the field. Finally, we will compare the calibrations and areas of applicability of this model with the Corps of Engineers hydraulic model (7) located at Matapeake, Maryland.

NUMERICAL MODEL

The Chesapeake Bay Circulation Model being developed for EPA is a two-dimensional, layered, finite-element model. It will include spatially and temporally varying forcing due to tides, wind, density currents, atmospheric pressure gradients, river inflows, Coriolis accelerations, etc. and simulate transport of salt, temperature and two conservative constituents. The solution technique (12) differs from traditional finite-element estuarine models (4, 9, 16) in that it uses an explicit, lumped-mass approach for economy. Also, for stability considerations, a wave equation is modeled instead of the usual continuity equation, giving the model system of equations, as

Wave equation: $$\left.\frac{\partial w}{\partial t}\right|_1^2 + \nabla.\frac{\partial}{\partial t}(\overline{v}h) = 0 \tag{1}$$

Momentum equation: $$\frac{\partial}{\partial t}(\rho\overline{v}h) + \nabla.(\rho h\overline{v}\overline{v}) + \nabla p + \overline{f}\wedge\overline{v}\rho h = \overline{\tau} + \nabla(\overline{N}.\nabla\rho h\overline{v}) \tag{2}$$

Transport equation: $$\frac{\partial}{\partial t}(ch) + \nabla.(\overline{v}ch) = \nabla\ (\overline{E}.\nabla ch) + hr_p \tag{3}$$

Equation of state: $$\rho = f(s,T) \tag{4}$$

[1]Associate Engineer, Camp Dresser & McKee Inc., Springfield, VA
[2]Research Oceanographer, Chesapeake Bay Model, WES, Matapeake, MD
[3]Vice President, Camp Dresser & McKee Inc., Springfield, VA

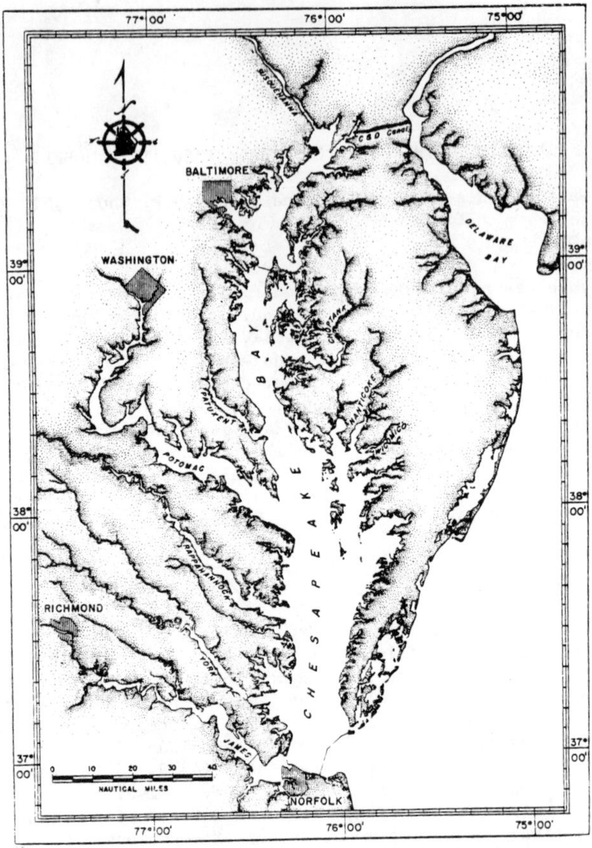

FIGURE 1. - MAP SHOWING CHESAPEAKE BAY

in which w = vertical velocity, (L/T); $\overline{v}$ = horizontal velocity vector, (L/T); h = depth, (L); p = pressure, (M/LT^2); ρ = density (M/L^3); f = Coriolis parameter, (1/T); $\overline{\tau}$ = stress vector, (M/LT^2); $\overline{N}$ = momentum transfer coefficient vector, (L^2/T); c = concentration; $\overline{E}$ = dispersion coefficient vector, (L^2/T); r_p = source/sink term, (1/T); s = salinity; and T = temperature.

The solution procedure is to solve the two-dimensional, layer-averaged momentum equations in each layer, and calculate vertical flows and surface displacements using the wave equation. Mass transport of salt, temperature and two conservatives is simulated at a larger time step, and weakly coupled to the wave and momentum equations through an equation of state.

Given observed data for the hydrodynamic and meteorological forcing functions, the major calibration parameters are the bottom and interficial friction terms, the dispersion coefficients, $\bar{E}$, and the momentum transfer coefficients, $\bar{N}$. Of the later two groups, the vertical coefficients, E_z and N_z are generally considered to be the most important.

PHYSICAL MODEL

The Corps of Engineers' physical model of Chesapeake Bay is an 8.6 acre fixed-bed, distorted scale concrete model housed in a 14 acre shelter located in Matapeake, Maryland. The model is accurately constructed from the off-shore Atlantic Ocean boundary to the heads of tide for all tributaries emptying into Chesapeake Bay (Figure 1). This model is different from most other physical models in that it has a higher degree of automation. Both freshwater inflows and tide generators are capable of complete computer control. Inflows are computer controlled through a feedback mechanism which allows extremely accurate control of discharges from drought to hurricane conditions. Any desired hydrograph can be programmed for each of the 21 inflow points on the model. For site specific studies watersheds can be further subdivided to additional inflow points. The tide generators are computer controlled as well, and are capable of simulating any desired tide sequencing including but not limited to a lunar-month of variable tides producing both neap and spring variations.

Economics and hydrodynamics were considered in choosing the 1:1000 horizontal and 1:100 vertical scales. Time (1:100), velocity (1:10), and discharge (1:1,000,000) scales result from Froude model laws. Stainless steel roughness strips projecting from the model bottom serve to increase the model Reynolds number to simulate the turbulent prototype. These strips are raised or lowered during calibration to adjust the longitudinal and lateral flow characteristics. An induced-mixing bubble line of compressed air running up the axis of the bay and major tributaries is utilized in reducing the degree of stratification believed resulting from the lack of meteorological mixing in the physical model.

DATA BASE

The numerical models employed on either the whole or part of Chesapeake Bay, have used a variety of calibration and verification data sets. These range from specially collected data, and the tide tables, to a compendium of tidal wave characteristics in the Bay dating from 1844 (10). These data sets were generally adequate for the models that used them, but were not always so for a model of the entire Bay region including tributaries.

By far the single most comprehensive data set for the Chesapeake Bay system is that collected for the Corps of Engineers Chesapeake Bay hydraulic model. These data were collected over a four year period from 1970 to 1973, and represent measurements of velocity, conductivity, temperature, tidal elevation, freshwater inflow, and surface weather observations. Due to the immense size of the bay and numerous tributaries, a non-synoptic sampling scheme was required to gather information from all areas of the estuarine system (Table 1). Equipment availability, equipment failures, and scheduling problems encountered can be expected in any large data collection effort, but added to the non-synoptic and spotty nature of the prototype data obtained.

TABLE 1. - TEMPORAL DISTRIBUTION OF DATA SET COLLECTED FOR HYDRAULIC MODEL

Zone	1970	1971	1972	1973
James River		X		
York River				X
Rappahannock River	X			
Potomac River	X	X		
Patuxent River			X	
Upper Bay (Above Patuxent R.)		X	X	X
Lower Bay (Below Patuxent R.)				X

Approximately two-thirds of the 75 tidal stations sampled by the National Oceanic and Atmospheric Administration (N.O.A.A.) are suitable for a one year harmonic analysis. Seasonal records are available for most of the other stations. Chesapeake Bay Institute (CBI), Chesapeake Biological Lab (CBL), and Virginia Institute of Marine Science (VIMS) were contracted to collect velocity and salinity information. Non-synoptic velocity data were collected at 770 depths distributed among the 205 stations monitored during the 28 discrete sampling periods. Continuous velocity measurements were generally on the order of five days duration, with an attempt to collect a longer duration record for a single station within a sampling region as a reference station. Salinity information was collected at 780 depths among 199 stations. In general, half hourly data were obtained during 13 consecutive daylight hours for three days concurrent with the velocity sampling periods at each station. In addition to these salinity data, several slack water cruises, conducted on the main bay and the Potomac, James, York, and Rappahannock Rivers, are available to provide a "synoptic" view of time history salinity variations. Wind data in the form of surface weather observations are also available from the National Climate Center of N.O.A.A. for varying periods of time from 24 weather stations located on the western shore of the bay.

Although this is the most extensive data set available for Chesapeake Bay, it is not ideal and complicates an overall calibration of any hydrodynamic model either numerical or physical. Synoptic data would be preferable, however, a large quantity of synoptic data for calibration would not be economically or physically feasible.

This desirability for a synoptic data set, has led EPA to conduct a Bay-wide hydrodynamic and nutrient sampling program planned for July 1980, and to coincide with the "Superflux" program planned for an area around the mouth of the Bay. Approximately 50 bottom moored current meters will be located on strings predominantly in the Lower Bay to better understand the Bay's three-dimensional structure through measurements of velocity, conductivity, temperature, surface elevations and meteorologic parameters. This program will be the center of future discussions.

NUMERICAL MODEL CALIBRATION AND VERIFICATION PROCEDURES

To ensure consistency with the hydraulic model developed by the Corps of Engineers, the same data set will be used to calibrate the numerical model. Chesapeake Bay will be divided into seven zones consisting of the James, York, Rappahannock, Potomac and Patuxent Rivers, the Upper Bay above the Patuxent, and the entire Bay. Several synoptic data sets will be developed for each zone from the larger Bay-wide set representing 1-2 weeks of continuous measurements. Calibration will then proceed on a zone-by-zone basis, beginning with the rivers, then the upper Bay, and finally the entire system.

The major calibration parameters have been identified as the bottom friction coefficient, represented by Manning's n or Chezy's C, the interfacial shear coefficient, τ_i; the vertical momentum transfer coefficient, N_z and the vertical dispersion coefficient, E_z. To facilitate calibration procedures, it is helpful to know typical values of these parameters used for other models (Table 2).

TABLE 2. - COMPARISON OF MODEL COEFFICIENTS FOR CHESAPEAKE BAY

Authors	Model Type	Bottom Friction m or C	Momentum Transfer Coefficients (m^2/s)	Dispersion Coefficients (m^2/s)	Data Set
Blumberg (1)	x-z f.d	n=0.022	N_z=0-0.003	E_z=0-0.03	Nov 1974
Blumberg (2)	x-y f.d	C=63	-	-	Tide Tables
Boicourt (3)	1-D f.d	-	-	E=10^2-10^3	CBI 1968
Caponi(5)	3-D f.d	-	N_h=57-93 N_z=0.002	Same	Tide Tables
Chen(6)	x-y f.e	n=0.022-0.026	N_h=300-1500	-	Hicks (10)
Gardner and Pritchard (8)	1-D f.d	n=0.026	-	-	CBI Nov 1973 - Jan 1974
Rives(13)	1-D f.d	n=0.022	-	-	Tide Tables
Rives and Pritchard (14)	1-D f.d	n=0.02	-	-	CBI May 1975

From Table 2 it can be seen that there is a uniformity in the selection of the bottom friction coefficient, here represented by Mannings's n. This range is 0.02 - 0.026. This allows us a ready calibration to the bottom friction.

The forms of the vertical coefficients N_z, and E_z, are those used by Blumberg (1) in developing an x-z model of the Potomac River. Recognizing the relationship between these coefficients, he used,

$$E_z = N_z/(1 + R_i) \tag{5}$$

where R_i = Richardson number

$$\text{and } E_z = E_{zo}\, z^2(1-\frac{z}{h})^2 \left|\frac{\partial\bar{v}}{\partial x}\right| (1 - \frac{R_i}{R_{ic}})\, f(R_i/R_{ic}) \tag{6}$$

in which E_{zo} = vertical dimensionless dispersion coefficient; and R_{ic} = critical Richardson number. $f(R_i/R_{ic})$ is a function representing the inhibition of mixing due to salinity gradients. Thus, the one vertical coefficients to be calibrated is the vertical dimensionless dispersion coefficient, E_{zo}.

A review of previous numerical models in the Bay area show that maximum values of E_z are in the range 20 $\sim$ 30 cm^2/sec. Using this range, observed velocities, and Eqs. (5) and (6), it should be possible to readily determine model coefficients for E_z and N_z.

The major remaining calibration coefficient is the interfacial shear coefficient, τ_i. Its form is not readily understood, except to say that it is a term representing friction due to velocity differences in adjacent computational layers. Its form is expected to be model dependent, depending on layer thickness and location.

Once the model has been calibrated in this piecewise manner, it will be verified with the synoptic data set collected this summer. This procedure serves two purposes which are interrelated. Firstly, the model's accuracy and calibration can be checked. Secondly, it is hoped that areas of suspect data, particularly meteorologic forcing will be much better represented, and parameters can be further adjusted to match this more complete data set.

PHYSICAL MODEL CALIBRATION AND VERIFICATION PROCEDURES

Early analysis of data from Chesapeake Bay showed substantial amounts of energy in the bay system that was not of astronomical origin. This was noticed as unusually high co-spectral density function values at low frequences in the Chester River (17), and by extremely low reduction in variance values indicated in the one-year 37-constituent harmonic analyses performed by N.O.A.A. for the hydraulic model. A detailed analysis program including harmonic analyses, spectral analysis, digital filtering, and associated mathematical relationships for waves and wave movements was undertaken by Chesapeake Bay Model Branch (CBMB) personnel to further study and understand the prototype records.

A non-recursive low pass filter was developed to separate the tidal series into a low frequency component with a period of 36 hours or greater and a high frequency component. A correction factor was also employed to smooth the filter and produce a cleaner separation. Based on phase angles derived from power spectral densities and harmonic analyses, the Bay and main tributaries were found to be basically linear with respect to diurnal and semi-diurnal constituents. The M_2 constituent was found to be the overriding dominant force in the high frequency record. The long period energy, on the order of 100 hours, was found to be unrelated to any physical characteristics associated with astronomically generated tides. An excellent relationship was found between the low pass residuals (the non-astronomical component) from selected tide stations along the bay and the north-south wind components from selected weather station records. Since this non-astronomical component could not be adequately generated in the physical model, additional analyses were not pursued in this regard.

Digital filtering and harmonic analysis of the prototype tide records produced an excellent data set of M_2 phase, amplitude, and offset values. Due to the non-synoptic nature of the prototype records and the variations in the non-harmonic tidal component, calibration of the physical model was begun using an M_2 constituent approach. During the beginning phases of calibration, the model was filled with fresh water and an M_2 tide was repeatedly generated for the model tide control station. Many iterations of roughness strip adjustments were performed in a longitudinal progression up the bay and rivers before tidal propagation (time of arrival and amplitude) in the model roughly approximated that of the prototype.

Once tidal propagation was in reasonable agreement, the long-term average (for years of record) fresh water discharge was introduced at each of the 21 model inflow points. Iterations of roughness strip adjustments were again performed with additional attention placed on the cross-sectional M_2 mid-depth velocity distribution. Alterations in the lateral distribution of model roughness were made to bring the model lateral flow distribution into closer agreement with the prototype flow distribution; to maintain an accurate tidal adjustment, the total amount of roughness between tide gages was not significantly changed. It should be emphasized that the prototype velocity records were not ideally suited for an M_2 analysis in that their short durations were not adequate for distinguishing and totally separating out the non-harmonic constituents. During this stage of model adjustment, however, only the lateral distribution of velocity were considered and not the absolute magnitudes.

At this point in calibration, salt was finally introduced into the model ocean to maintain the desired source salinity. The model was operated with the M_2 tide and long-term average fresh water discharge, and allowed to reach a stabilized salinity distribution. Roughness strip adjustments proved ineffective in producing the desired model-to-prototype agreement with respect to vertical, lateral, and longitudinal distribution of salinity and velocity magnitudes. Prototype sampling periods were then examined to select more representative boundary conditions of fresh water discharge and tidal range for discrete sampling periods. Roughness strip adjustments under these steady state conditions also proved ineffective.

A two-year prototype hydrograph of two-week average fresh water discharge and a repetitive 27-lunar-day tide was employed for the final calibration of the model. Under these dynamic conditions, the model was still found to be too stratified when compared with the prototype records. Due to the low amplitude of the tidal currents in the Chesapeake Bay System, the roughness strips were unable to generate sufficient turbulence to reproduce prototype mixing characteristics. The induced-mixing bubble line finally employed in the model, was found sufficient to increase the turbulent mixing while not impeding the propagation of the tides and currents.

Two types of verification tests were performed on the physical model with the established roughness configuration and induced-mixing bubble line. An M_2 tide and long term average fresh water discharge were additional boundary conditions used for steady state model and prototype M_2 tide and velocity comparisons. To examine the dynamic salinity responses of the model compared with the dynamic prototype, the 27-lunar-day tide was continuously generated for two, 2½ year, two-week average fresh water discharge hydrographs reproducing the four year prototype collection period. In addition to salinity sampling corresponding to prototype hourly and same-slack collections, velocity and tide measurements were also collected at selected stations for model-to-prototype comparisons. The physical model was shown to be calibrated and verified for large-scale Bay-wide studies.

DISCUSSION

There is a good deal of hydrodynamic data that has been collected for a number of numerical and hydraulic models of Chesapeake Bay. The largest data set is a four year effort for the Corp's hydraulic model at Matapeake, Maryland.

This data is fine for calibrating models. However, a synoptic data set is highly desirable as a test of Bay-wide model verification. Such a program is being undertaken by EPA in July 1980.

This paper has illustrated the differences between numerical and hydraulic model calibration. Even though both models are calibrated to the same data, it is difficult to relate calibration coefficients between the analog and digital processes. The analog hydraulic model is constrained by the properties of the working fluid and in that it cannot model meteorological or Coriolis effects. It is, however, a continuous medium, whereas the numerical model requires some calibration coefficients reflecting the discretization used, such as interfacial shear, momentum transfer and dispersion coefficients, which introduce more unknowns into the problem.

The advantages of having two such different models of the Bay are obvious. In a bay the size of the Chesapeake, a combination of the two models should provide insight into the complex three-dimensional processes. The numerical model can simulate wind and Coriolis effects which produce lateral salinity gradients, whereas the physical model should better simulate tidal phases and subgrid eddies.

One of the eventual purposes of the numerical model is to develop an atlas of flow patterns under given Bay conditions. A complementary combination of both models would be highly desirable to produce accurate three-dimensional flow regimes. As a first step, simulation under similar and compatible forcing conditions will be undertaken to examine the compatibility of the results for each model.

ACKNOWLEDGEMENTS

The authors wish to thank Dick Bruno of the Corp's hydraulic model and John Klein of EPA's Annapolis Field office for their help and input into this paper.

APPENDIX I. - REFERENCES

1. Blumberg, A.F., "A Numerical Investigation into the Dynamics of Estuarine Circulation," Chesapeake Bay Institute TR 91, The Johns Hopkins University, Baltimore, MD, Oct., 1975.
2. Blumberg, A.F., "Numerical Tidal Model of Chesapeake Bay," Journal of Hydraulics Division, ASCE, Vol. 103, HYI, Jan., 1977, pp. 1-10.
3. Boicourt, W., "A Numerical Model of the Salinity Distribution in Upper Chesapeake Bay," Chesapeake Bay Institute, The Johns Hopkins University TR 54, Ref. 69-7, May, 1969.
4. Brebbia, C.A., and Partridge, P.W., "Finite Element Simulation of Water Circulation in the North Sea," App. Math Modelling, Vol. 1, Sep., 1976, pp. 101-107.
5. Caponi, E.A., "A Three-Dimensional Model for the Numerical Simulation of Estuaries," Ph.D. Dissertation, University of Maryland, Jul., 1974.
6. Chen, H.S., "A Storm Surge Model Study. Volume II - A Finite Element Storm Surge Analysis and its Application to a Bay-Ocean System," Report to FIA, HUD, from Virginia Institute of Marine Sciences, SR 189, Gloucester Point, VA, Sep., 1978.
7. Corps of Engineers, "The Chesapeake Bay Plan of Study," prepared for NSF by U.S. Corps of Engineers, Baltimore, MD, Jun., 1970.
8. Gardner, G.B., and Pritchard, D.W., "Verification and Use of a Numerical Model of the Chesapeake and Delaware Canal," Chesapeake Bay Institute, The Johns Hopkins University TR 87, Ref. 74-7, Jul., 1974.
9. Grotkop, G., "Finite Element Analysis of Long-Period Water Waves," Computer Methods in Applied Mechanics and Engineering, Vol. 2, No. 2., May, 1973.
10. Hicks, S.D., "Tidal Wave Characteristics of Chesapeake Bay," Chesapeake Science, Vol. 5., No. 3, Fall, 1964, pp. 103-113.
11. Leendertse, J.J., et al., "A Three-Dimensional Model for Estuaries and Coastal Seas: Volume II, Aspects of Computations," Rand Corporation, Santa Monica, CA, Jun., 1975.
12. Lynch, D.R., and Gray, W.G., "A Wave Equation Model for Finite Element Computations," Computer and Fluids, 1979.
13. Rives, S.R., "A Mathematical Model of the Tide in the Potomac River," Chesapeake Bay Institute, The Johns Hopkins University TR 80, Ref. 73-6, Feb., 1973.
14. Rives, S.R., and Pritchard, D.W., "Adaptation of J.R. Hunter's One-Dimensional Model to the C and D Canal System," Chesapeake Bay Institute, The Johns Hopkins University, SR 66, Ref. 78-6, Oct., 1978.
15. Walton, R., Brandes, R.J., and Shubinski, R.P., "Towards a Chesapeake Bay Circulation Model," to be presented at "Inland Waters '80" Specialty Conferences, University of Wisconsin, Green Bay, WS, by Water Resources Engineers/CDM, Springfield, VA, Jul., 1980.
16. Wang, J.D., and Connor, J.J., "Mathematical Modeling of Near Coastal Circulation," Ralph M. Parsons Lab., M.I.T. Report No. 200, Apr., 1975.
17. Westinghouse Electric Corporation, "Chester River Study," A Joint Investigation by the State of Maryland Department of Natural Resources and Westinghouse Electric Corporation, Nov., 1972.

APPENDIX II. - NOTATION

English Symbols

C = concentration

C = Chezy friction coefficient

$\overline{E}$ = dispersion coefficient vector, (L^2/T)

E_z = vertical dispersion coefficient, (L^2/T)

E_{zo} = vertical dimensionless dispersion coefficient

$\overline{f}$ = Coriolos parameter, $(1/T)$

h = depth, (L)

n = Manning's friction coefficient

$\overline{N}$ = momentum transfer coefficient vector, (L^2/T)

N_h = horizontal momentum transfer coefficients, (L^2/T)

N_z = vertical momentum transfer coefficient, (L^2/T)

p = pressure, (M/LT^2)

r_p = source/sink term, $(1/T)$

R_i = Richardson number

R_{ic} = critical Richardson number

s = salinity

T = temperature

$\overline{v}$ = horizontal velocity vector (L/T)

w = vertical velocity, (L/T)

Greek Symbols

ρ = density, (M/LT^2)

$\overline{\tau}$ = stress vector, (M/LT^2)

τ_i = interfacial shear coefficient

OVERLAND/INSTREAM CONTAMINANT MIGRATION MODELING

S. M. Brown[M,ASCE(1)] and Y. Onishi[M,ASCE(2)]

Throughout the production, use and disposal of chemical substances, contaminants can be deposited on the land surface. Once there, the fate of these contaminants is controlled by a number of transport, degradation, inter-media transfer, and biologic uptake mechanisms. The relative importance of these mechanisms is a function of environmental conditions and the characteristics of the contaminant. A modeling study of the agricultural pesticide alachlor showed that the timing of runoff events after application and instream sediment concentrations had a major effect on the fate of alachlor. The fate of most radio-contaminants coming from shallow land disposal sites, on the other hand, are controlled by the delivery and movement of sediment materials.

Introduction

Events of recent years have forced the chemical industry and Federal government to take a careful look at the benefits and risks associated with chemical production, use, and disposal. In a recent study commissioned by the Congress it was found that disasters such as Love Canal, PBB's, and Kepone are not isolated incidents. Toxic contamination has resulted in widespread, enduring losses of natural resources i.e., wildlife, waterways and aquifers (Library of Congress, 1980). While the benefits of chemical use are substantial, many feel that the liklihood of future resource losses will increase, especially given the increases in chemical production projected for the next decade.

EPA's current toxic chemical control strategy calls for the production of integrated risk assessments and critical documents to support future regulatory decisions (EPA, 1979). Effective risk assessments will not only have to account for the effects of a chemical substance, but also the degree and duration of probable exposure. To obtain the latter, scientifically sound methods must be developed to measure and predict concentrations of toxic substances in the environment. One type of method which is receiving considerable attention is mathematical models capable of predicting the behavior of toxic substances as they move from sources to receptor organisms.

---

1 Section Manager, Battelle, Pacific Northwest Laboratories, Richland, Washington, 99352.

2 Staff Engineer, Battelle, Pacific Northwest Laboratories, Richland, Washington, 99352.

In order to obtain accurate predictions of contaminant transport, mathematical models must include major transport and fate mechanisms. The fate of a chemical substance in the environment is controlled by 1) its transport through environmental pathways (see Figure 1), 2) its persistence in a given environmental media i.e., soil, water and air, 3) its transfer between environmental media, and 4) its uptake by biota. While the last item is of key importance in the assessment of exposure and probable risk, biological uptake will not be discussed in any detail herein. The relative importance of each of the above factors in determining fate is highly dependent upon environmental conditions and the characteristics of the chemical.

This paper will briefly describe the use of a coupled set of mathematical models to evaluate overland and instream contaminant migration. The first modeling study deals with the fate of an agricultural pesticide, alachlor, in an Iowa watershed. The second covers the migration of radiocontaminants from low-level waste disposal sites in New Mexico.

Overland and Intream Models

In order to predict the fate of toxic chemicals in the overland-instream pathway, the Agricultural Runoff Management, ARM, model (Donigian and Crawford, 1976) and SERATRA (Onishi and Wise, 1979) a sediment-contaminant transport model, were coupled.

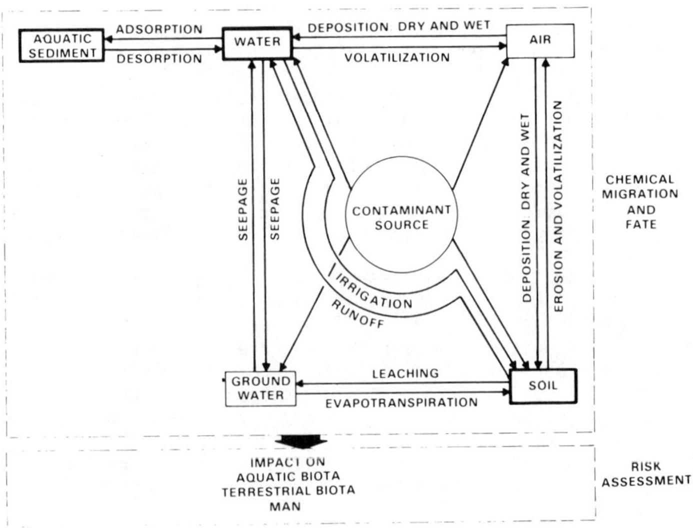

Figure 1. Multi-media Exposure Pathways

The ARM model used for this study consists of the following three components: The hydrologic component of ARM simulates the dominant physical processes which act to convert precipitation into runoff. The soil erosion component simulates sediment loss due to both sheet and rill erosion. Contaminant partitioning between the dissolved and particulate phases is simulated in the adsorption and removal component.

SERATRA is a finite element model which predicts time-varying longitudinal and vertical distributions of sediments and contaminants. The model consists of the following three submodels coupled to describe sediment-contaminant interaction and migration:

sediment transport for each sediment size fraction
dissolved contaminant transport
particulate contaminant (contaminants adsorbed by sediment) transport for each associated sediment size fraction.

The coupling of these models was achieved by breaking the bulk sediment loading predicted by ARM into different size fractions. Field data describing the grain size distribution of suspended sediments entrained in overland runoff are required to perform this operation. Once the bulk sediment has been divided into size fractions, total particulate contaminant loadings simulated by ARM can also be divided for each sediment size fraction by using a ratio of the partition coefficients associated with different sediment size fractions. Predicted runoff is input directly to SERATRA as a "tributary" inflow.

Agricultural Pesticide Fate

The coupled ARM and SERATRA models were applied to the Four Mile Creek watershed in Iowa to predict the fate of the pesticide, alachlor. This modeling study was part of a much larger program to evaluate the Chemical Migration and Risk Assessment (CMRA) methodology (Onishi et al. 1979) in which ARM and SERATRA form the chemical migration or exposure portion of the methodology.

Four Mile Creek watershed is located in northwestern Tama County, Iowa, and covers an area of 19.5 mi^2 (50.5 km^2). It is typical of the heavily cropped regions of Iowa. The major crops in the water shed include corn, soybeans, oats and hay.

Alachlor, a preemergence herbicide used to control most annual grasses and certain broadleaf weeds, is the most widely used herbicide in the watershed (Baker et al. 1979). When applied at the recommended rate, it has an effective soil persistence of 6 to 10 weeks, and is primarily degraded by microbial action and chemical breakdown (Weed Science Society of America, 1979). While alachlor will absorb to colloidal particles in the soil, it tends to be transported mainly in a dissolved form. Its half-life in water depends on the aerobic character of the water body and its sediments. Alachlor on a aerobic sediment has a half-life of less than two weeks. Anerobic sediment conditions speed the degradation process, resulting in a half-life of about four days (Onishi et al., 1979).

Over a three year period, from 1976 to 1979, Iowa State University (ISU) conducted a comprehensive field sampling program at the Four Mile Creek Watershed. The basic objectives of the program were: 1) to collect data for use in testing and refining ARM, and 2) to conduct complimentary model oriented research related to nonpoint pollutants from agricultural lands (Baker et al., 1979).

Two years of ISU data were used to calibrate ARM. The coupled set of models, ARM and SERATRA, were then applied to the lower portion of Four Mile Creek watershed (the shaded area in Figure 2). The transport of sediments, and dissolved and particulate alachlor were then calculated in Four Mile Creek and Wolf Creek, to its confluence with the Cedar River. The simulation period was June 1971 to May 1974. Soybeans were assumed to be the major crop type. The timing of plowing, planting, and cultivating operations was selected to be consistent with that in the rest of the watershed. Alachlor was applied the day after planting at a rate of 1.6 lb/acre; this is the application rate used by ISU on their field sites within the watershed.

On the land surface, contaminant transport is largely controlled by three major processes: runoff, infiltration/percolation, and soil erosion. Runoff acts as a solvent for a chemical substance and rapidly transports it overland. Through the mechanism of adsorption, contaminants attach to eroded soil particles which are also transported by overland runoff. The concentrations of a contaminant in each phase of the soil-water complex constantly change in an attempt to reach chemical equilibrium. This same mechanism is also at work when dissolved contaminants move vertically through the soil matrix. The availability of contaminants for transport is affected by a number of degradation mechanisms resulting from hydrolysis, photolysis, and oxidation and microbial activities. In addition to adsorption/desorption, the inter-media transfer of contaminants is also affected by volatilization.

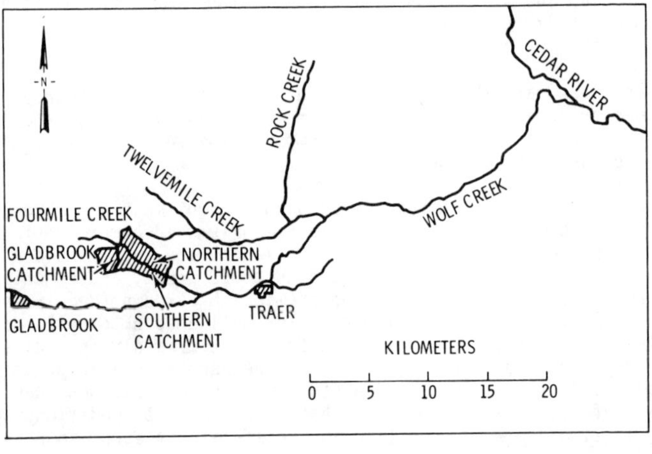

Figure 2. Four Mile And Wolf Creeks In Iowa

The modeling results showed that the fate of alachlor on the land surface is largely controlled by one single factor: the timing of significant runoff for overland transport by leaching it into the soil. Once there, it will degrade or continue to percolate to ground-water. Table 1 gives an annual summary of the "fate" of alachlor on each of the three subcatchments modeled with ARM (see Figure 2). Alachlor fate is presented in terms of its loss through the three dominant transport mechanisms (runoff, soil erosion and infiltration/percolation) and degradation. The differences in fate in each subcatchment are due to the timing of application, since the same sequence of meteorological events was used for each subcatchment.

Table 1. Alachlor Fate on the Land Surface (Percent of Total Applied).

		Transport Mechanisms			
Catchment	Year	Runoff*	Erosion	Infiltration Percolation	Degradation
Gladbrook	1971	1.89	0.002	0.005	98.10
	1972	1.78	0.001	0.02	98.20
	1973	1.39	0.001	0.01	98.60
Southern	1971	0.80	0.001	0.004	99.20
	1972	0.66	0.001	0.01	99.33
	1973	1.35	0.001	0.01	98.64
Northern	1971	1.00	0.001	0.004	99.00
	1972	0.60	0.001	0.01	99.40
	1973	2.84	0.010	0.004	97.15

* Summation of alachlor in surface runoff and interflow components.

From this table it is clear that only a limited quantity of alachlor was transported to Four Mile Creek, and that the bulk of that was dissolved. Even during periods of intense runoff less than three percent of the total applied alachlor was lost. During the period of simulation, an annual average of 1.4 percent of the total alachlor applied to the watershed was lost to Four Mile Creek. This percentage corresponds quite well with the values found by Wauchope in his survey of pesticide losses in runoff waters from agricultural lands (1978).

In stream systems many of the same factors affect contaminant fate. Dissolved contaminants delivered from the land surface or directly discharged into a stream are transported along with the flow. Adsorbed contaminants will also move downstream with the flow, although at a different rate mainly due to sediment deposition and resuspension. The chemical and biological degradation processes also act to attenuate contaminants in water bodies. In Four Mile and Wolf Creeks, the fate of alachlor was almost entirely controlled by flow and sediment transport mechanisms. With a degradation rate of 0.07 day^{-1} and an average travel time through the study reach of 2-3 days, little or no degradation occurred. Deposition of sediments in the study reach was also rather limited. Therefore, the bulk of the alachlor entering these creeks from the land surface was directly transported through the reach except those adsorbed by bed sediment directly from an overlying water.

One phenomena which is dramatically illustrated by the instream modeling results is the influence of sediment concentration on contaminant fate. It is believed that sediment transport processes have a major impact on contaminant fate when: 1) the chemical substance of concern has a high affinity for adsorption, 2) sediment concentrations are sufficiently high such that a significant fraction of the total contaminant load is sediment bound even if the substance has a low affinity for adsorption, and 3) sediments carrying a highly persistent substance are subject to deposition. The latter can result in a long-term source of contamination through subsequent desorption and/or resuspension.

With a relatively low partition coefficient 0.80 ft^3/lb (50 ml/gm) of alachlor, one would expect that sediment transport has little effect on the fate of alachlor. However, the modeling results show that sediment transport is important largely because of the high concentrations of suspended sediments in these creeks. Table 2 illustrates the impact of adsorption for the event having the highest instream alachlor concentrations over the three-year simulation period. As the alachlor migrated downstream, the percentage of dissolved alachlor to total alachlor was reduced and the particulate fraction increased. Figure 3 shows the distribution of alachlor over the entire study reach. The implications of this phenomena are: 1) that the alachlor available for direct uptake by aquatic biota i.e., the dissolved fraction, is reduced as a result of adsorption and 2) that alachlor will tend to accumulate in areas of sediment deposition.

Because alachlor is relatively non-persistent, however, the deposition of sediments is probably not likely to create a long-term source of contamination once introduction of alachlor to streams is terminated.

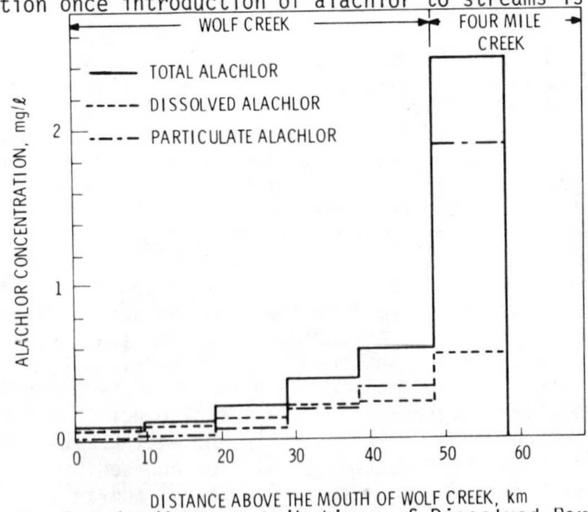

Figure 3. Longitudinal Distributions of Dissolved Particulate And Total Alachlor at 12 p.m. June 5, 1973

Table 2. Distribution of Alachlor at Different Locations - June 5, 1973 Storm Event

Location	Dissolved Fraction (%)	Particulate Fraction (%)
Overland Runoff	98	2
Mouth of Four Mile Creek	78	22
Mouth of Wolf Creek	65	35

Low-Level Radiocontaminant Fate

Considerable quantities of radioactive waste materials have been disposed to the ground at both commercial and Federal facilities across the U.S. Although current shallow land disposal practices provide an adequate short-term solution at many sites, long-term management strategies need to be developed.

Radiocontaminants can migrate from waste disposal sites through both biotic and abiotic pathways. Release mechanisms include exhumation by wind and water erosion, leaching, and transfer of the contaminants to the ground surface by ground-water movement, plants and animals. In general, these release mechanisms are relatively slow. They can, however, be significantly altered and/or changed in relative importance by acute natural events (e.g., earthquakes or flooding), and man's activities. In order to make decisions regarding the short-and long-term managements of buried wastes, especially over the long time frames these contaminants may remain in the environment, potential levels of exposure must be evaluated. Battelle, Pacific Northwest Laboratories and Los Alamos Scientific Laboratory (LASL) are jointly developing a methodology to predict the multi-media migration and fate of radiocontaminants. As a first step, ARM and SERATRA are being used to evaluate contaminant migration at the LASL site. Figure 4 shows a part of Mortandad Canyon, one of many canyons which dissect Pajarito Plateau in north-central New Mexico. The intermittent stream draining Mortandad Canyon is narrow and steep, with thin layers of sand bed sediment. Effluent discharges to the stream rapidly percolate into the alluvian within about 800-1200 m downstream of the outfall. During major summer thunderstorms, large surface flows may traverse about 3000 m below the outfall before disappearing in the alluvian (Hakonson et al. 1976). Area C is a small shallow land radionuclide disposal site located in a side-canyon within the Mortandad Canyon drainage.

Detailed modeling of the site is currently in progress. Comparison of dominant migration and fate mechanisms appeared in this contaminant fate problem and the one discussed previously indicates interesting differences between these two cases. On the land surface, the timing of runoff producing storm events and contaminant degradation are not important factors for this radionuclide migration case. Instead, the major factor controlling contaminant fate is the delivery of sediment materials

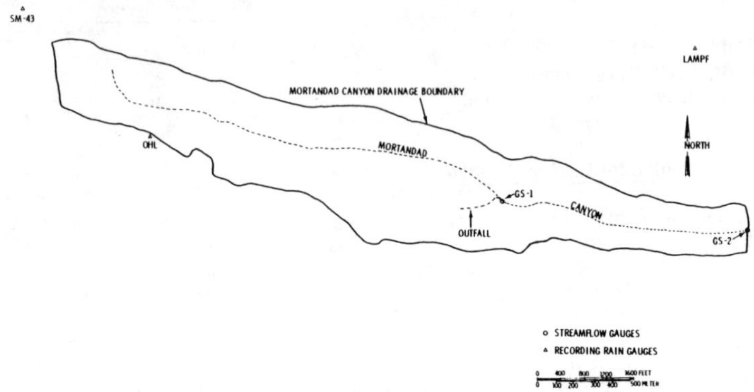

Figure 4. Mortandad Canyon Above Guaging Station GS-2

materials to the channel. This is largely due to the fact that these radionuclides $Pu^{239-240}$ and Cs^{137} have high affinities for adsorption to soil particles. Plutonium-239, Plutonium-240 and Cesium-137 are also having half lives of 24,390, 6,580 and 30 years, respectively. Therefore, the magnitude of storm events and their attendant capacities to transport soils will play a major role in contaminant fate.

Instream sediment transport processes will control contaminant fate. Suspended and bed sediments will tend to adsorb and carry the bulk of the contaminant load. Therefore, areas of sediment deposition will be of key importance because they will become long-term sources of contamination. This is in contract to the previous example, because of the high degree of persistence of these radiocontaminants.

Conclusions

Mathematical models capable of predicting the behavior of chemical substances in the environment will play a major role in assessing probable levels of exposure associated with chemical production, use, and disposal. Contaminant fate in a given environmental pathway is controlled by a complex set of mechanisms which affect contaminant transport, degradation, inter-media transfer, and biologic uptake. Given the characteristics of a contaminant and environmental conditions, mathematical models can be used to identify the key mechanisms controlling contaminant fate.

A modeling study of the pesticide alachlor showed that because it is relatively non-persistent on the land surface, the timing of runoff producing events had a major impact on alachlor delivery to the stream system. In stream it was found that even though alachlor does not have a high affinity for adsorption, a significant fraction of the total alachlor was sediment-bound because of high suspended sediment concentrations. As a result, sediment transport processes had a major impact on the instream fate of alachlor.

The fate of radiocontaminants in the overland-instream pathway was found to be dominated by a different set of mechanisms. Because these contaminants have high affinities for adsorption and are also highly persistent, sediment delivery to the stream channel and instream sediment transport control contaminant fate. Deposition of sediments is of particular concern because of the potential for creating a source of long-term contamination.

REFERENCES

1. Baker, J. L., H. P. Johnson, M. A. Borcherding and W. R. Payne, "Nutrient and Pesticide Movement From Field to Stream: A Field Study," in Best Management Practices for Agricultural and Silviculture, Proceedings of the 1978 Cornell Agricultural Waste Management Conference, Ann Arbor Science Publishers, Inc., Ann Arbor, MI, 1979.

2. Donigian, A. S. and N. H. Crawford, Modeling Pesticides and Nutrients on Agricultural Lands, EPA-600/2 76-043, U. S. Environmental Protection Agency, Washington, D. C., 1976.

3. Hakonson, T. E., et al., "Accumulation and Transport of Soil Plutonium in Liquid Waste Discharge Areas at Los Alamos," Transuranium Nuclides in the Environment, IAEA, Vienna, 1976.

4. Library of Congress, "Resource Losses from Surface Water, Ground Water, and Atmospheric Contamination: A Catalog." Prepared by the Environment and Natural Resources Policy Division, Congressional Research Service, Library of Congress for the U. S. Senate Committee on Environment and Public Works, Serial No. 96-9, U. S. Government Printing Office, Washington, D. C., March, 1980.

5. Onishi, Y. and S. E. Wise, "Mathematical Model, SERATRA, for Sediment-Contaminant Transport in Rivers and its Application to Pesticide Transport in Four Mile and Wolf Creeks in Iowa." Battelle, Pacific Northwest Laboratories, Richland, WA, August, 1979.

6. Onishi, Y., S. M. Brown, A. R. Olsen, M. A. Parkhurst, S. E. Wise and W. H. Walters, "Methodology for Overland and Instream Migration and Risk Assessment of Pesticides." Battelle, Pacific Northwest Laboratories, Richland, WA, August, 1979.

7. U. S. Environmental Protection Agency, "Research Outlook 1979," Office of Research and Development, Washington, D. C., EPA-600/9 79-005, February, 1979.

8. Wauchope, R. D., "The Pesticide Content of Surface Water Draining from Agricultural Fields - A Review," Journal of Environmental Quality, Vol. 7, No. 4, 1978.

9. Weed Science Society of America, Herbicide Handbook, Fourth Edition, Champaign, IL, 1979.

MASS TRANSPORT IN REGIONAL GROUNDWATER BASINS

Thomas G. Naymik[a]
Lewis D. Thorson[b]

ABSTRACT

Subsurface parameters affecting mass transport calculations in hydrogeologic environments are extremely heterogeneous. The uncertainty associated with the spatial variability of such input parameters as effective porosity and the distribution coefficient creates concern as to the predictive capability of mass transport models. A numerical model of a regional groundwater flow system (not site specific) was coupled with a mass transport model in order to study the influence of effective porosity and distribution coefficients in bedded media. The effects on model performance were discerned from long-term mass transport simulations. Model performance was based on initial breakthrough time, average breakthrough time, and the standard deviation of the breakthrough curve at a discharge surface. Large differences in model performance occurred when distribution coefficients were varied within reported ranges, while small differences resulted by varying effective porosity within naturally occurring ranges.

INTRODUCTION

Numerical models of aquifer systems are an accepted tool in water resource planning efforts. The models have value in predicting the effects of processes that occur over extended periods of time and are often the only available means of handling the complex boundary conditions, multiple inputs, and spatially distributed parameters of these systems.

Groundwater and mass transport models have also been used to study the behavior of dissolved constituents in groundwater flow regimes.[1,2] These studies, Pickens and Lennox (1976) and Schwartz (1975), are two among many dealing with waste migration in groundwater. This study[c] was carried out, in part to investigate geohydrological parameters important to the geological disposal of high level nuclear waste.

A generic model was developed for a large sedimentary basin. By varying parameter values in simulations, the effects of transport

---

[a] Hydrologist, Illinois State Water Survey, Urbana, IL
[b] Hydrologist, Lawrence Livermore Laboratory, Livermore, CA
[c] This work was performed at Lawrence Livermore Laboratory and was supported by the U.S. Nuclear Regulatory Commission under Interagency Agreement DOE 40-550-75 with the U.S. Department of Energy.

phenomena acting within the model were singled out. In this report, effective porosity, distribution coefficient, and source location were varied in long-term simulations of mass transport.

EQUATIONS AND ASSUMPTIONS

Groundwater flow. The velocity field of the model was calculated with a steady-state, finite element code originated by Taylor and Brown.[3,4] The code allows for nonhomogeneous and anisotropic layered material. Fluid flow is governed by Darcy's Law which can be put into the two dimensional form:

$$\bar{q} = -\bar{\bar{K}} \nabla \Phi \tag{1}$$

where $\bar{q}$ is the specific flux vector, $\bar{\bar{K}}$ is the hydraulic conductivity tensor, and Φ is the piezometric head. The steady-state head distribution is calculated using Darcy's Law in the continuity equation

$$\nabla \cdot (\rho \bar{q}) = 0 \tag{2}$$

where ρ is the fluid density. In this study three different boundary conditions were used:

no flow	$\bar{n} \cdot \bar{q} = 0$	(3)
recharge	$\bar{n} \cdot \bar{q} = \alpha$	(4)
water table	$P = 0$	(5)

where $\bar{n}$ is the normal to the surface, α is a constant that specifies the amount of fluid added by recharge, $\bar{q}$ is the specific flux vector, and P is the fluid pressure. The relationship between the specific flux vector and the average linear groundwater velocity in porous media is

$$\bar{v} = \frac{\bar{q}}{n} \tag{6}$$

where n is the effective porosity. The average linear velocities are used as part of the input to a material transport code that calculates mass concentration as a function of space and time.

Mass transport. The mass transport code used in this study was heavily modified from Duguid and Reeves.[5,6] The movement of the dissolved constituent is solved by

$$\nabla \cdot (\bar{\bar{D}} \cdot \nabla c) - \nabla \cdot (\bar{v} c) + \lambda R_d c = R_d \frac{\partial c}{\partial t} \tag{7}$$

where $\bar{\bar{D}}$ is the hydrodynamic dispersion tensor, c is the concentration of the dissolved constituent, $\bar{v}$ is the average linear groundwater velocity, λ is the radioactive decay constant and R_d is the retardation factor. In equation (7), R_d is used. In this study the distribution coefficient, K_d, was varied in multiple simulations. The relationship between the two is

$$R_d = 1 + \frac{\rho K_d}{n} \tag{8}$$

where ρ is the porous medium bulk density and n is the effective porosity. The derivation of equation (7) can be found in Bear.[7] In this study, equation (7) was solved in vertical cross section using a Galerkin finite-element approach with isoparametric elements.

Assumptions. Several assumptions were built into the model:

1) A nuclear waste repository in geologic media can be represented as a "point source" in a vertical cross section. The effect will be higher concentrations at the discharge surface of the model than would be expected in a three-dimensional model.

2) An isothermal situation was assumed which won't be the case near to the repository. The effect of high temperatures near the repository is to lower the groundwater viscosity, thereby increasing hydraulic conductivity.

3) The presence of a repository has no effect on the regional flow field. In reality, it will take time to resaturate and repressurize the mined out cavity. In this study a single pulse release of material takes place at the onset of simulations, thereby bypassing the problems of resaturation, repressurization, and dissolution of material at the source.

4) Trace amounts of material are in solution, thereby neglecting fluid density variations.

5) The hydrologic boundary conditions were assumed constant over very long periods of time. Intuitively this isn't the case. Research is under way in transient groundwater flow regimes.

6) The vertical cross section is assumed to be parallel to groundwater flow and all material movement occurs in the plane of the section.

7) The porous medium can be anisotropic and nonhomogeneous with respect to hydraulic conductivity, but is assumed isotropic with respect to dispersion.

8) The sorbed and solution phases of the material are in equilibrium.

DESCRIPTION OF THE BASIN MODEL

Groundwater flow. A conceptual model (not site specific) was developed for a large sedimentary basin (Fig. 1). The dimensions of the basin were 160 km by 240 km. A groundwater basin was assumed to be coincident with the geologic basin. A groundwater divide was assumed at the basin perimeter (point A), groundwater discharge takes place at the river (point B), and a vertical plane of symmetry was assumed to be present along and below the river at point B. The elevation at point A was defined at 1308 m to depict the marginal upland of the basin. The relief between points A and B was 560 m.

The stratigraphy was divided into five vertical hydrologic units, all of which had fairly low hydraulic conductivities (Fig. 2). The succession of units downward from the surface was: (1) aquifer, 170 m thick; (2) aquitard, 136 m thick; (3) aquiclude, 170 m thick; (4) aquitard, 136 m thick; and (5) aquifer, 136 m thick. These units outcropped along the horizontal axis between (1) 21 km and 80 km, (2) 15 km and 21 km, (3) 12 km and 15 km, (4) 8 km and 12 km, and (5) 0 km and 8 km.

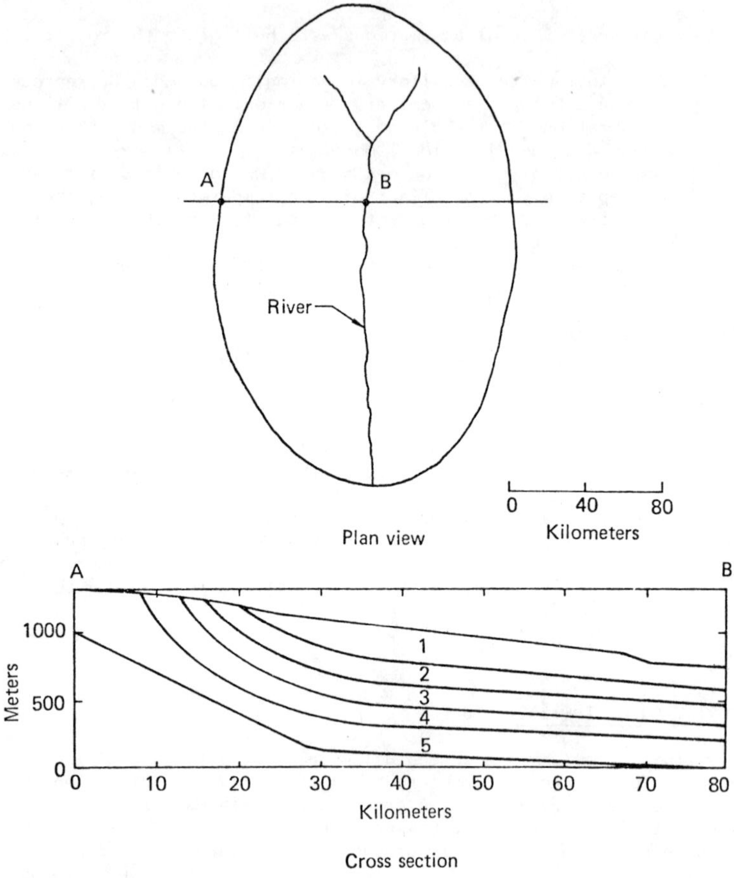

Fig. 1. Plan view of the generic sedimentary basin and cross section of half of the basin, AB.

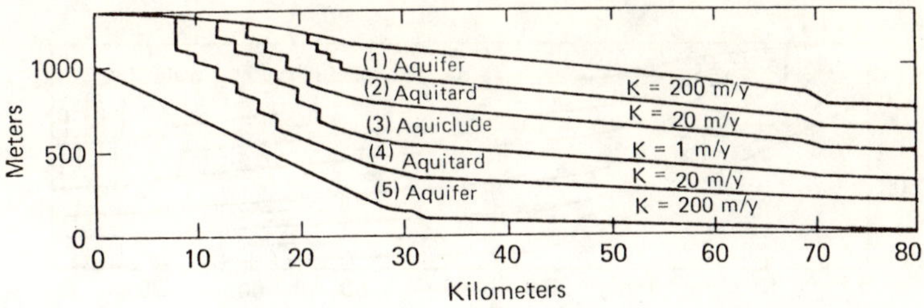

Fig. 2. Zones of hydraulic conductivities used in the model.

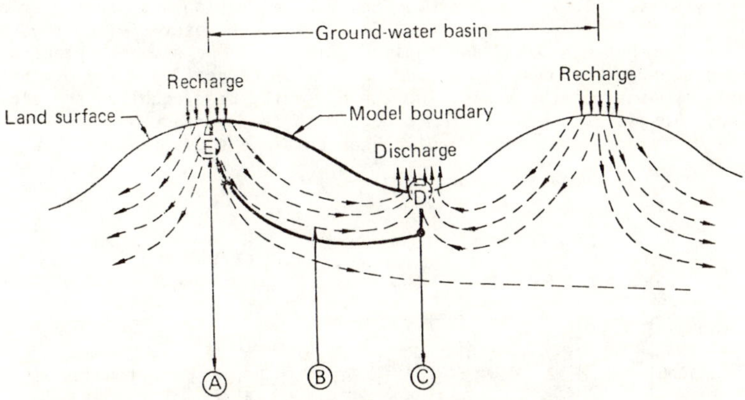

Fig. 3. Idealized hydrologic setting of the model (cross section). Line AE is the regional groundwater divide, point B is the divergence of regional and basin flow, and line CD is the symmetry plane for basin discharge.

The hydrologic setting of the basin (Fig. 3) contained three key features: (1) at line AE, a regional groundwater divide existed; (2) along line EC, the basin groundwater streamlines diverged from the regional flow pattern; and (3) line CD designated a symmetry plane for discharge in the basin flow regime. The basin was modeled most efficiently using these assumptions. Half of the basin was modeled because of the symmetry at the discharge region. The discharge region was treated as a broad flood plain. The groundwater divide was used as the vertical upland boundary, and the divergence between basin and regional flow was used as the lower boundary. The general flow regime is shown with arrows in Fig. 4.

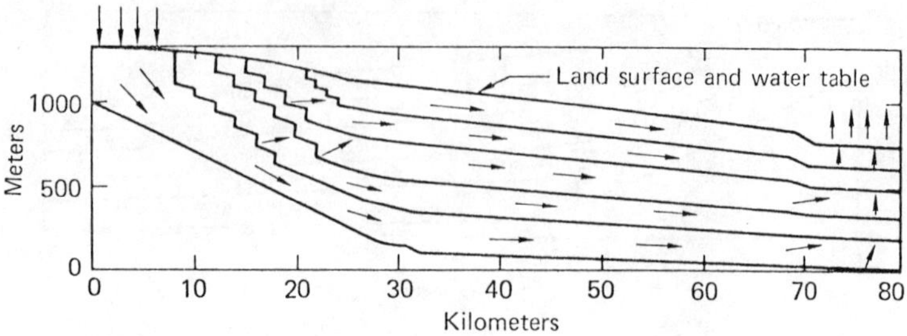

Fig. 4. Generalized groundwater flow pattern shown by arrows.

A finite-element model (1562 element) was designated to calculate steady state flow in the basin. The recharge region was treated as a distributed inflow boundary (Fig. 5). The groundwater divide, the divergence between basin and regional flow, and the symmetry plane at the discharge region were modeled as no-flow boundaries. The discharge surface was simulated by setting the hydraulic head equal to the elevation. The water table was coincident with land surface and was treated in 2 different ways, as hydraulic head equal to elevation and as a free surface.

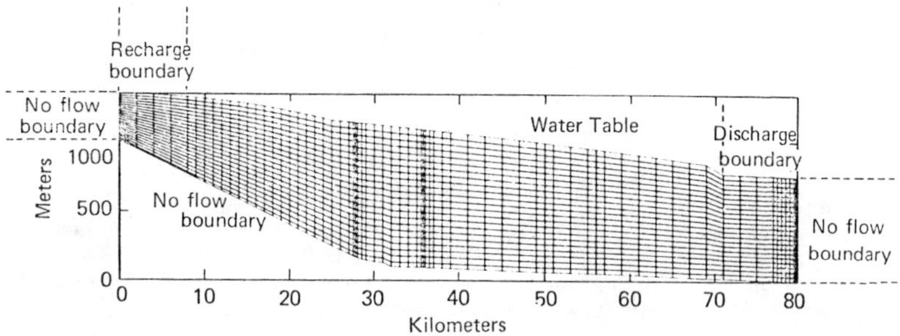

Fig. 5. Finite element mesh and model boundary conditions.

Mass transport properties. Given a velocity field, only the transport properties of the media and the location of the source remain to influence mass transport. Model performance was based on average breakthrough time and the standard deviation of the breakthrough curve at the discharge surface. In the cases of incomplete breakthrough curves, comparisons were based on the initial breakthrough time.

The basin cross section can represent a variety of stratified geologic materials by virtue of the mesh design and the flexibility of input parameters that control mass transport. The parameter values chosen (Table 1) were taken in part from Holdsworth et al.[8] The layers referred to in Table 1 are shown in Fig. 6 and are coincident with the hydrologic units.

TABLE 1. Mass transport properties held constant during simulations.

Layer No.	Approx. horizontal groundwater velocity, m/y	Approx. vertical groundwater velocity, m/y	Longitudinal dispersivity, m	Lateral dispersivity, m	Molecular diffusion, m^2/y	Tortuosity
1	1.3	0.001	50	5	10^{-5}	1.1
2	0.13	0.007	30	5	10^{-5}	1.1
3	0.008	0.001	30	10	10^{-5}	1.1
4	0.13	0.007	30	5	10^{-5}	1.1
5	1.3	0.01	50	5	10^{-5}	1.1

Location of the source. Ten source nodes were positioned to examine the movement of material from the source in the first 5000 years of simulation, as shown in Fig. 6. Not all sources were likely sites for a nuclear waste repository but, nevertheless, were used to check the model calculations.

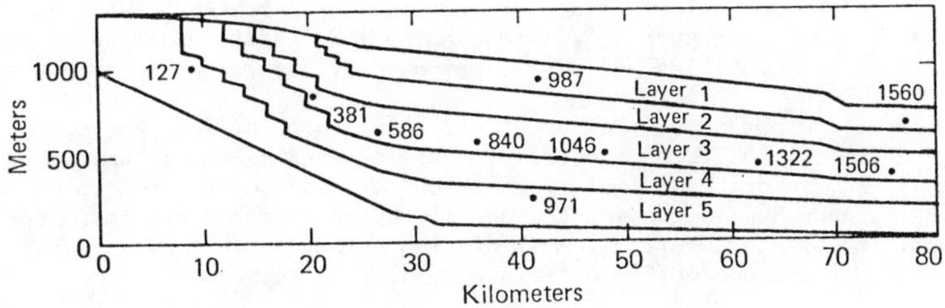

Fig. 6. Cross section of five layers used as material property zones and the position of ten source nodes.

Source node 127 was located in the lower aquifer ≈ 62 km from the discharge surface. The predominant flow direction was toward the discharge region with a slight upward component. Transport from this source was rapid via both the upper and lower aquifers. Node 381 was located at the bottom of layer 3 ≈ 51 km from the discharge surface. Material movement to the phreatic surface was rapid from this source. Source node 586 was near to the bottom of layer 3 ≈ 43 km from the discharge surface. The groundwater flow direction at this source was about 60° upward from the horizontal, toward the discharge surface. The steady-state flow rate at node 586 was about 0.16 m/y. Material migration from this source was slower than from the two upgradient sources. Source node 840 was also located in layer 3, but was ≈ 35 km from the discharge surface. The groundwater flow direction here was 3° upward from the horizontal, toward the discharge region. The flow rate was about 6 mm/y and transport was minimal. Of the ten sources, node 840 was positioned best to minimize material transport because in this section of the basin the equipotential lines were parallel and nearly vertical. Source nodes 971 and 987 were located in the lower and upper aquifers, respectively, ≈ 30 km from the discharge surface. These were positioned to study the transport behavior in faster flowing (1.3 m/y) regions of the model. The transport from sources 1046 and 1322 was similar to that of node 840 because the three were located in the same layer in similar flow regimes. Although transport was minimal from nodes 1046 and 1322, their proximity to the discharge surface, ≈ 24 km and ≈ 8 km, respectively, made them less suitable if one is trying to maximize breakthrough time. Source nodes 1506 and 1560 were located under the discharge surface. Node 1506 was positioned near the bottom of layer 3 and node 1560 near the bottom of the upper aquifer. These sources were used to study the transport phenomena at the downgradient end of the mesh where flow vectors were nearly vertical.

In summary, nodes 586 and 840 had the best locations for the purpose of maximizing breakthrough time. This statement is based on short-term simulations used to study the onset of transport. Source nodes 586 and 840 received additional study because of their long distance from the discharge surface and minimal transport in the first 5000 y of simulation.

SUMMARY OF MASS TRANSPORT SIMULATIONS

Mass transport simulations were carried our using nodes 586 and 840 as sources. The purpose was to examine model performance by varying effective porosities and distribution coefficients in the 5 layers.

Effective porosity. Values of effective porosity were varied in the 5 hydrologic units of the basin. Tables 2 and 3, in conjunction with Fig. 6, describe 9 simulations which were carried out twice: (1) using node 586 as the source, and (2) using node 840 as the source (identified by an "a" after the simulation number). In these simulations, distribution coefficients in all layers were held at zero.

TABLE 2. Effective porosity values of the five layers for simulations 1 through 9.

Layer number	Simulation number 1	2	3	4	5	6	7	8	9
1	.1	.02	.2	.1	.1	.1	.1	.1	.1
2	.05	.05	.05	.1	.01	10^{-3}	.05	.05	.05
3	.05	.05	.05	.05	.05	.05	.1	.05	10^{-3}
4	.05	.05	.05	.1	.01	10^{-3}	.05	.05	.05
5	.1	.02	.2	.1	.1	.1	.1	.1	.1

TABLE 3. The average breakthrough time and standard deviation of the breakthrough curves for simulations 1 through 9 (source at node 586) and simulations 1a through 9a (source at node 840).

Simulation number for node 586	Average breakthrough time, yrs	σ yrs	Simulation number for node 840	Average breakthrough time, yrs	σ yrs
1	5271	1283	1a	5486	1413
2	4677	1322	2a	5052	1461
3	5909	1193	3a	5671	1228
4	5661	1470	4a	5871	1495
5	4914	1111	5a	5145	1349
6	4831	1075	6a	5066	1338
7	5369	1333	7a	5829	1564
8	5163	1222	8a	5119	1250
9	5127	1206	9a	5028	1215

Node 586 was 8 km farther from the discharge surface than node 840. However, in more than half of the simulations, the average breakthrough time from source 586 was less than that from source 840. This was the result of the upward velocity component at 586 which readily moved material into the relatively fast flowing upper aquifer.

In the base cases (simulations 1 and 1a), the average breakthrough time from source 586 was less than that from source 840. When the effective porosities of the aquifers were decreased (simulations 2 and 2a), the mean arrival time at the discharge surface from both sources decreased. This occurred because material from source 586 was in the upper aquifer for most of its path to the discharge surface. When the effective porosities of the aquifers were increased (simulations 3 and 3a), the reverse of simulations 2 and 2a took place.

In simulations 4 and 4a, the effective porosities in layers 2 and 4 were increased. The result was increased mean arrival times. The delay of material crossing the discharge surface from source 840 was enhanced by the absence of artesian conditions in the vicinity of node 840. By decreasing the effective porosities in layers 2 and 4 (simulations 5, 5a, 6, 6a), mean arrival times decreased. Arrival time from source 840 was consistently greater. With nodes 586 and 840 as sources, the model was not sensitive to effective porosities below 10^{-3} in layers 2 and 4.

The effective porosity of layer 3 was increased in comparison to the base case for simulations 7 and 7a. In both simulations, average breakthrough times increased. Decreasing the effective porosity of layer 3 forced material to move more quickly away from the source. As a result, the mean arrival time decreased slightly (simulations 8, 8a, 9, 9a). The model was not sensitive to effective porosities below 10^{-3}, in layer 3, when either source node was used.

Distribution coefficient (K_d). The distribution coefficients in layers 1 through 5 were varied, while the effective porosities were held at 0.1, 0.05, 0.01, 0.05, and 0.1, respectively. Distribution coefficients were varied because of the high uncertainty associated with the parameters.[9] Table 4 describes seven simulations (10-16), each of which was then performed twice: (1) using node 586 as the source, and (2) using node 840 as the source. The results from these simulations were analyzed in terms of initial breakthrough time at the discharge surface. This was necessary because the maximum simulation time was 250,000 years, which was sufficient time in some runs to calculate the mean arrival time.

TABLE 4. Distribution coefficients (ml/g) of the five layers for simulations 10 through 16.

Layer number	Simulation number 10	11	12	13	14	15	16
1	0	10	100	0	0	10	100
2	0	10	100	0	0	0	0
3	0	10	100	10	100	0	0
4	0	10	100	0	0	0	0
5	0	10	100	0	0	10	100

Simulations 10 and 10a were the base cases for this set of runs using nodes 586 and 840, respectively, as source nodes. The initial breakthrough for these simulations was at about 2300 years. By setting K_d to 10.0 ml/g in all layers, the initial breakthrough came at 100,000 years in simulation 11 and at 85,000 years in 11a. In simulations 12 and 12a, the K_d equaled 100.0 throughout the model. The result was no breakthrough after 250,000 years of simulation.

In simulations 13 and 13a, layer 3 had a K_d of 10.0 ml/g, while in other layers it was set to zero. The initial breakthrough times increased only slightly from the base case because most of the travel-path length of the material was in layers 1 and 2. Similarly, in runs 14 and 14a, only a small increase in initial breakthrough occurred.

The distribution coefficients in the aquifers were increased to 10 ml/g (runs 15 and 15a) and 100 ml/g (runs 16 and 16a) and set to zero in the other layers in order to single out the effect from the aquifers. The result was initial breakthroughs of 38,000 years for run 15, 13,000 years for run 15a, and no breakthrough in 250,000 years for runs 16 and 16a. Clearly, the model was extremely sensitive to the K_d value used in the aquifers.

RECOMMENDATIONS

On the basis of the design and assumptions, it has been possible to make statements about the influence of effective porosity and distribution coefficient on long-term material transport simulations. The results lean heavily on the assumptions set forth at the beginning of this paper; they should be interpreted with care. If this type of modeling approach were applied to a field situation in order to predict region mass transport, the uncertainty in the field measurements themselves, in addition to the uncertainties caused by extrapolating input parameters over long distances, would more than likely mask the correct conclusion. Two concurrent tasks are recommended: (1) evaluate the usefulness of numerical models in predicting regional mass transport, and (2) review the geotechnical measurement and model parameter uncertainties inherent in regional mass transport models.

REFERENCES

1. J. F. Pickens and W. C. Lennox, "Numerical Simulation of Waste Movement in Steady Groundwater Flow Systems," Water Resources Research 12, (2) pp. 171-180 (1976).

2. F. W. Schwartz, "On Radioactive Waste Management: An Analysis of the Parameters Controlling Subsurface Contaminant Transfer," J. of Hydrology 27, pp. 51-71 (1975).

3. R. L. Taylor and C. B. Brown, "Darcy Flow Solutions with a Free Surface," Proceedings, American Society of Civil Engineers, Journal of Hydraulics Division 93 (HY2), pp. 25-33 (1967).

4. T. G. Naymik, User's Manual for a Two-Dimensional Ground-Water Flow Code on the Octopus Computer Network, Lawrence Livermore Laboratory, Livermore, CA, UCID-17908 (1978).

5. J. O. Duguid and M. Reeves, Material Transport Through Porous Media: A Finite-Element Galerkin Model, Oak Ridge National Laboratory, Oak Ridge, TN, ORNL-4928, Environmental Sciences Division Publication 733, p. 201 (1976).

6. T. G. Naymik and G. D. Mendez, User's Manual for a Material Transport Code on the Octopus Computer Network, Lawrence Livermore Laboratory, Livermore, CA, UCID-17986 (1978).

7. J. Bear, Dynamics of Fluids in Porous Media (American Elsevier Publishing Co. Inc., New York, 1971).

8. T. Holdsworth, D. Towse, D. J. Isherwood, T. F. Harvey, and R. A. Heckman, High-Level Waste Repository Site Suitability Study--Interim Report, Lawrence Livermore Laboratory, Livermore, CA, UCID-17879 (1978).

9. I. Y. Borg, R. Stone, H. B. Levy, and L. D. Ramspott, Information Pertinent to the Migration of Radionuclides in Groundwater at the Nevada Test Site, Lawrence Livermore Laboratory, Livermore, CA, UCRL-52078 (1976).

EFFECT OF DOCK RESISTANCE ON TANKER MOORING FORCES

by Daryl B. Simons,[1] F., ASCE, Yung Hai Chen,[2] M., ASCE
and Lawrence J. Swenson[3]

ABSTRACT

A physical hydraulic model study was conducted to investigate the source of extremely high hydrodynamic forces that were acting normal to the longitudinal axis of deep draft tankers while moored to a massive T-head dock in the Mississippi River, and to test the effectiveness of remedial structures placed in the river for the purpose of reducing the lateral mooring forces.

Remedial structures, installed for the purpose of either reducing the velocity of flow between the ship and bank or to deflect the current around the ship, were not effective in reducing the forces. The ship and the dock structure act in conjunction to alter the flow pattern around the ship with the result of increasing the effective angle of attack between the ship and the current. In order to eliminate the lateral mooring forces, either the ship must be rotated, bow towards the bank, through an angle equal to the effective angle of attack, or the velocity of flow on both sides of the ship must be made equal.

BACKGROUND

The growth in the dimensions and displacements of the oil tankers and ore carriers attempting to navigate and dock in the river environment, where the draft of the vessels occupies a significant portion of the channel depth, has led to a reexamination of the commonly used river dock design practices. A review of the literature on the subject of dock design indicates that if docks are constructed parallel to the current streamlines, the lateral forces acting to repel the moored ship away from the dock will be minimal. However, the experience of at least one owner of a deep draft river port terminal indicates that the magnitude of the hydrodynamic lateral mooring forces significantly exceeds the predicted values.

A physical hydraulic model study was conducted to investigate the source of the excessive lateral current forces and to test the

---

[1]Associate Dean of Research and Professor of Civil Engineering, Colorado State University, Fort Collins, Colorado

[2]Research Associate Professor of Civil Engineering, Colorado State University, Fort Collins, Colorado

[3]Graduate Research Assistant, Civil Engineering Department, Colorado State University, Fort Collins, Colorado

effectiveness of various current deflecting structures placed in the river upstream of the docking facilities for the purpose of reducing the mooring forces. The model study consisted of measuring the total hydrodynamic forces on a 100,000 DWT tanker model (Tanker 1) and a 50,000 DWT tanker model (Tanker 2), as functions of river stage and velocity, in the following four phases: 1) tankers moored to a low density dock structure (Dock 1), 2) tankers moored to a high density structure (Dock 2),

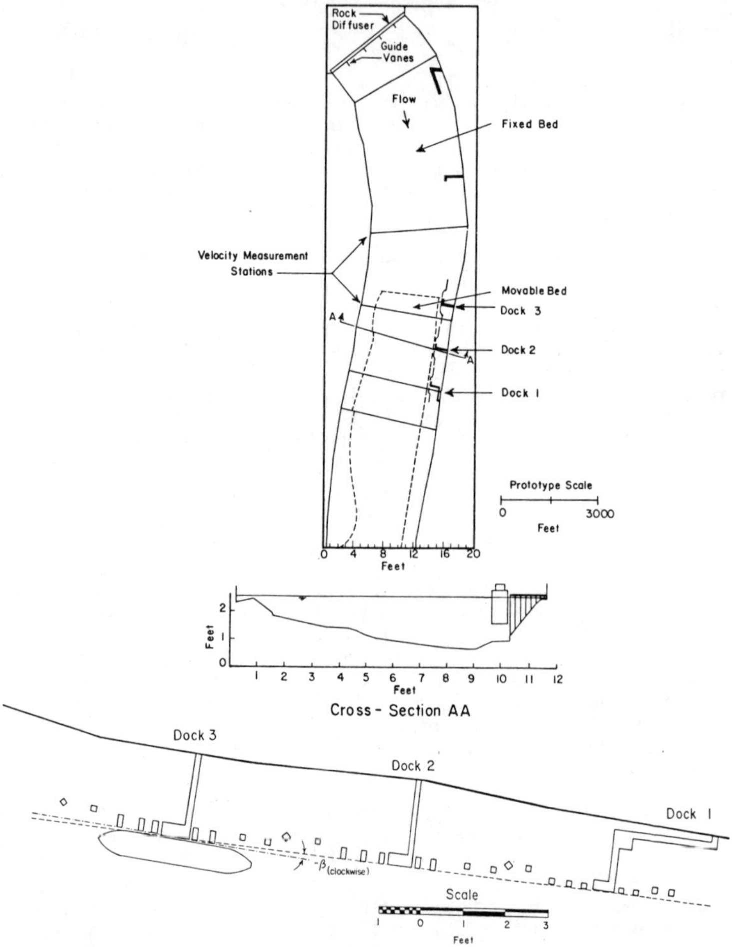

Figure 1. Schematic of Physical Model

3) tankers moored in the wake of current deflecting structures, and 4) given a particular dock structure design, (Dock 2), find the optimum angle between the dock structure and the current streamlines for minimum lateral mooring forces.

MODEL STUDY

The experimentation on the hydrodynamics of river docking was conducted in the River Mechanics Flume at the Engineering Research Center, Colorado State University. Figure 1 shows the geometric features of the River Mechanics Flume and channel reach. The horizontal scale was 1:225 and the vertical scale was 1:60. The vertical distortion of the model scales was mandated by the requirements of dynamic similarity for the sediment transport modeling portion of the study. The results of the sediment transport studies are outside the scope of this paper.

Two types of dock structures were tested. Dock 1, designed for tanker and barge traffic, is a steel T-head dock consisting of a main dock platform, with access bridge, and nine single pile dolphins. Dock 2 was constructed upstream of Dock 1, and consists of a massive T-head dock structure with a main platform, access bridge, five breasting dolphins and four mooring dolphins. The structures are shown in Figure 2. Dock 3, identical to Dock 2, was constructed upstream of Dock 2. The purpose of the fourth phase of the study was to determine the optimum orientation of Dock 3 with respect to the river current.

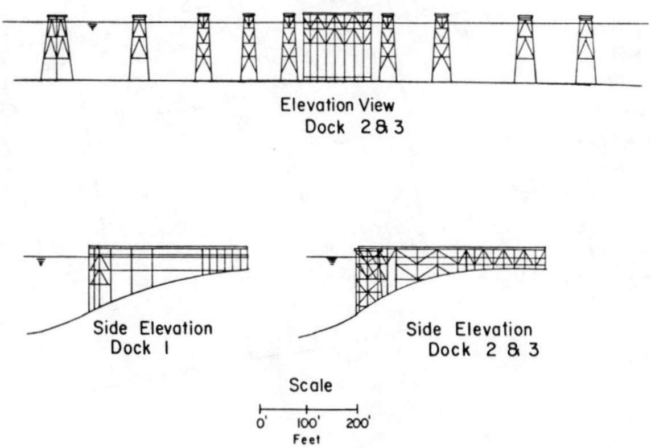

Figure 2. Comparison of Dock 1 and Dock 2 Structures

The river model was calibrated against the prototype channel for stage, discharge and current streamlines in the vicinity of the dock study area. Figure 3 shows the similarity of paths when the travel of floats placed in each was mapped. A miniature current meter was used to measure the distribution of velocity around the model ships and dock. Electronic force transducers were used to measure the forces acting upon the model ships. The transducer outputs were connected to chart recorders and an HP-1000 mini-computer to facilitate the acquisition and recording of data and the calculation of mean and root-mean-square force values. The transducers were mounted on the river banks and connected to the ships when the mooring forces were being measured during the first three phases of study. The fourth phase of the study required the construction of a rotatable platform on which the transducers and ships were connected. A schematic of the platform and a model ship is shown in Figure 4.

The Froude number of the model was equal to the Froude number of the prototype. Accordingly,

$$\frac{V_m}{\sqrt{g_m L_{v_m}}} = \frac{V_p}{\sqrt{g_p L_{v_p}}}$$

so that

$$\frac{V_m}{V_p} = \left(\frac{L_{v_m}}{L_{v_p}}\right)^{1/2}$$

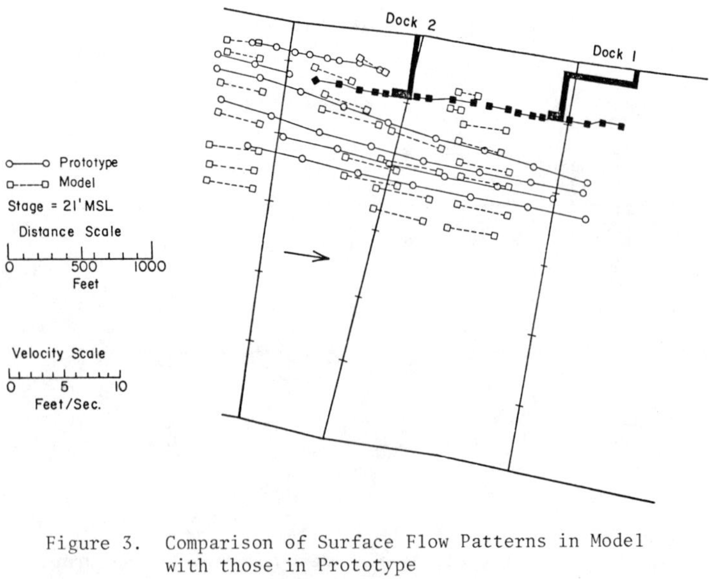

Figure 3. Comparison of Surface Flow Patterns in Model with those in Prototype

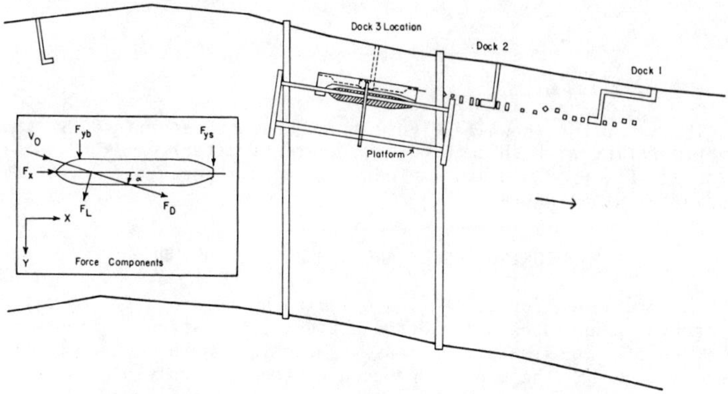

Figure 4. Rotating Platform Used to Measure Mooring Forces

The forces acting on the prototype and on the model were related using the ratio

$$\frac{F_m}{F_p} = \frac{\rho_m A_m V_m^2}{\rho_p A_p V_p^2} = \left(\frac{L_{H_m}}{L_{H_p}}\right)\left(\frac{L_{V_m}}{L_{V_p}}\right)\left(\frac{L_{V_m}}{L_{V_p}}\right) =$$

$$\left(\frac{1}{225}\right)\left(\frac{1}{60}\right)\left(\frac{1}{60}\right) = \frac{1}{810{,}000} \quad \text{or} \quad \frac{F_{m(lbs)}}{F_{p(tons)}} = \frac{1}{405}$$

The equations of the various force components (modified after Benham, Fang and Fetters, 1977) are as follows (see inset, Figure 4):

$$F_L = \frac{C_L \rho A_L V^2}{2} = F_y \cos\alpha - F_x \sin\alpha$$

$$F_D = \frac{C_D \rho A_D V^2}{2} = F_y \sin\alpha + F_x \cos\alpha$$

$$F_y = \frac{C_y \rho A_y V^2}{2}$$

$$F_y = F_D \sin\alpha + F_L \cos\alpha = F_{y_b} + F_{y_s}$$

$$C_y = C_D \sin\alpha + C_L \cos\alpha$$

$$A_L = LD = A_y$$

$$A_D = BD$$

The apparatus shown in Figure 4 was used to determine C_L, C_D and C_y by rotating each ship through an angle of attack of $\pm 20°$, while measuring F_x, F_{y_b} and F_{y_s}. The measurements were taken over a range of velocities and depths.

MOORING FORCES AND REMEDIAL STRUCTURES

The model bow forces (F_{y_b}) are scaled up to prototype magnitudes for each ship and dock and plotted against river stage in Figure 5. A bow force of approximately 45 tons was considered to be the threshold force above which the mooring procedure becomes unduly precarious and time consuming, when extreme care on the part of the tugs and line handlers is required.

The primary reason ships moored at Dock No. 2 experience very large forces directed upon the bow is that Dock No. 2 is located adjacent to the outside bank of a short radius curve of the river channel. As a result of this, the velocity and depth of flow, regardless of stage, are always near maximum values in the vicinity of Dock No. 2. In addition, the direction of flow immediately upstream of Dock No. 2 has a slight component normal to the starboard bow of tankers moored at Dock No. 2.

The following strategies were utilized for the selection of structures to reduce the forces on the tankers (see Figure 6):

(1) Install structures upstream of the bow which would cause high resistance to flow and thereby reduce the velocity of the current between the bow of the ship and the river bank (Alternative 1).

(2) Install structures that would deflect the high velocity flows away from the tanker thereby reducing velocities near the ship (Alternative 2).

(3) Install structures that would change the direction of flow and cause the river to flow parallel to the longitudinal axis of the ship. This would eliminate the normal component of the velocity vector (Alternative 3).

(4) Install structures that would change the direction of the ship to become parallel to the main flow direction (Alternative 4).

In addition, debris accumulation problems and effects of debris deflecting structures were investigated.

Evaluation of effects of Alternative 1, 2 and 3 Measures led to the following conclusions:

(1) High velocity fluid impinging on the starboard bow was not the only cause of the large transverse force.

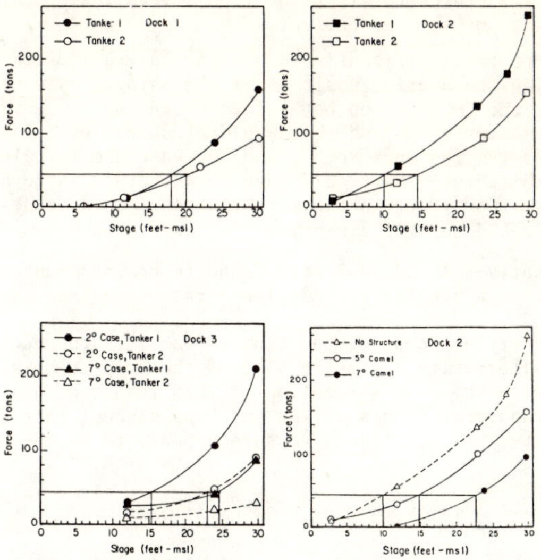

Figure 5. Force F_{y_b} versus River Stage

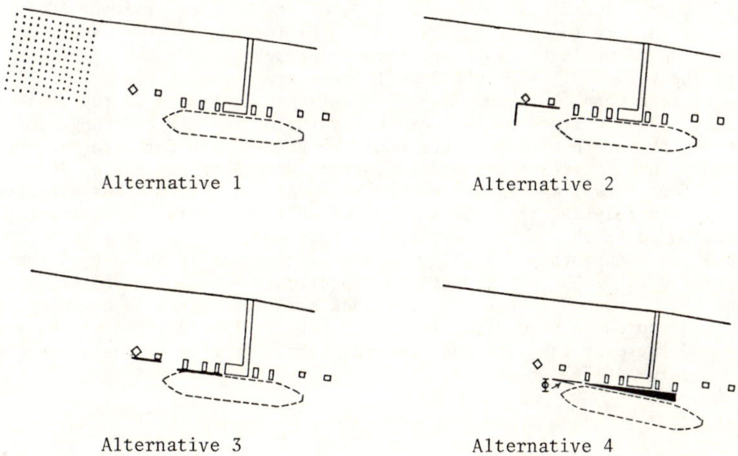

Figure 6. Mooring Force Reduction Structures

(2) Using structures to alter the flow streamlines parallel to the ship was not very effective.

(3) In order to significantly reduce the lateral forces without changing the mooring position of the ship, it was necessary to reduce the velocity on both sides of the ship. The structures required to accomplish this would block the exit lane of the tankers leaving Dock No. 2, and increase the turbulence acting upon the ship. They would have to withstand large hydraulic forces. Construction and maintenance of these types of structures would be very expensive.

(4) Alternatives 1, 2 and 3 were found to be inadequate and therefore a new force reduction strategy was required.

Instead of trying to alter the flow streamlines to become parallel to the ships, Alternative 4 Measure altered the ship to an alignment more parallel with the flow streamlines. This resulted in a significant reduction of transverse forces acting on ships without sacrificing the mooring stability. Alternative 4 Measure appears to be the most effective strategy.

OPTIMUM DOCK ORIENTATION

The purpose of the final phase of the model study was to determine the optimum orientation ($-\beta$, in Figure 1) of the longitudinal axis of the proposed Dock 3, so that the lateral mooring forces on a 100,000 DWT tanker would remain within acceptable limits.

Before Dock 3 was installed, Tanker 1 was connected to the rotating platform and placed upstream of Dock 2, during a river stage of 30 feet, as if it were in the mooring position. The forces F_x, F_{y_b} and F_{y_s} were measured while the ship was turned through the angles, β, of $\pm 9°$. The values of F_{y_b} obtained thereby are shown in Figure 7. F_y equaled zero when β equaled -3.5°. As shown in Figure 7, the angle of attack, α, was defined as zero at this value of β. At this angle, however, the stern of the tanker would be rotated too far into the navigation channel. It was decided to install Dock 3 at the angle β of -2°. After Dock 3 was installed at β equal to -2°, F_{y_b} was elevated to approximately 290 tons, at a stage of 30 feet. This value of F_{y_b} corresponded to that obtained when α equaled 10.5° in the free stream. Dock 3 was then rotated to -7° and moved approximately 80 feet closer to the river bank. F_y was reduced to approximately 150 tons (see Figures 7 and 5). It was apparent that the dock structure acted in conjunction with the moored ship to significantly alter the flow pattern around the ship and increase the resultant lateral mooring forces corresponding to an α value of 5.3°.

ANALYSIS

Strategies

The desirability of developing a methodology for predicting the magnitude of the lateral mooring forces, given the structural properties

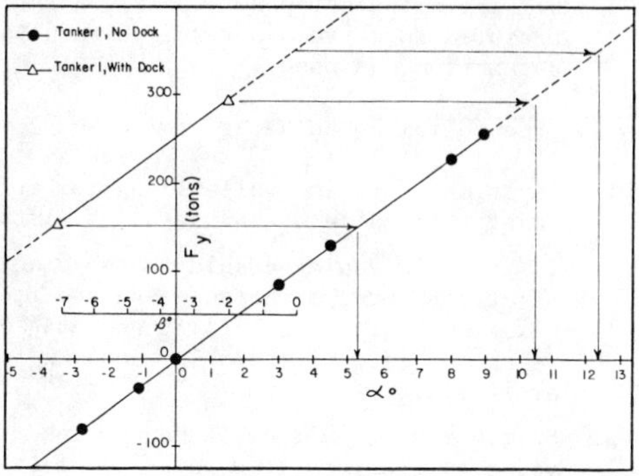

Figure 7. Effect of Dock 3 Structure on Mooring Forces

of the dock and the flow properties of the river, was recognized. The following strategies for obtaining an analytical solution were considered:

(1) Assume the dock resistance to flow results in a backwater effect and raises the water surface between the bank and the ship to an elevation higher than that on the river side of the ship. The difference in elevation causes a pressure imbalance tending to force the ship away from the dock.

(2) Assume the dock and ship act in conjunction to change the direction of river flow upstream of the bow of the ship which results in an angular difference between the flow streamlines and the longitudinal axis of the docks. The lateral forces could then be estimated by applying the laws of hydrodynamic lift and drag.

(3) Apply the fluid mechanics principles of force-momentum flux to estimate the quantity of flow that is diverted away from passing down the bankside of the ship to passing down the river side of the ship, causing a velocity gradient between the sides of the ship. Then, assuming the ship hull acts as a vertically oriented airfoil, apply the laws of circulation to calculate the lift, where, in the two-dimensional case:

$$F_L = \Gamma \ell \rho V$$

These strategies had certain drawbacks:

Strategy 1: The differences in water surface elevations in the model were too small to measure accurately.

Strategy 2: The angle between the axis of the ship and the direction of flow was very small, fluctuated dramatically in the turbulent flow, and depended upon where the angle was measured.

Strategy 3: Significant discrepancies between the calculated and measured force values resulted when the two-dimensional assumption was used.

The strategies had the following advantages:

Strategy 1: Empirical data is available on the effect of piers and constrictions on backwater and resistance to flow.

Strategy 2: If the dock could be said to cause an "apparent" angle of attack between the ship and the flow, then the classical equations of lift and drag could be applied using the data of ship C_L and C_D which are already available.

Strategy 3: Strategies 1 and 2 could be combined to calculate the velocity gradient from one side of the ship to the other and the two-dimensional circulation equation could be adjusted by a function of the draft to depth ratio to approximate the three-dimensional case.

APPARENT ANGLE OF ATTACK (α')

A methodology for estimating the total mooring forces caused by river current is being developed, and it is based on the concept of the "apparent angle of attack (α')". This angle is a fictitious angle between the oncoming flow and the longitudinal axis of the ship, which is created when a structure is placed in a current between a ship and a nearby bank, such that when that angle is used in the lift and drag equations with the coefficients of lift and drag obtained for the ship in the free stream, the forces calculated thereby will equal the forces experienced by the ship when moored to the structure.

Figure 8 shows the measured values of F_y for Tanker 1 at each of the docks as a function of stage. Also shown are the values of α' which when used with the free stream values of C_D and C_L in

$$F_y = F_D \sin\alpha + F_L \cos\alpha$$

yielded approximately the measured F_y values. Note that the sum of β and α' are nearly constant for the Dock 2 and Dock 3 cases as follows:

Stage	Tanker	Dock	β	α'	$\beta + \alpha'$
30	1	2	0	15	15
30	1	3	2	12.5	14.5
30	1	3	7	7	14

This was expected, as the design of Docks 2 and 3 are identical. The same trends were experienced with Tanker 2. Note that α' at Dock 1 ($\beta = 0$) equals 8°. The exposed structural area normal to the flow in Dock 1 is approximately one-half that of Docks 2 and 3.

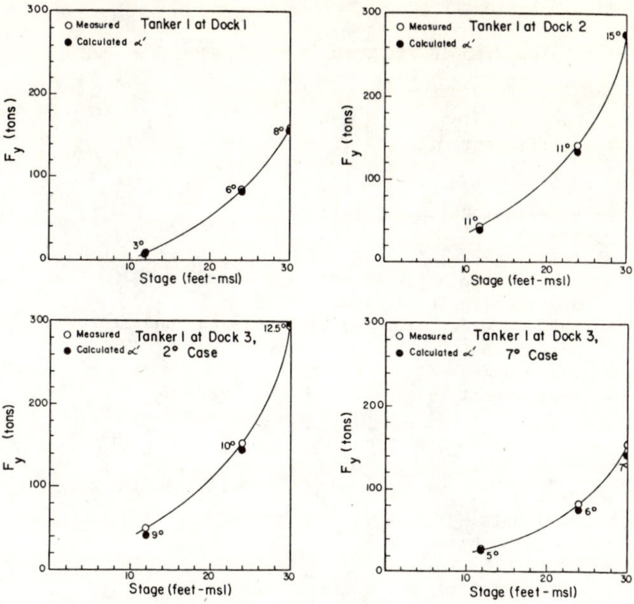

Figure 8. Apparent Angle of Attack (α') versus River Stage

CONCLUSIONS AND RECOMMENDATIONS

The most effective strategies for preventing high lateral forces on a ship moored in a river current are to: 1) maintain nearly equal velocities on both sides of the ship by minimizing the resistance to flow caused by the dock structure, or 2) if dock structural stability requirements preclude a low resistance structure, then the dock must be rotated towards the bank to generate a component of hydrodynamic lift nearly equal to the repelling forces caused by the structure.

The information provided in this paper is valid only for the docks modeled and is not intended to be sufficiently complete to calculate α' for any proposed dock structure. It is recommended that further research be conducted for the purpose of developing a methodology for estimating α', given the structural properties of a proposed dock, and the flow parameters of the river at the dock site.

Appendix I - Reference

1. Benham, F. A., Fang, S. S., and Fetters, C. H., "Wind and Current Shape Coefficients for Very Large Crude Carriers," Proceedings, Offshore Technology Conference, Houston, Texas, May, 1977.

Appendix II - Notation

The following symbols are used in this paper:

A = area
B = beam of ship
C = coefficient
D = draft
F = force
g = gravitational acceleration
L = length
V = velocity
α = free stream angle of attack
α' = apparent angle of attack
β = angle between proposed dock axis and undisturbed flow streamlines
Γ = circulation
ρ = mass density

<u>Subscripts</u>

D = drag
H = horizontal
L = lift
m = model
p = prototype
V = vertical
x = x direction
y = y direction

PHYSICAL HYDRAULIC MODEL STUDY OF AN ICE-COVERED RIVER

BY

R. P. Johnson*, Associate Member, ASCE
T. V. Kotras**, Member, ASCE

ABSTRACT

A physical model study was conducted to investigate possible river regulating policies which would maintain the section of the Mississippi River from the mouth of the Ohio River to St. Louis sufficiently clear of ice to minimize damage to river regulating works and to permit winter commercial tug-barge operation. This study was undertaken following the winter of 1976-77, during which ice on the Mississippi River caused damage to dikes and revetment in excess of $1.5 million between the confluence with the Ohio River and Chester, Illinois, river mile 0 to river mile 110. In addition, during the same winter, the Mississippi River above the Ohio was closed to navigation for 27 continuous days.

Using a unique modeling material which simulates the properly scaled physical and mechanical properties of ice, a series of hydraulic model tests were conducted of a typical bend in the Mississippi River. These tests examined: (1) the conditions necessary to initiate an ice jam; (2) the development of the jam and its effect on river hydraulics; and (3) the conditions under which a jam would collapse. During the tests, discharge and water levels were monitored along with the forces exerted by the ice on a single dike and on a series of dikes in different configurations and locations relative to the river bend.

Various schemes to "flush" the ice cover from the problem sections of the river by manipulating the river discharge and stage were simulated in the hydraulic model. These schemes failed to force the ice downstream until increases in the flowrate of 90% or more were obtained. It was concluded that such large, controlled variations in the flowrate were not feasible in this section of the river and that the most likely method of solving or minimizing ice problems was through prevention, using ice control or ice management systems rather than dealing with the problems after they occur.

* Consulting Engineer, ARCTEC, Incorporated, Columbia, Md.
** Vice President, ARCTEC, Incorporated, Columbia, Md.

BACKGROUND

The Mississippi River between the confluence with the Ohio River at Cairo, Illinois, and the confluence with the Missouri River at St. Louis, Missouri, does not typically experience serious river ice problems. However, during the particularly harsh winter of 1976-77, a rough, thick ice cover formed over an extensive section of the river (see Figure 1). The river ice conditions were described by the U.S. Army Corps of Engineers as follows:

> "...on 1 January 1977, dredging operations had to be halted because of heavy floating ice. Low flow and subzero temperatures produced the worst combination of ice and low water on the Middle Mississippi River in the memories of veteran rivermen. By the second week in January, ice began making navigation difficult at the upper end of the Chain of Rocks Canal and also at river mile 12 upstream of Cairo, Illinois...On 19 January, heavy ice conditions existed from river mile 12.5 to 54.1 and low temperatures indicated there would be no immediate improvement in the icing conditions. On this date, the Commander, Second U.S. Coast Guard District, closed the river between river mile 160.0 and Cairo, Illinois,...The Middle Mississippi River was officially closed for a period of 27 days. Reopening officially occurred at 1800 hours on 14 February following a breakup of the ice on 12 February which was accredited to warming temperatures." [4].

The closing of the river to commercial navigation slowed the southbound shipment of grain to New Orleans for export. Northbound commodities, such as steel, fuel, sugar and street salt, were effectively blocked. There were approximately 900 barges delayed at Cairo, Illinois due to the ice [4]. In addition to halting or delaying the transport of these vital commodities to communities in the midwest, the ice caused considerable damage to river training works; the moving ice eroded dike heads and carried away bank revetment. Extensive damage was reported between Cairo and Chester, Illinois. The cost to repair the damage from the ice was estimated, at 1977 prices, to be in excess of $1.5 million [4].

Since the conditions which arose during the winter of 1976-77 were not typical, it was not clear at the time what measures, if any, should be taken to alleviate the problem. To gain insight into the ice-river interaction and to test various river control strategies a series of ten physical ice-hydraulic model tests were conducted of a typical two mile section of the river. The objectives of this test program were to:

1. Demonstrate the capability of simulating ice-river interactions by use of physical models.

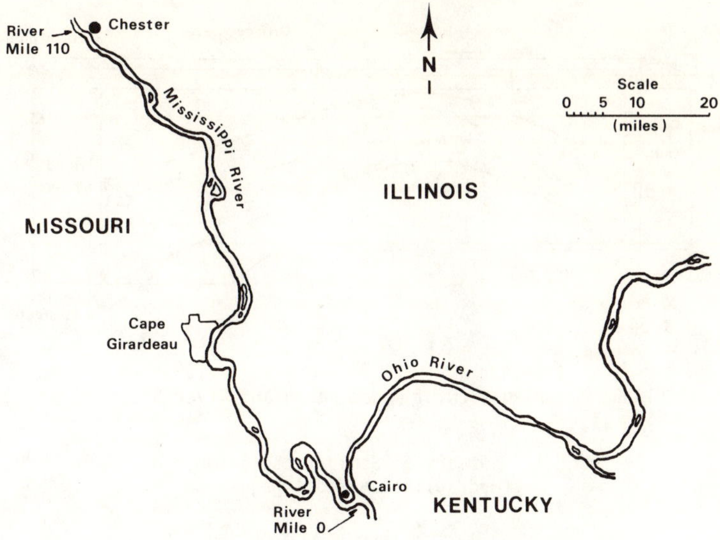

FIGURE 1. Ice Problem Area, Winter 1976-77, River Mile 0 to River Mile 110

2. Study ice arching and jamming mechanics in operation on a river.

3. Examine possible flow and stage manipulation strategies that may be used in encouraging the breakup of ice formations on a river.

4. Estimate forces on dikes in order to minimize structural damage due to ice and to study the effect dikes have on arching, jamming, and ice sheet failure.

MODEL DESCRIPTION

A fixed bed hydraulic model, as shown in Figure 2, was constructed of a typical two mile section of the Mississippi River between Cairo and Chester. The model was geometrically distorted with a horizontal prototype to model-scale ratio of 225:1 and a vertical scale ratio of 90:1, for a distortion ratio of 2.5. The important scale relationships for a properly scaled physical ice-hydraulic model have been developed by Kotras, et al. [2]. The scale relationships for this study are presented in Table 1.

Two different materials were used to simulate the river ice cover. Polypropylene plastic sheets were cut into 1, 3 and 6 inch squares and 6 inch irregular octagons to simulate 19, 56 and 113 foot

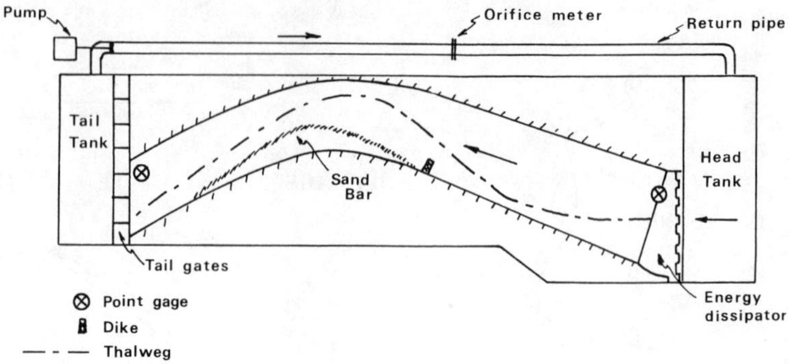

FIGURE 2: Plan View of Model Basin and River Section

TABLE 1. Summary of Scale Factors for the Hydraulic Ice Model [2]

	SYMBOL	SCALE RELATION	SCALE FACTOR
Horizontal Scale	λ	λ	225
Vertical Scale	β	β	90
Area, Horizontal	A_H	λ^2	50625
Area, Vertical	A_V	$\lambda\beta$	20250
Volume	∇	$\lambda^2\beta$	4556000
Time	T	$\lambda\ \beta^{1/2}$	23.7
Mass	M	$\lambda^2\beta$	4556000
Density	ρ	1.0	1.00
Water Velocity	V_w	$\beta^{1/2}$	9.49
Water Discharge	Q	$\lambda\beta^{3/2}$	192110
Water Depth	y	β	90
River Slope	S	β/λ	0.40
Bed Friction Factor (Darcy-Weisbach)	f_b	β/λ	0.40
Ice Cover Thickness	t	β	90
Ice Cover Friction (Darcy-Weisbach)	f_i	β/λ	0.40
Ice Cover Flexural Strength	σ_f	β	90
Ice Cover Crushing Strength	σ_c	β	90
Ice Cover Shearing Strength	σ_s	β	90
Ice Cover Elastic Modulus	E	β	90

square and 113 foot octagonal ice floes. These ice floe sizes span the range found in aerial photographs of the Mississippi River taken during the winter of 1976-77. While the plastic had the same density as fresh-water ice, it did not possess the hydrodynamic roughness or mechanical strength properties to scale according to the ratios presented in Table 1; they were too strong by several orders of magnitude and much too rigid. These plastic ice floes were used primarily to investigate geometry dependent phenomena, such as the occurrence of arching and the development of an unconsolidated ice cover into an ice jam.

The second material used to simulate the river ice cover was unique in that it could be controlled by varying the ratio of the components in the mixture to produce the properly scaled hydrodynamic roughness and strength properties, as well as the proper density. The material is a mixture of several chemicals and is used at room temperature. This model ice material was used in tests of consolidated level ice covers and in tests examining the development of ice jams composed of unconsolidated ice floes.

UNCONSOLIDATED ICE JAM TESTS

The initiation and development of a river ice cover may occur in several ways. One way is for the ice to grow out from the banks in smooth, level sheets that eventually meet in the middle. Another way is by arching of the ice floes which are moving downstream with the current. An arch is likely to occur when several floes try to pass through a surface constriction that is too narrow to accommodate all of them simultaneously. When a stable arch forms across the river, the incoming ice floes are blocked from continuing downstream and stack upstream from the arch, forming an unconsolidated ice cover.

Arching in the model occurred only with the 6 inch squares and the irregular octagons. The smaller simulated ice floes could not form and maintain a stable arch. Examination of aerial photographs of the Mississippi River taken during the winter of 1976-77 shows that arches did form at the tight river bends and that very large ice floes were involved.

In nature, an unconsolidated ice cover may thicken and become very rough, forming an ice jam, in two ways. The incoming ice floes may be submerged under the upstream edge of the ice cover. Or it may thicken by "shoving", which occurs when the cover is not strong enough to resist the drag forces exerted by the current. In this case, the cover will thicken by ice floes riding over one another until the forces between the floes become large enough to stabilize the cover and resist the drag forces [3].

Both of these ice cover thickening mechanisms were observed in the model. At the typical winter flowrate of 50,000 cfs and low water elevation, the smaller ice floes would submerge under the ice cover and cause it to thicken. As the flowrate was increased the unconsolidated cover "shoved". The presence of this thick unconsolidated ice cover, or ice jam, offered considerable resistance to the flow of water and caused an additional increase in head above the jam. This rough, thick cover would pose a serious problem to navigation and would exert large forces on river training structures.

ICE FLUSHING TESTS

Control strategies to manipulate the river discharge and stage were tested to evaluate their effectiveness in facilitating the breakup of ice formations on the river. Both unconsolidated ice jams and consolidated level ice covers were investigated. The river discharge control strategies that were tried included both raising and lowering the flowrate from a typical winter value of 50,000 cfs. The river stage control strategies included both raising and lowering the model tailwater elevation.

The unconsolidated ice jams were able to be flushed around the model river bend by increasing the discharge. The increase in discharge caused an increase in the water velocity which exerted more drag force on the broken ice cover and increased the local head drop across the jam. In the four ice jam flushing tests, the increase in head varied from 1.0 to 2.3 feet, full-scale. These conditions are in qualitative agreement with those observed on 12 February 1977 when a jam collapsed at approximately river mile 40 on the Mississippi River after an increase in stage of approximately 4 feet had been observed upstream from the jam [4].

The consolidated level ice was very resistant to flushing by manipulation of the river discharge and stage. In two of the five tests it was possible to move the level ice downstream by increasing the flowrate from 50,000 cfs prototype to 90,000 cfs and 95,000 cfs prototype. Such large, controlled increases in the flow are not feasible in this section of the Mississippi River during the winter months.

The effectiveness of raising or lowering the water level was also investigated. Even though the stage was lowered the full-scale equivalent of 6 feet and raised the equivalent of 9 feet, the ice sheet did not move downstream. It did crack away from the shore and rotate into a new position where the river banks could take up the forces exerted on the ice.

FORCES ON DIKES

Various configurations of stone dikes were tested, both singly and in a series, on both the inside and outside of the bend. The forces exerted by the ice on the model dikes were measured using a strain-gaged force block. Sloped, level,and notched dikes were tested to determine if there was a difference in the ice forces exerted upon them. For the typical winter low river stages that were modeled, a large portion of the dike was exposed, so that the ice action was essentially the same on the different types of dikes. There was no significant difference between the forces exerted upon the dikes located on the outside of the bend compared to those located on the inside of the bend.

In unconsolidated ice the force on the dike varied as shown in Figure 3. In one case, it was no greater than those recorded under similar hydraulic conditions in open water. In another case, the force on the dike was as much as 200 tons greater than the open water value. The large variability may be explained qualitatively by reference to Figures 4a and 4b. The force in the ice cover is transmitted to the river banks by arches of ice floes [3]. If the floes transmitting the force abut the dike, as shown in Figure 4a, the load on the dike may be very high. If no floes abut the dike, as shown in Figure 4b, the load on the dike will be very low.

Single dikes and a series of dikes were tested in consolidated level ice covers. As was to be expected, the upstream, or single, dike carried the largest ice loads. Figure 5 shows the forces on a single dike and the downstream dike in a pair. The maximum load recorded on the single dike was 1,000 tons, full-scale.

The dikes, whether located in the inside or outside of the bend, had no apparent effect on the formation of a stable arch at the model river bend. Furthermore, the dikes appeared to have no effect on the development of an ice jam of unconsolidated floes. In consolidated ice covers, the ice exerted large forces upon the dikes; however, when the load was removed by cutting the model ice away from the dike the ice cover did not move. Apparently, the dikes were not necessary to stabilize the ice cover because the load could be distributed to the river banks.

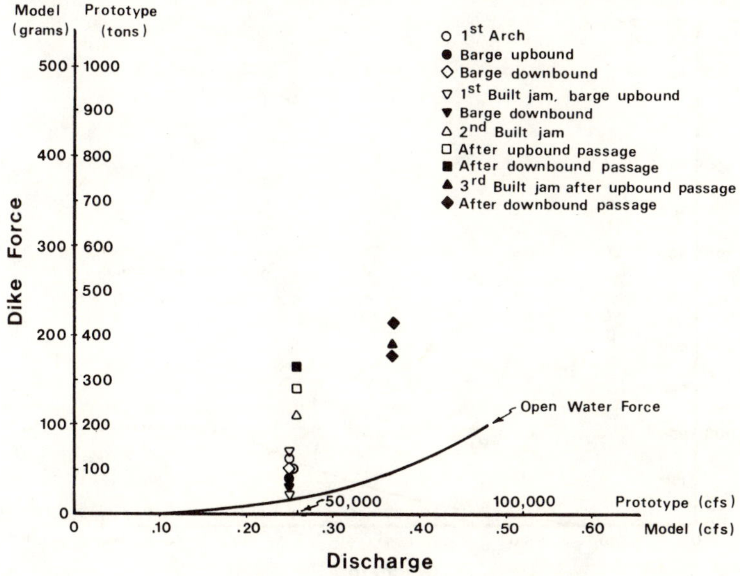

FIGURE 3. Dike Force Vs. Discharge
In Unconsolidated Ice

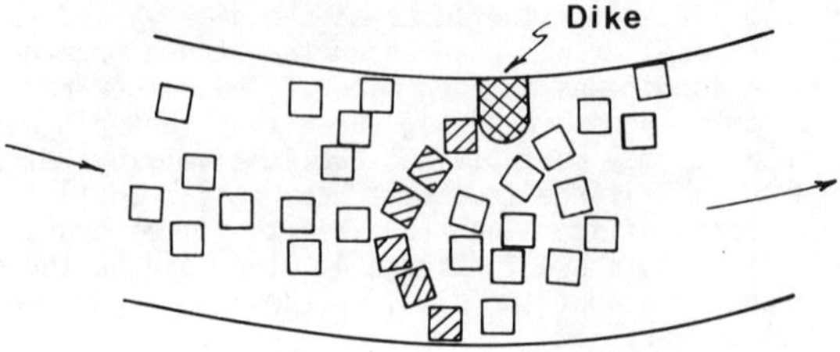

a. Arch Transmitting Large Force to Dike

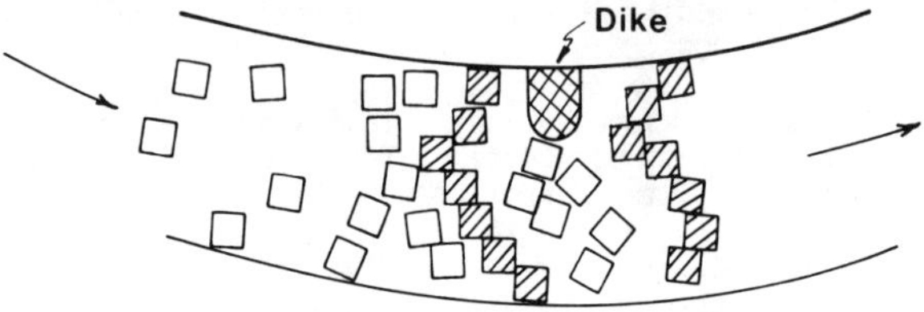

b. Arch Transmitting Little Force to Dike

FIGURE 4. Diagram of Arches in a Broken Ice Field Around Dike

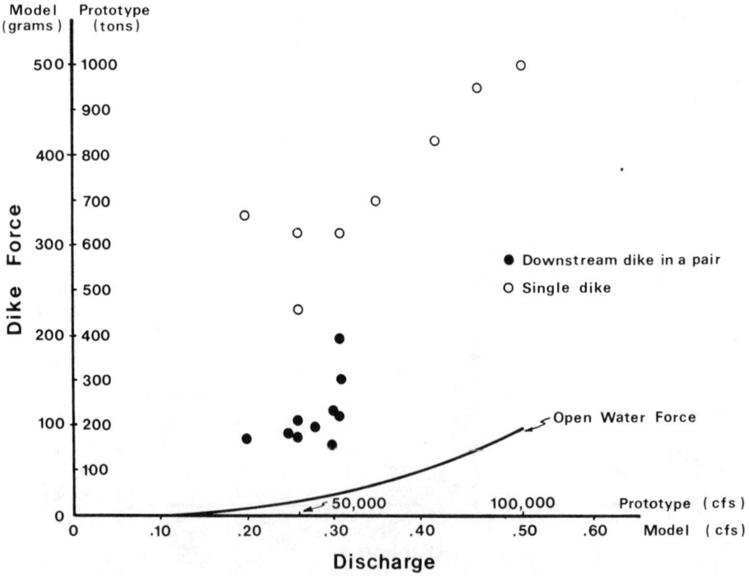

FIGURE 5. Dike Force Vs. Discharge in Consolidated Ice

CONCLUSIONS

The following major conclusions were drawn from this study:

1. The phenomenon of ice arching at a specific river section is dependent upon the ice flow concentration, ratio of the ice floe size to river opening, river geometry and floe shape. For the case studied, an arch would form at the bend only with the 113 foot ice floes. Sections of the river similar in geometry, flow, and stage characteristics to the model will be susceptible to jamming as ice pieces submerge under the upstream edge of the ice cover. The ice cover may thicken by shoving for river discharges greater than 50,000 cfs.

2. Once an unconsolidated ice jam occurred, the jam in the model could be forced to collapse due to an increase in the head drop across the jam. Unconsolidated jams in the model were broken up and flushed downstream under conditions qualitatively similar to those observed in the field. It is important to note, however, that the specific amount of head increase necessary to collapse the jam is dependent upon the stability of the jam and the degree of consolidation that has occurred, as well as upon river conditions. Because of this problem, relatively quick response to the formation of the jam is important; that is, the river discharge should be increased quickly in order to collapse the jam and flush it downstream before the jam consolidates and becomes stable. Also, there are no guarantees that increasing the head behind a jam will collapse it. In fact, the opposite can occur if the head is not increased sufficiently to break the jam; in this case conditions can be made worse by the submergence of additional ice underneath the upstream edge of the jam, which would make the jam longer, thicker and stronger.

3. The consolidated ice cover could not be flushed downstream by increasing the discharge or stage within reasonable limits. For the case studied, the discharge had to be increased 90% or more to flush the ice cover downstream. Again, it is important to note that increasing the flowrate may cause the problem to worsen as the ice tends to break, allowing larger pieces to submerge and raft. This would result in a thicker ice cover, with no assurance that the cover would be forced downstream.

4. The dikes experienced very large forces even when the ice did not move. In tests with a pair of dikes, the downstream dike experienced approximately half the force of a single dike. In unconsolidated ice covers, the force on the dike was dependent upon the random distribution of floes in the ice field, and the way in which they transmitted the force to the shore. The maximum dike force recorded was 1000 tons, full-scale.

ACKNOWLEDGEMENT

This study was conducted under contract to the University of Missouri at Rolla, Institute for River Studies, for the U.S. Army Corps of Engineers, St. Louis District. The authors wish to thank Dr. Glendon Stevens of the University and Mr. James T. Lovelace and Mr. Claude Strauser of the Corps of Engineers for their guidance and assistance in conducting this study.

REFERENCES

1. Johnson, R. P., T. V. Kotras, D. L. Benze, "Hydraulic Model Tests of a Typical River Section With Ice," submitted to the Institute of River Studies, Engineering Research Laboratory, University of Missouri at Rolla, ARCTEC, Incorporated Report No. 453C, August 1978.

2. Kotras, T. V., J. W. Lewis, R. A. Etzel, "Hydraulic Modeling of Ice Covered Waters," Fourth International Conference on Port and Ocean Engineering Under Arctic Conditions, Memorial University of Newfoundland, 1977, pp. 453-463.

3. Pariset, Hausser and Gagnon, "Formation of Ice Covers and Ice Jams in Rivers," Journal of the Hydraulics Division, ASCE, HY6, Nov. 1966, pp. 1-24.

4. "Report on Mississippi River Ice, 1976-1977," U.S. Army Engineer District St. Louis, Corps of Engineers, St. Louis, Missouri, Nov. 1977.

CALIBRATION OF NUMERICAL MODEL FOR HARBOR WAVES

Herman Bomze[1], Fellow ASCE

ABSTRACT

This paper describes the calibration of a numerical model for calculating the response of the water surface in a harbor subject to excitation by very long open-ocean waves. The numerical model is a Green's function/integral equation formulation of the Helmholtz equation in the nearshore domain, matched to an analytic solution for the open-ocean domain along a semi-circular boundary. The model was applied to the harbor of Acajutla, El Salvador, which had previously been model tested at the Hydraulics Research Station, Wallingford, England. Response amplitude operators (RAOs) were computed frequency-by-frequency at several points in the existing harbor and in several harbor layouts for a new container berth. Calibration factors were then obtained as the ratio of the physical model RAO to the computed RAO at each frequency.

Although the calibration factor at a given frequency varies from point to point in a harbor, the study demonstrated that the set of calibration factors derived for one harbor point produced a reasonable approximation to the significant wave heights at other harbor points. Furthermore, the set of calibration factors for the existing harbor led to realistic predictions of the harbor wave climate in different harbor layouts.

INTRODUCTION

The mathematics for describing the oscillations of a free water surface were developed over a century ago. However, practical applications of the theories were limited by inability to carry out the myriad of computations required. Solutions for harbor oscillation problems, therefore, had to be obtained by measurements in carefully controlled hydraulic model tests which simulated the real world.

The situation has changed with the advent of high speed electronic computers and the development of software for the solution of many problems in hydrodynamics. This makes mathematical modeling readily available and economical. However, most currently available computer programs for harbor oscillations are predicated on the simplifying assumptions of linear wave theory, and the only form of damping included in the solution is the energy in the wave radiating out of the harbor. These assumptions ignore real fluid effects such as viscosity, turbulence, eddy formation, flow separation, wave breaking, bottom friction, etc., all of which contribute to dissipation of energy and thus constitute additional forms of damping. Since damping causes a significant reduction in the amplitude of oscillations at resonant frequencies, a mathematical model based

[1] Assistant Vice President, PRC Harris, Inc., Lake Success, N.Y.

on linear wave theory will tend to predict resonant harbor oscillations several times larger than observed in nature. Linear wave models must therefore be calibrated by measurements either in the real harbor or a model harbor. Real world measurements are of course preferable since hydraulic model tests have their own built-in distortions due to scale effects and wave reflections from the model basin boundaries.

The technical literature abounds with mathematical predictions of high harbor waves at low natural frequencies of a harbor which could not be confirmed by real world or hydraulic model measurements. To cite a few examples: in a study of Long Beach Harbor, Hwang et al. (1) computed a root-mean-square amplification of ocean waves of 4.7 at the fundamental natural frequency; yet the measured amplification in the harbor ranged from about 0.9 to 2.9. Chen and Mei (2) and Ho and Bomze (3) studied numerically an offshore harbor formed by breakwaters designed to protect floating nuclear power plants. Although the mathematical formulations were different, both were based on linear wave theory and computed a root-mean-square amplification at the fundamental natural frequency of the breakwater enclosure of about 5; yet model tests by Dean et al. at the University of Florida and subsequently by Madsen and Vyas (4) at MIT showed very little amplification. Ito (5) has shown that flow separation at the end of a breakwater can totally suppress resonance at the fundamental natural frequency of a harbor. None of these studies attempted to correlate the computed results with the measurements.

Bomze (6) describes a mathematical model for the Port of Acajutla, El Salvador, and a procedure for calibrating the computed results against measurements at the Hydraulics Research Station, Wallingford, England (7). In (6) one harbor point was selected at which the response spectrum to six offshore wave spectra had been measured, and calibration factors were derived frequency-by-frequency to obtain a match between the computed response amplitude operators (RAOs) and the mean of the test RAOs. In order for calibration to be of practical value, it needs to be shown that (a) a set of calibration factors derived for one harbor point will lead to a reasonable computation of the significant wave height at other harbor points, and (b) the same set of calibration factors will also predict reasonable wave heights when the harbor configuration is modified. If these conditions exist, it would prove expeditious to conduct just enough measurements, in the real world or in a physical model, to calibrate a numerical model for harbor oscillations and to then use the calibrated numerical model for the study of alternate harbor layouts.

This paper applies the calibration procedure at a single point in Acajutla harbor described in (6) to two additional points. It establishes a valid range of frequencies between 0.001 and 0.020 Hz in which each of the three harbor points can be calibrated and demonstrates that the computation of a significant wave height at a point in the harbor may be performed with calibration factors determined at another point. Finally, it examines a modification of the existing harbor and demonstrates that the wave climate in the modified harbor can be predicted with reasonable accuracy by use of the calibration factors obtained for the existing harbor.

NSWICH PROGRAM FOR HARBOR OSCILLATIONS

NSWICH is a very efficient computer program, written in FORTRAN to run on a CDC 175 computer, for solving the Helmholtz equation which governs the oscillation of the free surface in a bounded region. The computational efficiency derives from combining a Green's function/integral equations solution in the near-shore region containing the harbor with an analytic solution for the open ocean region, the two solutions being matched on a semi-circular boundary. Figure 1 is a definition sketch which shows this boundary as well as additional boundaries for regions of different water depth inside the harbor. Up to four such regions can be accommodated by the NSWICH program. It can be shown (8) that the representation of a sloping harbor bottom by a series of vertical steps does not introduce significant errors into the solution. The principal causes of deviation of NSWICH results from the real world are the assumptions of totally reflective harbor boundaries (necessary for a Green's function formulation) and of linear wave theory which ignores all real fluid effects such as viscosity, bottom friction, wave breaking, turbulence, flow separation, eddy formation, etc.

The only form of damping considered by NSWICH is radiation damping of the wave exiting from the harbor. Since all other forms of damping have been ignored, and since a small amount of additional damping will cause a large reduction in resonant amplitudes, the computed resonant amplifications must be calibrated downward to match real world observations.

The input for the NSWICH program consists of the harbor boundary and bathymetry (in up to four increments of water depth); the incident wave direction; the wave frequencies of interest; and a grid of interior harbor points at which the wave height is to be calculated as a ratio of the incident wave height. This ratio is the RAO for the harbor wave height at the given point and wave frequency. The output is a tabulation of the RAO at each harbor grid point or special point of interest and the phase relationships between all the points. Water surface contours can be plotted as isobars of the RAOs at any desired phase of the harbor oscillation.

DERIVATION OF CALIBRATION FACTORS

For the reasons discussed earlier, the NSWICH RAOs need to be calibrated against real world or physical model measurements. The calibration procedure is illustrated on Figure 2 which shows three hypothetical wave spectra: an offshore spectrum Si; a measured spectrum So(T) at a point in the harbor; and a computed spectrum So(C) at the same point. By definition, the measured RAO is then $\sqrt{So(T)/Si}$, and the computed RAO is $\sqrt{So(C)/Si}$. The calibration factor is simply the ratio of the measured to the computed RAO. It will be seen that this ratio is identical to $\sqrt{So(T)/So(C)}$.

The measured RAO does not remain constant as the offshore wave spectrum changes. However, the variability was not found to be extreme. It appears reasonable to base a calibration factor on the mean of the measured RAOs. As a consequence the computed response spectra using the calibrated RAOs will not match the measured response spectra precisely, and some discrepancy is to be expected between the computed and the measured significant wave heights in the harbor.

VALID FREQUENCIES FOR CALIBRATION

Since the computed RAO appears in the denominator of the fraction which constitutes the calibration factor, a numerical instability arises at those wave

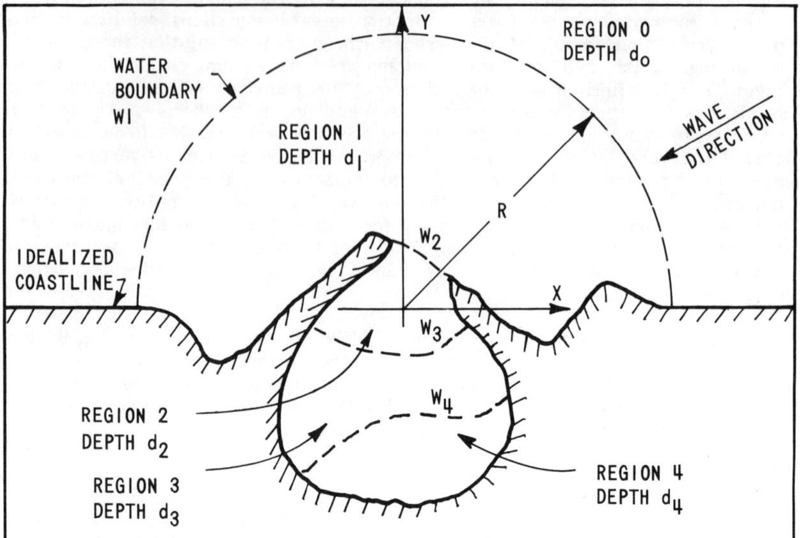

FIG. 1 - NSWICH PROGRAM DEFINITION SKETCH

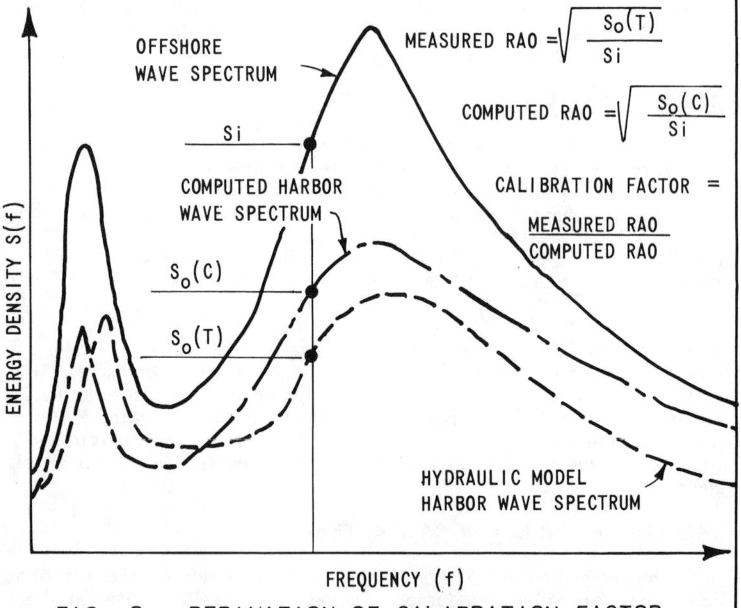

FIG. 2 - DERIVATION OF CALIBRATION FACTOR

frequencies which produce a node (RAO=0) near the point being calibrated. For the very long waves in the frequency range of .001 to .020 Hz studied at the Port of Acajutla, it was found that the calibration factor at a harbor point exceeded unity whenever the RAO at that point was less than about 0.5. As a practical matter for engineering evaluations, e.g., if measurements are available only at a single harbor point, one may have to use calibration factors in excess of unity at the few closely spaced frequencies at which they occur. The computed response spectrum will then show a much higher energy density at these frequencies than in the real world, and the significant harbor wave height computed from the response spectrum will be conservative. The degree of conservatism will depend on the shape of the offshore spectrum, but will generally be acceptable for engineering applications.

For a more rigorous appraisal, it should be recognized that a harbor wave RAO computed by linear wave theory must be an upperbound value, and that a calibration factor in excess of unity is therefore unrealistic and should be discarded. At those frequencies where it occurs, i.e., where the calibration point is located near a node of the harbor surface, another point further removed from the node should be selected for calibration.

NSWICH SOLUTION FOR EXISTING ACAJUTLA HARBOR

As an example of an NSWICH solution, Figure 3 presents a contour plot of the existing harbor surface at Acajutla for a unit wave height from S30°W at 0.006 Hz (166.7 second period), drawn in phase with a crest at Point B. The calculations were performed at intervals of 0.001 Hz up to a frequency of 0.094 Hz, above which the harbor response is insignificant. The frequencies of interest range only from 0.001 to 0.020 Hz because these induce relatively large ship motions in the surge, sway and yaw modes which, in turn, cause large mooring forces.

Model test harbor wave spectra were furnished by H.R.S. Wallingford at points B, C and D for the offshore wave spectrum shown on Figure 4. This information was used to calculate the measured RAO for each of the three points at each frequency (see Figure 2). The calibration factor is the ratio of the measured RAO to the RAO computed by NSWICH.

The calibration factors for points B, C and D are plotted on Figure 7. In accordance with the prior discussion of the validity of the calibration factors, values close to or in excess of unity should not be considered. For example, Figure 3 shows that Point D is close to a node at 0.006 Hz. At this frequency, Figure 7 shows a calibration factor of 1.13 which is overly conservative and should be discarded. Similarly, each of the three points contains a narrow frequency band in which its own calibration factor should be discarded in favor of the next lower calibration factor at another point.

Figure 5 shows response spectra computed for point B for the offshore spectrum of Figure 4, using three different sets of calibration factors, namely, those derived by calibrating points B, C, and D. The three spectra are similar in shape. They all indicate a peak energy density at 0.006Hz, although the peak is higher when calibration factors for points C and D are used than when the factors for point B are used. Since the significant wave height is proportional to the square root of the area under the spectrum, large differences in the peak energy density produce much smaller differences in significant wave heights. In the example of

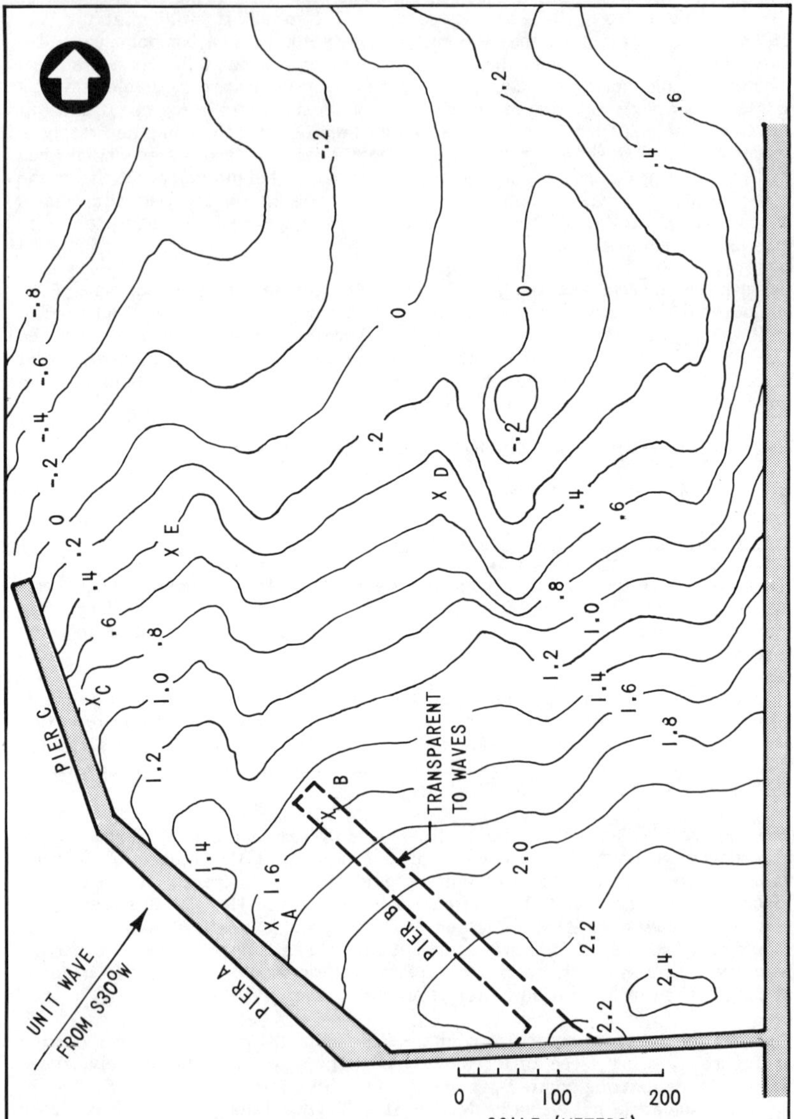

FIG. 3- EXISTING HARBOR WATER SURFACE AT 0.006 Hz
CONTOURS AS RATIO OF HARBOR TO INCIDENT WAVE HEIGHT IN PHASE WITH CREST AT POINT B.

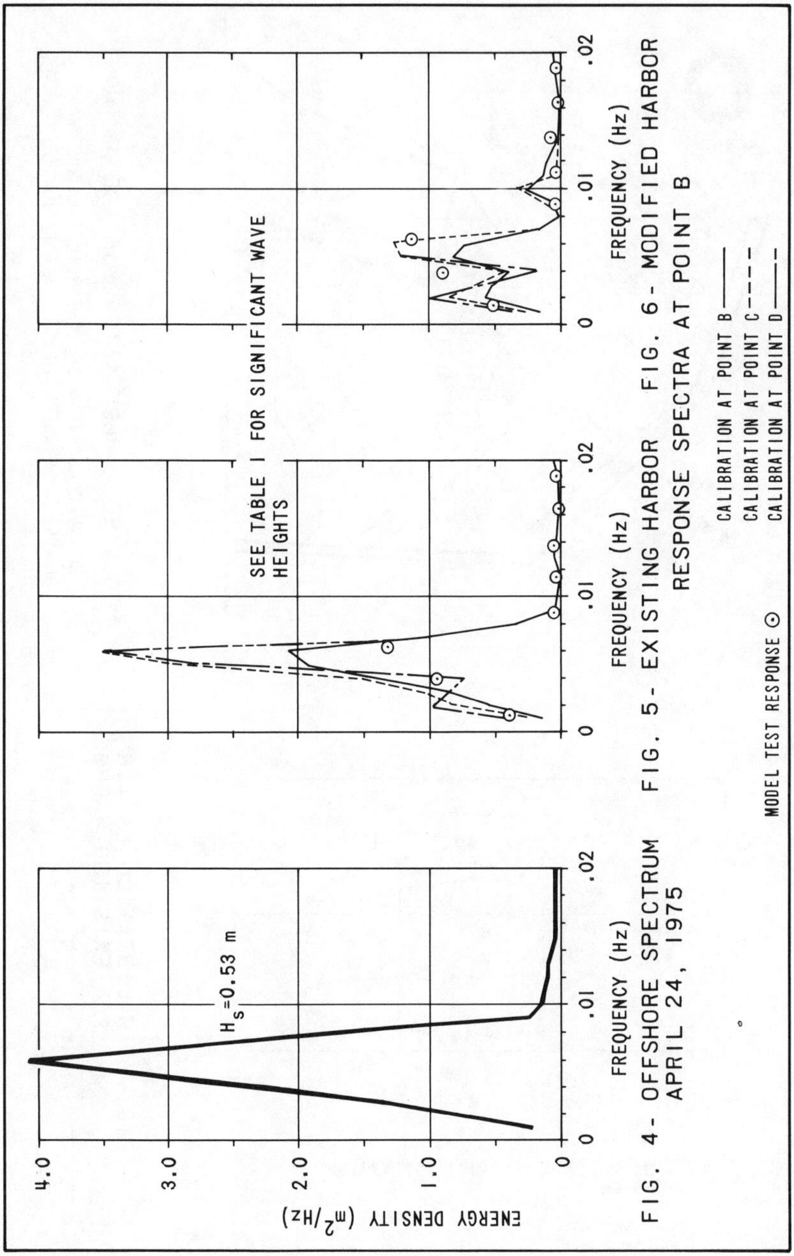

FIG. 4- OFFSHORE SPECTRUM APRIL 24, 1975

FIG. 5- EXISTING HARBOR FIG. 6- MODIFIED HARBOR RESPONSE SPECTRA AT POINT B

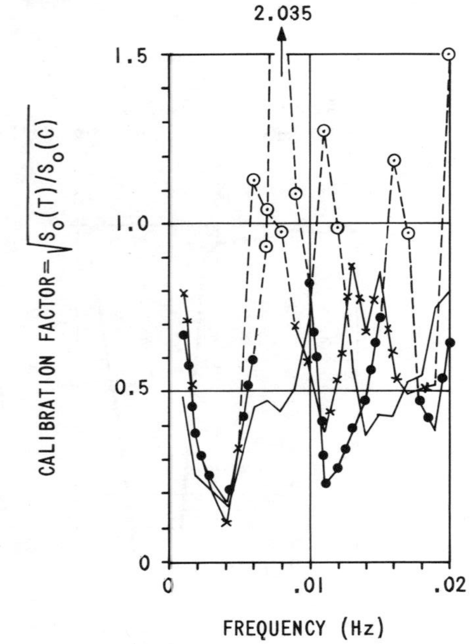

FIG. 7- CALIBRATION FACTORS FOR EXISTING HARBOR

AT POINT B
AT POINT D
AT POINT C
NOT VALID

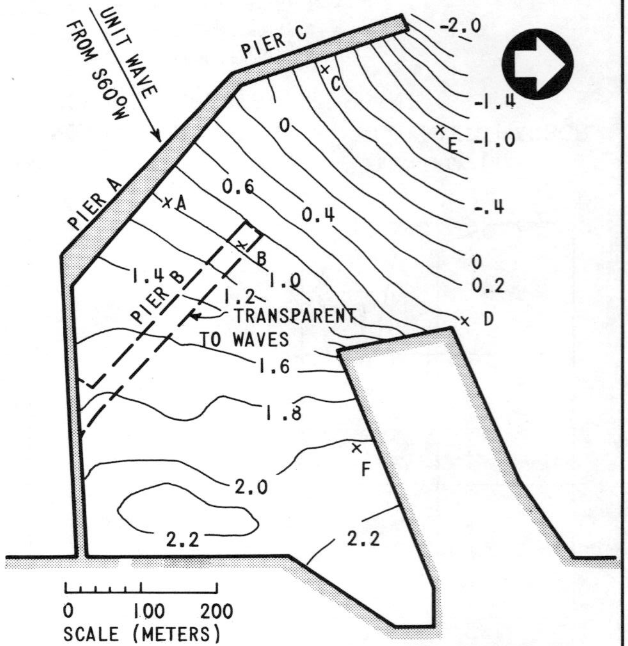

FIG.8-MODIFIED HARBOR WATER SURFACE AT .006 Hz

CONTOURS AS RATIO OF HARBOR TO INCIDENT WAVE HEIGHT IN PHASE WITH CREST AT POINT F

Figure 5, the ratio of the peak energies at point B is about 1.7, yet the ratio of the maximum to the minimum significant wave height is only 1.18. The response spectra at points C and D (not shown) are even less sensitive to which set of calibration factors is used.

Figure 5 also shows the model test response at existing harbor point B to the offshore spectrum of Figure 4. Energy densities were given in band widths of 0.0025Hz, which may be too wide to define the very narrow resonant peaks at low frequencies. This may acount for the test data showing a somewhat lower peak energy density than the computed response spectra, but the shape of the test response spectrum is quite similar to the computed shapes.

EXTRAPOLATION TO MODIFIED HARBOR

Figure 8 shows the harbor at Acajutla modified by a proposed container berth. It shows the contours of the water surface at the same 166.7 second period as was used for the existing harbor of Figure 3, but the contours are drawn in phase with a crest at point F. Model test response spectra were available at points B and F for the offshore spectrum shown on Figure 4. The question to be answered is whether the sets of calibration factors developed for points B, C and D in the existing harbor would be useful for computing the response spectra in the modified harbor.

Figure 6 shows the computed response spectra at point B in the modified harbor for the offshore spectrum of Figure 4, using the calibration factors plotted on Figure 7 for points B, C and D in the existing harbor. Paralleling the observations for the existing harbor (Figure 5), the spectra have similar shapes which for this harbor layout are triple-peaked. Because of the relatively wide band width (0.0025Hz) in the model test data it is impossible to confirm that the test response spectrum is also triple-peaked. However, the test data plotted on Figure 6 are reasonably close to the computed response spectra. In common with the existing harbor, the computed response spectra in the modified harbor appear to be quite insensitive to which set of calibration factors is used.

SUMMARY AND CONCLUSION

Table 1 presents a comparison of significant harbor wave heights at points B, C and D in the existing harbor and points B and F in the modified harbor. The harbor waves are in response to the offshore spectrum of Figure 4, using the different sets of calibration factors shown on Figure 7. The tabulation confirms that the significant harbor wave height Hs is relatively insensitive to which set of calibration factors is used, even though Figure 7 shows a variability in the calibration factors. The range of the Hs computed at any of the harbor points by the different sets of calibration factors falls within $\pm$11%.

The studies for the Port of Acajutla lend support to the view that numerical modeling based on linear wave theory is a useful tool for evaluating the wave climate in alternative harbor layouts, provided that the theoretical results can be calibrated by some real world or hydraulic model measurements. Once calibrated at a few harbor points, the numerical model will provide reasonable wave heights at other harbor points and for other harbor configurations.

REFERENCES

1. Hwang, L.S. et al., A Study of Wave and Ship Behavior At Long Beach Harbor, Tetra-Tech Report TC-182, February 1, 1972.

2. Chen, H.S. and Mei, C.C, "Oscillation and Wave Forces in a Manmade Harbor in the Open Sea", Tenth Symposium on Naval Hydrodynamics, June 1974.

3. Ho, R. and Bomze, H., "Basin Oscillations in an Offshore Harbor", Preprints, Paper No. OTC 2331, Offshore Technology Conference, Houston, Texas, May 1975.

4. Madsen, O.S. and Vyas, Y.Y., "Wave Motions in an Offshore Harbor", Symposium on Modeling Techniques, American Society of Civil Engineers, New York, N.Y., September 1975.

5. Ito, Y., "On the Effect of Tsumami - Breakwater", Coastal Engineering in Japan, Vol. XIII, December 1970, pp. 89-102.

6. Bomze, H., "Math Model for Harbor Waves/Ship Motions/Moorings", Proceedings, Ports '80 ASCE Specialty Conference, Norfolk, VA, May 1980.

7. Port of Acajutla, El Salvador - Random Wave Model Study, Hydraulics Research Station, Wallingford, England, Report No. EX739, July 1976.

8. Dean, R.G., "Long Wave Modification by Linear Transitions", ASCE Journal of the Waterways and Harbors Division, Vol. 90, No. WW-1, 1964, pp.1-30.

TABLE 1

EFFECT OF CALIBRATION POINT
ON COMPUTED LONG PERIOD HARBOR Hs (METERS)
FOR OFFSHORE SPECTRUM OF FIGURE 4

CALIBRATION PT. IN EXISTING HARBOR	EXISTING HARBOR Hs			MODIFIED HARBOR Hs	
	PT. B	PT. C	PT. D	PT. B	PT. F
B	0.37	0.29	0.27	0.25	0.46
C	0.44	0.31	0.29	0.29	0.53
D	0.42	0.32	0.29	0.29	0.52
MEAN Hs	0.41	0.31	0.28	0.28	0.50
RANGE (%)	+7.3 -9.8	+3.2 -6.5	+3.6 -3.6	+3.6 -10.7	+6.0 -8.0
PHYSICAL MODEL	0.33	0.30	0.33	0.30	0.36
MEAN Hs / PHYS. MODEL Hs	1.24	1.03	0.85	0.93	1.39

COMPUTER CONTROLLED PHYSICAL MODEL OF THE SACRAMENTO RIVER

J. Amorocho, F. ASCE[1], W. Hartman, A.M. ASCE[1],

and J. J. DeVries, M. ASCE[1]

ABSTRACT

A set of hydraulic models was developed for a portion of the Sacramento River in California, with the objective of studying in detail the behavior of sediments and the patterns of flow in the vicinity of a proposed major diversion structure for the Peripheral Canal of the state water system. One of these models, representing approximately 10 miles (16 km) of the river, was controlled automatically by a computer system to permit the simulation of unsteady flows, which include tidal action. The model, as well as the control and logging systems, are described in this paper.

INTRODUCTION

Two of the major components of the California Water System are the Federal Central Valley Project and the California State Water Project. Both involve engineering works provided to effect a redistribution of the water resources of the state by a large scale transfer of some of the excess water from the humid north to the fertile Central Valley and to the populous but dry south. The principal storage facilities regulate the flow of the Sacramento River, which joins the San Joaquin River forming a large network of deltaic channels before entering the chain of embayments of San Francisco Bay (see Fig. 1).

The main conveyance systems of water towards the south are the California Aqueduct and the Delta-Mendota Canal, which originate at the southern boundary of the Delta. To insure that an adequate proportion of the regulated flow of the Sacramento River reaches these conveyances, as well as to aid in the control of salinity of Delta waters and to protect and enhance fisheries and aquatic ecologies, the Department of Water Resources of California has planned the construction of the so-called Peripheral Canal. This large channel, which will be provided with a series of control structures, is intended to divert fresh water from the Sacramento River near the town of Hood and to convey it to the Clifton Court forebay at the entrance of the pumping plants of the Federal and State Aqueducts. Water releases will be made at the control structures along its route to provide fresh water

(1) Department of Land, Air and Water Resources, University of California, Davis.

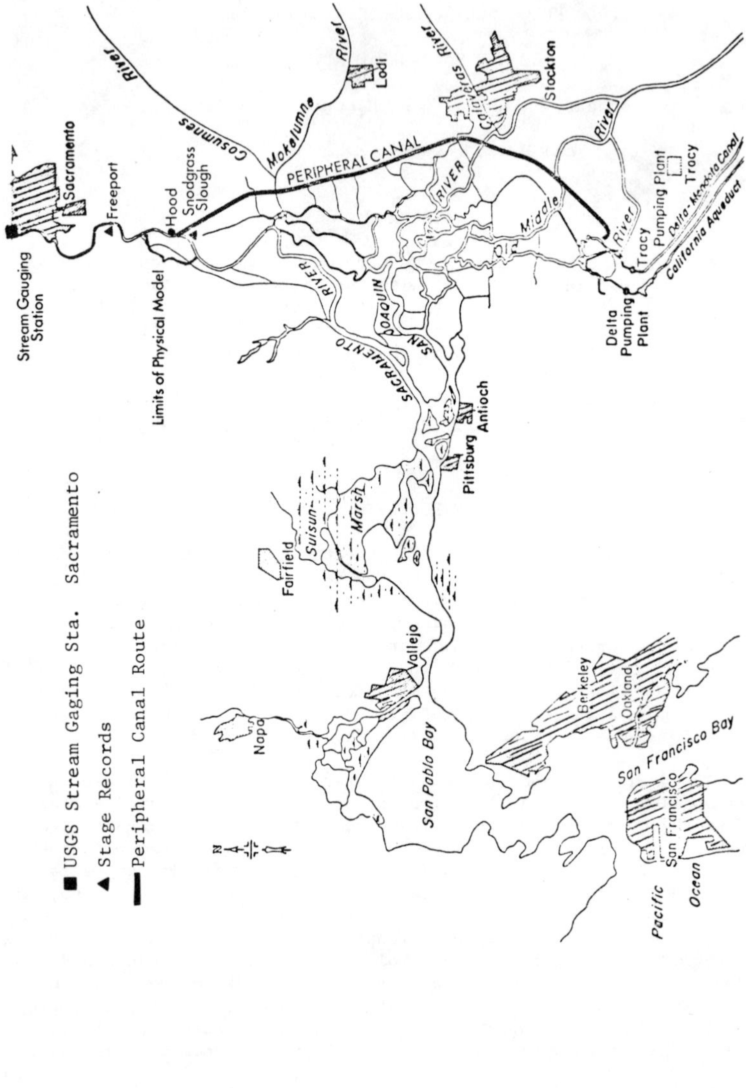

Figure 1. Location Map - Peripheral Canal Intake Model

circulation into the numerous delta branches, thus ensuring a positive forward flow.

The diversion structure for the Peripheral Canal on the Sacramento River must satisfy unusual requirements. In addition to preventing the diversion of excessive bed sediment loads under all conditions of operation, it must also avoid capturing migratory populations of small fish in various stages of development. The scheme as a whole should also minimize the development of conditions leading to undesirable sediment deposition in the river due to altered bedload concentrations downstream from the diversion, and should not stimulate localized predatory activity among fish species. The technical and scientific staff of the Department of Water Resources proposed several diversion configurations intended to satisfy the objectives. Two of these schemes are shown in Fig. 2. To test the performance of these schemes with respect to some of the constraints set for the structure, the Hydraulics Laboratory, Department of Land, Air and Water Resources, University of California, Davis undertook the performance of model tests of a portion of the Sacramento River in the vicinity of the general location selected for the diversion.

The objectives here were to observe the behavior of alternative structure configurations with respect to sediment capture and velocity distributions in the flow, and the selection of the most favorable specific location of the intake to satisfy the constraints imposed on the diversion.

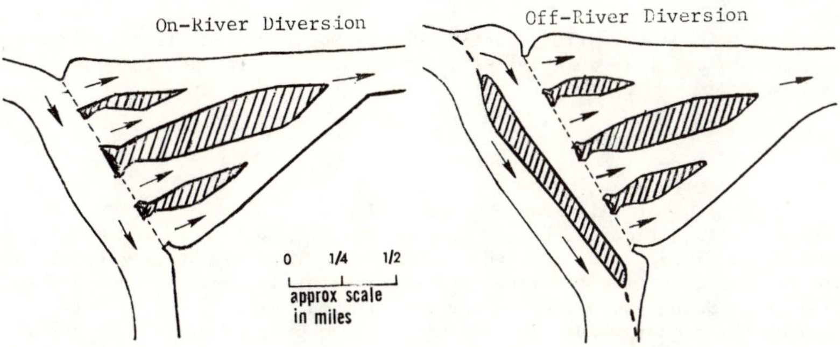

Fig. 2. On-River and Off-River Diversion Schemes

FUNCTIONS OF THE MODEL SYSTEMS

The value of reduced scale models for the study of conditions of flow and sediment transport depends on the faithfulness with which the unsteady flows of the river can be simulated in the area of the proposed diversion. Streamflow data originate at a USGS gaging station in Sacramento, approximately 34 kilometers (21 miles) upstream from the prospective diversion site. Continuous stage records are maintained at this station and at two other locations, one designated as Freeport, 12.9 km (8.0 miles) upstream from the town of Hood, and the other at Snodgrass Slough, shortly downstream from the same town.

For calibration of any physical model, as well as to permit the simulation of historical flows, proper discharge and stage data are necessary within the model boundaries. None of the three hydrographic stations mentioned were within these boundaries. Therefore, the data needed had to be obtained by computation. For this purpose, an auxilary mathematical model capable of calculating stages and discharges in the diversion area based on the data of the hydrographic stations was developed. This was designated as Model I.

For the selection of the best diversion structure configuration, as well as for the determination of the most appropriate point of diversion on the river and for a qualitative evaluation of possible sediment problems associated with the diversion operations, a distorted physical model to the horizontal scale of 1:240, designated as Model II, was built.

Finally, for the detailed study of velocity values in the immediate vicinity of alternative systems of fish screens and for the close examination of flow velocity patterns at the intake, a third model, designated Model III, was constructed at the scale of 1:50.

In this paper, brief descriptions of Models I and III are presented, and a more detailed description is given of Model II and its control systems.

MODEL I

The mathematical model employs an implicit solution of the St. Venant equations of motion proposed by Preissman reported by Ligget and Cunge (1975), and is in the form of a FORTRAN computer program. As mentioned, data from the recording stations at Sacramento, Freeport and Snodgrass Slough were used in the calibration, which involved determining the proper values of Manning's "n" to give results that corresponded to river stage measurements at the hydrographic stations.

A final simulation of a historical period (December 1974 - April 1975) was made with the computer model to verify its ability to accurately model the river. The data from this period also provided input for the calibration and verification tests of Model II. In general, these data consisted of the flow values required at the upper end of the physical model and the corresponding stages expected along its entire length during many steady and unsteady state conditions.

Simulation of diversions to the Peripheral Canal, both as a point diversion (for an off-river diversion scheme) and as a diversion over a length of river (for an on-river diversion) can be made using options available.

MODEL II

It was necessary to model approximately five miles of the river to allow flow patterns at the diversion to be properly represented. A horizontal scale of 1:240 was chosen on the basis of available laboratory space.

A vertical scale of 1:60 was chosen for this model, giving a distortion ratio of 1 to 4. If the model had been undistorted, the water depths would have been very small, making velocity measurements difficult, and possibly creating laminar flow conditions in some instances. In addition, it was desired to model the bed material. In an undistorted model, at the 1:240 scale, the bed shears would have been very small and extremely fine particles would have been needed to represent the bed material. This could have resulted in the sediment being transported in suspension. In the distorted model the bed shears (which are related to the hydraulic slope) are larger, allowing the use of bed materials of a larger size.

Layout and Details

The general layout of the model and its control systems are shown in Fig. 3. The simulated river bed was a concrete lined trapezoidal channel constructed on a sand base placed in a flume made of corrugated steel culvert halves, supported by a welded steel framework. The model was stable, due to its mass and rigidity, but was easy to level when required to correct for settlement of the floor slabs. The alignment of the stream bed in the flume was based on photogrammetric maps and cross-sections surveyed at approximately 1000 ft intervals. Movable bed material placed in the flume was shaped using templates cut to match the measured cross-sections.

Water was supplied to the model by a 20-hp pump, located in a tank beneath the laboratory floor and connected to a constant head tank. The rate at which water entered the model was controlled by an adjustable Sutro weir, designed so that the discharge was directly proportional to the head. The water surface in the weir tank was held constant and the weir was adjusted vertically to vary the discharge.

Additional water was supplied to the model system at its downstream end to simulate tidal action. This flow was measured by means of an orifice and was set to match the maximum incoming tide. Water leaving the model was controlled by another Sutro weir, positioned to maintain a specified water level at the downstream end.

The model was equipped with nine stilling wells connected to piezometer taps on both banks along its entire length. Water levels in these stilling wells could be read either visually by point gages, accurate to 0.001 ft or automatically with a scanner system. The water

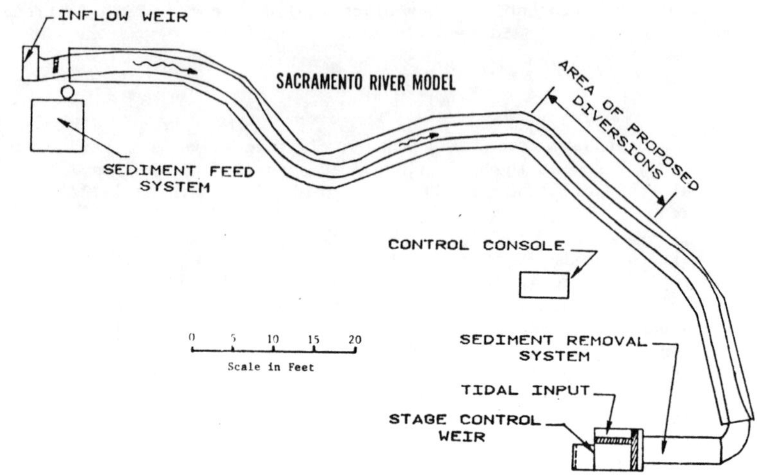

Fig. 3. General Model Layout

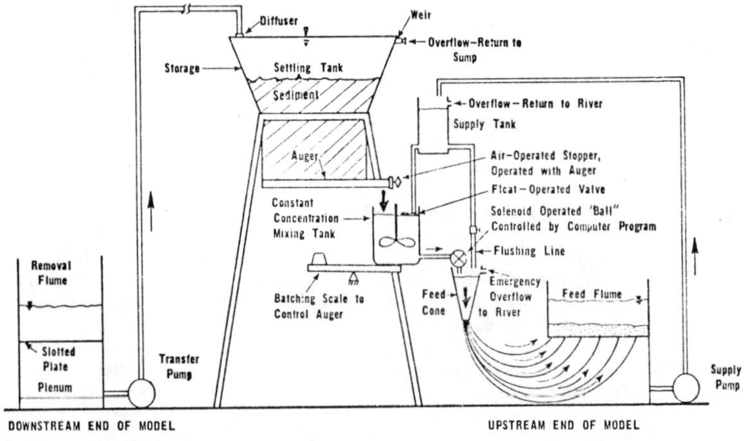

Fig. 4. Schematic Diagram of Sediment Feed System

surface slopes were determined from the elevations measured in the stilling wells.

The proper water surface slopes were obtained through trials with various artificial roughness elements consisting of chicken wire mesh and 1/4-inch mesh hardware cloth.

Manning's n for the final roughness configuration with the movable bed ranges from n_m = 0.0243 for a flow of 30,000 cfs to n_m = 0.0257 for a flow of 80,000 cfs, as determined from measured water surface slopes. The desired Manning's n values for the model, calculated from the relationship betweem model and prototype is n_m = 0.0244 for a flow of 30,000 cfs and n_m = 0.0250 for a flow of 80,000 cfs. Thus, there is good correspondence between the river and the model for Manning's n.

Sediment Supply System

The movable bed was placed in the model after the proper artificial roughness was established. A feed system (Fig. 4) was developed to supply sediment at the upstream end of the model.

The proper fed rate as a function of flow was determined by trial by adjusting it until the general bed elevations matched those measured in the field. Sediment was supplied to the model in the form of a slurry. This slurry was continuously mixed at a constant concentration, and was released into a feed cone in batches containing 0.133 lb of sediment. The times between releases were adjusted in proportion to the water inflow to the model to give the desired average sediment feed rate. A constant flow of wash water was used to dilute the slurry in the feed cone (see Fig. 4), from which the suspension was led into the bottom of the model at a nearly constant rate. The wash water used in the slurry as well as flows continuously maintained for washing the feed lines were pumped from the model just downstream from the upper flow control weir, so that no new water entered the model after the inflow was measured at the weir.

Sediment leaving the model at its downstream end was captured in a stilling chamber and pumped continuously back to the settling tank of the sediment supply system.

Control and Data Logging System

The main element of the model control and data logging system was a microcomputer located in the laboratory. This system's main function was the control of the upstream water inflow and the downstream stage in the model. It also had the purpose of handling a number of control and data logging tasks.

Figure 5 shows a schematic diagram of the control system. Programming was done in machine language. The system utilized feedback control to actuate various electromechanical devices such as weir positions and water levels for the operation of the model. The control of inflow and stage was effected either by using a table of historical

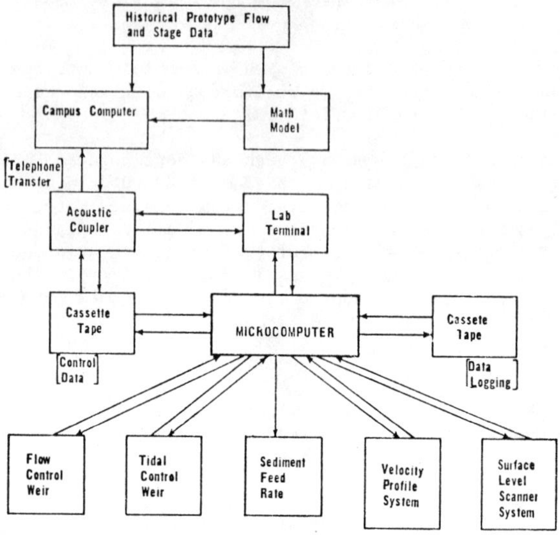

Fig. 5. Diagram of Control System.

records stored in the computer memory or any specified time dependent function. The system metered the sediment supply rates from the slurry storage in proportion to the incoming flow of water in the model. This flow was controlled by moving the upstream weir to calibrated positions on signals from the computer. The position of the weir was monitored by the computer by means of a potentiometer attachment. Downstream water levels were measured by means of a capacitance probe and controlled by the downstream weir.

Target values for inflows and stages in the model were specified at five second intervals (corresponding to about two and one-half minute increments in the river). Tidal gage readings from the river were available at 15-minute intervals. The model settings were interpolated between these readings to provide uniform variation.

Data Logging

Several model data logging and analysis duties were performed by the microcomputer system. All control values (inflow, stage, and control positions) were recorded at one or two minute intervals in real time and printed in both model and prototype time scales. All values of flow, stage, and velocities were converted to and recorded at prototype scale.

Water stages were measured using a commercial capacitance-type liquid-level probe (Robertshaw Level-Tel Model No. 158B2A1) and a water level scanning system. The nine stilling wells located along the model and two reference stilling wells, one maintained at a constant low water level and the other at a high water level, were connected to a 12 port fluid wafer-switch. The switch output was connected to a pressure transducer. The signal from the pressure transducer was sent to the control system, and the fluid switch was advanced from position to position by a rotary solenoid on command of the control system. A schematic diagram of the logging system is shown in Fig. 6.

In addition to the water level scans, a record was made of all control system positions, the model inflow, and the downstream stage. During the verification runs, these recordings were made each minute during the run. A sample output record is shown in Fig. 7.

During the testing period, water velocities were taken at selected cross-sections in the vicinity of the diversions. The data logger recorded the point velocities and cross-section position data using the subsystem described in the following section.

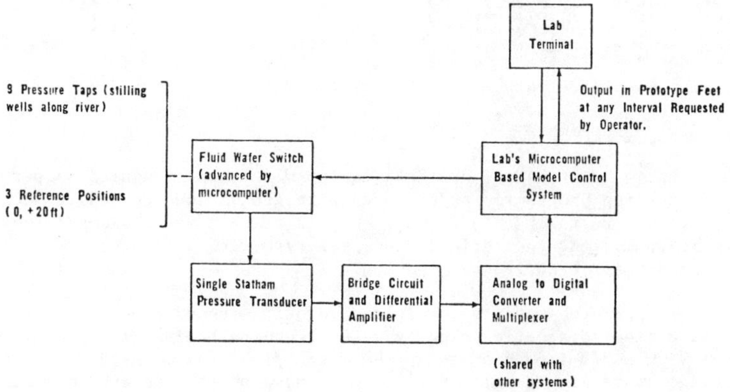

Fig. 6. Diagram of Logging System

Water level scan

1 14.5 2 14.2 3 13.9 4 13.5 5 13.2 6 12.7 7 11.7 8 11.6 9

Date & Time	Control codes				
0047:14:37:28	AC	AC	BF	BC	69
0047:15:08:28	AC	AC	BF	BC	66
0047:15:39:28	AC	AC	BD	BC	63
0047:16:10:28	AC	AB	BA	BC	63
0047:16:41:27	AC	AB	B8	BC	66
0047:17:12:28	AB	AB	B8	BC	6A
0047:17:43:27	AB	AB	B8	BC	6E
0047:18:14:28	AB	AA	BA	BC	72
0047:18:45:28	AA	AA	BB	BC	75
0047:19:16:28	AA	AA	BC	BC	77
0047:19:47:28	AA	AA	BC	BD	77
0047:20:18:27	AA	A9	BD	BD	77
0047:20:49:28	A9	A9	BE	BD	76
0047:21:20:28	A9	A9	BE	BD	74
0047:21:51:28	A9	A9	BE	BD	71
0047:22:22:28	A9	A9	BE	BC	6E
0047:22:53:27	A9	A9	BD	BC	6B
0047:23:24:28	A9	A9	BA	BB	69
0047:23:55:27	A9	A9	B9	BB	6B
0048:00:26:28	A9	A9	B8	BA	6E
0048:00:57:28	A9	A9	B9	BA	72
0048:01:28:28	A9	A9	B9	BA	74
0048:01:59:28	A9	A9	BA	BA	76
0048:02:30:28	A9	A9	BC	BA	74
0048:03:01:28	A9	A9	BC	BA	71
0048:03:32:27	A9	A9	BC	BA	6E
0048:04:03:28	A9	A8	BB	BA	6B
0048:04:34:28	A8	A8	B9	BA	6A
0048:05:05:28	A8	A8	B7	BA	6D
0048:05:36:28	A8	A8	B7	BB	71
0048:06:07:27	A8	A7	B9	BB	75
0048:06:38:28	A7	A7	B8	BB	78
0048:07:09:27	A7	A7	BA	BB	7C
0048:07:40:28	A7	A7	BC	BC	7D
0048:08:11:28	A7	A7	BE	BC	7D
0048:08:42:28	A7	A7	BE	BC	7A
0048:09:13:28	A7	A7	BE	BC	77
0048:09:44:27	A7	A7	BE	BC	74
0048:10:15:28	A7	A7	BD	BB	71

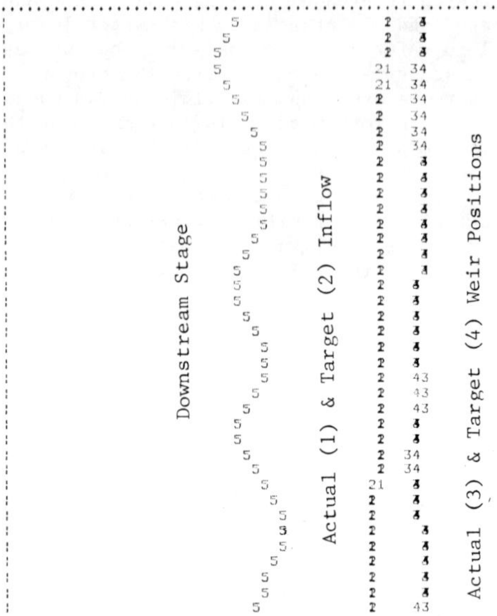

Fig. 7. Sample Output

Velocity Measurement

Point velocities in the flows in the model were measured as part of the testing program in order to compare the various designs. The velocities were taken on a 2 x 2 inch grid at selected cross-sections. This is equivalent to a grid in the prototype with a 10 ft vertical spacing and 40 ft horizontal spacing.

The apparatus used to obtain the velocity record consisted of a two-axis electromagnetic velocity meter (Cushing Engineering Model 602P) with a 3/8-inch diameter probe. The probe was mounted in a carriage traversing a horizontal track placed on tabs located over any selected cross-section. The carriage was equipped so the probe could be moved vertically and rotated about a vertical axis.

Only general velocity patterns were obtainable in this model, because the prove used was quite large in relation to the scale of the model.

Cross-Section Measurements

Due to the low specific gravity of the model sediment, any disturbance of the flow radically modified the bedforms and cross-section dimensions. A flow disturbance such as stopping the flow in order to measure the bedforms changed minor bedform features due to sediment slumping, and would have defeated the purpose of a continuous verification period. Dewatering the channel after tests to map bedforms had to be done slowly to prevent grading and erosion of the bars that had developed.

In order to minimize measurement effects on the cross-sections with the channel flowing, a multiprobed device for cross-section logging was developed.

The probes were raised to above water level, the frame was placed on reference tabs for the desired cross-section and a latch was released, allowing the probes to slide into the water. When the probes reached the bottom, the latch was fastened and the unit removed from the water for photographing. The probes were made of acrylic plastic, with a specific gravity of 1.2, and were lighter than the sediment; a small plate on the lower end of each probe prevented the probe from sinking into the sediment.

MODEL III

This physical model will be used to provide detailed data and verification of the design of the intake. It has sufficiently large size to permit very detailed velocity measurements. The vertical and horizontal scales of the model are 1:50. The space available at the laboratory, and other considerations, limit the model to encompass only a relatively short segment of the river channel upstream and downstream from the intake (2.1 miles or 3.5 km overall). The total length of the model along the river channel is 220 ft (67.1 m). The channel width varies between 14 and 16 ft (4.3 to 4.9 m). Only the off-river diversion, which was found to be most satisfactory from the studies with Model II, is simulated, but a number of screening schemes will be investigated. Very detailed measurements of point velocities are made by means of a laser-doppler system. The model is operated under steady states, only because it is not practical to make detailed velocity measurements during unsteady flows; however, the general pattern of secondary flows approaching the diversion area from upstream will be preserved by means of a system of veins and/or jets. To obtain data for these conditions, special river surveys were undertaken, in which measurements of velocities and flow direction were made at two cross-sections repeatedly over several hours to permit assessments of the instantaneous flow patterns in three dimensions.

Fig. 8. Overall View of 1:240 Model

Fig. 9. View of Computer System

REFERENCES

1. Hartman, W. J., J. Amorocho and J. J. DeVries, "Peripheral Canal Intake Studies 1. Distorted River Channel Model." Vols. 1 and 2. Department of Land, Air and Water Resources, University of California. Water Science and Engineering Paper 1064. June 1979.

2. Ligget, J. A. and J. A. Cunge, "Numerical Methods of Solution of the Unsteady Flow Equations," in Unsteady Flow in Open Channels, Water Resources Publications, Fort Collins, CO (1975).

SEDIMENT MODELING FOR THE SACRAMENTO RIVER DIVERSION TO THE PERIPHERAL CANAL

J. J. DeVries, M. ASCE, W. Hartman, A. M. ASCE,

and J. Amorocho, F. ASCE (1)

ABSTRACT

Similitude criteria for a 1:240 horizontal-scale, 1:60 vertical-scale movable bed river model were based on consideration of gravity and friction forces to duplicate the general hydraulic behavior of the river. To simulate scour and deposition in the model, similarity criteria were based on: 1) scaling of the bed shear stress; 2) matching the ratio of the particle fall velocity to the shear velocity; and 3) matching the bed forms in model and prototype. Even though it was not possible to satisfy all criteria simultaneously, the finely ground walnut shell material used for the model bed matched general scour and deposition patterns as shown through verification tests for a four-month prototype period.

INTRODUCTION

The Peripheral Canal is planned as part of the California State Water Project to convey water through the Sacramento-San Joaquin River Delta to pumping plants supplying the California Aqueduct of the California Water Project and the Delta-Mendota Canal of the Federal Central Valley Project. Water will also be released at a number of points in the Delta to improve water quality. The intake to the Peripheral Canal will be located near the town of Hood, California on the Sacramento River, about 21 miles (34 km) downstream from the city of Sacramento. The proposed intake works are designed to act as a barrier to fish which would otherwise be diverted with the river water. Elaborate fish screening mechanisms are being devised by the California Department of Water Resources to accomplish this fish protection. In addition, the river diversion works must function in a manner which minimizes undesirable sediment deposition in the river due to altered bedload transport conditions downstream from the diversion, and bedload material must be prevented from entering the diversion works.

A series of hydraulic model studies are being made at the Hydraulics Laboratory, Department of Land, Air and Water Resources, University of California, Davis to evaluate the performance of various intake designs

(1) Department of Land, Air and Water Resources, University of California, Davis.

for the California Department of Water Resources. Descriptions of the models and discussion of various aspects of the model study program are provided in another paper in these symposium proceedings (Amorocho et al., 1980).

This paper discusses similitude requirements for sediment modeling in a movable-bed model of 5 miles (8.0 km) of the Sacramento River, including the intake to the Peripheral Canal. The model length scale is distorted, the horizontal scale is 1:240 and the vertical scale is 1:60. By means of a computer-based control system, the inflow of the upstream end of the model and the downstream stage can be regulated as a function of time.

The successful performance of a movable-bed model requires that the combined action of the various hydraulic forces acting in the model bring about similitude with respect to bed movement as well as to fluid motion. The general procedure is to select scales and bed material which will result in bed movement of a nature similar to the prototype. Such a model will represent scour and deposition in a properly scaled manner.

SIMILITUDE CONSIDERATIONS FOR SEDIMENT MODELING

The purpose of the model study was to indicate where scour and deposition would occur in the river for various flow conditions simulated in the model. For sediment motion to be similar in model and prototype the model sediment must experience forces on the individual sediment particles which cause them to move in a manner similar to the movement of the particles in the prototype. The model must also, of course, duplicate the general hydraulic behavior of the prototype.

In any model representing an open channel flow, gravity forces are important. Therefore, to attain dynamic similitude, Froude numbers must be the same in model and prototype.

Viscous (or friction) forces are also significant in river flow situations. However, it is not possible to match the prototype Reynolds numbers in the model, and therefore a friction loss equation is used instead of the Reynolds number to provide a friction criterion (Graf, 1971). Friction provided by the unmodified model river banks or by the particles on the model bed is usually not sufficient to give adequate simulation of friction in the prototype, and additional roughness elements are usually required.

In movable-bed models in which scour and deposition are being studied, the similarity criteria which must be considered are (Einstein and Chien, 1954; Graf, 1971): 1) proper scaling of the bed shear stress; 2) matching of the ratio of sediment particle fall velocity w_f to the shear velocity u_* in model and prototype to permit sediment deposition and transport to be properly simulated; and 3) simulation of the viscous sublayer to correctly represent the fluid motion past individual bed particles. This last criterion may be neglected if there is turbulent flow around the bed particles in both model and prototype (Einstein and Chien, 1954).

Friction losses in rivers are affected strongly by the types of bed forms which exist. Different friction losses occur when the bed is covered by ripples than when there are dunes. Because the bed roughness affects the vertical velocity profile it would be desirable to match the prototype bed forms in the model, if possible.

No attempt was made to extrapolate the rate of sediment transport from the model to the prototype. The reader is referred to Graf (1971) for a discussion of some of the complications that arise when modeling the transport rate is attempted. The model sediment feed rate was determined in this study by trial to permit the supply of sediment at a rate equal to the mean transport rate at each given flow.

MODEL SCALES

Considerations of space and economy led to the selection of a horizontal length scale of 1:240. This was not an appropriate scale for the vertical dimensions of the model, however, because it would give very low model velocities, and the bed shear would be so small that extremely fine and light sediment material would be required for similitude of sediment movement. A vertical scale ratio of 1:60 was chosen to overcome these difficulties. The model thus had a distortion ratio of 4:1, which is within the generally accepted guidelines for river modeling.

Other distortions, in addition to the length scale distortions, are usually required in movable-bed models because it is not possible to satisfy all the scaling criteria simultaneously. For example, although the basic criterion for scaling for river models is the Froude criterion, models may be operated at Froude numbers which do not match those of the prototype in order to satisfy other criteria. In this model, however, the ratio of Froude numbers between model and prototype was kept at unity.

Froude Criterion

Setting model and prototype Froude number equal gives the following relation:

$$\left[\frac{V}{(g\ell_v)^{1/2}}\right]_m = \left[\frac{V}{(g\ell_v)^{1/2}}\right]_p \tag{1}$$

where

V = a representative velocity
g = acceleration due to gravity
ℓ_v = a representative vertical length
m = subscript denoting model, and
p = subscript denoting prototype.

Friction Criterion

It was assumed that Manning's equation was valid for expressing resistance to flow in both the model and the prototype. It can be shown (Graf, 1971) that the required ratio for Manning's n values is

$$\frac{n_p}{n_m} = \left(\frac{R_p}{R_m}\right)^{2/3} L_H^{-1/2} \tag{2}$$

where

n = Manning's roughness coefficient
L_H = horizontal scale for the model, and
R = hydraulic radius.

Shear Stress Criterion

To properly simulate bed shears in the model the dimensionless bedshear stress τ_* must be the same as in the prototype. The bed shears must also be large enough in the model to cause sediment motion under the same conditions which produce motion in the prototype. Conditions under which sediment motion occurs can be predicted from the Shields' diagram (Vanoni, 1975), which is a plot of τ_* versus the shear Reynolds number R_*.

$$R_* = \frac{u_* d_{50}}{\nu} \tag{3}$$

$$\tau_* = \frac{\tau_o}{(\gamma_s - \gamma) d_{50}} \tag{4}$$

where

d_{50} = median size of the sediment
τ_o = the bed shear
γ_s = weight per unit volume of sediment, and
γ = weight per unit volume of water.

Shields' diagram is given in Fig. 1. The solid line (Shields' curve) divides the region of sediment motion from the region of no motion. To insure similitude, the following relationship should exist:

$$\left[\tau_*\right]_m = \left[\tau_*\right]_p \tag{5}$$

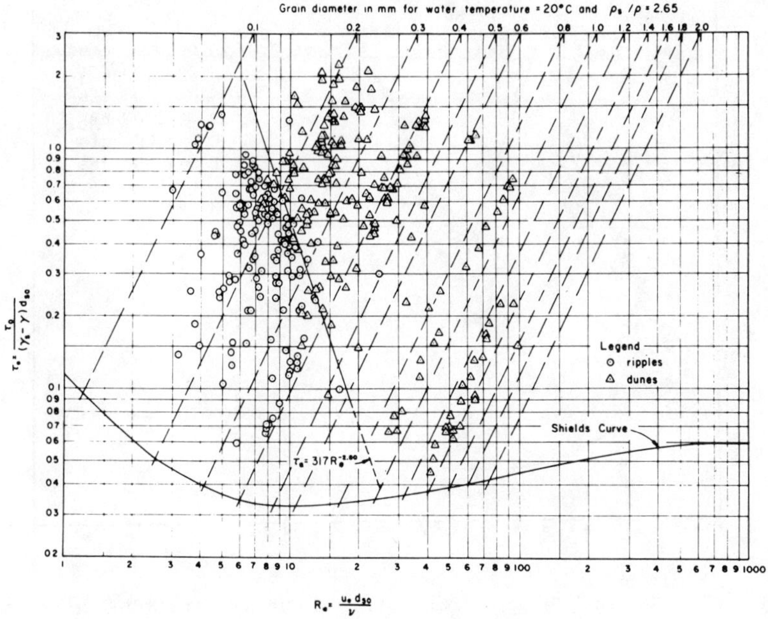

Figure 1. Shields' Diagram (after Vanoni, 1975).

Fall Velocity Criterion

The ratio of particle fall velocity w_f to shear velocity u_* should be the same in model and prototype so that deposition is properly simulated in the model. The following relationship should thus hold

$$\left[\frac{w_f}{u_*}\right]_m = \left[\frac{w_f}{u_*}\right]_p \tag{6}$$

Bed Form Criterion

At low Froude numbers (say below 0.5), the bed forms may be either ripples or dunes. The Shields' parameters τ_* and R_* can be used to predict the type of bed form which occurs (Vanoni, 1975). A plot of experimental data for flows with ripples and dunes from Vanoni (1975) is shown on the diagram in Fig. 2. The equation for the line which divides the two regions is

$$\tau_* = 317R_*^{-2.80} \tag{7}$$

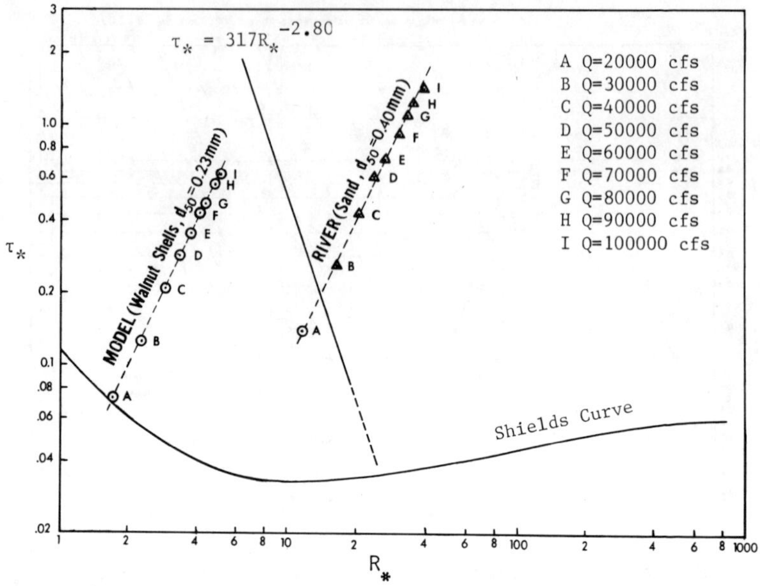

Figure 2. Shields' Diagram Illustrating Ripple and Dune Bed Form Regions.

MODEL SEDIMENT SELECTION

The criteria dealing with the dimensionless shear on the bed (τ_*) and the ratio of the fall velocity to the shear velocity (w_f/u_*) were considered to be the most important in the determination of the appropriate size of sediment for the model.

The consideration of possible types of sediment for the model indicated strongly that a light-weight material would be required. Ground walnut shells were chosen on the basis of their low specific gravity ($\gamma_s/\gamma = 1.33$), low cost, and availability. Walnut shell "flour" is commercially available in a range of sizes. However, there are a number of disadvantages to using walnut shells. The major problem is that the shells tend to decompose due to their organic nature. In addition, the shell material has to be soaked in water before use to remove very fine material (dust) and oils, both of which can be floated off.

Once the sediment material is chosen γ_s is fixed, and the values of (w_f/u_*) and τ_* can be computed as a function of particle size d_{50}. A plot of $[w_f/u_*]_p/[w_f/u_*]_m$ and $(\tau_*)_p/(\tau_*)_m$ versus d_{50} is given in Fig. 3. The plot shows that for a given set of model length scales and sediment material, it is not possible to satisfy both criteria simultaneously. As shown by the plot, it is not possible for both ratios to be equal to one.

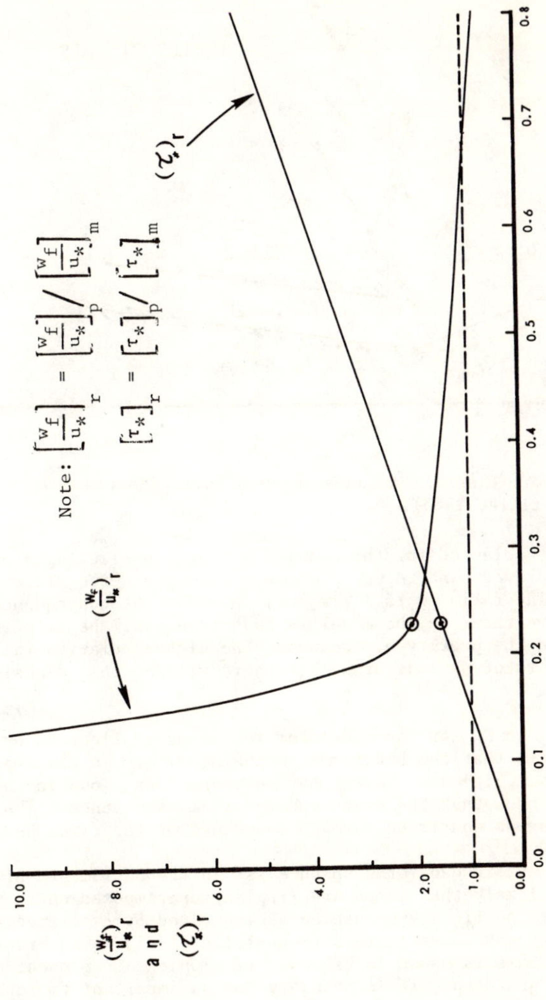

Figure 3. Prototype-to-Model Ratios for w_f/u_* and τ_* for $L_H = 240$, $L_V = 60$, and Walnut Shells with $d_{50} = 0.23$ mm.

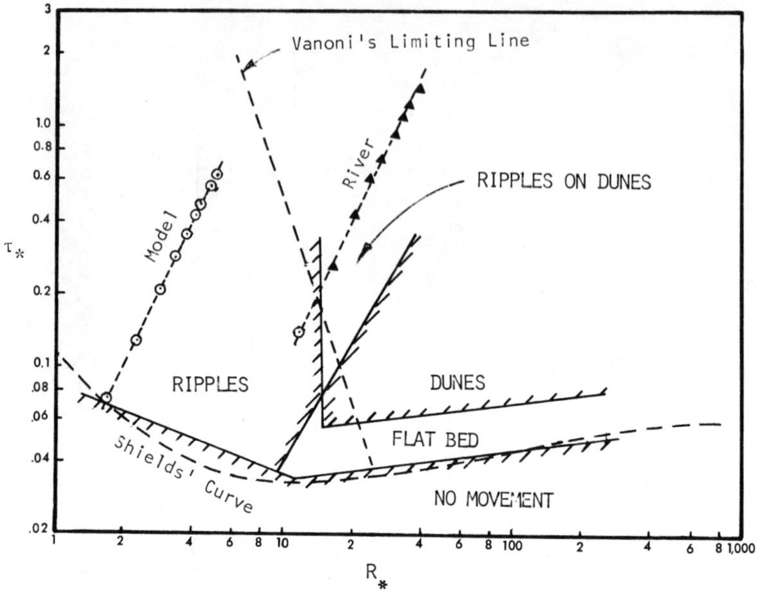

Figure 4. Bed Form Classification by French Investigators (after Bonnefille, 1965).

The sediment selected for the model has a median size (d_{50}) of 0.23 mm; for the river sand, d_{50} = 0.40 mm. This gives the ratio τ_{*p}/τ_{*m} = 1.5. The ratio $[w_f/u_*]_p/[w_f/u_*]_m$ = 2.0. The consequence of the lower relative shear in the model is that scour will be less in the model than in the prototype. However, the higher relative fall velocity in the prototype indicates that there will be less deposition in the model.

A plot of τ_* vs R_* for the model for the range of flows to be encountered indicates that the bed forms in the model are in the ripple classification (see Fig. 3). Values for corresponding flows in the Sacramento River show that the prototype bed forms are dunes. The presence of dunes was confirmed through soundings of the river bed.

Although the basic bed forms which exist in the Sacramento River are dunes, it is likely that there are ripples superimposed on these dunes for most of the flows encountered as indicated by experimental measurements by French investigators (Bonnefille, 1975). The French classification scheme is shown in Fig. 4. If ripples are present on the dunes in the prototype, it is probably not as important to reproduce the dunes in the model, because the ripples have a major effect in establishing the bed roughness.

It is not possible to produce dunes in the model without using a greatly different size and type of sediment. To do so would produce distortions in the τ_* and w_f/u_* ratios which would have large effects on the ability of the model to simulate sediment movement.

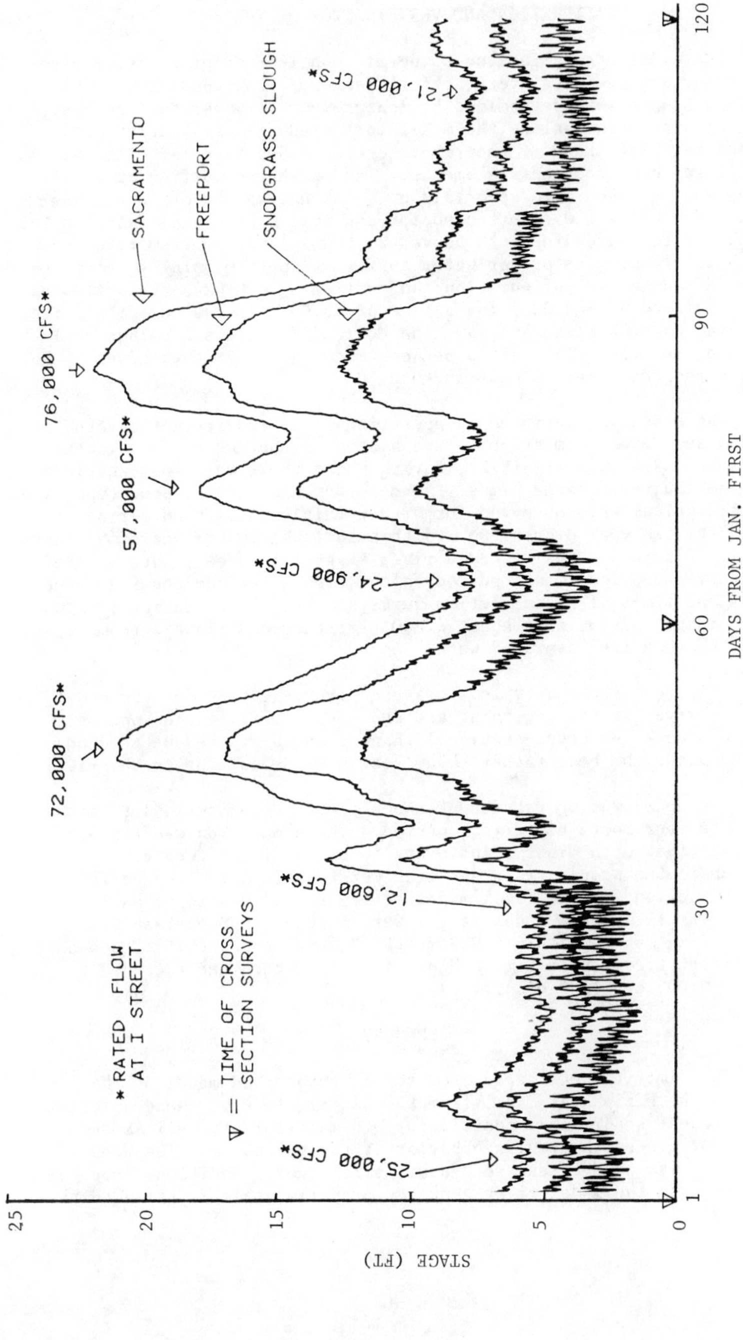

Figure 5. Sacramento River Stage Hydrographs for Four-Month Verification Period.

CALIBRATION AND VERIFICATION OF THE MODEL

The model was calibrated to permit duplication of observed river water surface profiles over the full range of flow conditions. Water surface slopes were determined by measurement of water surface elevation at nine points along the model for steady flows. Water surface slopes matching those of the river were obtained by trial using various artificial bank roughness elements. The roughness configuration selected was "chicken wire" overlaid by ¼-in. hardware cloth on each river bank. The chicken wire extended for the full bank height, whereas the height of the hardware cloth above the bed was varied from reach to reach to produce the proper water surface slope. Manning's n for the adopted roughness configuration ranged from $n_m = 0.0243$ for a flow of 30,000 cfs to $n_m = 0.0257$ for a flow of 80,000 cfs, as determined from measured water surface slopes. The desired Manning's n values, calculated from the relationship between model and prototype given by Eq. 2, are respectively: 0.0244 and 0.0250.

The model was verified by operating it to replicate the rated upstream and downstream stage over a historical period of four months (December 1974 to April 1975), during which the river cross-sections had been surveyed three times. No major changes in the prototype river cross-sections were observed for either this verification period or after the two-year drought period that followed (water years 1976 and 1977). Figure 5 shows recorded river stages at three points on the river over the four month period, along with notations showing rated flows and times of cross-section surveys. Mean flows during the two-year drought period were about 6,000 cfs; the peak flows were much lower than during "average" water years.

Figure 6 shows a typical measured cross-section for the model and the river at the beginning and end of the verification period. Some of the river cross-sectional changes shown may be due to dunes moving along the bed, rather than erosion or deposition in the river.

The model was operated continuously for 104 hr periods to simulate the four month historical period. The simulation was repeated several times with minor adjustments to the sediment feed rate equation until the proper cross-sections were obtained for the period. The simulation for the final model configuration was repeated to assure verification. Model stages were matched to Snodgrass Slough stages (Fig. 5) to within 0.002 ft (0.12 ft river) within 15 seconds of correct tidal cycle time. Model inflow matched the recorded to within 0.003 cfs (330 cfs river).

SUMMARY

The similitude criteria used for a movable-bed model of the intake to the Peripheral Canal from the Sacramento River were selected to represent scour and deposition of bed material, as well as duplication of general hydraulic behavior of the prototype. The Froude criterion was used to ensure dynamic similitude. Additional roughness was added to the model river banks to properly scale friction to the prototype.

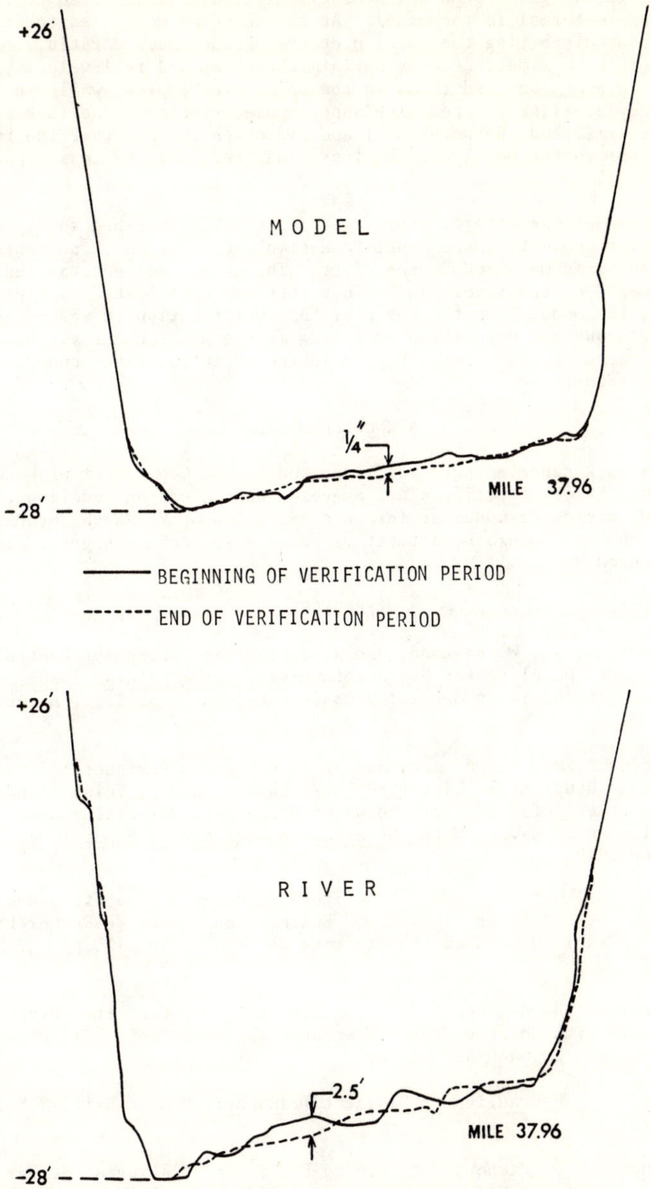

Figure 6. Measured Cross-Sections in River and Model.

A light-weight ground walnut shell material was selected to represent bed material in the model. At the selected model length scales, and without distorting the model/prototype Froude number ratio, it was not possible to simultaneously satisfy the shear and fall-velocity criteria. In addition, bed forms in the model are ripples, while in the river they are likely dunes with superimposed ripples. The lack of complete similitude between model and prototype is such that the relative scour in the model will be less, and there will be less deposition as well.

The model was able to reproduce the general river-bed configuration after subjecting the model to a simulated four-month sequence of flow conditions measured in the river. The model sediment was fed at a rate based on the river flow. This rate was established by experiment in the model. On the basis of this verification it was expected that bed scour and deposition occurring in the model under various test conditions reflect what will happen in the prototype river under similar conditions.

ACKNOWLEDGMENTS

The work reported here was supported by the Department of Water Resources, State of California. Suggestions concerning modeling criteria and review of model performance by Professor V. Vanoni, consultant to the California Department of Water Resources, are greatfully acknowledged.

REFERENCES

1. Amorocho, J., W. Hartman, and J. J. DeVries, "Computer Controlled Physical Model of the Sacramento River," Proceedings, 1980 Hydraulic Division Specialty Conference, Chicago, Ill., August, 1980.

2. Amorocho, J., W. Hartman, and J. J. DeVries, "Peripheral Canal Intake Studies 1. Distorted River Channel Model, Vols. 1 and 2, Department of Land, Air and Water Resources, University of California, Davis, Water Science and Engineering Paper 1064, June 1979.

3. Bonnefille, R., "Etude d'un critere de debut d'apparition des rider et des dunes fluviales," Bulletin du Centre de Recherches et d'Essais de Chatou, Electricite de France, No. 11, pp. 17-22, 1965.

4. Einstein, H. A., and N. Chien, "Similarity of Distorted River Models with Movable Beds," Transactions, ASCE, Vol. 121, Paper No. 2805, pp. 440-462, 1956.

5. Graf, W., Hydraulics of Sediment Transport, McGraw-Hill, N.Y., 1971.

6. Vanoni, V., "Factors Determining Bed Forms of Alluvial Stream, Journal of the Hydraulics Division, ASCE, Vol. 101, No. HY11, pp. 1435-1440, November 1975.

Subject Index

Author Index